PESTICIDES IN GROUND WATER

Distribution, Trends, and Governing Factors

Pesticides in Ground Water

Distribution, Trends, and Governing Factors

Jack E. Barbash, U.S. Geological Survey, Menlo Park, California
Elizabeth A. Resek, U.S. Environmental Protection Agency, Washington, D.C.

Volume Two of the Series
Pesticides in the Hydrologic System

Robert J. Gilliom, Series Editor
U.S. Geological Survey
National Water Quality Assessment Program

Ann Arbor Press, Inc.
Chelsea, Michigan

Library of Congress Cataloging-in-Publication Data

Barbash, Jack E.
     Pesticides in ground water : distribution, trends, and governing factors
     / Jack E. Barbash, Elizabeth A. Resek.
        p.   cm. — (Volume two of the series Pesticides in the hydrologic system)
     Includes bibliographical references and index.
     1. Pesticides—Environmental aspects—United States.
     2. Groundwater—Pollution—United States. I. Resek, Elizabeth A.
     II. Title.   III. Series: Pesticides in the hydrologic system : v. 2.
TD427.P35B37   1996
628. 1'6842—dc20   96-22170
ISBN 1-57504-005-0

This book represents information obtained from authentic and highly regarded sources. Reprinted material is quoted with permission, and sources are indicated. A wide variety of references are listed. Every reasonable effort has been made to give reliable data and information, but the author and the publisher cannot assume responsibility for the validity of all materials or for the consequences of their use.

Direct all inquiries to Ann Arbor Press, Inc., 121 South Main Street, Chelsea, Michigan 48118

No claim to original U.S. Government works
International Standard Book Number 1-57504-005-0
Library of Congress Card Number 96-22170
Printed in the United States of America 1 2 3 4 5 6 7 8 9 0
Printed on acid-free paper

INTRODUCTION TO THE SERIES

Pesticides in the Hydrologic System is a series of comprehensive reviews and analyses of our current knowledge and understanding of pesticides in the water resources of the United States and of the principal factors that influence contamination and transport. The series is presented according to major components of the hydrologic system—the atmosphere, surface water, bed sediments and aquatic organisms, and ground water. Each volume:

- summarizes previous review efforts;
- presents a comprehensive tabulation, review, and analysis of studies that have measured pesticides and their transformation products in the environment;
- maps locations of studies reviewed, with cross references to original publications;
- analyzes national and regional patterns of pesticide occurrence in relation to such factors as the use of pesticides and their chemical characteristics;
- summarizes processes that govern the sources, transport, and fate of pesticides in each component of the hydrologic system;
- synthesizes findings from studies reviewed to address key questions about pesticides in the hydrologic system, such as:

  How do agricultural and urban areas compare?

  What are the effects of agricultural management practices?

  What is the influence of climate and other natural factors?

  How do the chemical and physical properties of a pesticide influence its behavior in the hydrologic system?

  How have past study designs and methods affected our present understanding?

  Are water-quality criteria for human health or aquatic life being exceeded?

  Are long-term trends evident in pesticide concentrations in the hydrologic system?

This series is unique in its focus on review and interpretation of reported direct measurements of pesticides in the environment. Each volume characterizes hundreds of studies conducted during the past four decades. Detailed summary tables include such features as spatial and temporal domain studied, target analytes, detection limits, and compounds detected for each study reviewed.

Pesticides in the Hydrologic System is designed for use by a wide range of readers in the environmental sciences. The analysis of national and regional patterns of pesticide occurrence, and their relation to use and other factors that influence pesticides in the hydrologic system, provides a synthesis of current knowledge for scientists, engineers, managers, and policy makers at all levels of government, in industry and agriculture, and in other organizations. The interpretive analyses and summaries are designed to facilitate comparisons of past findings to current and future findings. Data of a specific nature can be located for any particular area of the country. For educational needs, teachers and students can readily identify example data sets that meet their requirements. Through its focus on the United States, the series covers a large portion of the global database on pesticides in the hydrologic system and international readers will find much that applies to other areas of the world. Overall, the goal of the series is to provide readers from a broad range of backgrounds in the environmental sciences with a synthesis of the factual data and interpretive findings on pesticides in the hydrologic system.

The series has been developed as part of the National Water-Quality Assessment Program of the U. S. Geological Survey, Department of Interior. Assessment of pesticides in the nation's water resources is one of the top priorities for the Program, which began in 1991. This comprehensive national review of existing information serves as the basis for design and interpretation of studies of pesticides in major hydrologic systems of the United States now being conducted as part of the National Water-Quality Assessment.

Series Editor

Robert J. Gilliom
U. S. Geological Survey

PREFACE

The poet Edna St. Vincent Millay (1892-1950) once wrote, "Upon this gifted age rains from the sky a meteoric shower of facts...they lie unquestioned, uncombined. Wisdom enough to leech us of our ills is daily spun, but there exists no loom to weave it into fabric." In the three decades since Rachel Carson's *Silent Spring* (1962) warned us of some of the potential ecological consequences of pesticide use, there has, indeed, been a downpour of information on the occurrence, transport, and fate of pesticides in ground water. However, while many authors have summarized various subsets of this information, efforts to interweave the results from all of the various areas of investigation of this broad subject into a single cloth have been nearly absent.

This book is an attempt to collect in one place the results from laboratory studies, field experiments, and well-sampling surveys ranging in scope from individual towns to the entire nation, and to summarize most of what is currently known across all scales of inquiry regarding the physical, chemical, and biological phenomena that govern the sources, transport, spatial and temporal distributions, and fate of pesticides and their transformation products in ground water. The topics also include an examination of the success with which mathematical simulations of pesticide movement in the subsurface and assessments of ground-water vulnerability to contamination have predicted actual pesticide occurrence in ground water. Although the emphasis is on ground water, pesticide detections in other subsurface media are also discussed, including soils, unsaturated-zone water, and tile drainage.

There are several excellent texts that summarize the basic principles of ground-water flow and solute transport in the subsurface. In order to minimize overlap with these earlier treatments, the present review is not intended as a pedagogic introduction to these subjects. Rather, it has been written as a reference guide to the existing literature on the sources, transport, distributions, and fate of pesticides and their transformation products in the subsurface. Readers are therefore expected to have at least a rudimentary familiarity with the fields of hydrogeology and environmental chemistry comparable to that provided by introductory courses at the undergraduate level.

The authors wish to express their great appreciation for the suggestions, reviews, and assistance provided by so many in the development of this book. George Hallberg, Dana Kolpin, Gordon Chesters, Robert Gilliom, and Michael Majewski provided peer reviews of the manuscript, as well as numerous additional discussions. We also wish to thank Elizabeth Behl, Mike Burkart, Gary LeMasters, Ed Brandt, Jack Kramer, JoAnn Gronberg, Lisa Nowell, Lehn Franke, Elisa Graffy, Steve Larson, Cathy Ryan, Dennis Erinakes, Lynette Seigley, Eric Vowinkel, Brian Katz, Leonard Gianessi, Paul Capel, Albert Leo, and Gail Thelin for providing copies of references and other materials, and for helpful discussions on various topics covered in the book. We also want to express our appreciation to Connie Haaser, Drew Klein, Dana Kolpin, and Pete Richards for sharing data from their respective studies.

The numerous maps presented are an important part of this book and are due in large part to the highly capable cartographic assistance provided by Naomi Nakagaki, Tom Haltom, and Donna Knifong. The help of Loreen Kleinschmidt and Will Fitzpatrick in providing many of the references used for this work, and that of Ed McCray in entering most of them into our bibliographic database, was also indispensable. We want to express our thanks to Susan Davis, Yvonne Gobert, and Glenn Schwegmann for the preparation of text, tables, and illustrations, and for the production of a high quality, camera-ready work. We also wish to thank the authors and publishers who have granted permission for the use of various figures from their publications. Finally, we are grateful to our technical editor, Tom Sklarsky, for his amazing patience and painstaking attention to detail in shepherding this book to its completion.

Jack E. Barbash
Elizabeth A. Resek

EDITOR'S NOTE

This work was prepared by the United States Geological Survey. Though it has been edited for commercial publication, some of the style and usage incorporated is based on the United States Geological Survey's publication guidelines (i.e., *Suggestions to Authors*, 7th edition, 1991). For example, references with more than two authors cited in the text are written as "Smith and others (19xx)," rather than "Smith, et al. (19xx)," and common-use compound adjectives are hyphenated when used as a modifier (e.g., quality-control procedures). Hyphenation is repeated when used in an original reference (e.g., State-Wide). For units of measure, the metric system is used except for the reporting of pesticide use, which is commonly expressed in English units. The original system of units is used when data are quoted from other sources.

Every attempt has been made to design figures and tables as "stand-alone," without the need for repeated cross reference to the text for interpretation of graphics or tabular data. There are a few exceptions, however, because of the complexity or breadth of the figure or table in individual instances. In some cases, a figure is shown just before its mention in the text to avoid continuity with unrelated figures or to promote effective layout. Some of the longer tables are at the end of the chapter to maintain less disruption of text. Some maps of the conterminous United States include insets of Alaska and Hawaii, though not spelled out. In most cases, the original style of notations on figures reprinted or redrawn from other sources is maintained (e.g., units of measurement).

As an organizational aid to the author and reader, chapter headings, figures, and tables are identified in chapter-numbered sequence. In the various summaries at the ends of sections and chapters, citations were kept to a minimum even though credit may be due, because of some repetitive discussion. The list of abbreviations and acronyms in the front of the book does not include chemical names, which are listed in the Appendix. Compounds that are given in terms of their chemical names in tables are not listed in the Appendix.

CONTENTS

FIGURES

TABLES

CONVERSION FACTORS

Multiply	By	To obtain
centimeter (cm)	0.3937	inch (in)
cubic meter (m^3)	35.31	cubic foot (ft^3)
gram (g)	0.03527	ounce, avoirdupois (oz)
hectare (ha)	2.469	acre
kilogram (kg)	2.205	pound, avoirdupois (lb)
kilometer (km)	0.6214	mile (mi)
liter (L)	0.2642	gallon (gal)
meter (m)	3.281	foot (ft)
square kilometer (km^2)	0.3861	square mile (mi^2)
square meter (m^2)	10.76	square foot (ft^2)

Multiply	By	To obtain
acre	0.405	hectare (ha)
cubic foot (ft^3)	0.02832	cubic meter (m^3)
foot (ft)	0.3048	meter (m)
gallon (gal)	3.7854	liter (L)
inch (in)	2.54	centimeter (cm)
mile (mi)	1.6093	kilometer (km)
ounce, avoirdupois (oz)	28.350	gram (g)
pound, avoirdupois (lb)	0.45359	kilogram (kg)
square foot (ft^2)	0.09290	square meter (m^2)
square mile (mi^2)	2.5900	square kilometer (km^2)

Temperature is given in degrees Celsius (°C), which can be converted to degrees Fahrenheit (°F) by the following equation:

$$°F = 1.8(°C) + 32$$

LIST OF ABBREVIATIONS AND ACRONYMS

Note: Clarification or additional information is provided in parentheses. Abbreviations for chemical compounds are included in the Appendix.

Monitoring and Other Studies

CPWTP, Cooperative Private Well Testing Program
MCPS, Midcontinent Pesticide Study
MMS, Metolachor Monitoring Study
NAWQA, National Water Quality Assessment
NAWWS, National Alachlor Well Water Survey
NPS, National Pesticide Survey
NUPAS, National Urban Pesticide Applicator Survey
PGWDB, Pesticides In Ground Water Database
SWRL, Iowa State-Wide Rural Well-Water Survey

Government and Private Agencies and Legislation

CDFA, California Department of Food and Agriculture
CERCLA, Comprehensive Environmental Response, Compensation, and
        Liability Act (Superfund)
DHS, Department of Health Services (Suffolk County, New York)
MDA, Minnesota Department of Agriculture
MDH, Minnesota Department of Health
NAS, National Academy of Sciences
NRC, National Research Council
OMOE, Ontario Ministry of the Environment (Canada)
OPP, Office of Pesticide Programs (U.S. Environmental Protection Agency)
PCPA, Pesticide Contamination Prevention Act (California)
RCRA, Resource Conservation and Recovery Act
RFF, Resources for the Future
SCS, Soil Conservation Service (now the Natural Resources Conservation Service)
USDA, U.S. Department of Agriculture
USDC, U.S. Department of Commerce
USEPA, U.S. Environmental Protection Agency
USFS, U.S. Forest Service
USGS, U.S. Geological Survey

Miscellaneous Abbreviations and Acronyms

cm/s, centimeter(s) per second
lb a.i., pound(s) active ingredient
lb a.i./yr, pound(s) active ingredient per year
kg/ha, kilogram(s) per hectare
kg/yr, kilogram(s) per year
mg/kg, milligram(s) per kilogram
mL/g, milliliter(s) per gram
ng/L, nanogram(s) per liter
ppb, parts per billion
ppm, parts per million

Miscellaneous Abbreviations and Acronyms—*Continued*

Θ, soil water content
μg, microgram(s)
$\mu g/g$, microgram(s) per gram
$\mu g/L$, microgram(s) per liter
$\mu g/m^2/yr$, microgram(s) per square meter per year
μm, micrometer(s)

AF, Attenuation Factor (Rao and others, 1985)
ARE, Application Rate Equivalent (Roy and others, 1993)
C_o, initial concentration
CPI, Corn Production Intensity (Pionke and others, 1988)
CWS, community water system wells
DAR, deethyl atrazine-to-atrazine concentration ratio
DOC, dissolved organic carbon
DOM, dissolved organic matter
DRASTIC, Depth to water; net Recharge; Aquifer media; Soil media; Topography; Impact of
        the unsaturated zone; and hydraulic Conductivity of the aquifer (Aller and
        others, 1987)
ELISA, enzyme-linked immunosorbent assay
f_{oc}, mass fraction of organic carbon in soil (organic-carbon content)
GWV, ground-water vulnerability
h, soil-water matric potential
H, Henry's Law constant
HA, health advisory
HAL, Health Action Level(s)
K, hydraulic conductivity
K_d, water-solids distribution coefficient
K_{oc}, soil-organic-carbon partition coefficient
K_{ow}, octanol-water partition coefficient
K_p, soil-water partition coefficient
LPI, Leaching Potential Index (Meeks and Dean, 1990)
MBAS, methylene-blue-active substances (surfactants)
MCL, Maximum Contaminant Level
MDL, Method Detection Limit
NOM, natural organic matter
PAL, Preventative Action Limit
pK_a, acid-base dissociation constant
RD, rural domestic wells
PRZM, Pesticide Root Zone Model (Carsel and others, 1985)
S_w, water solubility
TTR, total toxic residue
TU, Tritium Unit
VOC, volatile organic compound

PESTICIDES IN GROUND WATER

Distribution, Trends, and Governing Factors

Jack E. Barbash and Elizabeth A. Resek

ABSTRACT

A comprehensive review of published information on the distribution and behavior of pesticides and their transformation products in ground water indicates that pesticides from every chemical class have been detected in ground waters of the United States. Many of these compounds are commonly present at low concentrations in ground water beneath agricultural land. Little information is available on their occurrence beneath non-agricultural land, although use in such areas (on lawns, golf courses, rights of way, timberlands, etc.) is often comparable to, or greater than, agricultural use. Information on pesticides in ground water is not sufficient to provide either a statistically representative view of pesticide occurrence in ground water across the United States, or an indication of long-term trends or changes in the severity or extent of this contamination over the past three decades. This is largely due to wide variations in analytical detection limits, well selection procedures, and other design features among studies conducted in different areas or at different times. Past approaches have not been well suited for distinguishing "point source" from "nonpoint source" pesticide contamination.

Among the variety of natural and anthropogenic factors examined, those that appear to be most strongly associated with the intensity of pesticide contamination of ground water are the depth, construction, and age of the sampled wells, the amount of recharge (by precipitation or irrigation), and the depth of tillage. Approaches commonly employed for predicting pesticide distribution in the subsurface—including computer simulations, indicator solutes (e.g., nitrate or tritium), and ground-water vulnerability assessments—generally provide unreliable predictions of pesticide occurrence in ground water. Such difficulties may arise largely from a general failure to account for the preferential transport of pesticides in the subsurface. Significant improvements in understanding and predicting the occurrence and fate of pesticides in ground water are likely to depend on: (1) greater coordination of ground-water sampling across the nation to ensure consistency of study design, and thus comparability of results; (2) more extensive analyses for pesticide transformation products during ground-water monitoring studies; (3) substantially enhanced communication among investigators conducting laboratory experiments, small-scale field studies, and large-scale monitoring studies; and (4) more routine testing of predictions of pesticide behavior and ground-water vulnerability against actual field observations of pesticide occurrence in ground water.

CHAPTER 1

Introduction

Dramatic increases in the use of synthetic organic pesticides since World War II have helped to make the United States the largest producer of food in the world, but have also raised many concerns about potential adverse effects on the environment and human health. Approximately 1.1 billion pounds of pesticides are used each year in the United States to control many different types of weeds, insects, and other pests in a wide variety of agricultural and non-agricultural settings (Figure 1.1). The total amounts and number of different pesticides applied have grown steadily since the early 1960's, when the first reliable records were made. National use of herbicides and insecticides on cropland and pasture grew from 190 million pounds of active ingredient in 1964, to 560 million pounds in 1982 (Gilliom and others, 1985), and an estimated 660 million pounds in 1993 (Aspelin, 1994).

In most respects, the greatest potential for unintended adverse effects of pesticides is through the contamination of the Earth's hydrologic systems, which support both human and non-human ecosystems. Water is one of the primary media in which pesticides are transported from targeted applications to other parts of the environment. The application of pesticides thus provides the potential for their movement into and through all components of the Earth's hydrosphere (Figure 1.2), as demonstrated by their repeated detection in the surface waters (Wauchope, 1978; Gilliom and others, 1985; Thurman and others, 1992; Pereira and Hostettler, 1993; Larson, 1995), ground waters (U.S. Environmental Protection Agency [USEPA], 1990a, 1992a,b) and precipitation (Nations and Hallberg, 1992; Majewski and Capel, 1995) of the United States. Despite the extensive use of pesticides for several decades, however, many of the consequences of environmental contamination by these compounds and their transformation products remain unknown (Wauchope, 1987).

The detection of pesticides in ground water has been a public concern in the United States since at least 1980, by which time the contamination of subsurface water supplies by the insecticide aldicarb in New York and Wisconsin, and by the fumigant 1,2-dibromo-3-chloropropane (DBCP) in California, Arizona, South Carolina, and Maryland had been discovered (Cohen, 1990a). Contamination of ground water by pesticides is a major national issue because ground water is used for drinking water by about 50 percent of the nation's population. Concern about pesticides in ground water is especially acute in agricultural areas, where most pesticides are used and where over 95 percent of the population relies upon ground water for drinking water (Moody and others, 1988).

3

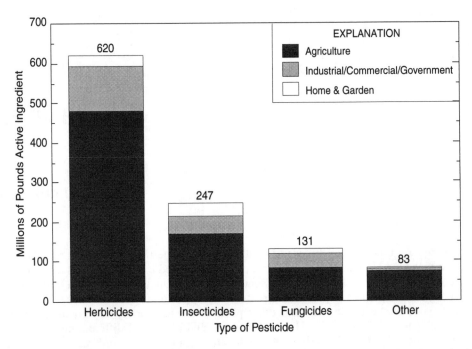

Figure 1.1. Total estimated mass of pesticides used in the United States during 1993 for agricultural, industrial/commercial/government, and home and garden applications (from Aspelin, 1994).

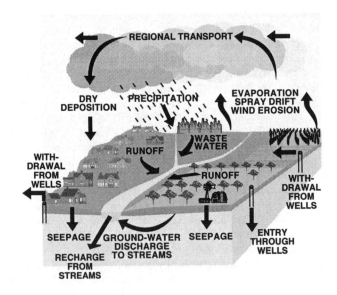

Figure 1.2. Pesticide movement in the hydrologic cycle (adapted from Majewski and Capel, 1995).

1.1 PURPOSE

The purpose of this book is to examine our present knowledge of the occurrence, spatial distributions, and temporal trends in the concentrations of pesticides in the ground waters of the United States, and to evaluate these observations within the context of our current understanding of natural and anthropogenic factors that affect the sources, transport, and fate of these compounds in the subsurface (i.e., in soils, the unsaturated zone, or the saturated zone). In relation to previous reviews, the present effort: (1) updates and expands on information analyzed by Hallberg (1989); (2) emphasizes the integration and analysis of results of studies conducted over a wide range of spatial and temporal scales, with particular attention to causal factors with large-scale significance, such as pesticide use patterns; (3) expands the scope of review to include non-agricultural land uses; and (4) provides a detailed geographic reference and tabular summary of all reviewed studies that examined pesticide occurrence in the subsurface. The maps of study locations and tabular summaries of study characteristics and results are designed to facilitate the acquisition and use of information on pesticides in ground water from specific investigations most appropriate to the objectives of future studies.

The most recent comprehensive review of pesticides in ground water (Hallberg, 1989) was built on the foundation laid by the earlier work of Cohen and others (1984, 1986) and Holden (1986). Similarly, the present book is an extension of the analysis by Hallberg (1989), and we emphasize results from monitoring studies published since that review was completed. Although the earlier monitoring studies will be included in this analysis where their results illuminate specific points, the reader is referred to earlier compilations for more complete summaries (e.g., Cohen and others, 1984, 1986; Holden, 1986; Hallberg, 1989; U.S. Environmental Protection Agency, 1992b; U.S. Environmental Protection Agency, 1995).

This book is one in a series of reviews of current knowledge of pesticide contamination of the hydrologic system being conducted as part of the Pesticide National Synthesis Project of the U.S. Geological Survey, National Water Quality Assessment (NAWQA) program. Other books in the series will focus on pesticides in the atmosphere, stream water, stream bed sediment, and aquatic biological tissues. These national topical books will complement the more detailed NAWQA studies of a broad range of water-quality issues in major national hydrologic basins (typically 10,000 to 30,000 mi^2, or 25,000 to 75,000 km^2) of the United States.

1.2 PREVIOUS REVIEWS

The extensive body of assessments and research on the occurrence, distribution, and behavior of pesticides in ground waters that has been published over the past decade has spawned a number of reviews. Table 1.1 (at end of chapter) cites 78 review publications and lists the general topics that they examined. Two general types of reviews are included: (1) overviews of data on the occurrence and distributions of pesticides in ground waters of the United States (and, in some cases, Canada and Europe); and (2) general reviews describing one or more aspects of the various processes that control the transport and fate of pesticides in the subsurface. The reviews span a wide range of detail and documentation. Although several of the reviews, such as those by Moye and Miles (1988), Nash (1988), Hallberg (1989), Smith (1990), and Rao and Alley (1993), include comprehensive bibliographies, many cited no previous publications at all. In addition to the studies in Table 1.1, a number of discussions of solute-transport simulations have also described the various physical and chemical processes governing pesticide behavior in the subsurface. Examples include reviews by van Genuchten and Wierenga (1976), Matthies (1987), Enfield and Yates (1990), Wagenet and Hutson (1990), and Wu and others (1994).

Reviews focusing on the distributions of pesticides in ground waters of the United States (Table 1.1) have generally been limited to compilations of statistics on the concentrations or frequencies of detection of individual pesticides within the area(s) of interest. Furthermore, many summaries of this type have involved the recapitulation of the conclusions reported in a relatively small number of publications on pesticide occurrence in ground waters of the United States; overviews by Cohen and others (1984, 1986) and Holden (1986) are among those cited most frequently in this regard. These summaries have provided an indication of the magnitude and spatial extent of ground-water pollution by pesticides in the nation, but offer relatively little direct insight regarding the influence of factors that may control the subsurface behavior and fate of these compounds.

A comprehensive review of monitoring studies by Hallberg (1989) represents a marked exception to this pattern. Building on the previous overviews by Cohen and others (1984, 1986) and Holden (1986), Hallberg (1989) provided summaries of pesticide detections that had been reported in the ground waters of individual states and three Canadian Provinces since the earlier overviews appeared, and used the results from the more recent studies to draw conclusions regarding the various factors that appear to govern the occurrence and behavior of pesticides in the subsurface.

Several publications have provided valuable comparisons among the study designs of different monitoring investigations (e.g., Williams and others, 1988; Barrett and Williams, 1989; U.S. Environmental Protection Agency, 1992a). In particular, Barrett and Williams (1989) summarized and compared 15 different monitoring studies in Iowa, Illinois, Kansas, and Nebraska with respect to the numbers, types, and depths of wells sampled, as well as the criteria employed for selecting wells and identifying times of year for sampling.

The reviews listed in Table 1.1 evaluated studies covering a broad range of spatial scales, from large, multistate regions to laboratory microcosms (soil columns or batch reactors). Between these extremes is the scale of individual agricultural fields, where most research on pesticide behavior in the subsurface has been carried out. Although numerous publications have discussed the principal physical, chemical, and microbiological processes that govern the transport and fate of pesticides in the subsurface, previous summaries of the results from field-scale studies have been largely empirical and site-specific, limited primarily to simple comparisons of observed concentrations of pesticides in the subsurface among different compounds, settings, and investigations (e.g., Hance, 1987; Hallberg, 1989), rather than integrated analyses of causal factors. This circumstance may be the result of the manner in which the results from field-scale studies are usually presented, with findings often reported in terms of "dissipation half-lives," depths of leaching, and specific concentrations for individual pesticides at particular locations in the subsurface. This approach has given rise to a body of information on pesticide behavior that is often highly site-specific, and hence, not easily transferred to other environmental settings, times, or agricultural management regimes, as discussed further in Chapter 2.

For example, a review by Nash (1988) organized the available data on rates of pesticide "dissipation" according to such parameters as compound class, pH, temperature, and soil type, but made little quantitative distinction between the influence of transformation, which modifies the chemical and physical properties of a pesticide, and of transport, which simply moves the pesticide to a different location (throughout this document, the term "transformation" will be used to denote any combination of biological and/or abiotic processes that lead to the modification of the chemical structure of a compound, giving rise to what will be referred to as either "transformation products" or "degradates"). The lack of distinction among the major influences on dissipation makes meaningful comparisons between studies difficult. In contrast, Boesten (1987) summarized the results from several field-scale studies in terms of the proportions of the applied compounds that were subsequently measured in the aqueous phase beneath the root zone. By "normalizing" the detected amounts to the quantity of compound applied during each of the field-scale studies of interest, Boesten (1987) was able to draw direct comparisons among the results from these studies to examine the influence of factors such as irrigation, soil type, and pesticide properties on the subsurface migration, and the persistence of the pesticides in question. Overall, however, relatively little effort has been directed toward examining the extent to which theoretical predictions have been confirmed or refuted by field observations.

Another subject that has not been extensively covered in previous reviews is the occurrence of pesticides in ground water associated with non-agricultural activities. Several publications have described the types of pesticides that have been detected in ground waters beneath specific non-agricultural areas, such as golf courses, pesticide handling facilities, disposal sites, residential areas, and ground-water recharge facilities, but no comprehensive overview of these observations is available.

1.3 APPROACH

This report focuses on pesticides in ground waters of the United States, although investigations conducted in Canada, Europe, and elsewhere are also considered for comparison with observations reported by studies from this country. The primary emphasis of this book is on observations of pesticides in ground water, but the distributions of pesticides among other environmental media are also considered to elucidate pathways of transport to ground water. In particular, pesticide concentrations in soils are included, because: (1) they provide information on the distribution and behavior of pesticides above the water table; (2) the majority of measurements of pesticide residues at the land surface have been conducted in whole soils, rather than in soil water alone; and (3) pesticide concentrations in soils provide a characterization of the mass and types of pesticide compounds that are available for leaching to greater depths. In addition to soils and ground water, other media of interest for particular issues are vadose-zone water, tile drainage, plant tissues, and surface water.

In addition to pesticides, selected discussions of other constituents in ground water are included to interpret findings for pesticides. Most significant is nitrate, which, like pesticides, is applied to the land in most agricultural settings—either as nitrate itself, or as reduced nitrogen species such as ammonia or urea, which undergo subsequent oxidation to nitrate. Nitrate has frequently been suggested as an inexpensive indicator for potential contamination of ground water by pesticides. In addition, nitrate is redox active, and its presence or absence can be used, in conjunction with measurements of dissolved oxygen and dissolved organic carbon (DOC), to help determine whether or not biogeochemical conditions are appropriate for denitrification.

The occurrence of pesticide transformation products and volatile organic compounds (VOCs) in ground water beneath agricultural areas is also examined to the limited degree possible. Pesticide transformation products are important for gaining a complete understanding of the ultimate fate of pesticides in the environment. Although fumigants, by design, are volatile compounds, VOCs are also of interest because they are currently among the "inert ingredients," or "adjuvants," contained within commercial pesticide formulations.

In many respects, the present understanding of the processes that govern the transport and fate of pesticides in the subsurface is reflected in the design of computer models used to simulate pesticide behavior in the subsurface and in the approaches employed to predict the susceptibility of ground water in different settings to pesticide contamination. Consequently, this book will also examine the capabilities of both solute-transport simulations and ground-water vulnerability assessments to predict pesticide occurrence in the subsurface.

The goal of the review process for monitoring studies was to locate and evaluate all significant investigations that have been published in an accessible format, including journal articles, federal and state reports, and university technical reports. It is important to note, however, that the monitoring studies examined do not encompass all of the data available on the occurrence of pesticides in ground waters of the United States. At the time of writing, many states were in the process of carrying out extensive monitoring programs designed to investigate the geographic distributions of pesticides and pesticide transformation products in their ground-water resources. Furthermore, the principal focus of this book is on the conclusions reached by authors of previous studies, rather than on an analysis or reanalysis of their original data on pesticide occurrence in ground water.

Although bibliographic sources similar to those used to locate monitoring studies were consulted to locate field-scale and laboratory investigations of pesticide behavior in the subsurface, it is not the intent of this report to summarize all of the myriad findings gathered during three decades of such studies. Rather, the results from field-scale and laboratory research are brought into the present discussion where they help to explain patterns of pesticide occurrence noted during the monitoring studies. These results are also addressed to answer specific questions regarding the factors that control pesticide behavior in the subsurface.

The studies examined for this book were assembled primarily through the combined use of bibliographic data bases, personal collections, and the reference lists from other publications. The electronic data bases searched for this purpose included AGRICOLA (U.S. Department of Agriculture), the National Technical Information Service (U.S. Department of Commerce), ChemAbstracts (American Chemical Society), and Selected Water Resources Abstracts (U.S. Geological Survey).

The findings of this review are presented in 10 parts. First, the principal design features of the field-scale investigations and monitoring studies reviewed are examined using a series of summary tables in Chapter 2. Tabulated features include geographic locations, sampling time frame, numbers and types of sampling points used, media sampled, and target analytes. The tables thus provide a comprehensive overview of and reference to the field-scale investigations and monitoring studies discussed in the book. These data serve as a basis for initial characterization of the nature, degree, and emphasis of the study efforts represented.

Second, a national perspective on the occurrence and geographic distribution of pesticides and their transformation products in ground water is developed in Chapter 3, based primarily on the observations reported by investigations carried out over regions encompassing more than one state. Although they are much less numerous than the state and local monitoring studies assembled for this book, the multistate studies are valuable because each provides data obtained with a consistent study design—and hence, the opportunity for direct comparisons of ground-water quality—across larger geographic areas. Chapter 3 also provides an overview of pesticide use, in both agricultural and non-agricultural settings across the United States, and examines the extent to which spatial patterns of pesticide detection in ground water correspond to geographic distributions of use. In addition to the focus on spatial variability, Chapter 3 includes a discussion of the temporal variability of pesticide concentrations and detection frequencies in ground water.

The third part of this book (Chapter 4) is a summary of the primary factors that affect pesticide concentrations in ground water (i.e., sources, partitioning, transport, and transformations). This provides a basis for understanding the patterns of pesticide occurrence that have been documented by the monitoring studies, and serves as a point of reference for addressing further questions regarding the nature of these factors and their relative significance among different compounds, and in different environmental and land-use settings.

In the remainder of this book, the results of research conducted at all spatial scales of interest are used to more closely examine the principal natural and anthropogenic phenomena that govern the causes and severity of ground-water contamination by pesticides (Chapters 5 through 8), to assess our ability to predict the likelihood of detecting pesticides in ground water (Chapter 9), and to evaluate the environmental significance of pesticides in ground water (Chapter 10). Because of the varying degrees to which these topics have been examined within the existing literature, the confidence with which they are addressed in this book is highly variable. Chapter 11 summarizes the major findings of this book and discusses their implications for future investigations of pesticide occurrence, movement, and fate in ground water.

Table 1.1. General features of previous overviews of pesticide occurrence and behavior in ground waters of the United States

[Scale(s) of principal Interest: L, laboratory; F, field; Reg, regional, if applicable. ‡, examined. Blank cells indicate no information applicable or available]

Reference (Pesticide[s] of primary interest, if applicable)	Primary Subjects Examined							Geographic Coverage (If applicable)	Number of References Cited	Scale(s) of Principal Interest
	Governing Factors				Transformation		Study Design			
	Occurrence and Distribution	Sources	Soils and(or) Hydrogeology	Partitioning and(or) Transport	Abiotic	Biological				
Reviews Focusing Primarily on Pesticide Occurrence in Ground Waters										
Croll, 1972	‡ (By compound class)							Selected areas of United States and Europe	37	Reg
Letey and Pratt, 1983	‡ (By state)			‡		‡		Selected areas of United States and England	21	Reg
Barles and Kotas, 1984	‡							Selected areas of United States	0	Reg
Cohen and others, 1984	‡							Nationwide	121	Reg
Bouwer and others, 1985	‡ (By state)	‡						Selected areas of United States	13	Reg
Cohen, 1986	‡ (By compound)		‡					California	28	Reg
Cohen and others, 1986	‡ (By compound)						‡	Nationwide	68	Reg
Hallberg, 1986	‡	‡						Selected areas of United States	86	Reg
Holden, 1986	‡ (By state)	‡						California, New York, Wisconsin, Florida	61	Reg
Krider, 1986	‡ (By state)							Selected areas of United States	11	Reg

Table 1.1 General features of previous overviews of pesticide occurrence and behavior in ground waters of the United States—*Continued*

Reference (Pesticide[s] of primary interest, if applicable)	Primary Subjects Examined							Geographic Coverage (If applicable)	Number of References Cited	Scale(s) of Principal Interest
	Occurrence and Distribution	Governing Factors					Study Design			
		Sources	Soils and(or) Hydrogeology	Partitioning and(or) Transport	Transformation					
					Abiotic	Biological				
U.S. Environmental Protection Agency, 1986	‡ (By state)	‡	‡	‡	‡	‡	‡	Nationwide	11	Reg
Brennan, 1987	‡	‡						Hawaii	1	Reg
Canter, 1987	‡ (By state)		‡	‡				Nationwide	63	L, F, Reg
Fairchild, 1987	‡ (By state)	‡						48 states	30	Reg
Hallberg, 1987	‡ (Selected studies)	‡	‡					Selected areas of United States	111	Reg
Schmidt and Sherman, 1987	‡ (Selected studies)							California	20	Reg
Lee and Nielsen, 1988	‡							Nationwide	1	Reg
Leistra, 1988	‡			‡	‡	‡		Selected areas in United States and Europe	17	Reg
Barrett and Williams, 1989 (Atrazine)	‡ (Selected states)		‡	‡		‡		Nationwide	31	Reg
Hallberg, 1989	‡ (Selected states)	‡	‡	‡		‡		Central and eastern (humid) United States	154	F, Reg
Shirmohammadi and Knisel, 1989	‡	‡	‡					Southeastern United States (7 states)	44	Reg
Felsot, 1990	‡	‡						Selected areas of United States	28	Reg

Table 1.1 General features of previous overviews of pesticide occurrence and behavior in ground waters of the United States—*Continued*

Reference (Pesticide[s] of primary interest, if applicable)	Primary Subjects Examined							Geographic Coverage (If applicable)	Number of References Cited	Scale(s) of Principal Interest
	Occurrence and Distribution	Governing Factors					Study Design			
		Sources	Soils and(or) Hydrogeology	Partitioning and(or) Transport	Transformation					
					Abiotic	Biological				
Lorber and others, 1990 (Aldicarb)	‡							Nationwide	45	F, Reg
Mackay and Smith, 1990	‡	‡	‡	‡		‡		California	14	Reg
Ritter, 1990	‡ (By compound)	‡		‡				Selected areas of United States	49	Reg
Cartwright and others, 1991	‡	‡						England	11	Reg
Nash and others, 1991	‡ (Selected studies)	‡		‡			‡	Selected areas of United States	25	F, Reg
Phillips and Birchard, 1991 (Organochlorines)	‡							Nationwide (by census division)	9	Reg
Wells and Waldman, 1991 (Aldicarb)	‡	‡		‡	‡	‡		Nationwide	125	Reg
Gustafson, 1993	‡	‡	‡	‡	‡	‡	‡	Selected areas of United States and Europe	451	F, Reg
Rao and Alley, 1993	‡ (Selected studies)					‡		Selected areas of United States	113	Reg
U.S. Environmental Protection Agency, 1995 (Triazines)	‡ (Selected studies)	‡	‡			‡		Selected areas of United States	109	Reg
Reviews Focusing Primarily on Processes and Matrix Distributions										
Robbins and Kriz, 1969		‡		‡					97	Reg
Le Grand, 1970			‡						18	

Table 1.1 General features of previous overviews of pesticide occurrence and behavior in ground waters of the United States—*Continued*

Reference (Pesticide[s] of primary interest, if applicable)	Primary Subjects Examined							Geographic Coverage (If applicable)	Number of References Cited	Scale(s) of Principal Interest
	Occurrence and Distribution	Governing Factors					Study Design			
		Sources	Soils and(or) Hydro-geology	Partitioning and(or) Transport	Transformation					
					Abiotic	Biological				
Chesters and Konrad, 1971		++	++	++	++	++			66	L,F
Haque and Freed, 1973				++	++	++			108	
Garrett and others, 1976			++	++			++		103	L, F
Riley, 1976				++					0	F
Roberts and others, 1982				++	++	++			32	
Jackson and Webendorfer, 1983a			++	++		++			0	
Jackson and Webendorfer, 1983b (Aldicarb)				++	++	++		Wisconsin	0	Reg
Rao and Jessup, 1983				++					84	L, F
Bishop, 1985				++		++			15	
Lavy and others, 1985	++			++	++	++		Selected areas of United States	196	L, F, Reg
Cheng and Koskinen, 1986		++		++	++	++			47	F
Jones, 1986 (Aldicarb)	++			++	++	++		Selected areas of United States	26	L, F
Wagenet, 1986a			++	++	++	++			0	
Boesten, 1987				++					29	F
Carsel and Smith, 1987		++		++					15	F, Reg
Erickson and Kuhlman, 1987	++	++	++	++		++		Kansas	0	F, Reg

Table 1.1 General features of previous overviews of pesticide occurrence and behavior in ground waters of the United States—*Continued*

Reference (Pesticide[s] of primary interest, if applicable)	Primary Subjects Examined							Geographic Coverage (If applicable)	Number of References Cited	Scale(s) of Principal Interest
	Occurrence and Distribution	Governing Factors					Study Design			
		Sources	Soils and(or) Hydro-geology	Partitioning and(or) Transport	Transformation					
					Abiotic	Biological				
Hance, 1987	‡ (Selected studies)			‡	‡	‡		Selected areas in United States and Europe	132	F, Reg
Helling, 1987				‡	‡	‡			26	
Jury and others, 1987a	‡			‡		‡		Selected areas of United States	119	L, F, Reg
Miller, 1987				‡	‡	‡		Selected areas of United States	271	L, F
Blodgett, 1988	‡			‡	‡	‡			0	
Brown, 1988				‡	‡	‡			0	
McVoy, 1988				‡		‡			0	
Moye and Miles, 1988 (Aldicarb)	‡ (By state)			‡	‡	‡		Selected areas of United States	118	L, F, Reg
Nash, 1988			‡	‡	‡	‡			137	L, F
Nicholls, 1988			‡	‡	‡				23	L, F
Abernathy, 1989				‡		‡			0	L
Chesters and others, 1989a (Alachlor and metolachlor)	‡	‡	‡	‡	‡	‡		Selected areas of United States and Canada	185	L, F, Reg
Hornsby, 1989				‡	‡	‡			5	
Mink and others, 1989 (Aldicarb)	‡ (By state)			‡	‡	‡		Selected areas of United States	45	L, F
Bouwer, 1990		‡		‡					51	F
Cohen, 1990a	‡			‡				United States, Canada, Western Europe	26	F
Fawcett, 1990		‡							0	F, Reg
Harper and others, 1990						‡			50	L, F

Table 1.1 General features of previous overviews of pesticide occurrence and behavior in ground waters of the United States—*Continued*

| Reference (Pesticide[s] of primary interest, if applicable) | Occurrence and Distribution | Governing Factors | | | Transformation | | Study Design | Geographic Coverage (If applicable) | Number of References Cited | Scale(s) of Principal Interest |
		Sources	Soils and(or) Hydrogeology	Partitioning and(or) Transport	Abiotic	Biological				
Marani and Chesters, 1990				++	++	++			90	L, F
Miller and Mayer, 1990	++			++		++	++	Selected areas of United States	655	L, F
Roberts, 1990				++	++	++			29	L, F
Smith, 1990	++		++	++	++	++	++	Selected areas of United States	126	F
Stover and Guitjens, 1990 (Aldicarb)	++			++	++	++		Selected areas of United States	43	L, F, Reg
Becker and others, 1991			++	++		++			0	F
Funari and others, 1991	++		++	++				Italy	42	Reg
Loch, 1991	++		++	++	++	++		The Netherlands	29	F, Reg
Chesters, 1992			++	++	++	++			80	L, F
Shestopalov and Molozhanova, 1992	++	++	++						6	F, Reg
Crutchfield and others, 1993	++	++	++						17	

CHAPTER 2

Characteristics of Studies Reviewed

Most of the studies reviewed fall into one of four categories, based on their objectives, their study design, and the spatial scale at which their measurements were made. These categories are: (1) "process and matrix-distribution studies;" (2) "state and local monitoring studies;" (3) "multistate monitoring studies;" and (4) compilations of monitoring data.

2.1 PROCESS AND MATRIX-DISTRIBUTION STUDIES

The principal objective of process and matrix-distribution studies is to gain a better understanding of the phenomena that control the sources, transport, and fate of pesticides within and among environmental compartments. Such investigations typically involve the application of one or more compounds at specified rates and under controlled conditions to an agricultural field or plot, or to greenhouse soil. This is followed by the monitoring of their subsequent movement away from the site(s) of application—either above or below the ground surface—in one or more environmental media. The media sampled by the studies examined for this report included ground water, lysimeter leachate, tile drainage, soils, plant tissues, and surface water. (Volatilization fluxes of pesticides during application were only rarely reported [e.g., Jury and others, 1984c; Clendening and others, 1990].) The spatial scope of these investigations has generally ranged from individual columns or containers of soil buried in the ground—typically a lysimeter, box, bag, or other vessel filled with soil with exposed surface areas of typically 1 m^2 or less—to small watersheds tens of hectares in size.

Tables 2.1 and 2.2 (at end of chapter) summarize the principal design features of 213 process and matrix-distribution studies of pesticide behavior in the subsurface. Table 2.1 lists the 55 reviewed studies that included measurements of pesticide concentrations in ground water; Table 2.2 lists the other 158 that only included measurements in related media, usually soils. The corresponding geographic locations of these two groups of investigations are shown in Figures 2.1 and 2.2, cross-referenced by the study numbers assigned in Tables 2.1 and 2.2. To avoid redundancy, compilations of the results of process and matrix-distribution studies (e.g., Jones 1986; Jones and others, 1986a) were excluded from Tables 2.1 and 2.2.

The degree to which results from process and matrix-distribution studies can be used to characterize the factors that control the movement and fate of pesticides in the subsurface is limited by the considerable spatial variability in physical, chemical, and biological properties exhibited by natural systems, even over scales of meters. As a result, many laboratory investigations of the partitioning, transport and(or) transformation of these compounds have been carried out using model systems (soil columns or batch reactors) under simulated environmental conditions. Although the emphasis of the present discussion will be on the process and matrix-distribution studies, the results from laboratory investigations will be drawn on selectively.

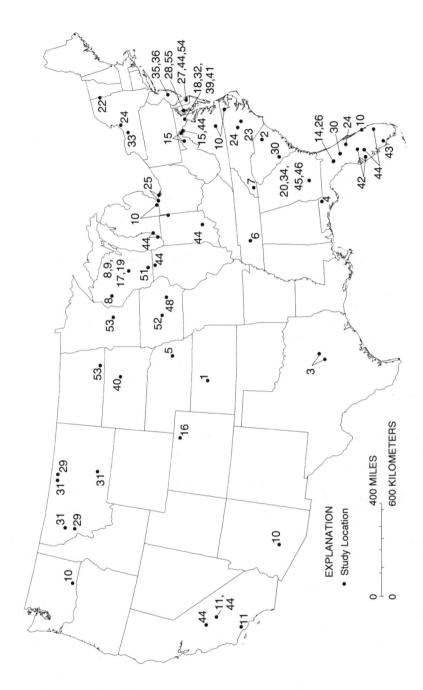

Figure 2.1. Process and matrix-distribution studies in the United States that included measurements of pesticide concentrations in ground waters. Study numbers correspond to those given in Table 2.1.

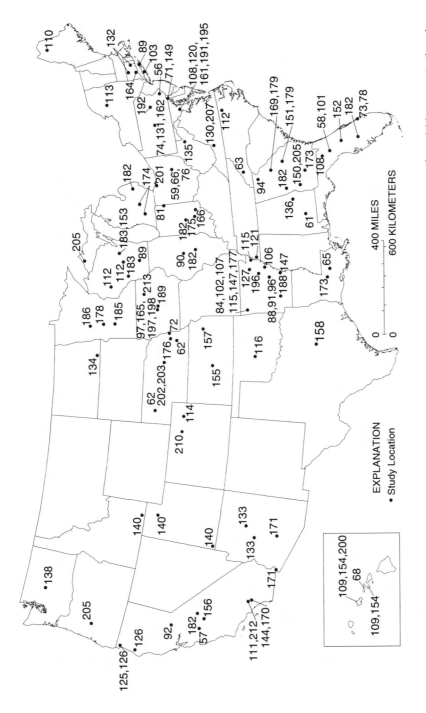

Figure 2.2. Process and matrix-distribution studies in the United States that did not include measurements of pesticide concentrations in ground waters. Study numbers correspond to those given in Table 2.2.

2.2 MONITORING STUDIES

The term "monitoring studies" is used to denote investigations that have focused on the determination of the spatial and(or) temporal extent of the occurrence of pesticides in ground waters sampled from existing wells over spatial scales ranging from subcounty areas to multistate regions. We examine the results from monitoring investigations in two categories of spatial scale: (1) "state and local studies," carried out over scales ranging from individual municipalities to entire states, and (2) "multistate studies." Investigations that involved the measurement of pesticide residues in wells located in and around individual agricultural fields were included among the process and matrix-distribution studies.

The principal features of the state and local monitoring studies and multistate monitoring studies reviewed for this book are summarized in Tables 2.3 (at end of chapter) and 2.4, respectively. The areas sampled for state and local monitoring studies are displayed in Figure 2.3, while those for the multistate investigations are shown in Figures 2.4 through 2.8. The state and local monitoring investigations carried out to date are more numerous than the multistate studies because most of the sampling for pesticide residues in ground water has been carried out by individual states, counties, universities, and pesticide manufacturers, rather than by federal agencies. In addition to the study designs of the multistate investigations, Table 2.4 provides an overview of their principal quantitative results.

2.3 COMPILATIONS OF MONITORING DATA

There have been two major attempts to assemble the available data on the detections of pesticides in the ground waters of the United States. The 1988 Survey of State Lead Agencies (Parsons and Witt, 1989) summarized data on the occurrence of pesticides in the ground waters of 35 states. The more extensive Pesticides in Ground Water Database, or PGWDB (U.S. Environmental Protection Agency, 1992b), contains data from 45 states. Table 2.4 includes a summary of the general features of both compilations.

The preparation of the 1988 Survey of State Lead Agencies and the PGWDB involved the collection of information from investigations with great variability in study design. Both summaries combined the results from studies that targeted known problems—such as 1,2-dibromo-3-chloropropane (DBCP) in the Central Valley of California and aldicarb in Long Island, New York—with those from more statistically representative, random surveys of ground-water quality. In addition, the monitoring studies included in these compilations tended to focus on the pesticides used most extensively within the areas of interest. Thus, the two compilations provide a broad picture of the spectrum of pesticides that have been investigated in ground waters and the levels of effort involved in different states. This picture, however, is biased toward the compounds and areas for which sampling has been most intensive, a point also noted by the authors of both compilations.

Of the two compilations, the PGWDB will be referred to more often because it: (1) is more recent and complete than the earlier summary, and (2) is part of an effort to catalogue pesticide detections across the nation on an ongoing basis, with periodic updates planned for the future (U.S. Environmental Protection Agency, 1992b). Despite its limitations, the PGWDB is the most comprehensive summary available on pesticide occurrence in the ground waters of the United States, with respect to both the range of compounds and the geographic coverage of the data that it encompasses.

Table 2.4. Overall study designs and results from five multistate monitoring studies and two data compilations on pesticides in ground waters of the United States

[Parameter: Targeted (T) sampling was usually directed toward areas with a specific land use, use of a particular pesticide, or potential contamination. Nontargeted (NT) sampling involved a stratified random well-selection process. MCL, maximum contaminant level; CWS, community water system wells; RD, rural domestic wells. NG, not given. USEPA, U.S. Environmental Protection Agency. <, less than]

Parameter	Monitoring Studies								Data Compilations	
	Midcontinent Pesticide Study		Cooperative Private Well Testing Program		National Pesticide Survey		National Alachlor Well Water Survey	Metolachlor Monitoring Study	Pesticides In Ground Water Database	1988 Survey of State Lead Agencies
	Pre-Planting	Post-Planting	Triazines	Acetanilides	CWS	RD				
Period of sampling	3/91-4/91	7/91-8/91	[1]1987-1993		4/88-2/90		4/88-5/89	4/88-6/89	1971-1991	NG
Targeted or nontargeted sampling	NT	NT (T for some compounds)	T		NT		NT	T	T, NT (varied among studies)	T, NT (varied among studies)
States sampled	12	9	17		50	38	26	4 (16 counties)	45	35
States with detections	[2]10	9	NG		NG		NG	4	42	33
Number of wells sampled for pesticide analytes	299	[3]45-100	14,044	12,539	[4]540	[4]752	1,430	240	68,824	NG (144,401 analyses)[5]
Percentage of sampled wells with analyte detections	20.4 percent	62 percent	9.8 percent	7.4 percent	10.2 percent	4.2 percent	12.95 percent	16.3 percent	24 percent	NG (4.17 percent of analyses)[5]
Percentage of samples with pesticides above MCL	0 percent	0.1 percent	0.4 percent (assumes 100 percent atrazine)[6]	1.5 percent (assumes 100 percent alachlor)[6]	<0.8 percent	0.6 percent	0.11 percent	0 percent	14 percent	NG (0.74 percent of analyses)[5]
Analytes examined—	13	63	(6)		126		5	1	302	169
Parents	11	55	(6)		101		5	1	258	167
Degradates	2	8	(6)		25		0	0	45	2

Table 2.4. Overall study designs and results from five multistate monitoring studies and two data compilations on pesticides in ground waters of the United States—*Continued*

Parameter	Monitoring Studies								Data Compilations	
	Midcontinent Pesticide Study		Cooperative Private Well Testing Program		National Pesticide Survey		National Alachlor Well Water Survey	Metolachlor Monitoring Study	Pesticides In Ground Water Database	1988 Survey of State Lead Agencies
	Pre-Planting	Post-Planting	Triazines	Acetanilides	CWS	RD				
Analytes detected in one or more wells	9	27	(6)		7	10	5	1	132	67
Parents	7	21	(6)		6	9	5	1	117	66
Degradates	2	6	(6)		1	1	0	0	16	1
Analytes detected above their MCL	0	1	(6)	(6)	NG	NG	2	0	36	17
Parents	0	1	(6)	(6)	NG	NG	2	0	33	17
Degradates	0	0	(6)	(6)	NG	NG	0	0	3	0
Reference(s)	Burkart and Kolpin, 1993; Kolpin and others, 1993; Kolpin and Goolsby, 1995; Kolpin and others, 1995		Baker and others, 1994		USEPA, 1990a, 1992a		Monsanto Agricultural Company, 1990; Holden and others, 1992	Roux and others, 1991a	USEPA, 1992b	Parsons and Witt, 1989

[1] Sampling period uncertain; inconsistent times given for commencement of sampling (1987 or 1988), and end of period (1993) inferred from date of publication.

[2] Separate results for the two sampling periods not given.

[3] Number of wells sampled depended upon compound analyzed.

[4] A total of 566 CWS and 783 RD wells were sampled during the NPS, but the data from only 540 of the CWS and 752 of the RD wells passed quality assurance requirements (U.S. Environmental Protection Agency, 1992a).

[5] Actual numbers of wells not given by authors; only the total number of analyses (i.e., "the sum of the number of wells in which each pesticide was analyzed" [Parsons and Witt, 1988]). Thus, wells for which larger numbers of pesticides were analyzed were over-represented in these estimates.

[6] Immunoassay used to detect pesticide analytes. Thus, the specific triazine or acetanilide herbicides responsible for the detections could not be determined (see Section 2.8.2).

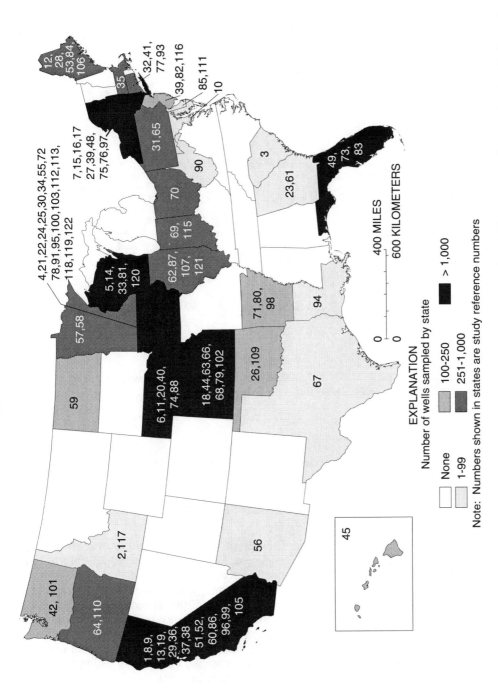

Figure 2.3. Number of wells sampled in each state for state and local monitoring studies of pesticide occurrence in ground water. Study numbers correspond to those given in Table 2.3.

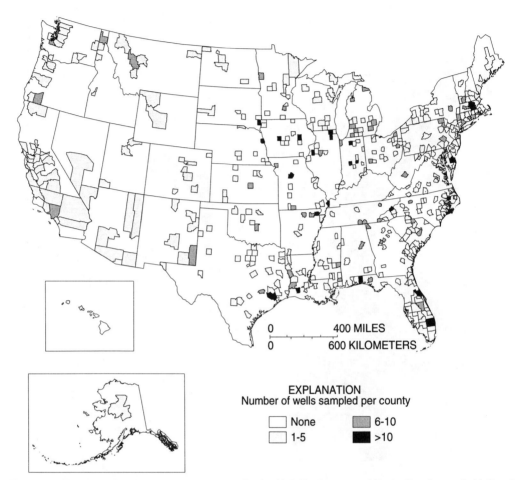

Figure 2.4. Number of wells sampled per county for the U.S. Environmental Protection Agency's National Pesticide Survey (data from Haaser, 1994).

2.4 GENERAL DESIGN FEATURES

Table 2.5 summarizes the general features of the process and matrix-distribution investigations and the monitoring studies examined for this book, based on information contained in Tables 2.1 through 2.4. Process and matrix-distribution studies are conducted in comparatively small areas to address specific questions and, when compared with monitoring studies, are more likely to involve a detailed characterization of the subsurface properties of the site(s) under investigation. Such studies often include high-quality information on pesticide application. In contrast, the focus on the larger areas examined by monitoring studies typically results in substantially less detailed descriptions of the sampled areas, frequently in terms of broad statistical summaries of individual parameters across the entire study region (e.g., well depth, well age, and distances to various land-use features of interest).

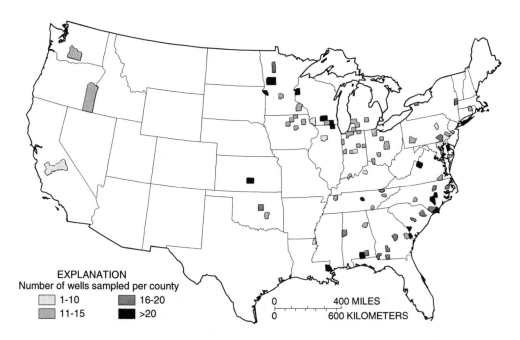

Figure 2.5. Number of wells sampled per county for the National Alachlor Well Water Survey (data from Klein, 1993).

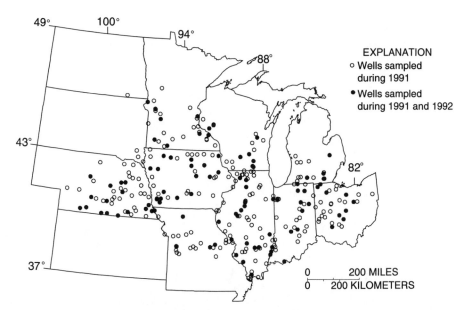

Figure 2.6. Locations of wells sampled during 1991 and 1992 for the Midcontinent Pesticide Study (redrawn from Kolpin and others, 1993).

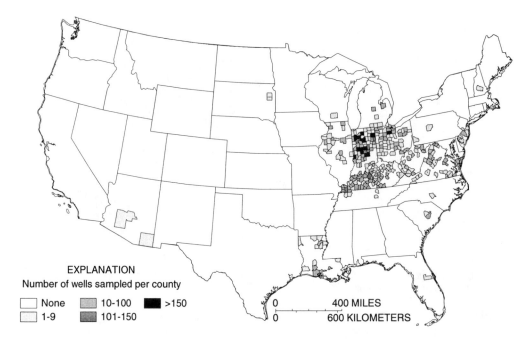

Figure 2.7. Number of wells sampled per county for triazine or acetanilide herbicides for the Cooperative Private Well Testing Program as of June 1994 (data from Richards, 1994).

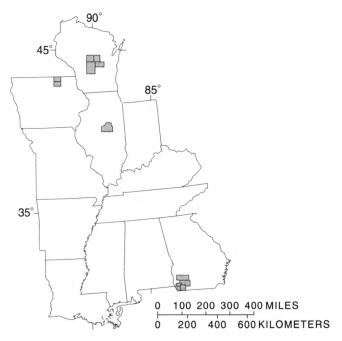

Figure 2.8. Counties sampled for the Metolachlor Monitoring Study (adapted from Roux and others, 1991a).

Table 2.5. General characteristics of the process and matrix-distribution and monitoring studies examined for this book, compiled from Tables 2.1 through 2.4

Study Characteristics	Number of Studies		
	Process and Matrix-Distribution Studies	State and Local Monitoring Studies	Multistate Monitoring Studies
Number of Studies	213	122	5
Sampled Matrices			
Ground water	55	121	5
Soil	148	0	0
Vadose-zone water	55	0	0
Tile drainage	31	7	0
Compound Classes Examined			
Triazines	90	79	4
Acetanilides	44	64	5
Carbamates and thiocarbamates	43	56	1
Ureas and sulfonylureas	12	19	1
Organochlorine compounds	16	45	2
Organophosphorus compounds	28	42	1
Chlorophenoxy acids	19	40	1
Fumigants	7	31	1
Transformation products	65	48	3
Others	75	58	2
Parameters Measured (Process and Matrix-Distribution Studies only—see Tables 2.1 and 2.2 for parameter definitions)			
Fmp only ($\pm f_{oc}$, $\pm T_{app}$)[1]	90		
Mgmt only ($\pm f_{oc}$, $\pm T_{app}$)[1]	42		
K_d, K_{ow}, K_{oc}, R_f or R	29		
Soil $t_{1/2}$ or k_1 (s,aq)	65		

[1]May or may not have included measurements of f_{oc} or T_{app}, denoted as $\pm$.

The high cost of chemical analyses of environmental samples for pesticides and their transformation products requires that investigators carefully balance the distribution and intensity of spatial and temporal sampling. Most observations of temporal variations in pesticide concentrations at a specific location are from process and matrix-distribution studies, rather than monitoring studies, which have concentrated mostly on large-scale spatial distributions. Temporal data are usually obtained for the aqueous phase (i.e., lysimeter leachate, tile drainage, or ground water), rather than for soils, and typically span periods ranging from several days to several years.

The monitoring studies examined for this book exhibited substantial variability in several design features, such as the nature of the well-selection process, the types and depths of the wells sampled, the detection limits employed, and the times of year when sampling was conducted (Tables 2.3 and 2.4). Such heterogeneity of design hampers efforts to integrate the results from these studies into a single, coherent picture of pesticide occurrence and behavior in ground waters across the nation.

2.5 GEOGRAPHIC DISTRIBUTION

The sampling of ground water for pesticides has been carried out over a range of environmental conditions that encompasses the majority of climatic, hydrogeologic, edaphic, and agricultural settings in which these compounds have been applied to the land in the United

States. The geographic scope of the investigations examined for this book is reflected in Figures 2.1 and 2.2 for the process and matrix-distribution studies, in Figure 2.3 for the state and local monitoring studies, and in Figures 2.4 through 2.8 for the multistate monitoring studies.

Statewide monitoring surveys have been conducted in a number of states across the country (Table 2.3). Among the multistate studies, only one—the National Pesticide Survey, or NPS (U.S. Environmental Protection Agency, 1990, 1992a)—involved sampling for pesticides in the ground waters of all 50 states. The four remaining multistate studies sampled ground waters in 4 to 26 states (Figures 2.5 through 2.8, Table 2.4). The areas of investigation for the multistate studies were chosen to encompass either: (1) a wide variety of environmental and agricultural settings (Roux and others, 1991a; Baker and others, 1994); (2) counties where a specific pesticide is sold (Holden and others, 1992); or (3) regions in which a particular suite of crops is grown (Kolpin and Burkart, 1991).

The areas of the United States where ground waters have been monitored most intensively for pesticides (Figures 2.1 through 2.8) are those regions of the country where agricultural activities and pesticide use are most extensive. Thus, the most extensive sampling of ground waters for pesticides has been carried out in California, Florida, New York (especially Long Island), most of the states in New England, the central Atlantic Coastal Plain, and the central and northern midcontinent. By contrast, sampling of ground waters for pesticides in the Rocky Mountain states and the arid southwest, where pesticide use is less extensive, has been relatively sparse.

The principal exceptions to the general geographic correspondence between pesticide use and ground-water monitoring efforts occur in many of the states in the southeast and southern midcontinent. With the exception of Florida, ground-water sampling for pesticides in these states has been limited, despite relatively extensive pesticide use.

2.6 MEDIA SAMPLED

The specific media in which pesticide residues have been measured have depended largely on the type of study involved. Nearly all monitoring studies examined for this report focused on the detection of pesticides in ground water, with only a minor degree of attention directed toward the measurement of residue levels in soils or vadose-zone waters (monitoring studies that reported the concentrations of pesticides in soils, but not in ground water [e.g., Mischke and others, 1985] were excluded from Table 2.3). By contrast, the process and matrix-distribution studies have examined the distributions of these compounds among a broader variety of media—including ground water, vadose-zone water, tile drainage water, soils, plant tissues, and surface water—but have focused primarily on residues in soils, rather than on subsurface waters. Thus, although 69 percent of the 213 process and matrix-distribution studies reviewed for this report measured pesticide concentrations in soils (148 studies), only 26 percent (55 studies) measured pesticide concentrations in ground waters. Measurements of pesticide residues in vadose-zone waters and in tile drainage were carried out in 26 and 15 percent of the process and matrix-distribution studies, respectively (Table 2.5).

The emphasis on pesticide residues in soils, rather than in ground water or other subsurface waters may result from a traditional focus on the degree to which pesticide residues persist in the root zone (or "rhizosphere"), rather than on their ultimate fate in the environment. This approach has been based on the common, implicit assumption that the environmental significance of pesticides decreases with diminishing concentrations of the parent compounds in soils (e.g., Garrett and others, 1975; Helling and others, 1988; Hall and others, 1989; Neary and Michael, 1989; Smith and others, 1990; Isensee and others, 1990; Bush and others, 1990; Kördel and others, 1991).

2.7 TARGET ANALYTES

2.7.1 PARENT COMPOUNDS

Most process and matrix-distribution studies involve the application of one or more pesticides under controlled conditions, and these compounds are usually the only ones for which chemical analyses are conducted. The principal objective of most monitoring studies, on the other hand, is to determine which pesticides are present in the ground waters in the area(s) of interest, thereby requiring a broad analytical capability.

Not surprisingly, the types of pesticides analyzed have been largely determined by the extent of use or concern at the time of sampling. Thus, decreases in the use of organochlorine pesticides since the 1960's were accompanied by a decrease in the proportion of studies examining these compounds (Figures 2.9 through 2.11). Conversely, with increases in the use of triazine and acetanilide herbicides over the past three decades, more recent studies have

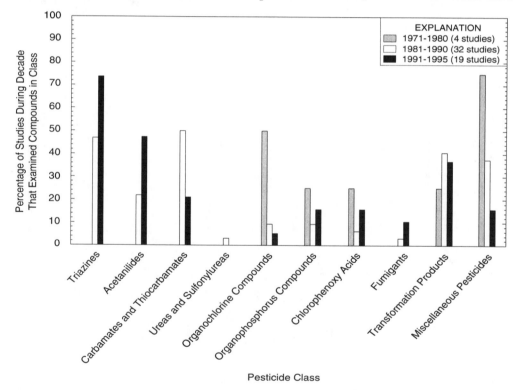

Figure 2.9. Percentage of process and matrix-distribution studies carried out in each decade from 1971 to 1995 that examined pesticides from individual chemical classes in ground water (data derived from Table 2.1).

increased the attention devoted to them. Ongoing concern over pesticides whose use has been discontinued, but that still persist in ground water where former use was heavy, is reflected in the considerable number of recent studies of the long-term subsurface fate of the fumigants DBCP and 1,2-dibromoethane (EDB).

In addition to these temporal trends, geographic patterns are also evident among the types of pesticides analyzed in ground waters in different regions of the United States, primarily reflecting the compounds used—and those of principal concern—within the region(s) of interest. For example, sampling for DBCP and EDB has been largely concentrated in California, Hawaii, Connecticut, Florida, Georgia, and Washington; investigations of aldicarb occurrence and behavior (Wells and Waldman, 1991) have focused on the sandy soils of the northern midcontinent, the northeast, the Atlantic Coastal Plain, California, and Prince Edward Island, Canada; and much of the sampling for triazine and acetanilide herbicides in ground waters has been carried out in the corn-and-soybean producing areas of the midcontinent.

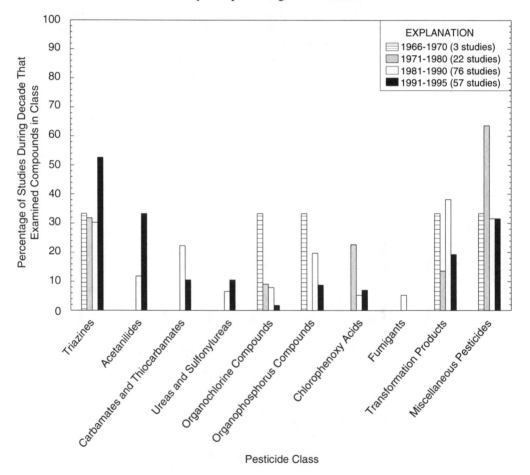

Figure 2.10. Percentage of process and matrix-distribution studies carried out in each decade from 1966 to 1995 that examined pesticides from individual chemical classes in subsurface media, excluding ground water (data derived from Table 2.2).

2.7.2 TRANSFORMATION PRODUCTS

Only a relatively small number of transformation products have been analyzed in ground-water monitoring studies to date, a point also noted by Hallberg (1989). Nevertheless, the few products that have been examined follow the historical shift noted for their parent compounds. Thus, although the transformation products of organochlorine insecticides (primarily DDD and DDE, from DDT, and heptachlor epoxide, from heptachlor) were of principal interest in many of the studies conducted in the late 1970's and early 1980's, more recent work has focused on the products generated from herbicides and insecticides in widespread use today, particularly those from atrazine, DCPA, alachlor, aldicarb, and carbofuran.

There are several possible reasons why pesticide transformation products have historically received comparatively little attention. First, as noted earlier, investigations of pesticide behavior in the subsurface have tended to focus more on measuring the time interval during which pesticide concentrations remain above biologically effective levels within the root zone—the domain within which most pesticides influence their target organisms—rather than on determining the ultimate fate of the parent compounds (e.g., Cooke, 1966; Brewer and others, 1982; and Randhawa and Gill, 1984). Second, most transformation reactions of organic

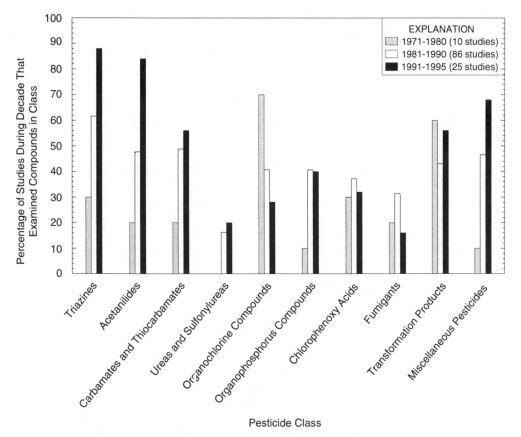

Figure 2.11. Percentage of state and local monitoring studies carried out in each decade from 1971 to 1995 that examined pesticides from individual chemical classes (data derived from Table 2.3).

molecules in soils and aqueous media increase the water solubility of the compound and the analytical methods are often more difficult than for the corresponding parent compounds. Finally, the more limited interest in tracking the environmental fate of pesticide transformation products in the subsurface, relative to the parent compounds, may be the result of a general lack of water-quality criteria for transformation products (Nowell and Resek, 1994), and hence, diminished regulatory attention toward them. However, some states, such as Wisconsin (Lawrence and others, 1993), have recently enacted water-quality standards that apply to pesticide transformation products, in addition to the parent compound.

Over the past five years, however, monitoring studies have devoted increasing attention to transformation products in ground waters, and some of these products have been found more frequently than their parent compounds. For example, deethyl atrazine and the demethylated products of DCPA hydrolysis ("DCPA acid metabolites")—rather than their parent compounds—were the analytes detected most often in ground waters during the first year of the study of herbicides in near-surface aquifers of the midcontinent by Burkart and Kolpin (1993, hereafter referred to as the Midcontinent Pesticide Study, or MCPS) and the NPS (U.S. Environmental Protection Agency, 1990a, 1992a), respectively. During the second year of sampling for the MCPS, alachlor ethanesulfonic acid (ESA) was detected in ground water nearly 10 times as frequently as its parent compound, alachlor (Kolpin and others, 1993).

To reduce the complexity or cost of the chemical analyses, the concentrations of transformation products are often quantified together with those of the parent compound. This has been common for studies of aldicarb, fenamiphos, and phorate, for which the sulfoxide and sulfone product concentrations are often summed with that of the parent and reported as "total toxic residue," or TTR. While reducing analytical costs, such an approach limits interpretation of the resulting data because transformation products usually exhibit substantially different toxicological, physical, and chemical properties from their parent compounds. The interpretation of results from studies employing immunoassay analyses (discussed below) poses similar difficulties.

2.8 ANALYTICAL METHODS AND DATA REPORTING

A wide variety of analytical methods have been employed to measure the concentrations of pesticides and their transformation products in environmental media. For example, according to a summary by Moye and Miles (1988), the detection and quantification of TTR for aldicarb has involved the use of at least 18 different analytical approaches since 1968. Such proliferation of analytical methods is typical of initial studies of relatively new or unfamiliar chemical species. As analyses for the parent compound and its transformation products become more routine for a given pesticide, however, procedures become more standardized.

2.8.1 DETECTION LIMITS AND ANALYTE RECOVERIES

Data on analytical uncertainties, detection limits, and analyte recoveries make it possible to determine whether results from different investigations may be compared, even if different methods were employed. Many of the process and matrix-distribution studies listed in Tables 2.1 and 2.2, however, did not include detection limits for the analytes of interest, and only occasionally provided data on analyte recoveries. Hallberg (1989) has noted similar deficiencies among such studies. Although some studies provided extensive detail on analytical methods, analyte recoveries, and detection limits (e.g., Lutz and others, 1973; Sirons and others, 1973; Muir and Baker, 1976; Neary, 1983; Segawa and others, 1986; Watson and others, 1989; Frank and others, 1990; Miles and others, 1990; Kladivko and others, 1991; Koterba and others, 1993; and Newton and others, 1994), many process and matrix-distribution studies and monitoring studies provided no information on analytical methods (e.g., Whitehead, 1974; Plumb, 1985; Carsel and others, 1986; Detroy, 1986; Jury and others, 1986a; Kelley and Wnuk, 1986; Schmidt, 1986a,b; Wagenet and Hutson, 1986; Detroy and others, 1988; Druliner, 1989; Eckhardt and others, 1989a,b; Shirmohammadi and others, 1989; Clendening and others, 1990; Kördel and others, 1991; Libra and others, 1991; Plumb, 1991; Smith and others, 1991a,b; and Richards, 1992a). Similarly, most discussions of minimum recommended requirements for the design of process and matrix-distribution studies (e.g., Kimball, 1988; DeMartinis, 1989; and Jones, 1990) have under-emphasized the need for consistent data reporting and adequate quality control in the analytical laboratory.

Monitoring studies have more frequently reported detection limits for target analytes than have the process and matrix-distribution studies, probably because of the greater emphasis placed by the former on frequencies of analyte detection in ground water. The inclusion of detection limits is of particular importance because variations in analytical detection limits directly influence the frequency of analyte detections (Gilliom and others, 1985; Barrett and Williams, 1989; Burkart and Kolpin, 1993; Kolpin and others, 1995), a point discussed further in Section 6.6.

Studies using field-spiked samples for determining analytical detection limits (e.g., Frank and others, 1990) or for detecting losses during storage (e.g., Weaver and others, 1983; Troiano and others, 1987; Marade and Segawa, 1988; Cavalier and others, 1991; Shaffer and Penner, 1991; and Ando, 1992) are uncommon. The need for data on pesticide recoveries, however, was underscored by a study of the analytes included in the NPS (Munch and Frebis, 1992). The authors measured the recoveries of 147 pesticides from well-water samples that had been spiked, biologically inhibited, and stored at 4°C. Despite the use of methods approved by the U.S. Environmental Protection Agency, up to 100 percent loss was observed after 14 days of storage for 26 out of the 147 compounds. Several of these 26 compounds have nevertheless been the subject of some of the studies listed in Tables 2.1 through 2.3; these compounds include methyl parathion, terbufos, azinphos-methyl, fonofos, prometon, and phorate.

2.8.2 USE OF IMMUNOASSAYS

Enzyme-linked immunosorbent assay (ELISA) has long been used for microbiological and serological analyses in the health sciences (Hallberg, 1995). More recently, this technique has been adapted for environmental applications, providing a rapid, convenient method for measuring concentrations of a variety of pesticide analytes in the field, using reasonably low levels of detection. Although use of ELISA methods is becoming more widespread (particularly among monitoring studies), of the investigations examined for this book, only a relatively small number employed them (Bushway and others, 1992; Krapac and others, 1993; Libra and others, 1993; Baker and others, 1994; Glanville and others, 1995), most notably the Cooperative Private Well Testing Program, or CPWTP (Baker and others, 1994).

ELISA methods provide a comparatively inexpensive way to detect the presence of one or more analytes in a particular chemical class, but they do not currently identify individual compounds within the class of interest (e.g., Wittmann and Hock, 1991; Bushway and others,

1982; Wallrabenstein and Baker, 1992; Baker and others, 1993, 1994; Meulenberg and others, 1995). Consequently, interference from unknown levels of related parent compounds and transformation products in environmental samples usually causes the ELISA technique to overestimate the concentration of the specific analyte to which the test is keyed, relative to the concentration measured by more compound-specific, chromatographic methods (Bushway and others, 1992; Baker and others, 1993; Krapac and others, 1993; Libra and others, 1993). The agreement is quite close, however, when ELISA results are compared with those from chromatographic analyses using known concentrations of specific compounds (Lawruk and others, 1992; Baker and others, 1993; Gascón and others, 1995; Gruessner and others, 1995). The ELISA method may also exhibit relatively high degrees of variability among replicate analyses (Lawruk and others, 1992).

Each ELISA method exhibits varying degrees of sensitivity to the different compounds to which it responds. For example, Baker and others (1994) found that the ELISA test for acetanilides was one-fifth as sensitive to ESA—the principal alachlor transformation product detected by the procedure—as it was to alachlor itself. The acetanilide concentrations that they reported may, therefore, have been underestimates of the "true" values for the sum of all of the acetanilide herbicides and transformation products during the CPWTP.

Despite these limitations, ELISA is a valuable, low-cost screening tool for identifying ground waters in which the pesticides of interest are not detectable, and thus may be used to reduce the number of samples requiring more elaborate analysis in the laboratory (Baker and others, 1994). It may also prove to be useful for examining long-term changes in spatial patterns of contamination in systems where the parent compounds and transformation products of interest are well known. In the laboratory, ELISA has been used as a relatively inexpensive detector for pesticides and their transformation products, such as ESA (Aga and others, 1994).

2.8.3 USE OF BIOASSAYS

Many field and laboratory studies employed bioassays, rather than chemical analyses, to measure pesticide concentrations in soils (e.g., Rodgers, 1968; Buchanan and Hiltbold, 1973; Hotzman and Mitchell, 1977; Majka and Lavy, 1977; Marley, 1980; Brewer and others, 1982; Randhawa and Gill, 1984; García and others, 1984; Braverman and others, 1986; Schmitz and Witt, 1988; van Biljon and others, 1988; Wadd and Drennan, 1989; van Biljon and others, 1990; and Vicari and others, 1994) or subsurface leachate (e.g., Hotzman and Mitchell, 1977). These tests typically infer pesticide concentrations on the basis of growth yields or levels of phytotoxicity observed for one or more sensitive plant species.

The technological difficulties associated with the chemical analysis of pesticides may have required this type of approach in the past, and may still require it where the requisite analytical instrumentation is lacking. However, plant bioassays systematically underestimate the concentrations of pesticides in soils relative to instrumental analyses, as has been observed for several compounds, including atrazine (Hall and Hartwig, 1978), pendimethalin (Barrett and Lavy, 1983), and oxadiazon (Barrett and Lavy, 1984). The difference between the pesticide concentrations measured by bioassay and those measured by chemical extraction has been characterized as the fraction of pesticide that is biologically inactive (Barrett and Lavy, 1983, 1984), but this has not been rigorously demonstrated, nor have the specific physical, chemical, or biological mechanisms behind such inactivation been elucidated.

2.8.4 USE OF RADIOLABELED COMPOUNDS

A relatively common method employed for the analysis of pesticides and their transformation products in environmental media during process and matrix-distribution studies has been the use of radiolabeled compounds in conjunction with scintillation counting (e.g., Majka and Lavy, 1977; Basham and others, 1987; Kubiak and others, 1988; Barnes and others, 1989; Goetz and others, 1990; Kördel and others, 1991; Demon and others, 1994; and Lee and others, 1994). This method makes it possible to follow the movement of relatively low analyte concentrations through the subsurface. However, despite its advantages, this approach, like the ELISA technique, is hampered by a lack of compound specificity. In the absence of procedures designed to separate individual analytes—typically through chromatography or pH-based fractionation—the method does not distinguish among the different compounds in which the ^{14}C label might be located. Recent work by Gu and others (1995) also suggests that the use of radiolabels to track solute movement in the subsurface may lead to considerable experimental error if the labeled solute exhibits significant volatility, due either to the presence of nonvolatile impurities or the potential influence of the ^{14}C label on diffusion and volatilization. This method is therefore of limited utility in elucidating transformation pathways (e.g., Kördel and others, 1991) or tracking the environmental fate of specific reaction products.

2.9 EFFECTS OF DIFFERENT AGRICULTURAL PRACTICES

Among the 213 process and matrix-distribution studies reviewed, 63 (30 percent) examined the influence of modifications in agricultural practices on the transport and fate of pesticides in the subsurface (Tables 2.1 and 2.2). Only ten, however, directly assessed impacts on ground water, most of the studies having focused on soils. Table 2.6 (at end of chapter) summarizes the topics investigated in each study. Of the practices examined, the influence of different degrees and methods of tillage received greatest attention (27 studies), followed by the effects of varying the rates at which water or the pesticide are applied to the land (19 and 12 studies, respectively). Other topics included the effects of variations in the timing and placement of pesticide applications, the timing of recharge, the nature of the pesticide formulation, and crop rotation sequences. In addition to the process and matrix-distribution studies, a few of the monitoring investigations evaluated the effects of different agricultural practices—particularly irrigation—on pesticide occurrence in the subsurface (see Chapter 5).

2.10 ANCILLARY PARAMETERS

Tables 2.1, 2.2, and 2.5 list the parameters measured most frequently during process and matrix-distribution studies to characterize pesticide behavior in the subsurface (monitoring investigations typically do not include the measurement of these parameters). The most common approach adopted by process and matrix-distribution studies has been to report experimental results solely in terms of pesticide concentrations in one or more subsurface media, or their "depth of leaching," following application. Of the 213 process and matrix-distribution investigations reviewed, 90 (42 percent) were of this type (Table 2.5). Such data are highly site-specific, and hence, not easily compared with analogous data from other locations, times, or agricultural management regimes. Other process and matrix-distribution studies have reported data characterizing the persistence or partitioning of pesticides—and in some cases their transformation products—under a variety of agricultural management practices (Table 2.6). Among the ancillary parameters provided, the most widely used and commonly reported are soil-water partition coefficients and soil dissipation half-lives.

2.10.1 SOIL-WATER PARTITION COEFFICIENTS

As organic compounds, pesticides have an affinity for the natural organic matter present in soils and other subsurface solids. Indeed, sorption to soil organic matter constitutes the principal mechanism by which the migration of pesticides is slowed relative to the rate of water movement through the subsurface. The stronger the affinity exhibited by a given compound for soil or aquifer solids, the more its migration through the subsurface is retarded relative to that of water. Among the 213 process and matrix-distribution studies reviewed, however, only 25 (12 percent) reported values of partitioning parameters for specific pesticides or transformation products (Table 2.5). The partition coefficients provided by these studies were keyed to the specific soil (K_p), soil organic carbon (K_{oc}), or octanol (K_{ow}). Other, related measurements included retardation coefficients (R) and soil thin-layer chromatography mobilities (R_f). The partitioning parameters will be defined, and their significance to pesticide transport discussed, in Chapter 4.

2.10.2 SOIL DISSIPATION HALF-LIVES

Among the 213 process and matrix-distribution studies reviewed, 65 (31 percent) reported one or more measures of pesticide persistence in subsurface media (Table 2.5). Most common among the parameters given was the "soil dissipation half-life," provided by 61 of the 65 studies. A dissipation half-life is an inherently site-specific parameter, reflecting the rate at which the concentration of a given compound within a specific medium decreases at a particular location, typically above the water table. As such, it represents the overall rate of depletion of the compound of interest by several simultaneous chemical and physical processes, including transformation, volatilization, and transport away from the site of measurement—in either the aqueous or gas phases—by advection and molecular diffusion. With few exceptions (Smith and others, 1991a; Yen and others, 1994), none of the studies reporting dissipation half-lives attempted to distinguish the relative importance of transformation and transport in reducing pesticide concentrations in situ.

The utility of a dissipation half-life value, in any reaction medium, is substantially reduced if the temperature of measurement is unknown or not specified. Despite a recognition of the importance of temperature (Nash, 1988), soil temperatures are seldom reported in conjunction with half-life values (e.g., Jones, 1986; Nash, 1988). Among the 61 reviewed investigations reporting soil dissipation half-lives, only 21 provided temperatures of measurement in situ.

A dissipation half-life is also of limited value if it pertains to a group of compounds simultaneously, rather than a single compound. In their efforts to simulate the distribution of aldicarb and its transformation products (aldicarb sulfoxide and aldicarb sulfone) in ground water, for example, Jones and others (1988) used a single, "total toxic residue degradation rate" to describe the reduction of aldicarb TTR concentrations in situ. Data summarized by Moye and Miles (1988), however, indicate that the second-order rate constants for the hydrolysis of aldicarb sulfone and aldicarb sulfoxide are an order of magnitude higher than that for aldicarb under alkaline conditions (temperature not given).

2.10.3 USE OF STANDARDIZED PARAMETERS

The spatial and temporal complexities of natural systems introduce a large amount of uncertainty in the environmental behavior of pesticides in the subsurface, thus reducing the confidence with which results from one site may be applied to another. The use of standardized parameters can control for some of this variability by "normalizing" each property of interest to a common reference point (e.g., Roberts and others, 1982). Efforts to express the results from process and matrix-distribution investigations in more standardized terms, however, have been limited. The most widely-used parameter of this type is the soil organic matter partition coefficient, or K_{oc} (discussed in Chapter 4). Of the 25 process and matrix-distribution studies that reported one or more measures of partitioning, however, only six provided estimates of K_{oc}. The reporting of pesticide concentrations in specific environmental compartments as proportions of the total mass of the parent compound applied (e.g., Boesten, 1987) provides a standardized measure of pesticide distributions that can be compared among different studies. Similarly, expressing either irrigation rates (Wyman and others, 1985; Troiano and others, 1990) or drainage rates (Radcliffe and others, 1988; Kolberg and others, 1989) as proportions of the local rate of evapotranspiration facilitates comparisons among the rates of water flux in different study locations.

2.11 SUMMARY

Tables 2.1 through 2.6 demonstrate the considerable number of studies and reviews that have examined the occurrence and behavior of pesticides in ground water and other subsurface media. The results from many of these studies have been reported in ways that limit the extent to which they can be generalized to other environmental settings, agricultural management regimes, and compounds. Despite these difficulties, however, the large volume of data on pesticide detections in ground waters of the United States provides a wealth of opportunities to examine patterns of occurrence, and test hypotheses regarding the processes that control the distribution and fate of pesticides in the subsurface.

Table 2.1. Principal characteristics of process and matrix-distribution studies that included measurements of pesticide concentrations in ground water

[Site Location(s): Sites placed within brackets indicate probable study locations, since exact locations were not given. Site Area(s): For studies that did not provide areas, estimates derived from study maps are given in brackets. ha, hectare. Soil Type(s): Includes textures of soils at all study sites, given for surficial soils only. Abbreviations taken from U.S. Department of Agriculture (1991). For studies where soil type was not provided, inferred soil type is given in brackets. Analytes Examined During Study: All analytes were not necessarily examined at all sites for a given study. Abbreviations used for chemical parameters: TTR, total toxic residue, consisting of the summed concentrations of the parent and one or more of its transformation products (used most frequently for aldicarb); alk, alkalinity; DO, dissolved oxygen; TDS, total dissolved solids; SC, specific conductance; NH_4, ammonium; NO_3, nitrate; NO_2, nitrite; SO_4, sulfate; PO_4, phosphate; E_H, oxidation-reduction potential; $\delta^{18}O$, δ^2H are isotope ratios for $^{18}O/^{16}O$ and $^2H/^1H$, respectively. (Note: This list is not intended to be exhaustive; additional chemical parameters were measured by several of the studies cited, but those listed here are the variables deemed to be most relevant for understanding pesticide transport and fate.) Media Sampled: Media sampled and analyzed for pesticides during field component of study. Abbreviations: PT, plant tissue (either standing plants or leaf litter); S, soil (any depth, analyzed either by chemical extraction or crop bioassay); V, vadose-zone water, obtained from lysimeters or buried soil columns; TD, tile drainage water; GW, ground water, obtained from wells or springs; SW, surface water. Parameter(s) Measured: Parameter definitions: aq. $t_{1/2}$, half-life for transformation in aqueous solution; f_{oc}, soil organic carbon content; fmp, field mobility and persistence in one or more of the media investigated; fmp (model test), study conducted for comparison with modeling results (see Table 9.1); K_d, distribution coefficient for soil (not normalized to organic carbon; measured in laboratory unless otherwise indicated); K_{ow}, octanol-water partition coefficient, measured in laboratory; mgmt, effects of different agricultural management practices (e.g., tillage, irrigation, or timing of pesticide applications) on movement and persistence of pesticides in the subsurface; P/P, product-to-parent concentration ratio(s) for selected pesticides; R, retardation factor; soil $t_{1/2}$, dissipation half-life in soil, measured in the field unless otherwise indicated; T_{app}, air or soil temperature measured during pesticide application. (Note: This list is not intended to be exhaustive. Additional parameters were measured by several of the studies cited, but those listed here were encountered most frequently among different studies.) Blank cells indicate no information applicable or available. NG, not given]

Study No.	Reference	Site Location(s)	Site Area(s) (ha, unless stated otherwise)	Soil Type(s)	Analyte(s) Examined During Study — Parent Pesticide(s)	Transformation Product(s)	Others	PT	S	V	TD	GW	SW	Parameter(s) Measured
1	Knutson and others, 1971	Kansas—Ellis County	11.5	sicl	Diazinon, parathion, methyl parathion, heptachlor, aldrin, endrin	Heptachlor epoxide, dieldrin			•			•		fmp, P/P, f_{oc}
2	LaFleur and others, 1973	South Carolina—Florence	0.01	[ls][1]	Toxaphene, fluometuron				•			•		fmp, soil $t_{1/2}$
3	Bovey and others, 1975	Texas—Riesel, Temple	3.1	c, c	2,4,5-T, picloram				•	•		•		fmp

Table 2.1 Principal characteristics of process and matrix-distribution studies that included measurements of pesticide concentrations in ground water—*Continued*

No.	Reference	Site Area(s) (ha, unless stated otherwise)	Site Location(s)	Soil Type(s)	Parent Pesticide(s)	Transformation Product(s)	Others	P/T	S	V	T/D	G/W	S/W	Parameter(s) Measured
4	Hebb and Wheeler, 1978	0.015	Florida—[Marianna]	s	Bromacil							•		fmp, f_{oc}
5	Wehtje and others, 1981, 1984	50	Nebraska—Bellwood	ls, s, sl	Atrazine		Br⁻ (tracer), NO_3		•	•		•		fmp, K_d, f_{oc}
6	Neary, 1983	35.2 100	Tennessee—Centerville Coleman Hollow Jenkins Hollow	crl crl	Hexazinone	Two unspecified products, presumably "metabolites A and B"[2]						•		fmp
7	Neary and others, 1985	4	North Carolina—Nantahala National Forest	stl	Picloram				•	•		•	•	fmp, soil $t_{1/2}$, f_{oc}
8	Wyman and others, 1985	0.014 0.048	Wisconsin—Hancock Cameron	s sl	Aldicarb (as TTR, along with products)	Aldicarb sulfoxide and sulfone (as TTR)	pH	•	•	•			•	mgmt, f_{oc}, T_{app}
9	Anderson, 1986	130	Wisconsin—Central Sand Plain	[s]	Aldicarb	Aldicarb sulfoxide and sulfone (as TTR)						•		fmp (model test)
10	Jones and others, 1986b	0.92 0.06 0.05 12.2 5.0, 14.2 0.07 0.3	Arizona—Maricopa North Carolina—Harrellsville Virginia—Blackstone Michigan—Blissfield Indiana—Bluecast Washington—Pasco Florida—Fort Pierce	[sl][1] [sl][1] [cl][1] [scl][1] [cl][1] [sl][1] [s][1]	Aldicarb (as TTR, along with products)	Aldicarb sulfone and sulfoxide (as TTR)	pH		•			•	•	fmp, soil $t_{1/2}$, f_{oc}

Table 2.1 Principal characteristics of process and matrix-distribution studies that included measurements of pesticide concentrations in ground water—*Continued*

No.	Reference	Site Location(s)	Site Area(s) (ha, unless stated otherwise)	Soil Type(s)	Parent Pesticide(s)	Transformation Product(s)	Others	P/T	S	V	S/D	T/D	G/W	S/W	Parameter(s) Measured
									Media Sampled						
11	Welling and others, 1986	California— Tulare County Ventura County	NG 4.1	1 1	Diuron, simazine Bromacil, diuron, simazine					•					fmp, soil $t_{1/2}$, f_{oc}
12	Krawchuk and Webster, 1987	Canada—Portage la Prairie, Manitoba	[220]	ls	Atrazine, 2,4-D, dicamba, chlorothalonil, MCPA, bromoxynil, decamethrin, disulfoton, mancozeb, EPTC, carbofuran, azinphos-methyl, diclofop-methyl, metribuzin, chlorpyrifos, metiram, propanil, paraquat, aldicarb, trifluralin	Carbofuran phenol						•	•		fmp, K_{ow}, f_{oc}
13	Priddle and others, 1987, 1989	Canada—Prince Edward Island	[15-25]	sifs till	Aldicarb (as TTR, along with products)	Aldicarb sulfone and sulfoxide (as TTR)	pH, E_H, DO, SC, NO_3, NH_3					•	•		fmp, soil $t_{1/2}$, T_{app}, P/P
14	Segal and others, 1987	Florida—Alachua County	15	s	Sulfometuron-methyl		NO_3, NH_4, PO_4, K, Ca, Mg					•	•	•	fmp
15	Weaver and others, 1987	West Virginia—Romney (Hampshire County); Berkeley and Jefferson Counties	12	NG	Endrin, napropamide, terbacil					•			•	•	fmp

Table 2.1 Principal characteristics of process and matrix-distribution studies that included measurements of pesticide concentrations in ground water—*Continued*

No.	Reference	Site Location(s)	Site Area(s) (ha, unless stated otherwise)	Soil Type(s)	Parent Pesticide(s)	Transformation Product(s)	Others	P/T	S	V	T/D	G/W	S/W	Parameter(s) Measured
16	Wilson and others, 1987	Colorado—Weld County	800	sl, ls	Atrazine							•	•	fmp
17	Fathulla and others, 1988	Wisconsin—Plover; Bancroft; Plainfield; Hancock	130; 130; 65; 260	ls; ls; s; ls, s, sl	Aldicarb (as TTR, along with products)	Aldicarb sulfone and sulfoxide (as TTR)	Major ions, SC, alk, NO$_3$, SO$_4$, pH					•		fmp, T$_{app}$, f$_{oc}$
18	Isensee and others, 1988	Maryland—Beltsville	0.0023	sil (Helling and others, 1988)	Alachlor, atrazine, cyanazine							•		fmp
19	Jones and others, 1988	Wisconsin—Hancock	195	s, sl	Aldicarb							•		fmp (model test), soil t$_{1/2}$, f$_{oc}$
20	Leonard and others, 1988	Georgia—Tifton	0.23	s	Aldicarb (as TTR, along with products), atrazine, butylate, EDB and fenamiphos (as TTR, along with products)	Aldicarb sulfone and sulfoxide (as TTR), fenamiphos sulfone and sulfoxide (as TTR)			•			•		fmp, soil t$_{1/2}$, f$_{oc}$
21	Maathuis and others, 1988	Canada—Outlook, Saskatchewan "Konst site" "Pederson site" "Sask. Irr. Dev. Center"	1,100; 16.7; 80	si, s, c; sic, cl; fl, si	Bromoxynil, 2,4-D, dicamba, diclofop-methyl, MCPA, trifluralin, triallate		Major ions, tritium, δ^{18}O, δ^{2}H				•			fmp

Table 2.1 Principal characteristics of process and matrix-distribution studies that included measurements of pesticide concentrations in ground water—*Continued*

No.	Reference	Site Location(s)	Site Area(s) (ha, unless stated otherwise)	Soil Type(s)	Parent Pesticide(s)	Transformation Product(s)	Others	PT	S	V	TD	GW	SW	Parameter(s) Measured
						Analyte(s) Examined During Study				Media Sampled				
22	Steenhuis and others, 1988, 1990	New York—Willsboro	0.21 - 0.29	scl, cl	Alachlor, atrazine, carbofuran, glyphosate		Br⁻ (tracer), NO_3, NO_2		•		•	•		mgmt
23	Garrison and others, 1989	North Carolina—Wayne County	1	s, sl, scl, ls	Fomesafen	Fomesafen amine		•	•			•		fmp, soil $t_{1/2}$, f_{oc}
24	Jones and others, 1989	North Carolina—Clayton (Johnston County)	0.08	ls	Thiodicarb	Methomyl	pH	•				•		fmp, soil $t_{1/2}$, f_{oc}
		New York—Palermo (Oswego County)	0.4	grls										
		Florida—Oviedo (Seminole County)	0.5	fs										
25	Keim and others, 1989	Ohio—Ottawa County	4.9	sic	Alachlor, atrazine, cyanazine, metolachlor, metribuzin, simazine		Pb, Zn, Cd (from applied sludge)				•	•		fmp
		Lucas County	6.5	sl										
26	Neary and Michael, 1989	Florida—Alachua County	10	s	Sulfometuron-methyl						•	•	•	fmp, f_{oc}, T_{app}
27	Ritter and others, 1989	Delaware—Georgetown	0.5	ls	Alachlor, atrazine, carbofuran, cyanazine, dicamba, metolachlor, simazine							•		mgmt
28	Shirmohammadi and others, 1989	Maryland—Queenstown Conventional till No till	5.8 8.9	sl	Atrazine, carbofuran, cyanazine, dicamba, metolachlor, simazine							•		mgmt, f_{oc}

Table 2.1 Principal characteristics of process and matrix-distribution studies that included measurements of pesticide concentrations in ground water—*Continued*

No.	Reference	Site Location(s)	Site Area(s) (ha, unless stated otherwise)	Soil Type(s)	Parent Pesticide(s)	Transformation Product(s)	Others	P T	S	V	T D	G W	S W	Parameter(s) Measured
29	Watson and others, 1989	Montana—Missoula Fort Missoula / North Fork, Elk Creek	0.24 / 2.6	l / sl	Picloram			•				•	•	fmp
30	Bush and others, 1990	Florida—Hughes Island / South Carolina—Barnwell	10 / [30]	s / ls, s	Hexazinone							•	•	fmp
31	Clark, 1990	Montana—Havre / Ronan / Huntley	4, 6 / 28, 18 / 8	l, sl / lfs, sil, fsl, sil / sic, cl	Atrazine, cycloate, prometon, simazine, dicamba, MCPA, picloram, silvex, 2,4-D, 2,4-DB, aldicarb, carbaryl, carbofuran	Aldicarb sulfone and sulfoxide, 3-hydroxy-carbofuran	SC, pH, alk, NO$_3$, Cr, Cu, Fe, Li, Mn, Mo, Sr, Ti, Zn	•			•			fmp
32	Isensee and others, 1990	Maryland—Beltsville	0.25 - 1.69	sil	Alachlor, atrazine, carbofuran, cyanazine	Deethyl atrazine						•		mgmt, f$_{oc}$
33	Porter and others, 1990	New York—Phelps	0.065	grsl	Aldicarb (as TTR, along with products)	Aldicarb sulfone and sulfoxide (as TTR)	pH	•			•		•	mgmt, soil t$_{1/2}$, f$_{oc}$
34	Smith and others, 1990	Georgia—Tifton	0.7	s	Atrazine, alachlor		Br$^-$ (tracer)	•	•					fmp
35	Winnett and others, 1990a	New Jersey—Chatsworth	NG	[s]	Parathion	Paraoxon						•	•	fmp
36	Winnett and others, 1990b	New Jersey—Chatsworth	NG	[s]	Chlorothalonil			•					•	fmp

Table 2.1 Principal characteristics of process and matrix-distribution studies that included measurements of pesticide concentrations in ground water—*Continued*

No.	Reference	Site Location(s)	Site Area(s) (ha, unless stated otherwise)	Soil Type(s)	Parent Pesticide(s)	Transformation Product(s)	Others	PT	S	V	TD	GW	SW	Parameter(s) Measured
37	Frank and others, 1991a	Canada—Ottawa, Ontario	14	cl	Atrazine (as TTR, along with deethyl atrazine)	Deethyl atrazine (as TTR)			•			•	•	fmp, soil $t_{1/2}$, T_{app}, f_{oc}
38	Frank and others, 1991b	Canada—Ottawa, Ontario	14	cl	Cyanazine, metolachlor				•			•	•	fmp, soil $t_{1/2}$, T_{app}, f_{oc}
39	Gish and others, 1991b	Maryland—Beltsville	1.28	sil	Alachlor, atrazine, carbofuran, cyanazine							•		mgmt, f_{oc}
40	Goodman, 1991	South Dakota—Breakins, Hamlin, and Kingsbury Counties	4-32	sil	Alachlor, atrazine, chloramben, chlorpyrifos, cyanazine, 2,4-D, dicamba, endrin, fonofos, lindane, methoxychlor, metolachlor, metribuzin, parathion, pendimethalin, phorate, picloram, propachlor, terbufos, toxaphene, trifluralin		NO_3, NO_2, NH_4, PO_4, DO, SC, Cl, SO_4, TDS			•		•		mgmt, f_{oc}, T_{app}
41	Isensee, 1991	Maryland—Beltsville	Buried, solution-filled vessel	Not applicable	Alachlor	Hydroxy-alachlor						•		fmp, $aq.t_{1/2}$, f_{oc}, T_{app}
42	Norris and others, 1991	Florida—Polk County, Manatee County	1.7, 0.1	fs, fs	Ethoprop			•				•		fmp, soil $t_{1/2}$, f_{oc}

Table 2.1 Principal characteristics of process and matrix-distribution studies that included measurements of pesticide concentrations in ground water—*Continued*

No.	Reference	Site Location(s)	Site Area(s) (ha, unless stated otherwise)	Soil Type(s)	Parent Pesticide(s)	Transformation Product(s)	Others	P/T	S	V	T/D	G/W	S/W	Parameter(s) Measured
43	Obreza and Ontermaa, 1991	Florida—Collier County	8	s	1,3-dichloropropene	3-chloroallyl alcohol		•				•	•	fmp, T_{app}, f_{oc}
44	Roux and others, 1991b	California— Tulare County site 1	6	sg	Simazine							•		fmp
		Tulare County site 2	32	fs										
		Fresno County	7	sg										
		Delaware—Sussex County	19	s										
		Florida—Hardee County	4	fs										
		Palm Beach County	4	s										
		Illinois—Winnebago County	6	s										
		Indiana—Jackson County	3	s										
		Michigan—Berrien County	8	sg										
		Van Buren County	11	s										
		West Virginia—Jefferson County	47	sil										
45	Smith and others, 1991a	Georgia—Plains	0.81	s	Alachlor, atrazine, carbofuran		Br⁻ (tracer)	•	•			•		fmp (model test), soil $t_{1/2}$
46	Smith and others, 1991b	Georgia—Tifton	0.7	s	Alachlor, atrazine		Br⁻ (tracer)	•	•			•		fmp (model test)

Table 2.1 Principal characteristics of process and matrix-distribution studies that included measurements of pesticide concentrations in ground water—*Continued*

Study		Site Location(s)	Site Area(s) (ha, unless stated otherwise)	Soil Type(s)	Analyte(s) Examined During Study			Media Sampled						Parameter(s) Measured
No.	Reference				Parent Pesticide(s)	Transformation Product(s)	Others	P T	S	V	T D	G W	S W	
47	van de Weerd and van der Linden, 1991	The Netherlands—Schaijk	Recirculating wells	s	Atrazine, 1,3-dichloropropene, dinoseb, mecoprop, propachlor		pH, SC, Cl⁻ (tracer)					•		fmp, R, soil $t_{1/2}$, T_{app}
48	Kalkhoff and others, 1992	Iowa—Iowa County	0.03	[sil][1]	Alachlor, atrazine, cyanazine		NO₃, NO₂		•			•		fmp
49	Priddle and others, 1992	Canada—Augustine Cove, Prince Edward Island	[4]	[glacial till]	Aldicarb (as TTR, along with products)	Aldicarb sulfone and sulfoxide (as TTR)	pH, E_H, DO, NO₃, major ions, alk	•						fmp
50	Waite and others, 1992	Canada—Regina, Saskatchewan	2,800	c	2,4-D, dicamba, bromoxynil, diclofop-methyl, triallate, carbofuran, carbaryl, chlorpyrifos, dimethoate, deltamethrin		Major ions					•	•	fmp
51	Levy and others, 1993	Wisconsin—Dane County	410	sil	Atrazine	Deethyl, deisopropyl and didealkyl atrazine						•		fmp, P/P
52	Jayachandran and others, 1994	Iowa—Ames	4	l	Atrazine	Deethyl atrazine, deisopropyl atrazine				•		•		fmp, P/P

Table 2.1 Principal characteristics of process and matrix-distribution studies that included measurements of pesticide concentrations in ground water—*Continued*

Study		Site Location(s)	Site Area(s) (ha, unless stated otherwise)	Soil Type(s)	Analyte(s) Examined During Study			Media Sampled						Parameter(s) Measured
No.	Reference				Parent Pesticide(s)	Transformation Product(s)	Others	P T	S	V	T D	G W	S W	
53	Komor and Emerson, 1994	Minnesota—Princeton, North Dakota—Oakes	0.02, 0.02	lfs, fsl	Atrazine	Deethyl atrazine, deisopropyl atrazine	Br⁻ (tracer), $\delta^{18}O$, δ^2H		•			•		fmp, f_{oc}
54	Ritter and others, 1994	Delaware—Georgetown	1.5	s	Atrazine, simazine, cyanazine, alachlor, metolachlor				•			•		mgmt
55	Sadeghi and Isensee, 1994	Maryland—Coastal Plain	NG	sil	Atrazine				•			•		mgmt

[1] Soil classification not given. Texture inferred from grain-size distribution data using textural classification given by Birkeland (1974).
[2] Based upon chemical names given by Neary and others (1983); "metabolites A and B" represent hydroxylated and demethylated degradates of hexazinone, respectively.

Table 2.2. Principal characteristics of process and matrix-distribution studies that did not include measurements of pesticide concentrations in ground water

[Site Location(s): Sites placed within brackets indicate probable study locations, since exact locations were not given. FRG, Federal Republic of Germany. Site Area(s): For studies that did not provide areas, estimates derived from study maps are given in brackets. ha, hectare. Soil Type(s): Includes textures of soils at all study sites, given for surficial soils only. Abbreviations taken from U.S. Department of Agriculture (1991). For studies where soil type was not provided, inferred soil type is given in brackets. Analyte(s) Examined During Study: All analytes were not necessarily examined at all sites for a given study. Abbreviations used for chemical parameters: TTR, total toxic residue, consisting of the summed concentrations of the parent and one or more of its transformation products (used most frequently for aldicarb). SC, specific conductance; NO_3,nitrate; PO_4, phosphate. (Note: This list is not intended to be exhaustive; additional chemical parameters were measured by several of the studies cited, but those listed here are the variables deemed to be most relevant for understanding pesticide transport and fate.) Media Sampled: Media sampled and analyzed for pesticides during field component of study. Abbreviations: PT, plant tissue (either standing plants or leaf litter); S, soil (any depth, analyzed either by chemical extraction or crop bioassay); V, vadose-zone water, obtained from lysimeters or buried soil columns; TD, tile drainage water; GW, ground water, obtained from wells or springs; SW, surface water (includes ponds). Parameter(s) Measured: Parameter definitions: aq. $t_{1/2}$, half-life for transformation in aqueous solution, measured in the laboratory; D_h, coefficient of hydrodynamic dispersion; (f,l,s,b), parameter measured in field, laboratory, soil or biomass, respectively; f_{oc}, organic carbon content; fmp, field mobility and persistence in one or more of the media investigated; fmp (model test), study conducted for comparison with modeling results (see Table 9.1); $k_1(s)$, first-order rate constant for compound disappearance in soil; K_d, distribution coefficient for soil, normalized to organic carbon (measured in laboratory unless otherwise indicated); K_{oc}, distribution coefficient for soil, normalized to organic carbon (measured in laboratory unless otherwise indicated); mgmt, effects of different agricultural management practices (e.g., tillage, irrigation, or timing of pesticide applications) on movement and persistence of pesticides in the subsurface; P/P, product-to-parent concentration ratio(s) for selected pesticides; R, retardation factor; R_f, mobility on soil thin-layer chromatography plates, measured in the laboratory; S_w, water solubility; soil $t_{1/2}$, dissipation half-life in soil, measured in the field unless otherwise indicated; soil $t_{1/2}$ (pH), pH-dependence of soil dissipation half-life, measured in the field; soil $t_{1/2}(T)$, temperature dependence of soil dissipation half-life, measured in the laboratory, unless otherwise indicated; T_{app}, air or soil temperature measured during pesticide application; v_p, pore-water velocity. (Note: This list is not intended to be exhaustive. Additional parameters were measured by several of the studies cited, but those listed here were encountered most frequently among different studies.) Blank cells indicate no information applicable or available. NG, not given]

Study		Site Location(s)	Site Area(s)	Soil Type(s)	Analyte(s) Examined During Study			Media Sampled						Parameter(s) Measured
No.	Reference		(ha, unless stated otherwise)		Parent Pesticide(s)	Transformation Product(s)	Others	PT	S	V	TD	GW	SW	
56	Cooke, 1966	Pennsylvania—Ambler	0.001	sil	Amiben, dinoben			•						mgmt
57	Johnston and others, 1967	California—San Joaquin Valley	40.5	NG	Aldrin, BHC, chlordane, DDT, dieldrin, heptachlor, lindane, TCBC[1], toxaphene, parathion	DDD, DDE, heptachlor epoxide			•	•				fmp
58	Rodgers, 1968	Florida—Gainesville	0.0080	fs	Ametryne, prometryne				•					fmp, f_{oc}
59	Edwards and Glass, 1971	Ohio—Coshocton	0.0008	sil	Methoxychlor, 2,4,5-T		Cl⁻ (tracer)		•			•		fmp

Table 2.2. Principal characteristics of process and matrix-distribution studies that did not include measurements of pesticide concentrations in ground water—*Continued*

No.	Reference	Site Location(s)	Site Area(s) (ha, unless stated otherwise)	Soil Type(s)	Parent Pesticide(s)	Transformation Product(s)	Others	Media Sampled P/T	S	V	T/D	G/W	S/W	Parameter(s) Measured
60	Osgerby, 1972	France—Avignon Bordeaux Versailles FRG—Geisenheim	NG NG	NG NG	None	2,6-dichloro-benzamide (2,6-dichloro-thiobenza-mide transformation product)			•					fmp (model test)
61	Buchanan and Hiltbold, 1973	Alabama—Central and southern Alabama	0.0026	fsl, sl	Atrazine, simazine (oat bioassay)				•					mgmt, soil $t_{1/2}$
62	Lavy and others, 1973; Lavy and Fenster, 1974	Nebraska—Lincoln Alliance	0.0006 0.0006	sicl sl	Atrazine, 2,4-D				•					fmp, T_{app}, soil $t_{1/2}$, f_{oc}
63	Lutz and others, 1973	North Carolina—Haywood County	0.016	cl, fsl, l	Picloram, 2,4,5-T				•					mgmt, T_{app}, f_{oc}
64	Sirons and others, 1973	Canada—Exeter, Ontario	0.022	cl	Atrazine, cyanazine	Deethyl atrazine, deisopropyl atrazine, cyanazine amide			•					fmp
65	Willis and Hamilton, 1973	Louisiana—Baton Rouge	0.045	sicl	Endrin				•		•		•	mgmt
66	Glass and Edwards, 1974	Ohio—Coshocton	0.0008	sil	Picloram		Cl⁻ (tracer)		•		•		•	fmp, f_{oc}
67	Liu, 1974	Puerto Rico—Río Piedras	0.0003	c	Diuron, fluometuron				•		•		•	fmp, f_{oc}

Table 2.2. Principal characteristics of process and matrix-distribution studies that did not include measurements of pesticide concentrations in ground water—*Continued*

No.	Reference	Site Location(s)	Site Area(s) (ha, unless stated otherwise)	Soil Type(s)	Parent Pesticide(s)	Transformation Product(s)	Others	P T	S	V	T D	G W	S W	Parameter(s) Measured
68	Rao and others, 1974	Hawaii—Molokai	0.0008	sic	Picloram				•					mgmt, R, f$_{oc}$
69	Smith and others, 1975	Canada—Saskatchewan	Irrigation ditches	NG	Atrazine, bromacil, monuron, simazine			•					•	fmp, soil t$_{1/2}$
70	Muir and Baker, 1976	Canada—Macdonald College, Quebec	5	sl, ls	Atrazine, cyprazine, cyanazine, metribuzin	Deethyl atrazine, deisopropyl atrazine, cyanazine amide, deethyl cyanazine					•			fmp
71	Hotzman and Mitchell, 1977	Delaware—Newark	0.046	Various	Dicamba, Vel-4207 (via bioassay)	Dicamba			•					mgmt, f$_{oc}$
72	Majka and Lavy, 1977	Nebraska—Irvington	0.01	sicl	^{14}C-cyanazine, ^{14}C-diuron			•						fmp, K$_d$, R$_f$, soil t$_{1/2}$ (T), f$_{oc}$, T$_{app}$
73	Mansell and others, 1977	Florida—Fort Pierce	20	s	Chlorobenzilate, 2,4-D, terbacil		NO$_3$; PO$_4$				•		•	mgmt, f$_{oc}$
74	Hall and Hartwig, 1978	Pennsylvania—Centre County	0.0029	sicl, cl	Atrazine			•		•				mgmt, f$_{oc}$
75	Fryer and others, 1979	England—Oxford	0.008	sl	Picloram			•						fmp

Table 2.2. Principal characteristics of process and matrix-distribution studies that did not include measurements of pesticide concentrations in ground water—*Continued*

No.	Reference	Site Area(s) (ha, unless stated otherwise)	Soil Type(s)	Parent Pesticide(s)	Transformation Product(s)	Others	P T	S	V	T D	G W	S W	Parameter(s) Measured
76	Glass and Edwards, 1979	0.0004	sil	Dicamba, picloram, 2,4,5-T			•	•				•	fmp, f_{oc}
77	Deleur and others, 1980	Buried soil columns	1	^{3}H-neburon, ^{3}H-chlortoluron, ^{3}H-metoxuron			•	•					fmp, f_{oc}
78	Mansell and others, 1980	NG	fs	Terbacil		K, NO_3		•					mgmt
79	Marley, 1980	0.02	c	Picloram (via bioassay)				•					mgmt
80	Zandvoort and others, 1980	0.0028	s	Bromacil				•					fmp, f_{oc}
81	Bottcher and others, 1981	17	sic, sil	Alachlor, carbofuran		NO_3, total N, PO_4, total P				•			fmp
82	Gusmao Helene and others, 1981	Buried soil columns	c	^{14}C-aldrin				•					fmp, f_{oc}
83	Jernlås and Klingspor, 1981	0.3	s, c	Trichloroacetic acid						•			mgmt
84	Rogers and Talbert, 1981	0.0006	sil	Metriflufen (via bioassay)				•					mgmt
85	Torstensson, 1981	0.3	s, c	Trichloroacetic acid	Cl^-			•					fmp, soil $t_{1/2}$

Table 2.2. Principal characteristics of process and matrix-distribution studies that did not include measurements of pesticide concentrations in ground water—*Continued*

No.	Reference	Site Location(s)	Site Area(s) (ha, unless stated otherwise)	Soil Type(s)	Parent Pesticide(s)	Transformation Product(s)	Others	P/T	S	V	T/D	G/W	S/W	Parameter(s) Measured
86	Wegman and others, 1981	The Netherlands—no town given	Green-houses	s, c	Methyl bromide	Br⁻					•			fmp, soil $t_{1/2}$
87	Basile, 1982	Italy—Bari	Buried soil columns	Various	1,3-dichloropropene, 1,2-dichloropropane			•						fmp
88	Brewer and others, 1982	Arkansas—Stuttgart	0.0048, 0.0096	sil	Fluchloralin, profluralin, trifluralin			•						mgmt, soil $t_{1/2}$ (f,l)
89	Enfield and others, 1982	New York—Cutchogue Wisconsin—Madison	2.8 NG	s sil	Aldicarb (as TTR, along with products) DDT	Aldicarb sulfone and sulfoxide (as TTR)		•						fmp (model test), f_{oc}
90	Felsot and others, 1982	Illinois—Central	0.5	sil	Terbufos			•						fmp, soil $t_{1/2}(T)$, K_{oc}, f_{oc}
91	Barrett and Lavy, 1983	Arkansas—Stuttgart	NG	sil	Pendimethalin			•						mgmt, soil $t_{1/2}$; T_{app}
92	Lear and others, 1983	California—Davis	0.082	NG	Methyl bromide	Br⁻		•	•					fmp
93	Nakamura and others, 1983	Japan—Kumagaya	Soil columns	Alluvial, volcanic	Chlornitrofen, benthiocarb, molinate, simetryne			•	•					mgmt
94	Neary and others, 1983	Georgia—Chattahoochee National Forest (near Clarksville)	0.85-1.09	sl	Hexazinone	"Metabolites A and B"[2]		•	•			•		fmp
95	Schmaland, 1983	FRG—Berlin	Buried lysi-meters	s, l	Toxaphene, lindane, butonate, trichlorfon, methyl parathion, dimethoate				•					fmp

Table 2.2. Principal characteristics of process and matrix-distribution studies that did not include measurements of pesticide concentrations in ground water—*Continued*

No.	Reference	Site Location(s)	Site Area(s) (ha, unless stated otherwise)	Soil Type(s)	Parent Pesticide(s)	Transformation Product(s)	Others	P T	S	V	T D	G W	S W	Parameter(s) Measured
96	Barrett and Lavy, 1984	Arkansas—Stuttgart	NG	sil	Oxadiazon			•						mgmt, soil $t_{1/2}$ (f,l)
97	Junk and others, 1984	Iowa—Ames	Buried 110-L cans filled with water and soil	sl	Alachlor, atrazine, carbaryl, 2,4-D ester, parathion, trifluralin	1-naphthol (from hydrolysis of carbaryl), 2,4-D acid		•						fmp
98	Leistra and others, 1984a	The Netherlands—Naaldwijk Honselersdijk Aalsmeer	Green-houses	s, sl, l	Aldicarb	Aldicarb sulfoxide, aldicarb sulfone		•		•			•	fmp (model test), $k_1(s)$, f_{oc}
99	Leistra and others, 1984b	The Netherlands—Naaldwijk Honselersdijk Wateringen Roelofarendsveen	Green-houses	s, sl, NG, NG	Diazinon, parathion, tetrachlorvinphos, triazophos			•		•			•	fmp, K_d, f_{oc}
100	Randhawa and Gill, 1984	India—Ludhiana	NG	ls	Atrazine, metoxuron, simazine			•						fmp
101	Bilkert and Rao, 1985	Florida—Gainesville	Buried soil columns	s, sl, sicl	Aldicarb, oxamyl, fenamiphos					•				fmp (model test), K_d, f_{oc}
102	Bouchard and others, 1985	Arkansas—Fleming Creek	11.5	grfsl	Hexazinone			•					•	fmp, soil $t_{1/2}$ (T)(f,l), aq.$t_{1/2}$
103	Carsel and others, 1985	New York—Long Island	NG	sl	Aldicarb (as TTR, along with products)	Aldicarb sulfone and sulfoxide (as TTR)		•						fmp (model test)

Table 2.2. Principal characteristics of process and matrix-distribution studies that did not include measurements of pesticide concentrations in ground water—*Continued*

Study		Site Location(s)	Site Area(s) (ha, unless stated otherwise)	Soil Type(s)	Analyte(s) Examined During Study			Media Sampled						Parameter(s) Measured
No.	Reference				Parent Pesticide(s)	Transformation Product(s)	Others	P T	S	V	T D	G W	S W	
104	Copin and others, 1985	Belgium—Tirlemont	0.006	s	Carbofuran, ethofumesate, metamitron		pH		•					fmp
105	Jernlås, 1985	Sweden—Björnstorp	2	cls	2,3,6-trichlorobenzoic acid (2,3,6-TBA)					•				mgmt, K_d, v_p, D_h, f_{oc}
106	Akkari and others, 1986	Arkansas—Marianna	0.0045-0.0105	sil	Methylarsonic acid, monosodium salt (MSMA)	Arsenate, dimethyl arsenate	pH, total arsenic		•					fmp, soil $t_{1/2}$ (T) (f,l), f_{oc}
107	Braverman and others, 1986	Arkansas—[Fayetteville]	NG	sil	Metolachlor				•					fmp, K_d, soil $t_{1/2}$ (T), f_{oc}
108	Carsel and others, 1986	Florida—Lake City Maryland—Beltsville	3.5 0.6	fs fsl	Metalaxyl				•					fmp, k_1(s)
109	Green and others, 1986	Hawaii—Oahu, Maui	NG	c	DBCP				•					fmp (model test), K_d, f_{oc}
110	Jones and others, 1986a	Maine—Aroostook County	0.017	stl	Aldicarb (as TTR, along with products)	Aldicarb sulfone and sulfoxide (as TTR)			•					mgmt, soil $t_{1/2}$, f_{oc}
111	Jury and others, 1986a,b	California—Etiwanda	0.64	ls	Napropamide, bromacil, prometryn		Cl⁻, Br⁻ (tracers)		•					fmp, K_d, K_{oc}, f_{oc}
112	Lorber and Offutt, 1986	North Carolina— Hertford County Wisconsin— Cameron Hancock	NG NG NG	sl sl ls	Aldicarb (as TTR, along with products)	Aldicarb sulfone and sulfoxide (as TTR)			•					fmp (model test), f_{oc}

Table 2.2. Principal characteristics of process and matrix-distribution studies that did not include measurements of pesticide concentrations in ground water—*Continued*

| Study | | Site Area(s) (ha, unless stated otherwise) | Site Location(s) | Soil Type(s) | Analyte(s) Examined During Study | | | Media Sampled | | | | | | Parameter(s) Measured |
No.	Reference				Parent Pesticide(s)	Transformation Product(s)	Others	P/T	S	V	T/D	G/W	S/W	
113	Wagenet and Hutson, 1986	0.13	New York—Phelps	s	Aldicarb (as TTR, along with products)	Aldicarb sulfone and sulfoxide (as TTR)		•						fmp (model test), f_{oc}
114	Anderson and Humburg, 1987	0.0032	Colorado—Akron (Central Great Plains)	sl, l, sil	Chlorsulfuron			•						fmp
115	Basham and others, 1987	3×10^{-5}	Arkansas—Fayetteville, Keiser	sil, sic	^{14}C-imazaquin		^{36}Cl^{-} (tracer)	•						fmp, K_d, soil $t_{1/2}$, f_{oc}, T_{app}
116	Dao, 1987	NG	Oklahoma—El Reno	sil, l	BAY SMY 1500 (herbicide)			•						mgmt, soil $t_{1/2}$, K_d, f_{oc}
117	Jarczyk, 1987	NG	FRG—Brenig/Bonn	Podzol	Azinphos-ethyl, dichlorprop, ethiofencarb, metamitron, methabenzthiazuron, oxydemeton-methyl, triadimefon, trichlorfon			•						fmp
118	Albanis and others, 1988	Buried soil columns	Greece—Ioannina	c, l, sil	Methyl parathion, lindane, atrazine					•				fmp
119	Bowman, 1988	Buried soil columns	Canada—London, Ontario	s	Aldicarb, metolachlor	Aldicarb sulfoxide, aldicarb sulfone		•		•				mgmt, K_d, $k_1(s)$, f_{oc}
120	Helling and others, 1988	0.0023	Maryland—Beltsville	sil	Alachlor, atrazine, cyanazine		Br^{-} (tracer)	•						fmp, f_{oc}

Table 2.2. Principal characteristics of process and matrix-distribution studies that did not include measurements of pesticide concentrations in ground water—Continued

No.	Reference	Site Location(s)	Site Area(s) (ha, unless stated otherwise)	Soil Type(s)	Parent Pesticide(s)	Transformation Product(s)	Others	P/T	S/V	T/D	G/W	S/W	Parameter(s) Measured
121	Klaine and others, 1988	Tennessee—Shelby County	18	sil	Atrazine			•				•	fmp, soil $t_{1/2}$
122	Kubiak and others, 1988	FRG—Merzenhausen	0.0025, Buried soil columns	Clayey loess	^{14}C-metamitron, ^{14}C-methabenzthiazuron	Desamino metamitron, methyl benzthiazolyl urea, unspecified metabolites (via ^{14}C)		•					fmp, T_{app}
123	Ramanand and others, 1988	India—Cuttack	0.0015	scl	Carbofuran			•					fmp
124	Schiavon, 1988	France—[Nancy]	Buried soil columns	Brown soil (ochrepts)	^{14}C-atrazine	Hydroxy-, deethyl, deisopropyl and diamino ^{14}C-atrazine		•					fmp
125	Weaver and others, 1988a	California—Del Norte County	[1,000]	grcl, sil, fsl	Fenamiphos	Fenamiphos sulfone and sulfoxide		•					fmp, f_{oc}
126	Weaver and others, 1988b	California—Del Norte County Humboldt County	0.0002 0.0002	grl l	Ethoprop, phorate	Phorate sulfoxide, phorate sulfone	pH	•					mgmt, f_{oc}
127	Barnes and others, 1989	Arkansas—Northeastern (no towns given)		sil, c	^{14}C-imazaquin			•					fmp, K_d, soil $t_{1/2}$, f_{oc}
128	Bowman, 1989	Canada—Ontario	Buried soil column	s	Atrazine, metolachlor, terbuthylazine				•				fmp, f_{oc}

Table 2.2. Principal characteristics of process and matrix-distribution studies that did not include measurements of pesticide concentrations in ground water—*Continued*

No.	Reference	Site Location(s)	Site Area(s) (ha, unless stated otherwise)	Soil Type(s)	Parent Pesticide(s)	Transformation Product(s)	Others	P T	S	V	T D	G W	S W	Parameter(s) Measured
129	Feng and others, 1989	Canada—Grande Prairie, Alberta	5	sic	Hexazinone									fmp
130	Foy and Hiranpradit, 1989	Virginia—Blacksburg	0.0022	sil	Atrazine				•				•	mgmt
131	Hall and others, 1989	Pennsylvania—Centre County	0.26	sicl	Atrazine, cyanazine, simazine, metolachlor				•	•				mgmt, f_{oc}
132	Huang and Frink, 1989	Connecticut—Simsbury Franklin	NG NG	ls sl	Alachlor, atrazine, metolachlor, simazine				•					fmp, f_{oc}
133	Johnsen and Morton, 1989	Arizona—Prescott, Flagstaff	0.005 - 3.1	sl, l, cl	Tebuthiuron				•					mgmt
134	Kolberg and others, 1989	North Dakota—Oakes	0.0006	lfs	Carbofuran, terbufos	3-hydroxy and 3-keto carbofuran, terbufos sulfone, terbufos sulfoxide			•	•	•			mgmt
135	Lavy and others, 1989	West Virginia—Parsons	10	sil	Hexazinone			•	•	•			•	fmp, K_d (s,b), soil $t_{1/2}$ (s,b), T_{app}, f_{oc}
136	Michael and others, 1989	Alabama—Tuskegee National Forest	1.2 - 4.1	ls	Picloram				•	•			•	fmp
137	O'Neill and others, 1989	Canada—Northwestern New Brunswick	3.0 - 10.4	NG	Dinoseb		Major ions, nutrients				•			fmp

Table 2.2. Principal characteristics of process and matrix-distribution studies that did not include measurements of pesticide concentrations in ground water—*Continued*

Study No.	Reference	Site Location(s)	Site Area(s) (ha, unless stated otherwise)	Soil Type(s)	Parent Pesticide(s)	Transformation Product(s)	Others	P/T	S	V	T/D	G/W	S/W	Parameter(s) Measured
138	Peryea, 1989	Washington— Wenatchee Orondo Cashmere Manson	NG NG NG NG	gsl, fsl, sl, l sl NG NG	Lead arsenate				•					fmp
139	Reiml and others, 1989	FRG—Neuherberg	0.36-m boxes[2]	s	^{14}C-buturon	Three products (^{14}C-labeled)[3]			•					fmp, P/P, f_{oc}
140	Summit and others, 1989	Utah— Hill Air Force Base Shivwitz Idaho—Malad	NG NG NG	sl fls sil	Tebuthiuron				•					fmp, soil $t_{1/2}$
141	Wadd and Drennan, 1989	England—Sonning Farm, Reading	0.0048	sl	Chlorsulfuron, metsulfuron-methyl				•					mgmt, soil $t_{1/2}$, f_{oc}
142	Bergström, 1990	Sweden— Kjettslinge Bulstofta	Buried soil columns	l, fs s, c	Chlorsulfuron, metsulfuron-methyl					•				fmp, f_{oc}
143	Blair and others, 1990	England—Higham, Bury Saint Edmunds: Experiment 1 Experiment 2 Lidgate, Suffolk	0.012 0.0072 0.0072	scl c	Isoproturon		pH		•					fmp, soil $t_{1/2}$(T), f_{oc}
144	Clendening and others, 1990	California—[Riverside]	0.0016	sl	Atrazine, bromacil, EPTC, prometon, triallate				•					fmp, soil $t_{1/2}$
145	Feng and Thompson, 1990	Canada—Port Alberni, Vancouver Island, British Columbia	0.0025	Variable	Glyphosate	Aminomethyl phosphonic acid (AMPA)			•	•				fmp, f_{oc}, soil $t_{1/2}$

Table 2.2. Principal characteristics of process and matrix-distribution studies that did not include measurements of pesticide concentrations in ground water—*Continued*

Study			Site Area(s)	Soil Type(s)	Analyte(s) Examined During Study			Media Sampled						Parameter(s) Measured
No.	Reference	Site Location(s)	(ha, unless stated otherwise)		Parent Pesticide(s)	Transformation Product(s)	Others	P T	S	V	T D	G W	S W	
146	Freitag and Scheunert, 1990	FRG—Neuherberg	Buried, soil-filled box	NG	^{14}C-monolinuron	six products (^{14}C-labeled)		•	•					fmp
147	Goetz and others, 1990	Arkansas—Fayetteville Rohwer	0.0042 0.0042	sil sic	^{14}C-imazethapyr	$^{14}CO_2$			•					fmp, f_{oc}, soil $t_{1/2}$ (T) (f, l)
148	Herrchen and others, 1990	FRG—Hildesheimer Börde	Buried soil columns	cl	^{14}C-BAS 263 I (carbamate insecticide)	$^{14}CO_2$			•	•				fmp, f_{oc}
149	Lee and others, 1990	Delaware—greenhouse soil	Steel cylinders	sl	^{14}C-monocrotophos	Various products (^{14}C-labeled)		•	•					fmp, R_f, soil $t_{1/2}$ (l), aq. $t_{1/2}$(T, pH), f_{oc}
150	Leonard and others, 1990	Georgia—Tifton	0.34	ls	Fenamiphos	Fenamiphos sulfoxide and sulfone			•					fmp (model test)
151	Minton and others, 1990	Georgia—Tifton	0.004	fl, l	Fenamiphos (as TTR, along with products)	Fenamiphos sulfoxide and sulfone (as TTR)	pH	•						mgmt, soil $t_{1/2}$, f_{oc}
152	Pennell and others, 1990	Florida—Davenport	1.6	s	Aldicarb (as TTR, along with products)	Aldicarb sulfone and sulfoxide (as TTR)			•					fmp (model test), mgmt, f_{oc}
153	Sauer and others, 1990	Wisconsin—Hancock	0.267	s	Atrazine, metolachlor				•					mgmt (model test), K_d, soil $t_{1/2}$, f_{oc}

Table 2.2. Principal characteristics of process and matrix-distribution studies that did not include measurements of pesticide concentrations in ground water—*Continued*

No.	Reference	Site Location(s)	Site Area(s) (ha, unless stated otherwise)	Soil Type(s)	Parent Pesticide(s)	Transformation Product(s)	Others	PT	S	V	TD	GW	SW	Parameter(s) Measured
154	Schneider and others, 1990	Hawaii—Central Oahu Lanai	0.2, 0.8 0.04	sic (Ultisols and Oxisols)	Fenamiphos (as TTR, along with products)	Fenamiphos sulfoxide and sulfone (as TTR)	pH	•						fmp, f_{oc}
155	Sophocleous and others, 1990	Kansas—Great Bend Prairie	NG	sicl, ls	Atrazine	Hydroxy-, deethyl and deisopropyl atrazine	Br⁻ (tracer), Cl⁻, pH	•		•				fmp, soil $t_{1/2}$, f_{oc}
156	Troiano and others, 1990	California—Fresno	0.0084	ls	Atrazine		Cl⁻, Br⁻ (tracers)	•						mgmt, f_{oc}
157	Adams and Thurman, 1991	Kansas—Topeka	0.011	sicl, sil	Atrazine	Deethyl and deisopropyl atrazine	Cl⁻ (tracer)	•		•				fmp, P/P
158	Bovey and Richardson, 1991	Texas—Riesel	0.74	c	Clopyralid, picloram						•			fmp
159	Bowman, 1991	Canada—London, Ontario	Buried soil columns	s	Isazofos, atrazine	CGA 17193 (isazofos product), deethyl atrazine		•						mgmt, soil $t_{1/2}$, f_{oc}
160	Brooke and Matthiessen, 1991	England—Herefordshire	5.5, 6.0	[sil][4]	Mecoprop, simazine		SC, pH				•		•	fmp (model test), f_{oc}
161	Gish and others, 1991a	Maryland—Beltsville	0.11, 0.033	sil	Atrazine, cyanazine			•		•				mgmt, soil $t_{1/2}$
162	Hall and others, 1991	Pennsylvania—Centre County	0.26	sicl	Atrazine, cyanazine, simazine, metolachlor					•			•	mgmt, f_{oc}

Table 2.2. Principal characteristics of process and matrix-distribution studies that did not include measurements of pesticide concentrations in ground water—*Continued*

No.	Reference	Site Location(s)	Site Area(s) (ha, unless stated otherwise)	Soil Type(s)	Parent Pesticide(s)	Transformation Product(s)	Others	P T	S	V	T D	G W	S W	Parameter(s) Measured
163	Hill and others, 1991	Canada—Lethbridge, Alberta	Buried, soil-filled pans	scl	Deltamethrin, λ-cyhalothrin				•					fmp, soil $t_{1/2}$, T_{app}
164	Huang and Ahrens, 1991	Connecticut—New Haven	0.0006	sl	Alachlor		pH		•					mgmt, f_{oc}
165	Jaynes, 1991	Iowa—Ames	0.0037	cl	Bromacil		Br$^-$ (tracer)			•				fmp, v_p, D_h, R
166	Kladivko and others, 1991	Indiana—Butlerville, Jennings County	[9]	sil	Alachlor, atrazine, carbofuran, chlorpyrifos, cyanazine, terbufos		Nutrients				•			mgmt, K_d, K_{oc}, f_{oc}
167	Klein, 1991; Kördel and others, 1991	FRG—Schmallenberg	Buried soil columns	s	^{14}C-cloethocarb, ^{14}C-bentazone	Unspecified metabolites (via ^{14}C)			•	•				fmp (model test), T_{app}
168	Leake, 1991	England—Essex	Buried soil columns	s, sl	Benazolin-ethyl	Benazolin, benzo-thiazolin			•	•				fmp, T_{app}, f_{oc}
169	Loague and Green, 1991	Georgia—Watkinsville	1.3	l, sl, scl	Atrazine					•				fmp (model test)
170	Neal and others, 1991	California—Riverside	0.008	cosl	Carbofuran, simazine, thiobencarb (ordram)				•					fmp, f_{oc}
171	Rice and others, 1991	Arizona— NG Maricopa Yuma	0.0013 0.62 0.0027	l sl s	Bromacil		Br$^-$ and four other tracers		•	•				fmp

Table 2.2. Principal characteristics of process and matrix-distribution studies that did not include measurements of pesticide concentrations in ground water—*Continued*

Study		Site Location(s)	Site Area(s) (ha, unless stated otherwise)	Soil Type(s)	Analyte(s) Examined During Study			Media Sampled						Parameter(s) Measured
No.	Reference				Parent Pesticide(s)	Transformation Product(s)	Others	P T	S	V	T D	G W	S W	
172	Rose and others, 1991	England—Brimstone farm Swavesey	0.2 500	Stagnogley Alluvial gley	Isoproturon, mecoprop						•			fmp
173	Sabbagh and others, 1991	Louisiana—Baton Rouge Georgia—Tifton	1.6 0.7	cl s	Atrazine, metolachlor atrazine			•					•	fmp (model test), f_{oc}
174	Shaffer and Penner, 1991	Michigan—East Lansing Hickory Corners	0.027 0.042	scl l	Alachlor, metolachlor			•						fmp
175	Sichani and others, 1991	Indiana—North Vernon	4	sil	Atrazine, carbofuran, cyanazine						•			fmp (model test)
176	Stahnke and others, 1991	Nebraska—Mead	0.0009	sicl	Pendimethalin			•						fmp, T_{app}, f_{oc}
177	Barnes and others, 1992	Arkansas—Fayetteville	0.0007	sil	Metolachlor			•						mgmt, K_d, soil $t_{1/2}$, T_{app}, f_{oc}
178	Clay and others, 1992	Minnesota—Westport	0.0001	sl	Alachlor		Br⁻, $^{15}NO_3$ (tracers), NO_3			•				mgmt
179	Mueller and others, 1992	Georgia—Midville Athens	0.0029 NG	ls sl	Alachlor, metribuzin, norflurazon			•						fmp (model test), soil $t_{1/2}$ (f,l), f_{oc}
180	Trevisan and others, 1993	Italy—Milano	3.1	l	Atrazine, metolachlor			•						fmp (model test), f_{oc}, T_{app}
181	Moyer and Blackshaw, 1993	Canada—Lethbridge, Alberta	0.0336	Chernozem	Atrazine, cyanazine			•						mgmt, soil $t_{1/2}$, f_{oc}

Table 2.2. Principal characteristics of process and matrix-distribution studies that did not include measurements of pesticide concentrations in ground water—*Continued*

| Study | | Site Location(s) | Site Area(s) (ha, unless stated otherwise) | Soil Type(s) | Analyte(s) Examined During Study | | | Media Sampled | | | | | | Parameter(s) Measured |
No.	Reference				Parent Pesticide(s)	Transformation Product(s)	Others	P T	S	V	T D	G W	S W	
182	Racke and others, 1993	California Florida Georgia Illinois Indiana Michigan	0.01–0.09 0.03–2 NG 0.01–1.33 0.03 0.01–0.09	l s, fs sl sil, lfs, fsl cl sl	Chlorpyrifos			•	•					mgmt, soil $t_{1/2}$, f_{oc}
183	Wietersen and others, 1993	Wisconsin—Lower Wisconsin River Valley (Greenhouse)	Soil columns	s	Alachlor, atrazine, metolachlor		Br⁻ (tracer)		•					mgmt
184	Boul and others, 1994	New Zealand—Mid Canterbury	0.09–0.1	stsil	DDT	DDD, DDE, p,p'-dichloroben- zophenone			•					mgmt, f_{oc}
185	Buhler and others, 1994	Minnesota—Westport	0.0120	sl	Alachlor, metolachlor, atrazine				•					mgmt
186	Burgard and others, 1994	Minnesota—Becker	0.0090	s	Metribuzin		Br⁻ (tracer)	•	•					fmp, K_d, K_{oc}, k(s), soil $t_{1/2}$ (T, pH)
187	Cooper and Zheng, 1994	French West Indies— Martinique France—Languedoc	0.01 0.01	c sic	Metolachlor				•					fmp, K_d, soil $t_{1/2}$, T_{app}, f_{oc}
188	Corbin and others, 1994	Arkansas—Clarkedale	0.0435	sil, sic	Fluometuron, trifluralin				•					mgmt, soil $t_{1/2}$
189	Czapar and others, 1994	Iowa—Kanawha	0.0036	sicl	Alachlor, cyanazine, pendimethalin		Rhoda- mine, WT, Br⁻, Cl⁻ (tracers)			•				fmp, f_{oc}

Table 2.2. Principal characteristics of process and matrix-distribution studies that did not include measurements of pesticide concentrations in ground water—*Continued*

No.	Reference	Site Location(s)	Site Area(s) (ha, unless stated otherwise)	Soil Type(s)	Parent Pesticide(s)	Transformation Product(s)	Others	P T	S V	T D	G W	S W	Parameter(s) Measured
190	Demon and others, 1994	France—[Nancy]	Outdoor soil columns	sic, sicl	^{14}C-atrazine	^{14}C-"chlorinated metabolites", ^{14}C-hydroxy-atrazine		•					fmp, f_{oc}
191	Gish and others, 1994	Maryland—Upper Marlboro	1	sl	Atrazine, alachlor			•					mgmt, soil $t_{1/2}$, T_{app}
192	Hall and Mumma, 1994	Pennsylvania—Centre County	0.6	sicl	Dicamba			•				•	mgmt, f_{oc}
193	Harris and others, 1994	England—Oxfordshire	Lysimeters	c (fractured)	Isoproturon, mecoprop			•				•	fmp, K_d, soil $t_{1/2}$, f_{oc}
194	Hassink and others, 1994	Germany—Schmallenberg	Outdoor soil columns	sil, sl, l	Diuron, methabenz-thiazuron, simazine			•					fmp, f_{oc}, soil $t_{1/2}$, T_{app}
195	Isensee and Sadeghi, 1994	Maryland—Beltsville	1.4	sil	Atrazine			•					mgmt, f_{oc}
196	Johnson and Lavy, 1994	Arkansas—Stuttgart	Buried, soil-filled mason jars	sil	Benomyl, carbofuran, thiobencarb, triclopyr	MBC[5], 3-keto carbofuran, 3-hydroxy carbofuran		•					fmp, f_{oc}, T_{app}
197	Kanwar and Baker, 1994	Iowa—Boone	2.4	NG	Atrazine		NO$_3$			•			mgmt
198	Kanwar and others, 1994	Iowa—Nashua	15	fl	Alachlor, atrazine, cyanazine, metribuzin		NO$_3$			•			mgmt

Table 2.2. Principal characteristics of process and matrix-distribution studies that did not include measurements of pesticide concentrations in ground water—*Continued*

Study		Site Location(s)	Site Area(s) (ha, unless stated otherwise)	Soil Type(s)	Analyte(s) Examined During Study			Media Sampled						Parameter(s) Measured
No.	Reference				Parent Pesticide(s)	Transformation Product(s)	Others	P/T	S	V	T/D	G/W	S/W	
199	Lee and others, 1994	Korea—Kakyung-dong, Cheong Ju	Stainless steel lysim-eters	l, sl, sl, ls	^{14}C-carbofuran			•	•					fmp, f_{oc}, T_{app}
200	Loague and others, 1994	Hawaii—Oahu	0.011	Oxisol	Chlorpyrifos, fenamiphos	Fenamiphos sulfone, fenamiphos sulfoxide	Br⁻ (tracer)		•					fmp, K_d, K_{oc}, f_{oc}, soil $t_{1/2}$, D_h
201	Logan and others, 1994	Ohio—Wood County	0.3152	sic	Atrazine, alachlor, metolachlor, metribuzin		NO_3				•	•		mgmt
202	Mills and Thurman, 1994a	Kansas—Topeka	0.013	sil	Atrazine	Deethyl atrazine, deisopropyl atrazine	Br⁻ (tracer)		•	•		•		mgmt, K_d, S_w, P/P
203	Mills and Thurman, 1994b	Kansas—Topeka	0.013	sil	Atrazine, propazine, simazine	Deethyl, deisopropyl, and didealkyl atrazine	Br⁻ (tracer)		•	•		•		fmp, P/P
204	Mueller, 1994	NG	NG	c, s	Bentazon, dichlorprop					•				fmp (model test)
205	Newton and others, 1994	Oregon—Corvallis Michigan—Chassell Georgia—Cuthbert	8.0 8.0 8.0	s, cl sl sl	Glyphosate	Aminomethyl phosphonic acid (AMPA)			•	•		•		fmp, T_{app}
206	Rouchaud and others, 1994	Belgium—Lubbeek	NG	sil	Imidacloprid				•					mgmt
207	Thomas and Robinson, 1994	Virginia—Blacksburg	0.000036 (soil beneath concrete slabs)	g, sil	Chlorpyrifos				•					fmp, f_{oc}

Table 2.2. Principal characteristics of process and matrix-distribution studies that did not include measurements of pesticide concentrations in ground water—*Continued*

Study		Site Location(s)	Site Area(s) (ha, unless stated otherwise)	Soil Type(s)	Analyte(s) Examined During Study			Media Sampled						Parameter(s) Measured
No.	Reference				Parent Pesticide(s)	Transformation Product(s)	Others	P T	S	V	T D	G W	S W	
208	Traub-Eberhard and others, 1994	Germany—Welver-Borgeln, Bad Sassendorf-Bettinghausen	5.0, 1.1, 0.7	lsi lsi	Chloridazon, metamitron, pendimethalin, isoproturon						•			mgmt, f_{oc}
209	Vicari and others, 1994	Italy—Emilia Romagna, Lombardy, Friuli, Apulia	0.03 - 0.06	c sic sl sl	Chlorsulfuron, metsulfuron				•					fmp, f_{oc}, soil $t_{1/2}$
210	Yen and others, 1994	Colorado—Windsor	Twelve 60-m long rows	cl	Alachlor		Br⁻ (tracer)		•					fmp, K_d, K_{oc}, $k_1(s)$, f_{oc}, soil $t_{1/2}$
211	Flury and others, 1995	Switzerland—Obfelden, Les Barges	0.001 0.001	l s	Atrazine, terbuthylazine, triasulfuron		Br⁻, Cl⁻, Brilliant Blue FCF (tracers)		•					fmp, mgmt, f_{oc}
212	Poletika and others, 1995	California—Etiwanda	Lysimeters	ls	Simazine		Br⁻ (tracer), MS-2 coliphage		•	•				fmp, K_d, f_{oc}, R, T_{app}
213	Weed and others, 1995	Iowa—Nashua	14.4	l	Alachlor, atrazine, metribuzin				•			•		fmp, mgmt, soil $t_{1/2}$, f_{oc}, T_{app}

[1] Trichlorobenzyl chloride.
[2] Based upon chemical names given by Neary and others (1983); "metabolites A and B" represent hydroxylated and demethylated degradates of hexazinone, respectively.
[3] Three different transformation products detected: 4-chloroaniline, methyl N-(4-chlorophenyl)-carbamate and "conjugated 4-chloroaniline."
[4] Soil classification not given. Texture inferred from grain-size distribution data using textural classification given by Birkeland (1974).
[5] Methyl-2-benzimidazole carbamate.

Table 2.3. Study designs for state and local monitoring investigations of pesticide occurrence in ground waters of the United States, Canada, and Europe

[Location(s): FRG, Federal Republic of Germany. SWRL, Iowa State-Wide Rural Well-Water Survey. Wells Sampled: Community, Used as water supply by 25 or more people (includes those designated as "municipal" wells and those serving utilities); Discontinued, No longer in use; Domestic, Used as private water supply for a single residence or farm (includes wells used for mixing agrichemicals on farms and those referred to as private or farmstead wells); Non-community, Used as water supply in public buildings (e.g., libraries, schools, post offices, etc.); Dom/irrig., Wells used for both domestic and irrigation purposes; Observation, Installed solely to monitor ground-water quality; Stock, Irrigation, Industrial, Drainage, Fire protection, Commercial, Non-potable and Agricultural are self-explanatory. Site Selection Strategy: Targeted (T) sampling was usually directed toward areas with a specific land use, use of a particular pesticide, or suspected contamination. Nontargeted (NT) sampling usually involved either a stratified random well-selection process (e.g., the National Pesticide Survey [U.S. Environmental Protection Agency, 1990a]), or complete surveys of all wells within a particular category, such as major municipal supplies (e.g., Richard and others, 1975; Cherryholmes and others, 1989). Pesticides Examined (compound class abbreviations): TRI, triazines; ACET, acetanilides; CARB, carbamates and thiocarbamates; UREA, Urea and sulfonylurea herbicides; OC, organochlorine herbicides; OP, organophosphorus compounds; PHEN, chlorophenoxy acids; FUM, fumigants (includes several compounds also used as industrial solvents, e.g., trichloroethylene, tetrachloroethylene, and carbon tetrachloride); TRANS, transformation products; MISC, miscellaneous compounds. (Note: For many of the studies, not all wells were analyzed for all of the pesticides listed.) S, Analytical screen conducted for this class of pesticides, but specific compounds not given. Blank cells indicate no information applicable or available. NG, not given; <, less than]

No.	Reference	Sampling Date(s)	Location(s) (State [Province, Country]—Town, County or Region)	Type(s)	Number	Site Selection Strategy	TRI	ACET	CARB	UREA	OC	OP	PHEN	FUM	TRANS	MISC
1	Tamblyn and Beck, 1968	9/63-12/64 1965 1966 1967	California—San Joaquin Valley	Tile drains	8 6 19 37	T	2				12				3	
2	Dion, 1971	1970	Idaho—Boise-Nampa area	NG	6	NG	Unspecified "herbicides and insecticides"									
3	Achari and others, 1975	NG	South Carolina—Georgetown County	NG	27	NG					3					
4	Richard and others, 1975	6/19/74 - 8/10/74	Iowa—Major cities	Community systems	9 systems	NT	1				1				1	

Table 2.3. Study designs for state and local monitoring investigations of pesticide occurrence in ground waters of the United States, Canada, and Europe—*Continued*

No.	Reference	Sampling Date(s)	Location(s) (State [Province, Country]—Town, County or Region)	Type(s)	Number	Site Selection Strategy	TRI	ACET	CARB	UREA	OC	OP	PHEN	FUM	TRAS	MISC
5	Hindall, 1978	11/73-12/75	Wisconsin—Central Sand Plain	Domestic / Observation / Discontinued observation / Community / Industrial / Irrigation	[1]127 / 48 / 5 / 8 / 3 / 19	NG					8	7	3		3	
6	Junk and others, 1980	6/78, 9/78	Nebraska—Buffalo and Hall Counties	Monitoring	35	NG	1	1			1					
7	Myott, 1980	1978	New York—Nassau County	Community	335	NG					9		3		1	
8	Oshima and others, 1980	11/27/79 11/29/79	California—Kern County Monterey County	NG / NG	6 / 6	T			1						2	
9	Peoples and others, 1980	5/79	California—Central Valley	Domestic / Community / Irrigation / Dom/irrig. / NG	123 / 61 / 63 / 13 / 2	NG					9			2	3	1
10	Pinto, 1980	8/2/79-1/21/80	Maryland—Wicomico County	Irrigation, domestic, community, and commercial	36	T								2		
11	Spalding and others, 1980	8/78	Nebraska—Buffalo and Hall Counties	Irrigation	14	T	1	1	1		7		1		2	
12	Bushway and others, 1982	6/29/81-8/11/81	Maine—Hancock and Washington Counties	NG	<7	T						1			1	

Table 2.3. Study designs for state and local monitoring investigations of pesticide occurrence in ground waters of the United States, Canada, and Europe—*Continued*

No.	Reference	Sampling Date(s)	Location(s) (State [Province, Country]—Town, County or Region)	Type(s)	Number	Site Selection Strategy	TRI	ACET	CARB	UREA	OC	OP	PHEN	FUM	TRANS	MISC
13	Maddy and others, 1982	NG	California—San Joaquin and Salinas Valleys	Community; Domestic	48; 6	T					13	20		1	3	1
14	Rothschild and others, 1982	12/80-8/81	Wisconsin—Central Sand Plain (Wood County)	Monitoring; Irrigation; Domestic	67; 7; 25	T			1						2	
15	Baier and Robbins, 1982a	1981	New York—Eastern Suffolk County, North Fork	Domestic; Observation	44; 23	T			1					1	2	
16	Baier and Robbins, 1982b	1982	New York—Eastern Suffolk County, South Fork	Observation	34	T			1					1	2	
17	Zaki and others, 1982; Zaki, 1986	4/80-6/80	New York—Suffolk County	Domestic; Non-community supply; Community; NG	8051; 274; 68; 11	T			1					3	2	
18	Spruill, 1983	1976-1981	Kansas—Statewide	Domestic, community, stock, and irrigation	766	NG					5		2			1
19	Weaver and others, 1983	5/27/82-7/8/82	California—Santa Maria Valley; Salinas Valley; Upper Santa Ana Valley; San Joaquin Valley	Domestic, community, agricultural, and industrial	7; 21; 23; 166	NT	2		1					2		
20	Wehtje and others, 1983	1980-1981	Nebraska—Buffalo and Hall Counties	Observation	41	T	1									

Table 2.3. Study designs for state and local monitoring investigations of pesticide occurrence in ground waters of the United States, Canada, and Europe—*Continued*

	Study			Wells Sampled			Pesticides Examined									
No.	Reference	Sampling Date(s)	Location(s) (State [Province, Country]—Town, County or Region)	Type(s)	Number	Site Selection Strategy	TRI	ACET	CARB	UREA	OCP	OP	PHEN	FUM	TRANS	MISC
21	Hallberg and others, 1984	10/27/81-1/4/84	Iowa—Big Spring basin (Clayton County)	Spring	1	Not applicable	2	2			1	1				
22	Libra and others, 1984	12/82-12/83	Iowa—Floyd and Mitchell Counties	Domestic Spring	19 1	T	3	2			1	1				
23	Marti and others, 1984	10/81-8/83	Georgia—Donalsonville (Seminole County)	Irrigation	3	NG	1	1				1		2		2
24	Thompson, 1984	8/24/83-8/26/83	Iowa—Upper Des Moines River	Community, commercial, and domestic Observation	4 9	NG	2	1	1		S	1				1
25	Kelley, 1985	5/84-3/85	Iowa—Statewide	Community	70	T	3	2			14	5	1	7	5	4
26	Smith and others, 1985	1979-1984	Oklahoma—Fort Cobb, El Reno, and Woodward	Observation	6	T					S	S	S			
27	Soren and Stelz, 1985	1980-1982	New York—Eastern Suffolk County	Domestic Fire protection Irrigation Observation	3 1 7 31	NG			1						2	
28	Williams and Tolman, 1985	NG	Maine—South-central	Observation	4	T			1							1
29	California Dept. of Health Services, 1986	1985	California—Statewide (58 Counties)	Community	2947 (807 systems)	NT	2							5		

Table 2.3. Study designs for state and local monitoring investigations of pesticide occurrence in ground waters of the United States, Canada, and Europe—*Continued*

No.	Reference	Sampling Date(s)	Location(s) (State [Province, Country]—Town, County or Region)	Type(s)	Number	Site Selection Strategy	TRI	ACET	CARB	UREA	OC	OCP	PHEN	FUM	TRANS	MISC
				Wells Sampled			**Pesticides Examined**									
30	Detroy, 1986	NG	Iowa—Iowa River alluvium	Community, domestic, and observation	NG	NG	3	2								1
31	Fishel and Leitman, 1986	9/82-10/83	Pennsylvania—Upper Conestoga River Basin (Lancaster County)	Domestic Spring	42 1	NG	2	2								
32	Frink and Hankin, 1986	NG	Connecticut—Statewide	Community	[1]82	NT			1		16	5		2	4	3
33	Harkin and others, 1986	12/80-9/85	Wisconsin—Central Sand Plain	Observation	123	T			1						2	
34	Kelley and Wnuk, 1986	5/85	Iowa—Little Sioux River	Community	25	NT	3	2			13	6	1	6	6	4
35	Scarano, 1986	5/83-10/85	Massachusetts—Statewide	Domestic Community	311 16	T			1						2	
36	Schmidt, 1986	NG	California—Fresno/Dinuba metropolitan area	Observation	21	T								1		
37	Segawa and others, 1986	2/86	California—Willows (Glenn County)	Domestic	137	NT	3	2	S		S	S				
38	Welling and Nicosia, 1986	7/85 9/85	California—Yolo County / Solano County	Non-community Domestic	6 8	T		2								

Table 2.3. Study designs for state and local monitoring investigations of pesticide occurrence in ground waters of the United States, Canada, and Europe—Continued

| Study | | | | Wells Sampled | | | Pesticides Examined | | | | | | | | | |
No.	Reference	Sampling Date(s)	Location(s) (State [Province, Country]—Town, County or Region)	Type(s)	Number	Site Selection Strategy	TRI	ACET	CARB	UREA	OCC	OPP	PHEN	FUM	TRANS	MISC
39	Barton and others, 1987	3/11/85-4/19/85	New Jersey—Potomac-Raritan-Magothy outcrop (Middlesex and Mercer Counties)	Community, industrial, domestic, irrigation, non-community, commercial, and discontinued	65	NT	8				11	7		4	4	
40	Chen and Druliner, 1987	1984	Nebraska—High Plains aquifer	Irrigation, domestic, community, and stock	57	NT	8									
41	Droste, 1987	1985, 1986	Connecticut—Simsbury	Observation Domestic	18 72	T								1		
42	Ellingson, 1987	9/84	Washington—Vantage (Grant County)	Monitoring	5	T					S	S	S			
43	Frank and others, 1987a	11/81; 7/82 11-12/84	Canada—Southern Ontario	Domestic Domestic	11 91	T	4	3	8	4	1	13	4			8
44	Koelliker and others, 1987	12/85-2/86	Kansas—Statewide	Domestic	103	T	2	3			9	3	3	9	1	2
45	Oki and Giambelluca, 1987	NG 1982-83 NG	Hawaii—Oahu	NG Community Springs	[1]95 10 [1]8	T							3			

Table 2.3. Study designs for state and local monitoring investigations of pesticide occurrence in ground waters of the United States, Canada, and Europe—*Continued*

Study				Wells Sampled			Pesticides Examined									
No.	Reference	Sampling Date(s)	Location(s) (State [Province, Country]—Town, County or Region)	Type(s)	Number	Site Selection Strategy	TRI	ACET	CARB	UREA	OC	OP	PHEN	FUM	TRANS	MISC
46	Ontario Ministry of Environment, 1987a	1985	Canada—Ontario (Province-wide)	Domestic	351	T	5	2					1		2	1
47	Ontario Ministry of Environment, 1987b	1986	Canada—Ontario (Province-wide)	Domestic Community	37 5	T	5	2							2	
48	Pacenka and others, 1987	1980, 1983, 1984	New York—Long Island (selected areas)	Domestic	169	T			1						2	
49	Rutledge, 1987	4/84-3/85	Florida—Ocala National Forest, Orlando, Windermere, Bartow	Drainage, irrigation, community	[2]29	T	8				13	7	4	9	3	
50	Simmleit and Herrmann, 1987a,b	1/84-3/84	FRG—Upper Franconia	Springs	2	Not applicable					2					
51	Troiano and others, 1987	NG	California—Central Valley Fresno County San Joaquin County Kern County	NG	24 12 5	T						1			2	
52	Troiano and Segawa, 1987	5/86	California—Tulare County	NG	122	NT	3			1		S	S			1
53	Williams and others, 1987	1985-1986	Maine—Aroostook County	Domestic, observation	85	T	5	1	6	1	5	8	3		1	8

Table 2.3. Study designs for state and local monitoring investigations of pesticide occurrence in ground waters of the United States, Canada, and Europe—Continued

No.	Reference	Sampling Date(s)	Location(s) (State [Province, Country]—Town, County or Region)	Type(s)	Number	Site Selection Strategy	TRI	ACET	CARB	UREA	OC	OP	PHEN	FUM	TRANS	MISC
54	Capodaglio and others, 1988	10/86-2/88	Italy—Pavia Province	Domestic, community	77	NG	1		1							
55	Detroy and others, 1988	1985-1987	Iowa—Statewide	Community	355	T	3	2	2			5	2			3
56	Ellingson and Redding, 1988	7/30/86-9/19/86; 1/87	Arizona—Statewide (All counties except Apache, Graham, and Gila)	Community	40	NT			1		1			8		
57	Klaseus and others, 1988 (Minnesota Dept. of Agriculture)	Spring, 1986-Spring, 1987	Minnesota—19 Counties	Observation / Domestic / Irrigation / Tile drains	65 / 31 / 4 / 5	T	4	3	5	1	1	6	2		3	4
58	Klaseus and others, 1988 (Minnesota Dept. of Health)	7/85-6/87	Minnesota—77 counties	Community / Non-community	224 / 176	T	5	3	6	1	1	7	4			4
59	Lym and Messersmith, 1988	1985 / 1986	North Dakota—Ten counties	Domestic	144 / 44	T / T										1
60	Marade and Segawa, 1988	9/85	California—Central Valley	Domestic, irrigation, community	169	NT			2						2	
61	McConnell, 1988	11/87	Georgia—Donalsonville (Seminole County)	Observation / Domestic	5 / 4	T							1			

Table 2.3. Study designs for state and local monitoring investigations of pesticide occurrence in ground waters of the United States, Canada, and Europe—*Continued*

No.	Reference	Sampling Date(s)	Location(s) (State [Province, Country]—Town, County or Region)	Type(s)	Number	Site Selection Strategy	TRI	ACET	CARB	UREA	OC	OP	PHEN	FUM	TRANS	MISC
62	McKenna and others, 1988	12/8-7/87	Illinois—Mason County	Observation / Irrigation / Domestic	21 / 11 / 19	T	3	2	1	1	1				2	1
63	Perry and others, 1988	1985-1986	Kansas—Kansas River Basin, south-central Kansas	NG	56	T	5	2								1
64	Pettit, 1988	6/85-12/87	Oregon—Statewide	Domestic and community	216	T	5	3	10	1	8	11	3	6		1 / 3
65	Pionke others, 1988; Pionke and Glotfelty, 1989	12/10/85-4/14/87	Pennsylvania—Mahantango Creek Watershed	NG / Springs	21 / 2	T	3	2	1		3	3	1			1
66	Snethen and Robbins, 1988	11/86-12/87	Kansas—Eastern Counties	Domestic	84	T	1	1			3		1		1	1
67	Stecher and Rainwater, 1988	1982	Texas—City of Lubbock	Non-community / Community / NG	18 / 10 / 3	NG					9	4	2		4	
68	Steichen and others, 1988	12/85-2/86	Kansas—Statewide	Domestic	103	NT	2	3			9	2	2	5	2	2
69	Turco and Konopka, 1988	1987-1988	Indiana—Newton, Jasper, Tippecanoe and Jennings Counties	Domestic	47	NT	2	2								1
70	Baker and others, 1989	4/88; 3/89	Ohio—Statewide	Domestic	610	T	4	2		1						

Table 2.3. Study designs for state and local monitoring investigations of pesticide occurrence in ground waters of the United States, Canada, and Europe—*Continued*

No.	Reference	Sampling Date(s)	Location(s) (State [Province, Country]—Town, County or Region)	Type(s)	Number	Site Selection Strategy	TRI	ACET	CARB	UREA	OC	OP	PHEN	FUM	TRANS	MISC
				Wells Sampled			Pesticides Examined									
71	Cavalier and others, 1989	1985-1987	Arkansas—Central and northeastern Counties, Ouachita National Forest	Domestic, irrigation, community, and springs	119	T	3	2	2	3			2			6
72	Cherryholmes and others, 1989	11/86-11/87	Iowa—Statewide (all counties)	Community	856	NT	3	3	3		10	8	2		4	2
73	Choquette and Katz, 1989	1983-87	Florida—Polk County, Jackson County, Highlands County	Domestic and community	3024 / 2039 / 988	NT								1		
74	Druliner, 1989	1984, 1985, 1987	Nebraska—High Plains aquifer	Irrigation, domestic, stock, and community	159	NT	1									
75	Eckhardt and others, 1989a	1978-1984	New York—Nassau and Suffolk Counties	Community and observation	903	NG		2	2		2			3	1	
76	Eckhardt and others, 1989b	6/87-10/87	New York—Nassau and Suffolk Counties	NG	90	NT		S	S		S			1		
77	Grady, 1989	4/85-9/87	Connecticut—Four stratified-drift aquifers	Observation	83	NT	4	1	1					3		
78	Hallberg and others, 1989	10/83-9/87	Iowa—Big Spring basin (Clayton County)	Spring	1	Not applicable	3	2	1		8	4	3			4
79	Helgesen and Rutledge, 1989	1987	Kansas—High Plains aquifer	Domestic, stock, and irrigation	82	NT	1					1				

Table 2.3. Study designs for state and local monitoring investigations of pesticide occurrence in ground waters of the United States, Canada, and Europe—Continued

Study				Wells Sampled			Pesticides Examined									
No.	Reference	Sampling Date(s)	Location(s) (State [Province, Country]—Town, County or Region)	Type(s)	Number	Site Selection Strategy	TRI	ACET	CARB	UREA	OC	OP	PHEN	FUM	TRANS	MISC
80	Lavy, 1989	1985-87	Arkansas—Selected areas	Irrigation Community Non-community	78 20 10	NG	3	2	2	3			2			5
81	LeMasters and Doyle, 1989	8/88-2/89	Wisconsin—Statewide (dairy farms)	Domestic	534	NT	5	2	2	1	13	11			3	5
82	Louis and Vowinkel, 1989	1986, 1987	New Jersey—Coastal Plain	Irrigation Domestic Community Non-potable	37 25 17 2	T	7	2	5	1	9	5	3	5	7	3
83	Miller and others, 1989	1984	Florida—Statewide	Community	>700	T			1						2	
84	Neil and others, 1989	1985-1987	Maine—Statewide	Domestic and observation	95	T	5	1	6	1	6	8	2		1	8
85	Ritter and others, 1989	6/84; 3/85	Delaware—Appoquinimink Watershed	Observation	24	T	1		1							
86	Sitts, 1989	12/88-4/89	California—Central Valley	Domestic	190	T	3		2	1			1			2
87	Voelker, 1989	1984-87	Illinois—Statewide	Community	330	NT	3[3]	2[3]								
88	Exner and Spalding, 1990	1975-1989	Nebraska—Statewide	Domestic, observation, irrigation, stock, and community	2260	NT	6[3]	3[3]				2[3]	1[3]		1[3]	3[3]
89	Frank and others, 1990	1986 1987	Canada—Ontario	Domestic	103 76	NT	4	2	3	2	2	15	6		1	0

Table 2.3. Study designs for state and local monitoring investigations of pesticide occurrence in ground waters of the United States, Canada, and Europe—Continued

No.	Reference	Sampling Date(s)	Location(s) (State [Province, Country]—Town, County or Region)	Type(s)	Number	Site Selection Strategy	TRI	ACET	CARB	UREA	OC	OP	PHEN	FUM	TRANS	MISC
90	Hogmire and others, 1990	5/85-10/85	West Virginia—Eastern Panhandle	Domestic	20	T					1					1
91	Kross and others, 1990	1988-1989	Iowa—Statewide (SWRL)	Domestic	686	NT	3	3	2		1	9	2		5	7
92	Ontario Ministry of Environment, 1990	1987	Canada—Ontario	Domestic Community	41 1	T	8	2		13		13	4		2	2
93	Pignatello and others, 1990	1986	Connecticut—Simsbury	Observation Domestic	18 90	T								1		
94	Stuart and Demas, 1990	10/1/83 - 9/30/88	Louisiana—Statewide	NG	65	NG	9	2			11	7	4	5	4	1
95	Thompson, 1990	9/85-10/86	Iowa—West Fork, Des Moines River	Observation	9	T	3	2	2			1				1
96	Troiano and Sitts, 1990	2/87 5/87	California—Merced County	Domestic	30 29	NT	1	2								
97	Walker and Porter, 1990	1985-1987	New York—Statewide	Domestic, irrigation, stock, tile drains, and observation	73	T	3	2	2						1	
98	Dehart and others, 1991	1988, 1989	Arkansas—Northwestern Counties	Springs	25	NG	2	2		1			1			
99	Domagalski and Dubrovsky, 1991; 1992	1984-1990	California—San Joaquin Valley	Community Observation Tile drains	183 123 46	NT	11	1	4		1	12	3	7	2	4

Study | Wells Sampled | Pesticides Examined

Table 2.3. Study designs for state and local monitoring investigations of pesticide occurrence in ground waters of the United States, Canada, and Europe—*Continued*

No.	Reference	Sampling Date(s)	Location(s) (State [Province, Country]—Town, County or Region)	Type(s)	Number	Site Selection Strategy	TRI	ACET	CARB	UREA	OC	OP	PHEN	FUM	TRANS	MISC
100	Libra and others, 1991	12/87-9/89	Iowa—Big Spring basin (Clayton County)	Spring	1	Not applicable	3	2	1		1	8	2			5
101	Mayer and others, 1991	NG	Washington—Whatcom County	NG	107	T								1		
102	Perry and Anderson, 1991	1987	Kansas—Northwest, north central, southwest and south-central	Irrigation	111	NT	5	3	1			2				2
103	Seigley and Hallberg, 1991	1987-1989	Iowa—Bluegrass Creek watershed	Domestic Tile drains	7 12	NG	3	3	1			9				2
104	Trevisan and others, 1991	NG	Italy—Lomellina, Vercelli, Novara	NG	24	NT										1
105	Ando, 1992	8/6/90-8/23/90	California—Central Valley	Domestic	60	T	3			1					2	2
106	Bushway and others, 1992	5/90-11/90	Maine—Central and Southern Counties	Domestic	58	NG	1	1	1							
107	Goetsch and others, 1992	3/91-4/92	Illinois—Statewide	Domestic	337	NT	5	3	5	2	3	2	2		8	8
108	Rudolph and others, 1992	10/91-3/92	Canada—Ontario (Provincewide)	Domestic Observation	1292 160	NT T	3	2							1	
109	Christenson and Rea, 1993	4/88-7/89	Oklahoma—Oklahoma City urban area	Domestic	143	NT	10	4	11		13	10	3	8	7	7
110	Istok and others, 1993	8/87-9/87	Oregon—Ontario	NG	42	T									2	1

Table 2.3. Study designs for state and local monitoring investigations of pesticide occurrence in ground waters of the United States, Canada, and Europe—*Continued*

No.	Reference	Sampling Date(s)	Location(s) (State [Province, Country]—Town, County or Region)	Type(s)	Number	Site Selection Strategy	TRI	ACET	CARB	UREA	OC	OCP	PHEN	FUM	TRANS	MISC
				Wells Sampled			**Pesticides Examined**									
111	Koterba and others, 1993	1988–1990	Delaware/Maryland/Virginia—Delmarva Peninsula	Domestic, community, and observation	100	NT	10	4	11				4		4	8
112	Libra and others, 1993	6/91	Iowa—Statewide (SWRL resampling)	Domestic	65	NT	3	3	1			9			3	2
113	Rex and others, 1993	10/90	Iowa—Statewide (SWRL resampling)	Rural domestic	60	NT	3	2	1							1
114	Rudolph and others, 1993	6/92–8/92	Canada—Ontario (Provincewide)	Domestic; Observation	1237; 160	NT; T	3	2							1	
115	Risch, 1994	12/85–4/91	Indiana—Statewide	Domestic; Community; Non-community; Observation; Springs	304; 106; 73; 36; 2	T	12	4	7		15	21	5	6	12	1; 2
116	Szabo and others, 1994	1991	New Jersey—Coastal Plain	Observation; Irrigation; Community	15; 29; 7	T	3[3]	3[3]	3[1]						3[2]	3[3]
117	Boyle, 1995	6/93–9/94	Idaho—Boise area	Domestic, irrigation, and monitoring	22	NT	6	3	2		5	13			2	1
118	Glanville and others, 1995	5/93–6/93; 5/94–6/94	Iowa—Nine counties	Domestic; Monitoring	88; 9	T	S	S								

Table 2.3. Study designs for state and local monitoring investigations of pesticide occurrence in ground waters of the United States, Canada, and Europe—*Continued*

No.	Reference	Sampling Date(s)	Location(s) (State [Province, Country]— Town, County or Region)	Type(s)	Number	Site Selection Strategy	TRI	ACET	CARB	UREA	OC	OOP	PHEN	FUM	TRANS	MISC
119	Kalkhoff and Schaap, 1995	10/88-9/91	Iowa—Deer Creek watershed (Clayton County)	Monitoring Lysimeter Tile line	4 7 1	T	2	2								
120	LeMasters and Baldock, 1995	5/16/94- 11/2/94	Wisconsin—Statewide	Domestic	289	NT	3	2							4	
121	Mehnert and others, 1995	3/90-2/91	Illinois—Selected areas	Domestic	240	NT	4	3	5	1	7	4	2		4	8
122	Seigley, 1993	1990-1992	Iowa—Bluegrass Creek watershed	Active Inactive Tile lines	10-31 25-35 60-102	NG	3	2								2

[1]Numbers of wells estimated from maps or tables describing monitoring network.
[2]Number of "analyses" (number of wells sampled not provided).
[3]Non-detected pesticides not given.

Table 2.6. Agricultural management practices examined by the process and matrix-distribution studies

[Pesticide Analyte(s) Examined: All analytes were not necessarily examined at all sites for a given study. TTR, "total toxic residue," consisting of the summed concentrations of the parent and one or more of its transformation products (used most frequently for aldicarb). Cropping: Includes effects of cropped versus fallow land, different crops, and different crop rotations. Irrigation/Precipitation: Investigations of influence of different irrigation methods (e.g., furrow, basin, drip, and sprinkler) included under "rate." Pesticide Application, Placement: Includes investigations of the effects of incorporation versus surface application, as well as effects of application in different locations (e.g., ridge versus furrow; broadcast versus banding). MISC, Miscellaneous topics include the influence of tile line spacing (Kladivko and others, 1991), chemigation (Barnes and others, 1992), and fertilizer applications (Boul and others, 1994; Rouchaud and others, 1994). Media Sampled represents media sampled and analyzed for pesticides during field component of study. Abbreviations: PT, plant tissue (either standing plants or leaf litter); S, soil (any depth, analyzed either by chemical extraction or crop bioassay); V, vadose-zone water, obtained from lysimeters or buried soil columns; TD, tile drainage water; GW, ground water, obtained from wells or springs; SW, surface water]

Studies that Included Measurements of Pesticide Concentrations in Ground Water

Reference	Study Site(s)	Parent Pesticide(s)	Transformation Product(s)	Cropping	Tillage	Irrig./Precip. Rate	Irrig./Precip. Timing	Pest. App. Rate	Pest. App. Timing	Pest. App. Placement	Pest. App. Formulation	MISC	Media: PT	S	V	TD	GW	SW
Wyman and others, 1985	Wisconsin—Hancock Cameron	Aldicarb (as TTR, along with products)	Aldicarb sulfoxide and sulfone (as TTR)					•	•	•			•				•	
Steenhuis and others, 1988; 1990	New York—Willsboro	Alachlor, atrazine, carbofuran, and glyphosate			•									•	•	•		
Ritter and others, 1989	Delaware—Georgetown	Alachlor, atrazine, carbofuran, cyanazine, dicamba, metolachlor, and simazine			•			•									•	

Table 2.6. Agricultural management practices examined by the process and matrix-distribution studies—*Continued*

Reference	Study Site(s)	Pesticide Analyte(s) Examined — Parent Pesticide(s)	Pesticide Analyte(s) Examined — Transformation Product(s)	Cropping	Tillage	Irrigation/Precipitation — Rate	Irrigation/Precipitation — Timing	Pesticide Application — Rate	Pesticide Application — Timing	Pesticide Application — Placement	Pesticide Application — Formulation	MISC	PT	SV	TD	GW	SW
Shirmohammadi and others, 1989	Maryland—Queenstown	Atrazine, carbofuran, cyanazine, dicamba, metolachlor, and simazine			•											•	
Isensee and others, 1990	Maryland—Beltsville	Alachlor, atrazine, carbofuran, and cyanazine	Deethyl atrazine		•											•	
Porter and others, 1990	New York—Phelps	Aldicarb (as TTR, along with products)	Aldicarb sulfone and sulfoxide (as TTR)						•					•	•	•	
Gish and others, 1991b	Maryland—Beltsville	Alachlor, atrazine, carbofuran, and cyanazine			•											•	

Table 2.6. Agricultural management practices examined by the process and matrix-distribution studies—*Continued*

Reference	Study Site(s)	Pesticide Analyte(s) Examined — Parent Pesticide(s)	Transformation Product(s)	Cropping	Tillage	Irrigation/Precipitation — Rate	Irrigation/Precipitation — Timing	Pesticide Application — Rate	Pesticide Application — Timing	Pesticide Application — Placement	Pesticide Application — Formulation	MISC	Media Sampled P/T	S/V	T/D	G/W	S/W
Goodman, 1991	South Dakota—Brookings, Hamlin, and Kingsbury Counties	Alachlor, atrazine, chloramben, chlorpyrifos, cyanazine, 2,4-D, dicamba, endrin, fonofos, lindane, methoxychlor, metolachlor, metribuzin, parathion, pendimethalin, phorate, picloram, propachlor, terbufos, toxaphene, and trifluralin			•		•		•					•		•	
Ritter and others, 1994	Delaware—Georgetown	Atrazine, simazine, cyanazine, alachlor, and metolachlor			•								•			•	
Sadeghi and Isensee, 1994	Maryland—Coastal Plain	Atrazine			•		•						•			•	
Studies that did not Include Measurements of Pesticide Concentrations in Ground Waters																	
Cooke, 1966	Pennsylvania—Ambler	Amiben and dinoben				•											
Buchanan and Hiltbold, 1973	Alabama—Central and southern Alabama	Atrazine and simazine (oat bioassay)						•	•	•	•		•				

Table 2.6. Agricultural management practices examined by the process and matrix-distribution studies—*Continued*

Reference	Study Site(s)	Parent Pesticide(s)	Transformation Product(s)	Cropping	Tillage	Irrigation/ Precipitation – Rate	Irrigation/ Precipitation – Timing	Pesticide Application – Rate	Pesticide Application – Timing	Pesticide Application – Placement	Pesticide Application – Formulation	MISC	Media – P/T	Media – S	Media – V	Media – D	Media – T/D	Media – G/W	Media – S/W
Lutz and others, 1973	North Carolina—Haywood County	Picloram and 2,4,5-T						•					•						
Willis and Hamilton, 1973	Louisiana—Baton Rouge	Endrin											•		•		•		•
Rao and others, 1974	Hawaii—Molokai	Picloram				•	•		•					•	•				
Hotzman and Mitchell, 1977	Delaware—Newark	Dicamba and Vel-4207				•								•	•				
Mansell and others, 1977	Florida—Fort Pierce	Chlorobenzilate, 2,4-D, and terbacil			•												•		•
Hall and Hartwig, 1978	Pennsylvania—Centre County	Atrazine						•	•				•	•	•				
Mansell and others, 1980	Florida—Fort Pierce	Terbacil			•										•				
Marley, 1980	England—Warwick	Picloram (via bioassay)						•					•						
Jernläs and Klingspor, 1981	Sweden—Plönninge	Trichloroacetic acid				•		•								•			
Rogers and Talbert, 1981	Arkansas—Fayetteville	Metriflufen (via bioassay)							•				•						
Brewer and others, 1982	Arkansas—Stuttgart	Fluchloralin, profluralin, and trifluralin				•	•	•					•			•			

Table 2.6. Agricultural management practices examined by the process and matrix-distribution studies—*Continued*

Reference	Study Site(s)	Parent Pesticide(s)	Transformation Product(s)	Cropping	Tillage	Irrigation/ Precipitation Rate	Irrigation/ Precipitation Timing	Pesticide Application Rate	Pesticide Application Timing	Pesticide Application Placement	Pesticide Application Formulation	MISC	PT	SS	SV	TD	GW	WW
Barrett and Lavy, 1983	Arkansas—Stuttgart	Pendimethalin				•	•			•			•					
Nakamura and others, 1983	Japan—Kumagaya	Chlornitrofen, benthiocarb, molinate, and simetryne			•	•							•		•			
Barrett and Lavy, 1984	Arkansas—Stuttgart	Oxadiazon				•							•					
Jernlås, 1985	Sweden—Björnstorp	2,3,6-trichloro-benzoic acid (2,3,6-TBA)						•								•		
Jones and others, 1986a	Maine—Aroostook County	Aldicarb (as TTR, along with products)	Aldicarb sulfone and sulfoxide (as TTR)					•	•	•			•					
Dao, 1987	Oklahoma—El Reno	BAY SMY 1500 (herbicide)			•								•					
Bowman, 1988	Canada—London, Ontario	Aldicarb and metolachlor	Aldicarb sulfoxide and sulfone			•	•						•		•			
Weaver and others, 1988b	California—Del Norte and Humboldt Counties	Ethoprop and phorate	Phorate sulfoxide and sulfone								•		•					
Foy and Hiranpradit, 1989	Virginia—Blacksburg	Atrazine			•								•				•	

Table 2.6. Agricultural management practices examined by the process and matrix-distribution studies—*Continued*

Reference	Study Site(s)	Parent Pesticide(s)	Transformation Product(s)	Cropping	Tillage	Irrigation/Precipitation Rate	Irrigation/Precipitation Timing	Pesticide Application Rate	Pesticide Application Timing	Pesticide Application Placement	Pesticide Application Formulation	MISC	PT	SV	TD	GW	SW
Hall and others, 1989	Pennsylvania—Centre County	Atrazine, cyanazine, simazine, and metolachlor			•		•						•	•			
Johnsen and Morton, 1989	Arizona—Prescott, Flagstaff	Tebuthiuron						•					•				
Kolberg and others, 1989	North Dakota—Oakes	Carbofuran and terbufos	3-hydroxy and 3-keto carbofuran; terbufos sulfone and sulfoxide			•							•	•	•		
Wadd and Drennan, 1989	England—Sonning Farm, Reading	Chlorsulfuron and metsulfuron-methyl		•				•					•				
Minton and others, 1990	Georgia—Tifton	Fenamiphos (as TTR, along with products)	Fenamiphos sulfoxide and sulfone (as TTR)		•								•				
Sauer and others, 1990	Wisconsin—Central Wisconsin	Atrazine, carbofuran, chlorpyrifos, and metolachlor			•								•				
Troiano and others, 1990	California—Fresno	Atrazine				•							•				

Table 2.6. Agricultural management practices examined by the process and matrix-distribution studies—*Continued*

Reference	Study Site(s)	Parent Pesticide(s)	Transformation Product(s)	Cropping	Tillage	Irrig. Rate	Irrig. Timing	App. Rate	App. Timing	Placement	Formulation	MISC	P T	S V	T D	G W	S W
Bowman, 1991	Canada—London, Ontario	Isazofos and atrazine	CGA 17193 (isazofos degradate) and deethyl atrazine			•					•		•				
Gish and others, 1991a	Maryland—Beltsville	Atrazine and cyanazine			•								•				
Hall and others, 1991	Pennsylvania—Centre County	Atrazine, cyanazine, simazine, and metolachlor			•								•				•
Huang and Ahrens, 1991	Connecticut—New Haven	Alachlor									•		•				
Kladivko and others, 1991	Indiana—Butlerville, Jennings County	Alachlor, atrazine, carbofuran, chlorpyrifos, cyanazine, and terbufos										•		•			
Barnes and others, 1992	Arkansas—Fayetteville	Metolachlor										•	•				
Clay and others, 1992	Minnesota—Westport	Alachlor				•				•					•		
Moyer and Blackshaw, 1993	Canada—Lethbridge, Alberta	Atrazine and cyanazine		•									•				

Management Practice(s) Examined columns: Cropping, Tillage, Irrigation/Precipitation (Rate, Timing), Pesticide Application (Rate, Timing, Placement, Formulation), MISC.

Media Sampled columns: P T, S V, T D, G W, S W.

Table 2.6. Agricultural management practices examined by the process and matrix-distribution studies—*Continued*

Reference	Study Site(s)	Pesticide Analyte(s) Examined — Parent Pesticide(s)	Transformation Product(s)	Cropping	Tillage	Irrigation/Precipitation Rate	Irrigation/Precipitation Timing	Pesticide Application Rate	Pesticide Application Timing	Pesticide Application Placement	Pesticide Application Formulation	MISC	Media Sampled P/T	Media Sampled S/V	Media Sampled D/W	Media Sampled T/G	Media Sampled W/W	Media Sampled S/S
Racke and others, 1993	California, Florida, Georgia, Illinois, Indiana, and Michigan	Chlorpyrifos		•				•		•			•					
Wietersen and others, 1993	Wisconsin—Lower Wisconsin River Valley (Greenhouse)	Alachlor, atrazine, and metolachlor				•					•			•				
Boul and others, 1994	New Zealand—Mid Canterbury				•	•						•	•					
Buhler and others, 1994	Minnesota—Westport	Alachlor, metolachlor, and atrazine											•					
Corbin and others, 1994	Arkansas—Clarkedale	Fluometuron and trifluralin			•								•					
Gish and others, 1994	Maryland—Upper Marlboro				•						•		•					
Hall and Mumma, 1994	Pennsylvania—Centre County	Dicamba			•									•				•
Isensee and Sadeghi, 1994	Maryland—Beltsville	Atrazine			•	•	•		•				•					

Table 2.6. Agricultural management practices examined by the process and matrix-distribution studies—*Continued*

Reference	Study Site(s)	Parent Pesticide(s)	Transformation Product(s)	Cropping	Tillage	Irrigation/Precipitation Rate	Irrigation/Precipitation Timing	Pesticide Application Rate	Pesticide Application Timing	Placement	Formulation	MISC	PT	SV	TD	GW	SW
Kanwar and Baker, 1994	Iowa—Boone	Atrazine			•					•					•		
Kanwar and others, 1994	Iowa—Nashua	Alachlor, atrazine, cyanazine, and metribuzin		•	•										•		
Logan and others, 1994	Ohio—Wood County	Atrazine, alachlor, metolachlor, and metribuzin		•	•	•									•		•
Mills and Thurman, 1994a	Kansas—Topeka	Atrazine	Deethyl atrazine and deisopropyl atrazine								•		•	•			•
Rouchaud and others, 1994	Belgium—Lubbeek	Imidacloprid										•	•				
Traub-Eberhard and others, 1994	Germany—Welver-Borgeln, Bad Sassendorf-Bettinghausen	Chloridazon, metamitron, pendimethalin, and isoproturon							•						•		
Flury and others, 1995	Switzerland—Obfelden, Les Barges	Atrazine, terbuthylazine, and triasulfuron				•							•				
Weed and others, 1995	Iowa—Nashua	Alachlor, atrazine, and metribuzin		•	•								•		•		

CHAPTER 3

Overview of Pesticide Occurrence and Distribution in Relation to Use

Existing data on the occurrence and geographic distribution of pesticides in ground waters of the United States are voluminous. The most extensive compilation of this information is the Pesticides In Ground Water Database, or PGWDB (U.S. Environmental Protection Agency, 1992b). This database contains results from the sampling of 68,824 wells in 45 states for pesticides and some of their transformation products from 1971 to 1991 (Table 2.4). As noted previously, however, the perspective provided by the PGWDB on pesticide contamination of ground waters of the United States is not necessarily representative of ground-water quality throughout the country because of the inherent biases in the many studies included (U.S. Environmental Protection Agency, 1992b; Cohen, 1993; Haaser and Waldman, 1994).

Among the monitoring studies that have been conducted (Table 2.3), only the National Pesticide Survey, or NPS (U.S. Environmental Protection Agency, 1990a, 1992a), sampled wells in all 50 states. The spatial and analytical scope of the NPS was the most extensive of any monitoring investigation carried out to date for the investigation of pesticides in ground water (Table 2.4). Nevertheless, the data from the NPS constitute sparse geographic coverage across the nation (Figure 2.4). Furthermore, "the focus of the NPS [was] on the quality of drinking water in wells rather than on the quality of ground water" (U.S. Environmental Protection Agency, 1990a). The NPS, therefore, did not account for the influence of well characteristics (e.g., construction, depth, and age) on the chemistry of the waters sampled.

Despite the value of the data from the PGWDB and the multistate monitoring studies, a statistically representative picture of pesticide occurrence in ground waters across the United States still does not exist. Furthermore, because of variability in study design, the state and local monitoring investigations, when viewed in aggregate, provide an even less reliable picture of pesticide occurrence in ground waters across multistate regions than do the individual multistate studies. The discussion to follow therefore relies on data sources that fall short of a complete picture of the types and geographic distributions of pesticides detected in ground waters across the nation, but nevertheless provide many useful insights.

3.1 PESTICIDE-RELATED COMPOUNDS DETECTED IN GROUND WATER

Given the high cost of chemical analyses for pesticides, individual ground-water monitoring studies typically examine only a small number of compounds, usually those that are of greatest concern or are used most extensively within the area of interest. Data that are currently available therefore, give a better indication of the compounds most likely to be detected in ground waters than of those less likely to be encountered.

3.1.1 PARENT COMPOUNDS

One or more compounds from every major chemical class of pesticides have been detected in the ground waters of the nation. Table 3.1 and Figure 3.1 summarize the total number of pesticides and transformation products from each class that have been reported in ground waters of the United States by all of the monitoring studies and data compilations reviewed. The pesticide classes with the largest number of compounds detected are the organochlorines and organophosphorus compounds, carbamates and thiocarbamates, and fumigants. Among the multistate monitoring studies, the groups that have been examined most frequently are the triazine and acetanilide herbicides.

Figure 3.2 displays the pesticides and pesticide transformation products detected in 100 or more wells in the United States between 1971 and 1991, based on the PGWDB. In general, these are the compounds that have received the most attention during monitoring studies, either because of their extensive use (triazines and acetanilides), or their widespread detection in ground waters (e.g., carbamates and fumigants). Because reporting limits varied considerably among different compounds for individual studies, as well as among different studies for the same compound, the data in Figures 3.1 and 3.2 are not accurate reflections of the relative frequencies of detection of these compounds in ground water. The influence of reporting limits on detection frequencies will be examined in detail in Section 6.6.2.

Table 3.2 (at end of chapter) expands on Table 3.1 by listing all pesticides and transformation products detected in ground waters of the United States during: (1) the NPS (U.S. Environmental Protection Agency, 1990a, 1992a); (2) any of the monitoring studies included in the PGWDB (U.S. Environmental Protection Agency, 1992b); or (3) any of the monitoring investigations reviewed for this report (Table 2.3). Just as important, Table 3.2 also lists the compounds for which analyses were conducted, but that were not detected. The table focuses on the compounds that have (and have not) been detected in ground water as a result of typical agricultural practices (see Section 5.3.1). It thus excludes the results from process and matrix-distribution studies, which are often conducted in a manner designed to increase the likelihood of detecting pesticides in subsurface waters (e.g., through increased application or irrigation rates).

The PGWDB specifically excludes data from the NPS (U.S. Environmental Protection Agency, 1992b). Thus, Table 3.2 demonstrates the degree to which compounds detected in ground waters as a result of a wide variety of study designs (PGWDB) have also been encountered in wells sampled by a statistically representative national design (NPS). Comparisons between the data from the PGWDB and those from the monitoring studies are less informative because several of the latter were included in the PGWDB. In general, the compounds for which detections were reported by both the NPS and the PGWDB are among those that have been used most extensively. The most prominent examples include atrazine, simazine, alachlor, bentazon, prometon, chlordane, dibromochloropropane (DBCP), and ethylene dibromide (EDB).

Although a number of organophosphorus pesticides have been detected in wells across the nation (Table 3.1), compounds from this chemical class represent a large proportion of the pesticides that have been looked for, but not detected in ground waters (Table 3.2). The low persistence of some of these compounds is likely to be a contributing factor to this scarcity, but such low detection rates may also have resulted from the comparatively low rates at which organophosphorus pesticides are typically applied. Many of the other pesticides that have been looked for but not detected in wells are compounds that are not widely used.

Table 3.1. Pesticides reported in ground waters of the United States by the monitoring studies and data compilations examined for this book, listed according to chemical or use class

[Pesticide Class: "Other nitrogen-containing herbicides" include amides (other than the acetanilides), amines, uracils, Bentazon, Dinoseb, and Paraquat. Miscellaneous Insecticides include pyrethroids, arsenic, copper, and xylenes (o-, m-, and p-). State and Local Studies: Compilation of all compounds detected during state and local monitoring studies reviewed. National Pesticide Survey: CWS, community water system wells; RD, rural domestic wells. USEPA, U.S. Environmental Protection Agency. Blank cells indicate no information applicable or available]

										Data Compilations	
		Numbers of Compounds Within Each Class Detected by Each Study									
		Monitoring Studies									
Pesticide Class	Cooperative Private Well Testing Program	Midcontinent Pesticide Study			National Pesticide Survey		National Alachlor Well Water Survey	Metola- chlor Monitoring Study	State and Local Studies	Pesticides In Ground Water Database	1988 Survey of State Lead Agencies
		Pre- Planting	Post-Planting								
			1991	1992	CWS	RD					
Herbicides											
Triazines	(1)	5	5	5	3	3	3		9	10	7
Degradates	(1)	2	2	3					3	2	
Acetanilides	(2)	2	2	3		1	2	1	4	3	3
Degradates	(2)			2					4	1	
Chlorophenoxy acid derivatives				2					8	7	4
Ureas and sulfonylureas				1		1			4	5	2
Chlorobenzoic acid derivatives				1					3	4	3
Degradates				1	1	1			1	1	
Other nitrogen-containing herbicides				6	1	1			13	14	5
Insecticides											
Carbamates and thiocarbamates				2					12	15	6
Degradates									5	[3]6	

Table 3.1. Pesticides reported in ground waters of the United States by the monitoring studies and data compilations examined for this book, listed according to chemical or use class—*Continued*

Pesticide Class	Numbers of Compounds Within Each Class Detected by Each Study									
	Monitoring Studies								Data Compilations	
	Cooperative Private Well Testing Program	Midcontinent Pesticide Study		National Pesticide Survey		National Alachlor Well Water Survey	Metolachlor Monitoring Study	State and Local Studies	Pesticides In Ground Water Database	1988 Survey of State Lead Agencies
		Pre-Planting 1991	Post-Planting 1992	CWS	RD					
Insecticides— Continued Organochlorine compounds				1	1			23	22	20
Degradates			1					5	4	1
Organophosphorus compounds			1					17	16	12
Degradates									4[1]	
Miscellaneous								6	3	
Fungicides								3	3	
Degradates								1	1	
Fumigants				1	2			10	11	4
Miscellaneous Pesticides								1	2	
Totals	9	9	28	7	10	5	1	132	131	67
References	Baker and others, 1994	Kolpin and others, 1993; Kolpin and Goolsby, 1995		USEPA, 1990a, 1992a		Monsanto Agricultural Company, 1990; Holden and others, 1992	Roux and others, 1991a	See Table 2.3	USEPA, 1992b	Parsons and Witt, 1989

[1]Specific compounds not discernible with the immunoassay procedure used (see Section 2.8.2). Compounds to which the triazine immunoassay responds include atrazine, cyanazine, simazine, "and several other triazine herbicides, as well as some triazine breakdown products" (Wallrabenstein and Baker, 1992).

[2]Specific compounds not discernible with the immunoassay procedure used (see Section 2.8.2). Compounds to which the acetanilide immunoassay responds include alachlor, metolachlor "and various alachlor breakdown products" (Wallrabenstein and Baker, 1992).

[3]Includes 1-naphthol, which was presumed to have been a degradate of carbaryl (Majewski, 1993).

[4]Pertains to 4-nitrophenol, a degradate of methyl parathion. Because it is also used as a fungicide, this compound was included in both categories, following the approach adopted by the authors of the Pesticides In Ground Water Database.

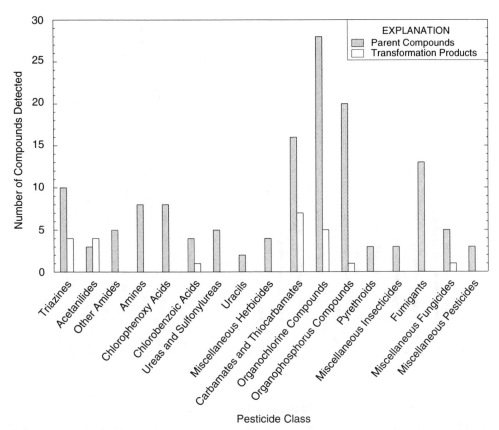

Figure 3.1. Numbers of pesticide compounds (parents and transformation products) from each chemical or use class that have been detected in ground waters of the United States during the monitoring studies and data compilations reviewed.

The pesticides that have been detected in ground water (Table 3.2) include several that are relatively hydrophobic, particularly the organochlorine compounds. Due to their affinity for natural organic matter, these compounds have traditionally been assumed to remain associated with surficial soils and subsurface solids, rather than subsurface waters (see Section 4.3.1). Indeed, the organochlorine pesticides are among those most commonly encountered in recent fluvial and lacustrine sediments in agricultural areas. Nevertheless, Figure 3.2 indicates that two of these compounds, DDT and pentachlorophenol, were among the 21 pesticides detected in more than 100 wells across the country between 1971 and 1991. Furthermore, the PGWDB reports detections of three other organochlorine insecticides—hexachlorobenzene, chlordane, and lindane (γ-hexachlorocyclohexane)—that were also among the pesticides detected in wells during the NPS (U.S. Environmental Protection Agency, 1990a). The frequent detections of these compounds in well waters indicate that affinity for organic matter is an unreliable indicator of whether an individual pesticide will reach ground waters, probably because of the effects of nonequilibrium transport (see Section 4.3.3) or colloid transport (Section 4.3.1).

Table 3.3 summarizes the frequencies with which individual pesticides and transformation products were detected in ground waters of the United States during four of the multistate monitoring programs reviewed. (Data from the Cooperative Private Well Testing

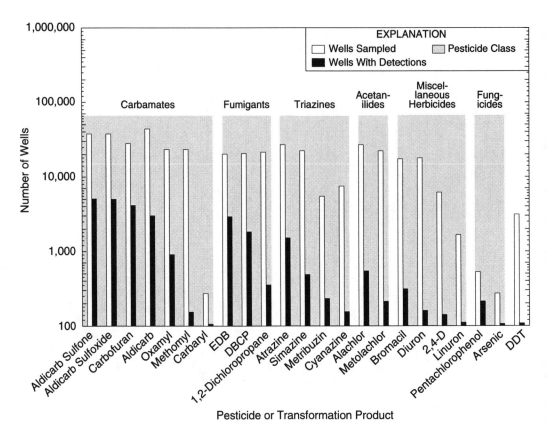

Figure 3.2. Pesticides and transformation products (grouped by pesticide class) detected in at least 100 wells in the United States, based on the Pesticides in Ground Water Database (data from U.S. Environmental Protection Agency, 1992b).

Program, or CPWTP [Baker and others, 1994], were not included in the table because this study has focused on two general groups of pesticides—the triazine and acetanilide herbicides—rather than on specific compounds). Despite the fact that all four of these studies were conducted within a 5-year period (Table 2.4), and with a considerable degree of geographic overlap (Figures 2.4, 2.5, 2.6, and 2.8), there are substantial differences in the frequencies with which individual pesticides were detected in ground water by the different studies. As noted earlier for the PGWDB, many of these differences may have arisen from variations in analytical detection limits and study design. The frequencies of detection for a given analyte generally increase when lower detection limits are used, or when sampling is targeted toward areas where ground-water contamination by the compound is considered to be more likely (see Section 6.6).

As indicated in Table 2.4, pesticides have occasionally been detected in ground waters at concentrations exceeding Maximum Contaminant Levels (MCLs). However, none of the reviewed studies reported pesticide concentrations in relation to standards for the protection of aquatic ecosystems—criteria that have been summarized most recently by Nowell and Resek (1994). Such standards are of interest in areas where contaminated ground waters may discharge to surface waters through natural discharge zones or from tile lines or ditches draining agricultural fields. Relations between pesticide concentrations measured in ground water and water-quality standards will be examined in greater detail in Chapter 10.

Table 3.3. Individual pesticide analytes detected in ground waters of the United States by the multistate sampling studies examined for this book

[Pesticide or Transformation Product: Transformation products are indented. Well Selection Procedure: Targeted (T) sampling was usually directed toward areas with a specific land use, use of a particular pesticide, or potential contamination. Nontargeted (NT) sampling involved a stratified random well-selection process. Estimated percentages given for each analyte for the National Pesticide Survey and National Alachlor Well Water Survey. CWS, Community water system wells; RD, Rural domestic wells. ND, not detected. Blank cells indicate no information applicable or available. <, less than. µg/L, micrograms per liter]

Frequencies of Detection

Wells with detections as a percentage of all wells sampled (upper number)

Analyte detection or reporting limits (µg/L) given in italics (lower number)

Pesticide or Transformation Product	Midcontinent Pesticide Study			National Pesticide Survey		National Alachlor Well Water Survey	Metolachlor Monitoring Study
	Pre-Planting	Post-Planting 1991	Post-Planting 1992	CWS	RD		
Atrazine	14.7 *0.05*	20.4 *0.05*	43.0 *0.005*	1.7 *0.12*	0.7 *0.12*	11.68 *0.03*	
Deethyl atrazine	15.4 *0.05*	21.1 *0.05*	31.0 *0.015*	ND *2.2*	ND *2.2*		
Deisopropyl atrazine[1]	4.0 *0.05*	7.5 *0.05*	18.2 *0.050*				
DCPA "acid metabolites"			15.6 *0.010*	6.4 *0.10*	2.5 *0.10*		
Prometon	4.0 *0.05*	6.1 *0.05*	9.0 *0.010*	0.5 *0.15*	0.2 *0.15*		
Metolachlor	3.0 *0.05*	2.5 *0.05*	11.0 *0.002*	ND *0.75*	ND *0.75*	1.02 *0.03*	16.3 *0.1*
Simazine	0.7 *0.05*	1.4 *0.05*	13.0 *0.005*	1.1 *0.38*	0.2 *0.38*	1.60 *0.03*	
Metribuzin	0.7 *0.05*	1.4 *0.05*	1.0 *0.005*	ND *0.60*	ND *0.60*		
Alachlor			5.0 *0.002*	ND *0.50*	<0.1 *0.50*	0.78 *0.03*	
Alachlor ethanesulfonic acid (ESA)			47.0 *0.100*				
2,6-Diethylaniline			16.0 *0.003*				
Cyanazine	0.3 *0.05*	1.1 *0.05*	3.0 *0.008*	ND *2.4*	ND *2.4*	0.28 *0.1*	

Table 3.3. Individual pesticide analytes detected in ground waters of the United States by the multistate sampling studies examined for this book—*Continued*

Pesticide or Transformation Product	Frequencies of Detection — Wells with detections as a percentage of all wells sampled (upper number) / Analyte detection or reporting limits (μg/L) given in italics (lower number)						
	Midcontinent Pesticide Study			National Pesticide Survey		National Alachlor Well Water Survey	Metolachlor Monitoring Study
	Pre-Planting	Post-Planting		CWS	RD		
		1991	1992				
Cyanazine amide			9.3 *0.05*				
Hexachlorobenzene				0.5 *0.060*	ND *0.060*		
1,2-Dibromo-3-chloropropane (DBCP)				0.4 *0.010*	0.4 *0.010*		
1,2-Dibromoethane (EDB)				ND *0.010*	0.2 *0.010*		
Lindane (γ-HCH)				ND *0.043*	0.1 *0.043*		
Ethylene thiourea (ETU, degradate of EBDC fungicides[2])				ND *4.5*	0.1 *4.5*		
Bentazon			ND *0.10*	ND *0.25*	0.1 *0.25*		
Dinoseb				<0.1 *1.3*	ND *1.3*		
2,4-D			6.7 *0.010*				
DDE			6.4 *0.006*				
Dicamba			4.4 *0.010*				
Picloram			4.4 *0.010*				
Chlorpyrifos			4.2 *0.004*				
Ethalfluralin			3.2 *0.004*				

Table 3.3. Individual pesticide analytes detected in ground waters of the United States by the multistate sampling studies examined for this book— *Continued*

Pesticide or Transformation Product	Frequencies of Detection Wells with detections as a percentage of all wells sampled (upper number) Analyte detection or reporting limits (µg/L) given in italics (lower number)						
	Midcontinent Pesticide Study			National Pesticide Survey		National Alachlor Well Water Survey	Metolachlor Monitoring Study
	Pre-Planting	Post-Planting		CWS	RD		
		1991	1992				
2,4,5-T			2.2 *0.010*				
EPTC			2.1 *0.002*				
Trifluralin			2.1 *0.003*				
Triallate			2.1 *0.001*				
Benfluralin			1.1 *0.004*				
Napropamide			1.1 *0.003*				
Pendimethalin			1.1 *0.008*				
Propachlor			1.1 *0.007*				
Tebuthiuron			1.1 *0.010*				
Well Selection Procedure	NT	NT (T for some compounds)		NT		NT	NT
References	Kolpin and others, 1993; Kolpin and others, 1995			U.S. Environmental Protection Agency, 1990a, 1992a		Monsanto Agricultural Company, 1990; Holden and others, 1992	Roux and others, 1991a

[1]May also be derived from cyanazine (Thurman and others, 1994).
[2]Includes Mancozeb, Maneb, Metiram, Zineb, and Ziram.

3.1.2 TRANSFORMATION PRODUCTS

The environmental significance of pesticide transformation products has only recently been recognized. Although the nationwide compilation by Parsons and Witt (1989) provided data for only one pesticide transformation product in ground waters of the United States (the widely-studied DDT degradate, DDD), the PGWDB included occurrence data for 45 transformation products. Recent acknowledgment of the importance of transformation products was also inherent in the design of the NPS, which included 25 degradates in its list of 126 target analytes (Table 2.4).

Table 3.4 lists transformation products that have been detected in ground waters of North America and Europe, their respective (and often inferred) parent compounds, and the locations where they were encountered. A total of 55 different transformation products are listed, derived from 33 parent compounds. A similar list presented earlier by Somasundaram and Coats (1991) contained 14 degradates, derived from 11 parent compounds. Among the transformation products listed in Table 3.4, those most frequently detected in ground water are deethyl and deisopropyl atrazine (DEA and DIA, respectively), alachlor ethanesulfonic acid (ESA), aldicarb sulfoxide and aldicarb sulfone. The common detection of these compounds in ground water is probably related to their stability, their relatively simple chemical analysis, the widespread use and occurrence of their parent compounds, and regulatory demand.

Pesticide transformation products detected during ground-water monitoring studies have often been detected with greater frequency than their parent compounds. Table 3.5 compares detection frequencies of several transformation products with those of their respective parent compounds, based on data from monitoring studies that analyzed for one or more parent/product combinations in ground water. The relatively high rates of detection for several of the degradates listed in Table 3.5 suggest that monitoring for parent compounds alone may fail to account for the behavior and fate of a substantial proportion of the pesticides used.

Pesticides that have been detected in ground waters substantially less frequently than one or more of their transformation products during monitoring studies include alachlor, DCPA, and aldrin. In contrast, atrazine has generally been detected in ground waters at frequencies comparable to or greater than those for its transformation products. Although, as stated earlier, the PGWDB does not provide a statistically representative picture of pesticide occurrence in the ground waters of the United States, overall patterns of detection from the PGWDB (Table 3.5) appear to be largely consistent with these observations.

3.1.3 NONFUMIGANT VOLATILE ORGANIC COMPOUNDS

Fumigants, by design, are volatile organic compounds (VOCs). However, a substantial proportion of the "inert ingredients," or "adjuvants" commonly employed in pesticide formulations are also VOCs (e.g., U.S. Environmental Protection Agency, 1987; Mackay, 1988; Mackay and Smith, 1990; Rose, 1990; and Buchmiller, 1992). The limited data assembled to date indicate that nonfumigant VOCs have been detected in ground waters beneath several agricultural areas in the United States (e.g., Garrett and others, 1975; Kelley, 1985; Kelley and Wnuk, 1986; Hallberg, 1987; Steichen and others, 1988; Cherryholmes and others, 1989; Domagalski and Dubrovsky, 1991; and Kolpin and Thurman, 1995). Whether or not the presence of these compounds arose from their use as adjuvants, however, was not established. VOCs have also been detected in ground waters beneath agrichemical mixing-and-loading facilities (Long, 1989), as discussed in Section 8.1.5.

Table 3.4. Pesticide transformation products detected in ground waters of North America and Europe, and their (often inferred) parent compounds

[Reference: PGWDB, Pesticides In Ground Water Database: (U.S. Environmental Protection Agency, 1992b). Note that because the PGWDB is a compilation of studies, its results may include those from other studies listed in the table. DHS, Department of Health Services. NPS, National Pesticide Survey (U.S. Environmental Protection Agency, 1990a, 1992a). USEPA, U.S. Environmental Protection Agency]

Parent Compound (often inferred)	Transformation Product(s)	Where Product was Detected in Ground Water	Reference
Alachlor	Alachlor ethanesulfonic acid (ESA)	Midcontinental United States Wisconsin	Baker and others, 1993 Kolpin and others, 1993 LeMasters and Baldock, 1995
	Hydroxy-alachlor	Iowa	PGWDB (USEPA, 1992b)
	2,6-diethylaniline	Midcontinental United States Massachusetts	Kolpin and others, 1993 Potter and Carpenter, 1995
	19 other products[1]	Massachusetts	Potter and Carpenter, 1995
Aldicarb	Aldicarb sulfoxide; aldicarb sulfone	Various locations around the United States	Jones, 1986; PGWDB (USEPA, 1992b)
	Aldicarb sulfone	The Netherlands	Loch, 1991
Ametryn	Deethyl ametryn	Oahu, Hawaii	Miles and others, 1990
Atrazine	Deethyl atrazine (DEA); deisopropyl atrazine (DIA)	Quebec, Canada The Netherlands Iowa Iowa, Indiana Midcontinental United States New Jersey Wisconsin	Muir and Baker, 1976 Lagas and others, 1989 Kross and others, 1990 PGWDB (USEPA, 1992b) Burkart and Kolpin, 1993 Szabo and others, 1994 LeMasters and Baldock, 1995
	Deethyl atrazine (DEA)	Ontario, Canada Oahu, Hawaii	Frank and others, 1979; Rudolph and others, 1992 Miles and others, 1990
	Diamino atrazine	Wisconsin	LeMasters and Baldock, 1995
	"Dealkylated atrazine"	Illinois	Goetsch and others, 1992
Carbofuran	3-Keto-carbofuran	Iowa	Kross and others, 1990; PGWDB (USEPA, 1992b)
	3-Hydroxy-carbofuran	Iowa New York Iowa, New Jersey, New York, Rhode Island Andalucia, Spain	Kross and others, 1990 Walker and Porter, 1990 PGWDB (USEPA, 1992b) Barceló and others, 1995
Chlorothalonil	Unspecified degradates	Suffolk County, New York	Suffolk County DHS (1989)
Chlorpyrifos	3,5,6-Trichloro-2-pyridinol	Cape Cod, Massachusetts	Cohen and others, 1990
Cyanazine	Cyanazine amide; DIA[2]	Quebec, Canada Lawrence, Kansas Midcontinental United States	Muir and Baker, 1976 Meyer and Thurman, 1994 Kolpin and Goolsby, 1995

Table 3.4. Pesticide transformation products detected in ground waters of North America and Europe, and their (often inferred) parent compounds—*Continued*

Parent Compound (often inferred)	Transformation Product(s)	Where Product was Detected in Ground Water	Reference
Cyprazine	Decyclopropyl cyprazine (identical to DEA)[3]	Quebec, Canada	Muir and Baker, 1976
DDT	DDE, DDD	California, Connecticut, Indiana, Mississippi, New Jersey, South Carolina	PGWDB (USEPA, 1992b)
	DDE	Waste disposal sites Oklahoma Midcontinental United States	Plumb, 1991 Christenson and Rea, 1993 Kolpin and others, 1995
Dacthal (DCPA)	TPA[4]	California Midcontinental United States	Ando, 1992 Kolpin and others, 1995
	DCPA "acid metabolites" (unspecified)	United States (nation-wide)	NPS (USEPA, 1990)
		California, Massachusetts, Oregon	PGWDB (USEPA, 1992b)
		Oregon Suffolk County, New York	Monohan and Field, 1995 Suffolk County DHS,1989
Dichlobenil	2,6-Dichlorobenzamide (BAM)	The Netherlands	Lagas and others, 1989
1,3-Dichloro-propene	3-Chloroallyl alcohol	Del Norte County, California	Brown and others, 1986
Endosulfan	Endosulfan sulfate	Indiana, New York	PGWDB (USEPA, 1992b)
Endrin	Endrin aldehyde	Waste disposal sites in the United States	Plumb, 1991
Heptachlor	Heptachlor epoxide	Stanislaus and Fresno Counties, California Kansas Alabama, Illinois, Indiana, Kansas, Massachusetts, South Carolina, Virginia Illinois	Brown and others, 1986 Steichen and others, 1988 PGWDB (USEPA, 1992b) Goetsch and others, 1992
Isazofos	CGA17193[5]	Ontario, Canada	Bowman, 1991
Methiocarb	Methiocarb sulfone	Andalucia, Spain	Barceló and others, 1995

Table 3.4. Pesticide transformation products detected in ground waters of North America and Europe, and their (often inferred) parent compounds—*Continued*

Parent Compound (often inferred)	Transformation Product(s)	Where Product was Detected in Ground Water	Reference
Maneb/mancozeb/ metiram/zineb	Ethylene thiourea (ETU)	Maine The Netherlands California	Neil and others, 1989 Lagas and others, 1989 PGWDB (USEPA, 1992b)
Metam-sodium	Methyl isothiocyanate	The Netherlands	Lagas and others, 1989
Methyl bromide	Bromide ion (Br⁻)	The Netherlands	Wegman and others, 1981
Methyl parathion	4-nitrophenol	Mississippi	PGWDB (USEPA, 1992b)
Metribuzin	Metribuzin DADK metribuzin DA, DK, DADK	Illinois Wisconsin	Goetsch and others, 1992 Lawrence and others, 1993
Molinate	Molinate sulfoxide	California	Brown and others, 1986; PGWDB (USEPA, 1992b)
Monolinuron	N-(4-hydroxyphenyl)-N'-methoxy-N'-methylurea	Federal Republic of Germany	Freitag and Scheunert, 1990
Parathion	Paraoxon	Kern County, California	Brown and others, 1986
Pendimethalin	4-hydroxymethyl pendimethalin	Mead, Kentucky	Stahnke and others, 1991
Simazine	DIA[2]	The Netherlands	Lagas and others, 1989
Thiodicarb	Methomyl	Palermo, New York Oviedo, Florida	Jones and others, 1989
Triadimefon	Triadimenol	Ithaca, New York	Petrovic and others, 1994a

[1]Of the twenty alachlor transformation products detected, the structures of nine (including 2,6-diethylaniline) were identified by comparison with standards or other published mass spectral data. The remaining compounds were identified solely from their mass spectra (Potter and Carpenter, 1995).

[2]DIA has been found to be a product of transformation of cyanazine, as well as of atrazine, in soils (Beynon and others, 1972; Sirons and others, 1973) and the vadose zone (Thurman and others, 1994). It is also produced from the transformation of simazine.

[3]Structure identical to deethyl atrazine. May represent carry-over of atrazine transformation product from previous growing seasons, rather than cyprazine transformation product.

[4]TPA, tetrachloroterephthalic acid (Ando, 1992) or 2,3,5,6-tetrachloro-1,4-benzenedicarboxylic acid (Kolpin and others, 1995).

[5]CGA17193, 5-chloro-3-hydroxy-1-isopropyl-1H-1,2,4-triazole (Petrovic and others, 1994a).

Table 3.5. Frequencies of detection of transformation products in ground waters of the United States, compared with those for their respective parent compounds

[Data shown are those for parent/product combinations examined by one or more monitoring studies during which at least one of the compounds was detected. Pesticide or Transformation Product: Transformation products are indented. National Pesticide Survey: Estimated percentages given for each analyte. CWS, Community water system wells; RD, Rural domestic wells. SWRL, Iowa State-Wide Rural Well-Water Survey; PGWDB, Pesticides In Ground Water Database. USEPA, U.S. Environmental Protection Agency. ND, not detected; NG, not given. Blank cells indicate no information applicable or available. µg/L, micrograms per liter]

Pesticide or Transformation Product	Midcontinent Pesticide Study (Post-planting only)		National Pesticide Survey		Illinois State-Wide Survey	SWRL (Iowa)	PGWDB (45 States)
	1991	1992	CWS	RD			
Herbicides							
Alachlor		5.0 *0.002*	ND *0.50*	<0.1 *0.50*	0.7 *1.3*	1.2 *0.02*	2.0
Alachlor ethanesulfonic acid (ESA)		47.0 *0.100*					
2,6-Diethylaniline		16.0 *0.003*					ND
Hydroxy-alachlor						0.2 *0.10*	0.3
Atrazine	20.4 *0.05*	43.0 *0.005*	1.7 *0.12*	0.7 *0.12*	2.1 *0.43*	4.4 *0.13*	5.6
Deethyl atrazine	21.1 *0.05*	31.0 *0.015*	ND *2.2*	ND *2.2*		3.5 *0.10*	3.9
Deisopropyl atrazine[1]	7.5 *0.05*	18.2 *0.050*				3.4 *0.10*	3.5
Didealkyl atrazine					0.1 *1.4*		ND
Cyanazine[1]		3.0 *0.008*	ND *2.4*	ND *2.4*	ND *NG*	1.2 *0.12*	2.0
Cyanazine amide		9.3 *0.05*					

Table 3.5. Frequencies of detection of transformation products in ground waters of the United States, compared with those for their respective parent compound—Continued

Pesticide or Transformation Product	Frequencies of Detection Wells with detections as a percentage of all wells sampled (upper number) Analyte detection or reporting limits (µg/L) given in italics (lower number)						
	Midcontinent Pesticide Study (Post-planting only)		National Pesticide Survey		Illinois State-Wide Survey	SWRL (Iowa)	PGWDB (45 States)
	1991	1992	CWS	RD			
DCPA		ND *0.002*	ND *0.060*	ND *0.060*		0.4 *0.01*	0.2
DCPA "acid metabolites"		15.6 *0.010*	6.4 *0.10*	2.5 *0.10*			50
Metribuzin	1.3 *0.05*	1.0 *0.005*			0.1 *0.43*	1.9 *0.01*	4.3
Metribuzin DA					ND *NG*		ND
Metribuzin DK					ND *NG*		ND
Metribuzin DADK					0.3 *0.7*		ND
Insecticides							
Aldrin					0.3 *0.004*		2.0
Dieldrin[2]					1.6 *0.004*		2.7
Carbofuran		ND *0.01*	ND *1.2*	ND *1.2*	ND *NG*	ND *0.01*	14.7
3-Hydroxy-carbofuran			ND *1.1*	ND *1.1*		0.4 *0.02*	0.2
3-Keto-carbofuran						0.4 *0.02*	0.4
Carbofuran phenol			ND *21*	ND *21*	ND *NG*		ND
3-Keto-carbofuran phenol			ND *0.93*	ND *0.93*			

Table 3.5. Frequencies of detection of transformation products in ground waters of the United States, compared with those for their respective parent compound—*Continued*

Frequencies of Detection

Wells with detections as a percentage of all wells sampled (upper number)

Analyte detection or reporting limits (µg/L) given in italics (lower number)

Pesticide or Transformation Product	Midcontinent Pesticide Study (Post-planting only)		National Pesticide Survey		Illinois State-Wide Survey	SWRL (Iowa)	PGWDB (45 States)
	1991	1992	CWS	RD			
Endrin			ND *0.13*	ND *0.13*	0.8 *0.006*		1.0
Endrin aldehyde			ND *0.13*	ND *0.13*	ND *NG*		ND
Heptachlor			ND *0.060*	ND *0.060*	ND *NG*		1.7
Heptachlor epoxide			ND *0.060*	ND *0.060*	0.5 *0.004*		1.0
References	Kolpin and others, 1993; Kolpin and others, 1995		USEPA, 1990a, 1992a		Goetsch and others, 1992	Kross and others, 1990	USEPA, 1992b

[1]Deisopropyl atrazine found to be produced from transformation of cyanazine, as well as from atrazine (Meyer and Thurman, 1994).

[2]Although it is a product of aldrin transformation, dieldrin was also used as an insecticide.

3.2 PESTICIDE USE IN AGRICULTURE

Despite the large quantities of pesticides that are currently used in the United States, the actual amounts that have been applied to the land—for agricultural, as well as non-agricultural purposes—are not well known. Indeed, even the most recent estimates of agricultural use for areas larger than individual counties have been derived primarily from inference, rather than from application records.

Table 3.6 (at end of chapter) summarizes estimates of the use of agricultural pesticides applied in quantities of at least 8,000 pounds of active ingredient (a.i.) per year within the United States from 1987 to 1991 (Gianessi and Puffer, 1990; 1992a,b), as well as the corresponding estimates of agricultural pesticide use reported by the U.S. Department of Agriculture (USDA) for 1966 (Eichers and others, 1970) and 1971 (Andrilenas, 1974). One hundred pesticides (57 herbicides, 31 insecticides, and 12 fungicides) were applied in quantities greater than 500,000 pounds a.i. per year nationwide in 1987-91 for agricultural purposes.

The pesticide use estimates provided by Gianessi and Puffer (1991; 1992a,b) are not specific to any particular year, because they were assembled from data collected at different times during the period from 1987 to 1991. The use of a particular pesticide on a specific crop in a given county was estimated by multiplying the total area in the county planted with the crop by "use coefficients" representing: (1) the percentage of the crop in the county treated with the pesticide; and (2) the estimated rate at which the pesticide is applied to the crop in that area. The crop acreage data were obtained from the 1987 Census of Agriculture (U.S. Department of Commerce, 1990) and other sources, including records of grazing acreage on federal land. The application rates estimated for the herbicides generally represent use patterns for 1989-1990 (Gianessi and Puffer, 1990), while those employed for insecticides and fungicides were mostly obtained for 1989-1991 (Gianessi and Puffer, 1992a,b). Consequently, these estimates reflect general patterns of annual pesticide use from 1987 to 1991, rather than for any specific year during this period.

Any changes in crop acreage, pesticides used on particular crops, or application rates since the end of the index period of interest (1987-1991) will affect the accuracy of the use totals. For example, the recommended application rates for atrazine decreased considerably between 1990 and 1992, while the use of cyanazine reportedly increased by as much as 25 percent in the Midwest during this same period (Goolsby and others, 1994).

Figure 3.3 displays the type of locally derived data from which application rates may be estimated, using the example of information reported by Scarano (1986) on aldicarb use in individual communities in Massachusetts. The strength of the correlation between the amount of aldicarb applied and the number of treated acres in different communities ($r^2 = 0.80$) indicates that crop acreage can be a reliable predictor of pesticide use when the data for both parameters are examined on a local scale.

In addition to the use data, Table 3.6 also shows the compounds that have been detected in ground waters, surface waters, rain, and air in the United States. This summary indicates that pesticides from every major chemical class have been detected in surface and meteoric waters, as well as in ground waters, but that more pesticides have been detected in ground water than in any other part of the Earth's hydrosphere.

The information in Table 3.6 provides an overall summary of the total amounts of individual pesticides used for agricultural purposes in the United States, but for examining the potential influence of pesticide applications on water quality, some measure of the amounts of pesticide used per unit area is needed. At present, the finest spatial resolution at which data on pesticide use are available for the entire nation is on a countywide basis (e.g., Gianessi and Puffer, 1990, 1992a,b). Given the relatively short distances traversed by ground water within a growing

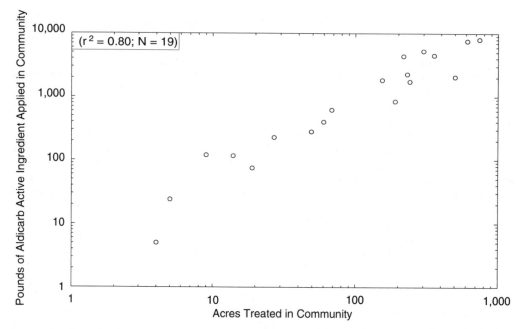

Figure 3.3. Pounds of aldicarb active ingredient applied, versus the number of acres treated with aldicarb in individual communities in Massachusetts (data from Scarano, 1986). Relation found to be statistically significant ($\alpha = 0.05$).

season (see Section 4.1), normalizing countywide pesticide use to the area of land to which the chemicals are actually applied is most appropriate for examining the effects of pesticide use on ground-water quality in agricultural areas. In contrast, normalizing countywide use to the area of an entire county (e.g., Battaglin and Goolsby, 1995) is more appropriate for studies of pesticide occurrence in surface waters (e.g., Larson, 1995) because surface-water flow can integrate water quality across entire watersheds over time scales of days to weeks.

Figures 3.4, 3.5, and 3.6 demonstrate that the amounts of herbicides, insecticides, and fungicides (respectively) applied per unit area of treated land vary substantially among different crops. For the insecticides and fungicides, application rates vary by two orders of magnitude. These data also indicate that the rates of pesticide use on turfgrass are higher than those for many agricultural crops, a point examined further in Section 3.3.

The pesticide use data summarized in Figures 3.4 through 3.6, as well as in Section 3.3.1, were provided by E. Brandt of the U.S. Environmental Protection Agency's Office of Pesticide Programs (Brandt, 1995). Brandt compiled these data from a variety of different sources to obtain information on the amounts of pesticides applied to agricultural crops (Gianessi, 1992) and turf (Kline and Company, 1990; Lucas, 1995), as well as information on the percentage of land treated with pesticides in agricultural areas (Doane Marketing Services, 1992), on golf courses (Maritz Marketing Research, Inc., 1992), and in areas where the chemicals are applied by homeowners and commercial applicators (Whitmore and others, 1992).

There is considerable geographic variability in pesticide use across the nation, caused largely by variations in: (1) pesticide use among different crops; (2) the types of agricultural products grown in different areas; and (3) pesticide application rates for a given crop in different regions. Figures 3.7a through 3.7d show the geographic distributions of agricultural pesticide use in the United States for the index years from 1987 to 1991, expressed in terms of the mass of active ingredient applied annually per unit area of agricultural land. Data for all agricultural

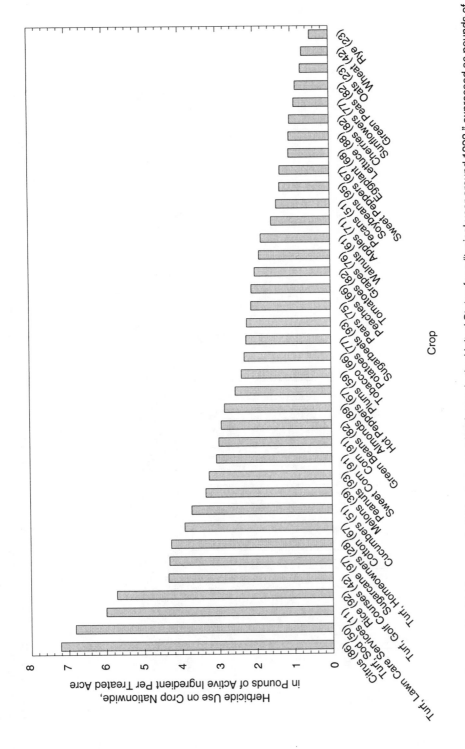

Figure 3.4. Annual herbicide use on individual crops and turfgrass in the United States for a "typical year around 1992," expressed as pounds of active ingredient applied per treated acre. Percentage of cropland treated with herbicides given in parentheses for each crop (data from Brandt, 1995).

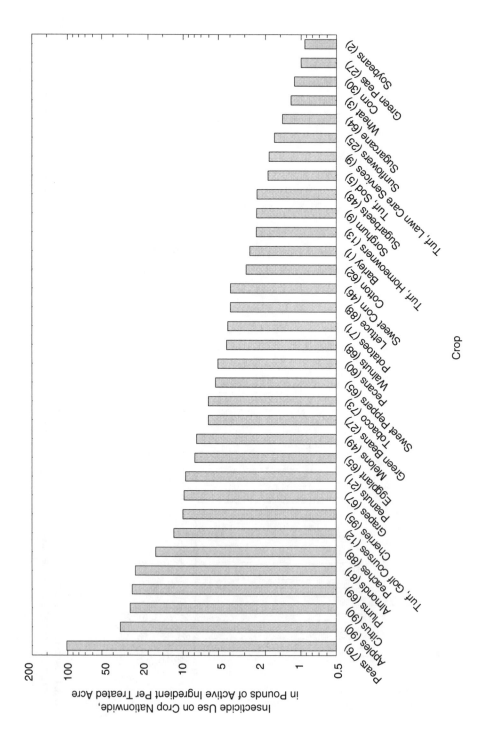

Figure 3.5. Annual insecticide use on individual crops and turfgrass in the United States for a "typical year around 1992," expressed as pounds of active ingredient applied per treated acre. Percentage of cropland treated with insecticides given in parentheses for each crop (data from Brandt, 1995). Note logarithmic scale.

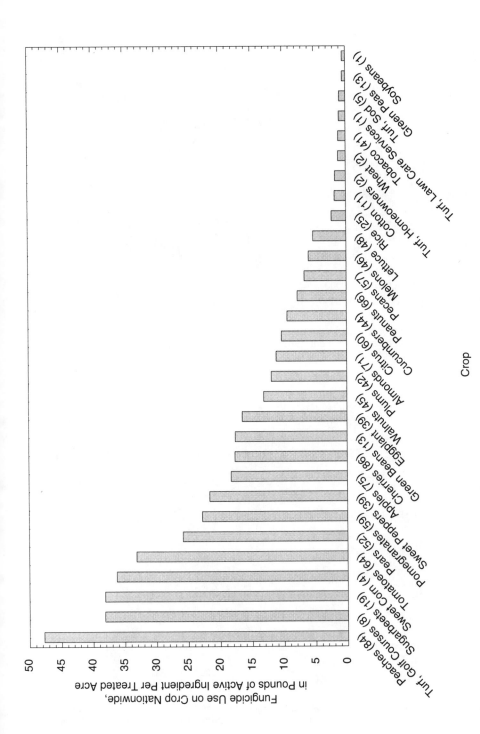

Figure 3.6. Annual fungicide use on individual crops and turfgrass in the United States for a "typical year around 1992," expressed as pounds of active ingredient applied per treated acre. Percentage of cropland treated with fungicides given in parentheses for each crop (data from Brandt, 1995).

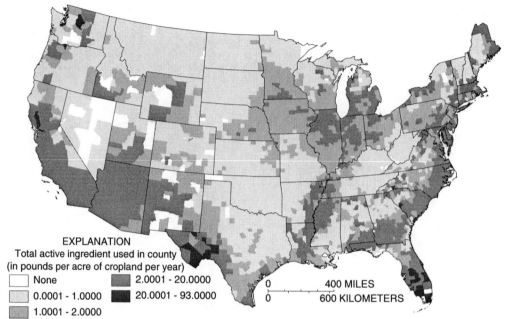

Figure 3.7a. Annual estimated pesticide use per acre of agricultural land in each county in the conterminous United States, based on the index years from 1987 to 1991. Data represent total use of all fungicides, herbicides, and insecticides reported by Gianessi and Puffer (1990, 1992a,b).

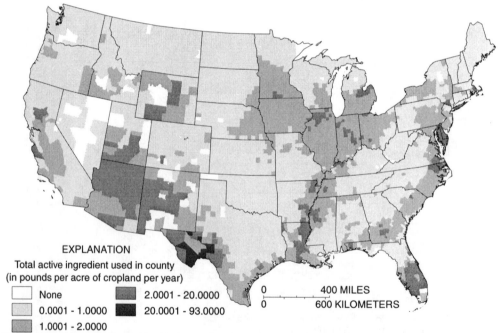

Figure 3.7b. Annual estimated herbicide use per acre of agricultural land in each county in the conterminous United States, based on the index years from 1987 to 1991. Data represent total use of all herbicides reported by Gianessi and Puffer (1990).

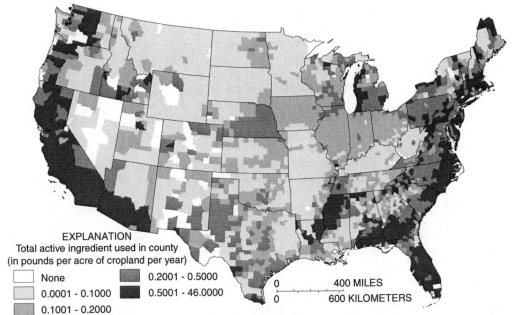

Figure 3.7c. Annual estimated insecticide use per acre of agricultural land in each county in the conterminous United States, based on the index years from 1987 to 1991. Data represent total use of all insecticides reported by Gianessi and Puffer (1992b).

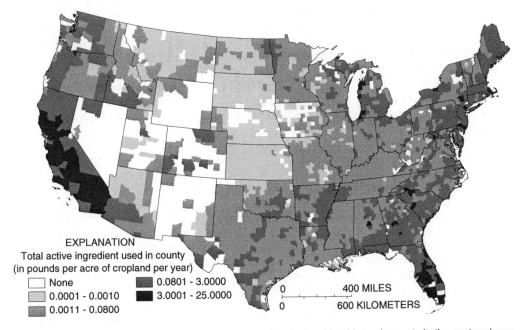

Figure 3.7d. Annual estimated fungicide use per acre of agricultural land in each county in the conterminous United States, based on the index years from 1987 to 1991. Data represent total use of all fungicides reported by Gianessi and Puffer (1992a).

pesticides are shown in Figure 3.7a (herbicides, insecticides, and fungicides combined), while Figures 3.7b, 3.7c, and 3.7d display the geographic distributions of use for herbicides, insecticides, and fungicides, respectively. The use of oil as an insecticide, and sulfur and copper as fungicides is not accounted for in the figures. For each county, these data were computed by dividing the total annual countywide use of all of the pesticides in the class(es) of interest (Gianessi and Puffer (1991; 1992a,b) by the area of cropland in the county. The latter parameter, obtained from the 1987 Census of Agriculture, represents "the sum of acreage for cropland harvested, cropland used for pasture or grazing, cropland used for cover crops, cropland on which all crops failed, cropland in cultivated summer fallow, and idle cropland" (U.S. Department of Commerce, 1990).

Although individual counties are shaded according to a single value of pesticide use per unit area of agricultural land in Figures 3.7a through 3.7d, each value, as noted above, pertains only to the cropland within a given county, and is therefore not representative of pesticide use in non-agricultural areas. This gives rise to some distortion in these maps, particularly in the western states, where counties are generally much larger than elsewhere in the nation. Many counties in the southwestern states, for example, contain areas of intensive, irrigated agriculture where pesticide use is substantial, as well as large expanses of desert with virtually no farming. While the actual use may be restricted to only a relatively small area within a given county, one shading is used for the entire county in Figures 3.7a through 3.7d.

The data presented in this chapter in relation to herbicide use per unit area of cropland (e.g., Figures 3.4 and 3.7b) may not accurately reflect the use of compounds applied to grazing land. Although, as noted earlier, Gianessi and Puffer (1990) employed estimates of grazing acreage on federal lands to compute the use of rangeland herbicides, the 1987 Census of Agriculture did not include federal lands in their estimates of cropland area. This suggests that data computed in the manner employed for Figure 3.7b may overestimate the actual use per unit area for rangeland herbicides, such as 2,4-D or picloram. During a recent field study in Kansas, however (Cress, 1994), the actual amount of 2,4-D applied to pastureland was found to be higher than that estimated for the herbicide on pastureland in Kansas by Gianessi and Puffer (1990). Thus, while the estimates cited in this chapter for the use of rangeland herbicides may be inaccurate, neither the direction nor the magnitude of these errors are known. In contrast, the potential difficulties associated with estimating application rates on rangeland are not likely to be as pronounced for insecticides or fungicides because their use in such settings is generally negligible (Gianessi and Puffer, 1992a,b).

Despite the limitations discussed above, several general observations can be made from Figures 3.7a through 3.7d regarding patterns of agricultural pesticide use across the United States. Pesticides are applied for agricultural purposes in virtually every county in the conterminous 48 states, the areas of most intensive use being in the northern midcontinent (the "corn belt"), Washington, the Southwest, the lower Mississippi River Valley, Florida, coastal areas of the Southeast, and parts of New England (Figure 3.7a). As shown earlier (Figure 1.1, Table 3.6), among the three major use categories (herbicides, insecticides, and fungicides), herbicides are used in the greatest quantities, and fungicides in the smallest quantities. Because herbicide use overshadows that of insecticides and fungicides in most areas, the geographic patterns of herbicide application (Figure 3.7b) closely approximate those for the use of all pesticides (Figure 3.7a) across most of the United States, major exceptions being the states on the west coast and in New England.

Areas of more intensive application of insecticides and fungicides (Figures 3.7c and 3.7d, respectively) are more widely distributed than those for the herbicides. Whereas herbicide use is concentrated most heavily within the midcontinent, the regions of highest use for insecticides and fungicides are predominantly the coastal areas and the lower Mississippi River Valley.

Furthermore, the geographic patterns of use are much more fragmented for insecticides and fungicides than for herbicides, because of the more substantial use of insecticides and fungicides on specialty crops in relatively small areas.

The geographic range of application varies widely among different pesticides in the same use category. Some compounds have widespread use in most areas of the United States, such as the herbicides 2,4-D, EPTC, and glyphosate, and the insecticides chlorpyrifos and carbofuran. Others, such as the herbicide molinate, and the insecticides oxamyl and methidathion, are used almost entirely in relatively small regions, because of their application to a few specialized crops grown in limited areas. Maps showing the geographic variations in use across the United States have been provided for these and most other high-use pesticides by Battaglin and Goolsby (1995), and Larson (1995).

3.3 NON-AGRICULTURAL PESTICIDE USE

In addition to their use in agriculture, pesticides are applied in a wide variety of non-agricultural settings. These uses include turf maintenance on residential lawns, golf courses, and commercial properties; insect control in homes and commercial buildings; the clearing of vegetation from railroad, transmission-line, and roadway rights-of-way; elimination of pests during the cultivation and processing of wood products; and pest control in private and public gardens. As an example, Figure 3.8 illustrates the variety of purposes for which pesticides were used, as well as the amounts applied, in non-agricultural settings in California during 1978. Non-agricultural uses of pesticides have also been summarized by Hodge (1993).

Although quantitative data on pesticide use in residential and other non-agricultural areas are much more sparse than those for pesticides used in agriculture, the available information indicates that pesticide use for agriculture is still considerably more extensive. In both 1990 and 1991, the total amounts of pesticide sold nationwide for agricultural use exceeded those sold for non-agricultural use (industry, commercial, government, home, and garden combined) by a factor of three (Aspelin, 1992). In California in 1978 (the example invoked in Figure 3.8), agricultural use of pesticides was more than 10 times greater than non-agricultural use (Litwin and others, 1983). Nationwide estimates of noncropland use of 68 herbicides, as well as comparisons between cropland and noncropland use for the nine highest-use herbicides in 1987, have been provided by Gianessi and Puffer (1990).

3.3.1 PESTICIDE USE IN URBAN AREAS

Pesticide use in urban settings has undergone major changes over the last several decades. The expansion of suburban areas, the growth of the lawn care industry, the development of new herbicides and insecticides, and the virtual replacement of organochlorine insecticides with alternative compounds, have influenced both the amount and type of pesticides applied in urban areas of the United States. The amounts applied are large—the professional applicator and consumer markets for pesticides were each estimated at $1.1 billion in sales, at the manufacturers' level, in 1991. For comparison, sales in the agricultural market were estimated to be $4.9 billion (Hodge, 1993). A 1981 survey of professional pesticide applicators identified 1,073 pesticide products containing 338 different active ingredients (a.i.). Total applications to lawns, trees, and structures by professional applicators in 1981 were estimated at 47 million pounds a.i. (Immerman and Drummond, 1984).

Pesticides are introduced into the urban environment in a variety of ways. Herbicides, insecticides, and fungicides are applied to lawns, gardens, golf courses, cemeteries, and some

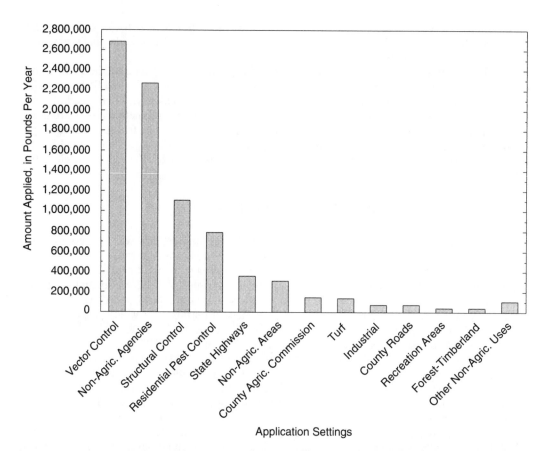

Figure 3.8. Non-agricultural pesticide use in California in 1978 (adapted from Litwin and others, 1983).

parks as liquid sprays, dusts, or granular solids. In many parts of the United States, building foundations and the soil surrounding them are routinely treated with insecticides to control termites and other destructive insects. Control of mosquitoes also has a high priority in some parts of the country, for the purposes of both public health and nuisance reduction. Algae and other plants that interfere with recreational uses of lakes and reservoirs, such as Eurasian watermilfoil, are removed by the application of herbicides in both urban and rural areas. In addition, the control of specific insect pests for agricultural purposes, such as the medfly in California, has involved the aerial spraying of insecticides over urban areas (Larson, 1995).

The *National Home and Garden Pesticide Use Survey*, conducted by the U.S. Environmental Protection Agency (USEPA), estimated household pesticide use across the United States from 1989 to 1990 (Whitmore and others, 1992). Rather than collecting quantitative information on the amounts of individual pesticides applied, however, the survey gathered data on the number of pesticide products on hand and the number of times the products were used in each surveyed household during the previous year. Results indicate that approximately 73 percent of households (69 million of 94 million total) used some type of pesticide during 1990, for either indoor or outdoor use (Whitmore and others, 1992). Other estimates indicate that home and garden pesticide use has been relatively stable in recent years, with 65-88 million pounds a.i. applied each year from 1979-1991 (Aspelin and others, 1992).

The types of pesticides used in urban and suburban areas differ considerably from those used in agriculture. In addition to the data for agricultural use, Table 3.6 lists the pesticides applied most often in and around homes and gardens, along with the reported number of outdoor applications in 1990, based on the Home and Garden Survey (Whitmore and others, 1992). An important point that emerges from a comparison of the fifty most widely used agricultural pesticides with the fifty most widely used home and garden pesticides is that there is only a 20 percent overlap between the two groups.

In 1981, the National Urban Pesticide Applicator Survey (NUPAS) estimated use by professional pest control firms in three sectors of the professional applicator industry—lawn care, tree care, and treatment of structures. Treatment of structures with insecticides, much of which may have been conducted indoors, accounted for more than 50 percent of the total amount of pesticides used by professional applicators in 1981 (Immerman and Drummond, 1984). Indeed, both the NUPAS and the Home and Garden Survey indicate that insecticide use accounts for the largest portion of urban use of pesticides. However, since the Home and Garden Survey did not estimate the actual mass used, it is difficult to determine which compounds were used most extensively by consumers. More recent estimates (1991) indicate that herbicides and insecticides now account for approximately 50 percent and 30 percent, respectively, of the professional applicators' market, but about 20 percent and 75 percent, respectively, of the consumer market (Hodge, 1993).

As noted earlier (Section 3.2), pesticide applications for turf maintenance are comparable to, or larger than, those employed for crop protection in many agricultural settings in the United States. Rates of herbicide application per unit area of turf treated on sod farms and on lawns treated by professional lawn-care services across the nation are exceeded only by the amounts applied to citrus crops (Figure 3.4). Herbicide application rates on golf courses and on lawns treated by private homeowners also rank among the top 10 crops grown in the United States, while rates of fungicide application on golf courses are second only to those employed for growing peaches (Figure 3.6). Table 3.7 summarizes estimates of pesticide use on lawns, golf courses, and sod farms, provided by the USEPA Office of Pesticide Programs (Brandt, 1995), and—like Figures 3.4 through 3.6—places these estimates within the context of agricultural use. For all turfgrass uses other than golf courses, application rates per unit acre are highest for herbicides and lowest for fungicides; golf course application rates exhibit the opposite trend. In terms of the total amounts used nationwide, herbicides, insecticides, and fungicides applied to turfgrass (lawns, golf courses, and sod farms combined) rank third, fifth, and sixth, respectively, among all crops grown in the United States.

What appear to be substantial rates of pesticide use on residential lawns may be explained in part by the fact that, as noted by Christenson and Rea (1993), "the cost of pesticides is a much less important economic constraint on a home owner than it is on a farmer." Consequently, pesticides applied for lawn maintenance are likely to pose less of a threat to ground-water quality in highly urbanized, industrialized, or rural areas than in suburban areas, where turfgrass is substantially more common. The potential significance of pesticide use on turfgrass in predominantly suburban areas is illustrated by the fact that, as of 1987, turfgrass accounted for 28 percent of the land area of Long Island, New York (Petrovic and Hummel, 1987).

3.3.2 PESTICIDE USE ON GOLF COURSES

By 1993, there were over 14,000 golf courses in the United States (Smith and others, 1993). As noted in previous sections, golf courses are treated with pesticides at rates that are among the highest of all crops. Figure 3.9 shows the rates of application for the pesticides used

Table 3.7. Estimated annual rates of pesticide use for turfgrass maintenance in the United States for "a typical year around 1992"

[Data from Brandt (1995). lb a.i., pounds of active ingredient; lb a.i./yr, pounds of active ingredient per year. NA, not applicable]

Turfgrass Use	Pesticide Class	Estimated Average Application Rate per Treated Acre		Total Estimated Annual Use	
		lb a.i.	Rank Among Crops	million lb a.i./yr	Rank Among Crops
Sod farms	Herbicides	6.82	2	0.5201	23
	Insecticides	1.90	26	0.01446	33
	Fungicides	0.97	28	0.00741	30
Lawns (professional services)	Herbicides	6.00	3	17.73	6
	Insecticides	1.86	27	4.332	9
	Fungicides	1.04	27	0.323	24
Golf courses	Herbicides	4.36	5	2.542	12
	Insecticides	12.03	7	1.401	17
	Fungicides	38.13	2	4.464	7
Lawns (private homeowners)	Herbicides	4.28	7	25.06	4
	Insecticides	2.38	23	6.215	6
	Fungicides	1.69	24	0.707	17
Turf (all applications)	Herbicides	NA	NA	45.85	3
	Insecticides	NA	NA	12.62	5
	Fungicides	NA	NA	5.502	6

most commonly on golf courses in the United States during 1982, the most recent year for which such detailed data appear to be available.

The potential effects of high pesticide application rates on ground-water quality beneath golf courses are exacerbated by the fact that the soils used for tees and putting greens, where application rates are highest (Cohen and others, 1990; Smith and Tillotson, 1993), are also designed to maximize infiltration rates. Furthermore, the incorporation of large amounts of sand into these soils to facilitate drainage (at least 85 percent by volume, or 97 percent by weight), coupled with the use of soil sterilization for controlling fungi and weeds, substantially reduces the capacity of the soil to either retain or transform the applied compounds (Smith and others, 1993). These potentially negative effects may be partially counteracted by the fact that the pesticides are surface-applied to turf (which may retain some of the pesticide mass), rather than being incorporated into bare soil (Hallberg, 1995).

3.3.3 PESTICIDE USE ON RIGHTS-OF-WAY

A variety of herbicides are used to control weeds and grasses along roadway, utility-line, railroad, and other rights-of-way for safety and aesthetic purposes, and as firebreaks in some areas. The particular compound applied to a given location is often a local choice based on the type of weeds and the weather conditions. Some of the most frequently used herbicides include 2,4-D and other chlorophenoxy acid herbicides, picloram, and triclopyr. Occasionally, insecticides, such as fonofos, are applied to roadsides to help control the movement of insects, such as grasshoppers, during infestations. The usual method of application is to spray the pesticides from a moving truck along the roadsides. Spraying by hand is employed for small areas, such as under guard rails and near bridges and overpasses. Some states have policies for

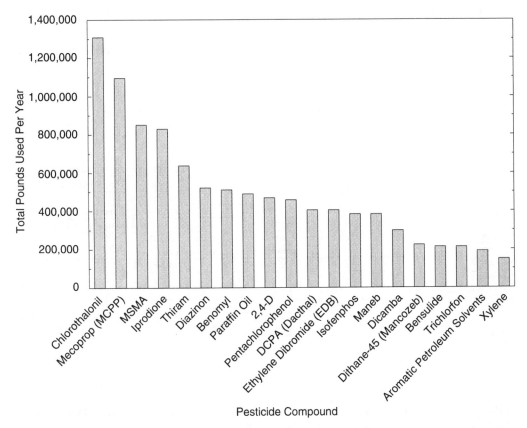

Figure 3.9. Use of pesticide chemicals on golf courses in the United States in 1982 (redrawn from Cox, 1991).

herbicide use such that the majority of the roadsides are sprayed each year, while other states have a more conservative approach to pesticide use, in which only locations with specific problems—perhaps 5 percent of the total roadside area—are sprayed. In some areas, particularly in several national forests, the application of herbicides along roadsides is prohibited (Larson, 1995).

The time of year when pesticides are applied to rights-of-way may significantly affect the risk of ground-water contamination by these compounds. In the San Joaquin Valley, California, for instance, the maximum rates of application of simazine to roadways are employed during the rainy season in late fall and winter (Domagalski and Dubrovsky, 1991), presumably because this is the period of most rapid weed growth. Because this is also the period of maximum recharge, however, the applied compounds may be carried into the subsurface in infiltrating waters relatively soon after application. As discussed in Section 5.1.3, previous work has suggested that the shorter the time interval between application and significant recharge, the greater the likelihood of ground-water contamination by the applied compounds.

An illustration of the variety and amounts of pesticides employed for these purposes is provided in Table 3.8, which lists the compounds used to clear rights-of-way in the San Joaquin Valley, California, during the mid-1980's (Domagalski and Dubrovsky, 1991). Although this

Table 3.8. Pesticides used on rights-of-way in the San Joaquin Valley, California during 1986

[Excerpted from Domagalski and Dubrovsky (1991). Application data for 1986 obtained from the California Department of Food and Agriculture (1986). lb a.i., pounds of active ingredient. Blank cells indicate no information applicable or available]

Pesticide	Total Application to San Joaquin Valley in 1986 (lb a.i.)	Pesticide Class	Other Applications
Acrolein	500,522	Herbicide	
Diuron	238,324	Herbicide	Oranges, alfalfa, asparagus
Glyphosate[1]	233,759	Herbicide	Orchards, row crops, nuts, vineyards
Simazine	190,377	Herbicide	Vineyards, orchards
2,4-D	182,169	Herbicide	Grain, corn, orchard floors
Diazinon	160,218	Insecticide	Residential/commercial lawns
Copper sulfate pentahydrate	120,675	Fungicide and algicide	Rice, fruit
Atrazine	51,936	Herbicide	Corn
Bromacil	26,426	Herbicide	Oranges
Amitrole	16,329	Herbicide	
Ammonium sulfamate	4,436	Herbicide	
Fenac	3,876	Herbicide	
Prometon	2,707	Herbicide	
Dalapon	1,281	Herbicide	
Tebuthiuron	676	Herbicide	
Copper sulfate, anhydrous	342	Fungicide and algicide	Residential/commercial lawns

[1]Given as "glyphosphate" in Domagalski and Dubrovsky (1991), but presumed to be "glyphosate."

summary is not universally applicable across the United States, these compounds have probably been used on rights-of-way in many other areas of the nation.

3.3.4 PESTICIDE USE IN FORESTRY

Pesticides serve a number of purposes in silviculture. Herbicides are used primarily for site preparation and conifer release. During site preparation, herbicides are used to eliminate competing vegetation in areas where replanting is to take place. Conifer release involves application of an herbicide several years after planting to protect the growing trees from competing, overtopping vegetation. With decreased competition for light and water, conifers can normally outgrow competing vegetation without further treatment. Herbicides are also applied in relatively minor amounts to maintain rights-of-way and to manipulate vegetation for wildlife management. Thus, herbicides are usually applied to replanted areas only once or twice in the 25- to 50-year period between planting and harvesting. Insecticides are used to control outbreaks of specific pests, such as the gypsy moth, the spruce budworm, bark beetles, and cone and seed insects. They are also used on rangelands to control grasshopper infestations. Fungicides and fumigants are used primarily on nursery stock (Larson, 1995).

Pesticides have been used in the forests of the United States for several decades. The compounds employed most extensively for silviculture during the 1950's and 1960's were chlorinated insecticides, such as DDT and endrin, whose use has since been banned in this country. During the 1970's and 1980's, new insecticides and biological agents replaced the organochlorines for insect control in forests, and the use of herbicides for vegetation control became more common. Since the early 1990's, use of pesticides, particularly herbicides, has apparently declined in some sectors of the forestry industry. The amount of pesticides applied in forestry, however, has always been a small fraction of the amount used in agriculture. Similarly, the area treated with pesticides each year is much smaller in forestry than in agriculture (Larson, 1995).

Data on the actual amounts of pesticides applied for silviculture, and the total areas involved, are difficult to obtain because of the varied ownership of forested land and the absence of a national database on non-agricultural pesticide use. Forests in the United States are owned or administered by the U.S. Forest Service (USFS), the Bureau of Land Management, states, counties, municipalities, farmers, individual land owners, and private companies. Figure 3.10, for example, shows the distribution of ownership of forested land in silviculture in Minnesota during the period from 1990-1991.

Complete data are available on pesticide use by the USFS, which administers the 191 million acres of national forest—24 percent of the approximately 800 million acres of forested land in the United States (U.S. Forest Service, 1993). Pesticides were used on less than 0.2 percent of national forest land in 1993, and on less than 1 percent each year since the mid-1970's, when detailed reporting began (U.S. Forest Service, 1978, 1985, 1989, 1990, 1991, 1992, 1993). It is not clear whether these data are representative of use on the remainder of forested land in the nation. Only two states, California and Virginia, have collected statistics on pesticide use on all forested land within their borders. These data are compared with information on pesticide use in the national forests in Table 3.9. Approximately 75,000 pounds a.i. of pesticides were applied to 0.2 percent of the forested land in California in 1991 (California Department of Food and Agriculture, 1991; Johnson, 1988). These figures apparently do not include treatment with bacterial or viral insecticides. In Virginia, approximately 79,000 pounds a.i. of herbicides were applied to 0.4 percent of the forested land in the state in 1993 (Artman, 1995). In addition, approximately 0.5 to 1.5 percent of forested land in Virginia has been treated for gypsy moth suppression each year from 1990-1993 (U.S. Forest Service, 1994).

The data in Table 3.9 indicate that the pesticides used on forested lands in California and Virginia are also used on USFS land, suggesting that pesticide use on national forest land is representative of use on the remainder of forested land in the nation. Thus, a very small percentage of the total forested area in the United States is treated with any type of pesticide in a given year. Pesticide use on 0.2 to 1.5 percent of forested land would imply that between 1.6 and 12 million acres (of about 800 million total acres) are treated with some type of pesticide in a given year nationwide. These are probably overestimates, as land receiving applications of more than one pesticide is counted more than once in the Forest Service data (Larson, 1995). For comparison, agricultural applications of atrazine and alachlor covered approximately 50 million and 27 million acres, respectively, in 1987-91 (Gianessi and Puffer, 1990).

Although pesticide use on forested land in the United States may be small compared to agricultural use, both in terms of the mass of pesticide applied and the acreage involved, consideration of this topic is important for several reasons. Forests provide vital habitat for wildlife and support a number of important fisheries. In addition, forested lands in the United States are often relatively pristine and highly valued for their aesthetic and recreational uses. The headwaters of most of the nation's major river systems are in forested areas. Many of the nation's national parks and wilderness areas border forested land that may be treated with pesticides. Forestry applications represent a large proportion of the total use of a number of pesticides, such as triclopyr, hexazinone, and diflubenzuron (dimilin), and so must be considered when evaluating the results of monitoring studies and research on pesticide occurrence in ground water (Larson, 1995). Finally, in comparison with grassland soils, which

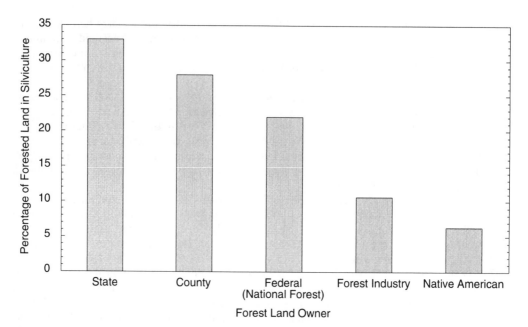

Figure 3.10. Ownership of forested land in silviculture in Minnesota, 1990-91, based on data from the Minnesota Environmental Quality Board (1992). Adapted from Larson (1995).

are similar to most agricultural soils, forest soils tend to be thinner, more acidic, less well-buffered, and have lower amounts of organic matter and clay (Birkeland, 1974). These differences in soil properties can influence the transport and fate of pesticides in the subsurface (see Chapter 4).

Figure 3.11 shows the changes in the total mass of herbicides, insecticides, fungicides, and fumigants applied to national forest land for the period from 1977 to 1993. While these values represent only a fraction of the total amount applied to forests in the United States, they indicate the general trends in silvicultural pesticide use over this period.

The herbicides with the highest use in silviculture during the late 1970's and early 1980's were 2,4-D, picloram, and hexazinone. Applications of these compounds have declined in recent years, however, and triclopyr is now the herbicide with highest use. Overall, there has been a significant decline in herbicide use over the last decade in the national forests. This is partially a result of a 1984 ban on aerial application of herbicides in national forests, and the increased costs of preparing the required environmental impact statements (Wehr and others, 1992). In fact, a number of national forests, particularly in the upper Midwest, have suspended all herbicide use for the past several years. Although data are scarce, the decline in herbicide use in the national forests has probably not occurred in other sectors of the forestry industry (Larson, 1995).

In comparison with herbicide applications, insecticide use in forestry is focused much more on controlling outbreaks of specific pests in localized areas, and is not a routine part of normal silvicultural practice. As in agriculture, there has been a dramatic change in the types of insecticides used in forestry over the last 30 years. DDT and other organochlorine insecticides were used extensively in the 1950's and 1960's. During the 1970's and 1980's, several organophosphate compounds (malathion, azinphos-methyl, trichlorfon, and acephate) and carbamate compounds (carbaryl and carbofuran) were used. More recently, a bacterial agent,

Table 3.9. Use of pesticides on forested lands

[Modified from Larson (1995). Data Source for National Forest Lands: U.S. Forest Service (1993). Data Source for California: California Department of Food and Agriculture (1991). Data Source for Virginia: Artman (1995). lb a.i., pounds of active ingredient. Blank cells indicate no information applicable or available]

Pesticide Compound	Pesticide Use (lb a.i. x 1000, unless otherwise indicated)		
	All National Forest Lands (1992)	California (1991)	Virginia (1993)
Herbicides			
Triclopyr	47	19	2
Hexazinone	18	20	8
Glyphosate	17	6	50
2,4-D	12	9	0
Picloram	8	0	0
Imazapyr	2	0	18
Fosamine	2	0	1
Dacthal (DCPA)	2	0	0
Dicamba	1	0	0
Diuron	1	0	0
Insecticides			
Bt (*Bacillus thuringiensis* var. *kurstaki*)	300,000 acres		25,000 acres
Carbaryl	97		
Malathion	5		
Diflubenzuron (Dimilin)	0.0018 (60 acres)		1.4 (46,000 acres)
Fungicides and Fumigants			
Dazomet	35	0	
Methyl bromide	40	16	
Borax	27	0	
Chloropicrin	27	3	
Chlorothalonil	5	0	

Bacillus thuringiensis var. *kurstaki* (Bt), has become the main insecticide used to control outbreaks of several major insect pests, including the gypsy moth, spruce budworm, and various cone and seed insects. Use of Bt has accounted for 25-90 percent of the total acreage treated with insecticides each year from 1984-1993 on USFS land. The apparent decline in insecticide use shown in Figure 3.11 is largely related to the replacement of many of the more traditional insecticides by Bt (amounts of Bt used are not included in Figure 3.11). Carbaryl is still used in relatively large quantities by the USFS, primarily for control of grasshoppers and crickets on rangeland (Larson, 1995). Diflubenzuron (dimilin) is also applied to control gypsy moths in the eastern United States (Artman, 1995), although use data from the USFS do not reflect this.

Fungicides and fumigants used most commonly in forestry include methyl bromide, dazomet, chloropicrin, and borax (Table 3.9). Use of these compounds, primarily on nursery stock and in seed orchards, has remained relatively stable in the national forests over the last 15 years (Figure 3.11). Whether this is true for other forested land is not known. Borax is also applied aerially to extinguish forest fires.

Pesticides are applied by several methods in forestry. Aerial application has been employed with liquid formulations of insecticides, including Bt, and with liquid and pelleted formulations of herbicides. Several variations of ground-based application are used for liquid or pelleted formulations of herbicides, including broadcast spraying from vehicles, manual spot spraying, single-stem injection, and banded spraying along tree rows in commercially owned

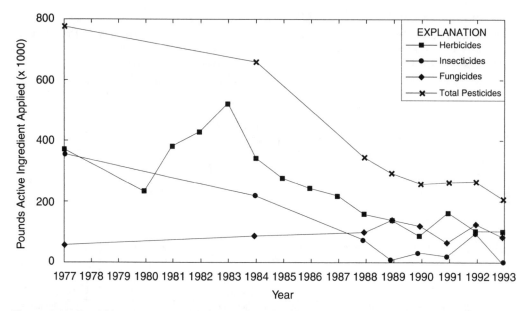

Figure 3.11. Pesticide use on national forest land, 1977-93. Insecticide amounts do not include use of *Bacillus thuringiensis* (Bt). Data from U.S. Forest Service, 1978, 1985, 1989, 1990, 1991, 1992, 1993, 1994. Redrawn from Larson (1995).

forests. Typical application rates for herbicides, inferred from national forest use data, range from approximately one pound per acre for triclopyr, glyphosate, and 2,4-D, to about two pounds per acre for hexazinone (Larson, 1995).

3.4 GEOGRAPHIC DISTRIBUTION IN RELATION TO USE

The majority of the pesticides detected in ground water—particularly the triazine and acetanilide herbicides—have been those used most extensively in agriculture, and have been found beneath predominantly agricultural areas. These general patterns have been observed over a wide range of spatial scales, including individual counties (e.g., Troiano and Segawa, 1987), states (e.g., Goetsch and others, 1992), multistate areas (Hallberg, 1989), and the entire nation (U.S. Environmental Protection Agency, 1990). Such observations are influenced, however, by the fact that most monitoring studies have focused on agricultural pesticides, as well as on regions dominated by agricultural activities. Because of the limited amount of quantitative data that are currently available on either pesticide use or occurrence in non-agricultural settings, the following discussion of relations between pesticide detection frequencies and use focuses primarily on agricultural areas.

The frequencies with which pesticides are detected in ground water within a given area are likely to increase with increasing application rates, but the extent of correlation between detections and use appears to diminish as the spatial scale of observation increases. In this section, the degree to which pesticide detections are related to use will be examined at three levels of specificity: (1) for all pesticides as a single class of contaminants in ground water; (2) for individual use classes (herbicides, insecticides, fumigants, and fungicides); and (3) for several of the individual pesticides for which published data on both use and occurrence are available.

Figure 3.12 shows the frequencies with which one or more pesticide residues were detected in individual states (expressed as the percentages of sampled wells with detections) during monitoring studies conducted between 1971 and 1991, based on the PGWDB (U.S. Environmental Protection Agency, 1992b). The total number of wells sampled in each state is also displayed. Geographic patterns of pesticide occurrence are not immediately evident from these data; states in each of the specified detection-frequency ranges appear to be more-or-less randomly distributed across the nation, with adjacent states in some cases exhibiting markedly different rates of pesticide detection, despite similar levels of agricultural activity (e.g., Indiana versus Michigan, North Carolina versus South Carolina or Virginia). The absence of discernible spatial patterns among the PGWDB data on pesticide detections across the nation is not surprising, given the difficulties associated with combining the results from studies that encompass such a wide variety of different objectives and designs (see Section 2.3).

Figure 3.13 displays the relation between overall pesticide detection frequencies and total pesticide use per acre of cropland in states with data from 100 or more wells in the PGWDB. No significant correlation was observed between the two variables (Spearman rank correlation, $\alpha = 0.05$). While states with low rates of agricultural pesticide use (less than 1 lb a.i./acre/yr) show comparatively infrequent detections of pesticides in ground water, states with high use show a considerably broader range of detection frequencies. This pattern is not unexpected. Although

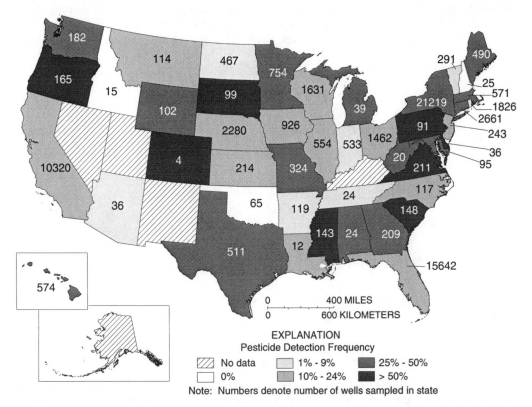

Figure 3.12. Number of wells sampled per state and percentage of sampled wells with pesticide detections in each state, based on the Pesticides In Ground Water Database, 1971-91 (data from U.S. Environmental Protection Agency, 1992b).

low detection rates are to be anticipated in areas of low use, high pesticide use does not necessarily ensure high rates of pesticide detection, because of the mitigating influences of climate, soils, hydrogeologic setting, and agricultural practices, as well as study design (see Chapters 5 and 6). Other authors have also called attention to the low proportions of wells with pesticide detections in some states with relatively high pesticide use rates (e.g., Hallberg, 1989; Burkart and Kolpin, 1993).

3.4.1 HERBICIDES

Figures 3.14 through 3.16 display the spatial distributions of herbicide detections reported by the two multistate monitoring studies that examined these compounds as a group. Figures 3.14 and 3.15 show the immunoassay results from the CPWTP for the triazines (atrazine, cyanazine, simazine, and, according to the authors, "several other triazine herbicides, as well as some triazine breakdown products") and acetanilides (alachlor, metolachlor, and "various alachlor breakdown products"), respectively (Baker and others, 1994). Figure 3.16 shows the distribution of herbicide detections in ground waters of the midcontinent in 1991 from the Midcontinent Pesticide Study, or MCPS.

Though triazines were analyzed for in more counties than acetanilides during the CPWTP, the spatial patterns of detection for the two herbicide groups are generally similar (Figures 3.14 and 3.15). An exception is northern Indiana, where the acetanilides have been detected more frequently than the triazines. Conversely, triazines have been detected at higher

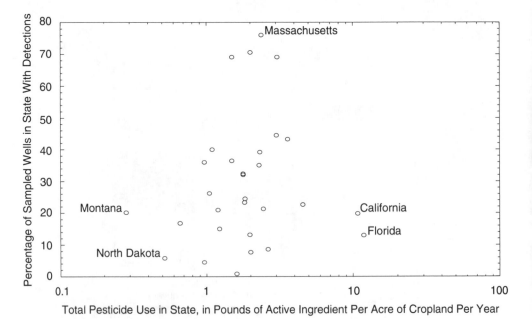

Figure 3.13. Percentage of sampled wells with pesticide detections in each state with 100 or more wells sampled, based on the Pesticides In Ground Water Database (U.S. Environmental Protection Agency, 1992b) versus statewide use of herbicides, insecticides, and fungicides combined, expressed as pounds of active ingredient (a.i.) per year (Gianessi and Puffer, 1990, 1992a,b), per acre of cropland (U.S. Department of Commerce, 1990).

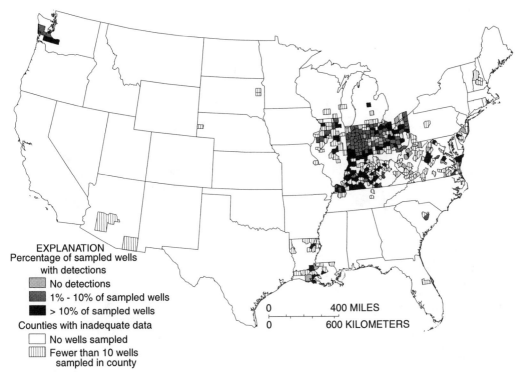

Figure 3.14. Frequencies of triazine herbicide detection (detection limit: 0.05 micrograms per liter) in counties with 10 or more wells sampled for the Cooperative Private Well Testing Program, as of February 1994 (data from Richards, 1994).

frequencies than the acetanilides in many counties in Kentucky, Louisiana, and Washington, but sampling for the triazines has also been more extensive in these states.

The study areas for the CPWTP and the MCPS exhibit their most extensive overlap in Illinois, Indiana, and Ohio, but the spatial patterns of herbicide occurrence from the two studies are most similar in Illinois. Differences between the results from the two studies become more pronounced moving eastward through Indiana and Ohio. In particular, herbicides were detected less frequently in Ohio during the 1991 MCPS sampling than during the CPWTP. However, the 1992 sampling for the MCPS, which involved considerably lower detection limits for several of the compounds of interest (Table 3.3), yielded pesticide detection patterns that more closely resembled those arising from the CPWTP. Indeed, in 1992, the MCPS detected pesticides in all of the wells sampled in Ohio, and in nearly all of those sampled in Indiana (Kolpin and others, 1993).

In some areas, the results from the multistate studies may be inconsistent with the observations of more localized investigations. For example, although herbicides were not detected at several sites in south-central Wisconsin by the MCPS in 1991 (Figure 3.16) or 1992 (Kolpin and others, 1993), this region was found during two successive statewide surveys (LeMasters and Doyle, 1989; LeMasters and Baldock, 1995) to exhibit the highest frequency of atrazine detection in the state, despite the use of a higher detection limit (0.15 µg/L) than that employed for the MCPS (0.05 µg/L in 1991, 0.005 µg/L in 1992). The reasons for these

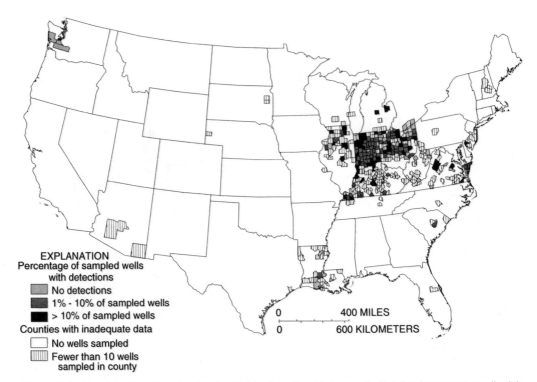

Figure 3.15. Frequencies of acetanilide herbicide detection (detection limit: 0.2 micrograms per liter) in counties with 10 or more wells sampled for the Cooperative Private Well Testing Program, as of February 1994 (data from Richards, 1994).

differences are often unclear, but they may be related to the much lower spatial density of sampling that characterizes the larger-scale studies.

The likelihood of detecting herbicides in wells shows considerable spatial variability (Figures 3.14 through 3.16), even among adjacent counties in uniformly high-use areas such as parts of the northern midcontinent (Figure 3.7a). This point is also reflected in Figures 3.17 and 3.18, where the herbicide detection frequencies from the CPWTP are plotted against the combined use of the herbicides of interest in counties where 10 or more wells were sampled. Because the CPWTP data are derived from enzyme-linked immunosorbent assay (ELISA) results, the compounds for which use data were compiled for these figures were only those to which the ELISA kits responded. Thus, the triazine data were computed for ametryn, atrazine, cyanazine, prometon, propazine, simazine, simetryn, terbuthylazine, and trietazine, while the acetanilide data were computed for alachlor, metalaxyl, and metolachlor. The use data for each herbicide were also scaled according to the sensitivity of the method to the compound, as reported in the product literature (Millipore Corporation, 1990).

Consistent with the pattern observed for the statewide data on pesticides in general (Figure 3.13), Figures 3.17 and 3.18 indicate that, for both the triazines and the acetanilides, counties with comparatively low herbicide use have relatively low rates of detection, while high-use counties have a wide range of herbicide detection rates. In some of the high-use counties,

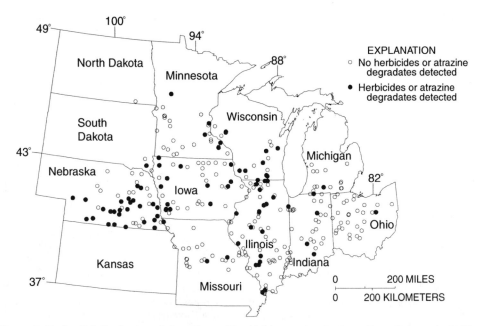

Figure 3.16. Spatial distribution of detections of herbicides or atrazine transformation products (deethyl or deisopropyl atrazine) in wells sampled during 1991 for the Midcontinent Pesticide Study, based on a reporting limit of 0.05 micrograms per liter for all compounds. Redrawn from Kolpin and others (1993).

triazine or acetanilide herbicides were detected in 100 percent of the wells sampled, while in others, they were not detected at all.

Areas where pesticide use is more intensive also tend to receive applications of a wider variety of pesticides. Consequently, the number of different pesticides detected in ground water has been used by some authors as an indicator of the severity of ground-water contamination by these compounds (e.g., Kross and others, 1990; Goodman, 1991). Figures 3.19 through 3.24 compare the number of different herbicides detected in ground waters in individual states during the period from 1971 to 1991, based on the PGWDB, to the use of those compounds, based on countywide data from Gianessi and Puffer (1990). In examining the data on the numbers of pesticides detected in individual states in Figures 3.19, 3.21 and 3.23 (and similar figures in Sections 3.4.2 and 3.4.3), it should be recognized that the number of pesticides looked for in ground water also shows considerable variability among different states and among different pesticide classes.

Among the herbicide classes examined, the numbers of triazines and acetanilides detected in individual states appear to show the closest relations with use (Figures 3.19 and 3.20). States in the northern midcontinent, along the Mississippi River Valley, and on the Atlantic Coast, where use of the triazine and acetanilide herbicides is relatively high (Figure 3.20), have reported detections of more of these compounds in ground water than the western states (Figure 3.19), where their use is lower. In contrast, less of a geographic correspondence between occurrence and use is apparent for the chlorophenoxy acid, urea (Figures 3.21 and 3.22), and miscellaneous herbicides (Figures 3.23 and 3.24). However, a relatively high number of chlorophenoxy acid and urea herbicides have been detected in the ground waters of California,

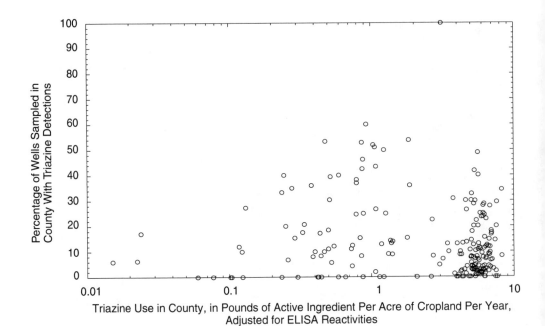

Figure 3.17. Percentage of sampled wells with triazine herbicide detections during the Cooperative Private Well Testing Program (data from Richards, 1994) versus use, expressed as pounds of active ingredient per year (Gianessi and Puffer, 1990) per acre of cropland (U.S. Department of Commerce, 1990), in counties with 10 or more wells sampled. Use of individual triazine herbicides scaled according to the sensitivity of the ELISA method for each compound, based on product literature (Millipore Corporation, 1990).

and at least 21 different herbicides have been detected in the wells of Virginia—more than in any other state. In addition, a large number of the triazine, acetanilide, and miscellaneous herbicides have been detected in the ground waters of Missouri, despite the relatively low use reported for these compounds in the state.

3.4.2 INSECTICIDES AND FUNGICIDES

For the insecticides, geographic relations between detections and use appear to be stronger in coastal states than in the Midwest (Figures 3.25 through 3.30). In particular, a relatively high number of organochlorine pesticides have been detected in the northern midcontinent, despite their limited use in this area. This may be largely a result of the fact that many of the organochlorine insecticides that are most commonly detected in ground waters (e.g., aldrin, DDT, endrin, and toxaphene) have been discontinued. Thus, Figure 3.29 primarily reflects past use patterns. As noted earlier for the herbicides, a comparatively large number of insecticides have also been detected in the ground waters of Missouri, despite their relatively low reported use in the state. Compared to the data for herbicides and insecticides, spatial relations between occurrence in ground water and use are even less apparent for fungicides (Figures 3.31 and 3.32), perhaps because fungicides are less frequently examined during monitoring studies.

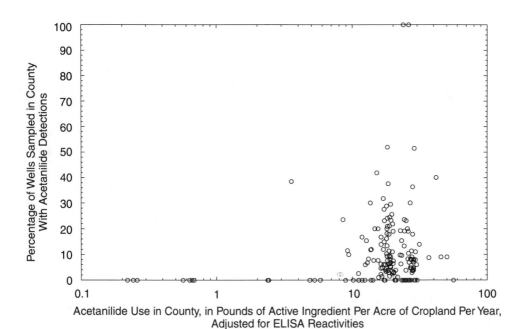

Figure 3.18. Percentage of sampled wells with acetanilide detections during the Cooperative Private Well Testing Program (data from Richards, 1994) versus use, expressed as pounds of active ingredient per year (Gianessi and Puffer, 1990, 1992a) per acre of cropland (U.S. Department of Commerce, 1990), in counties with 10 or more wells sampled. Use of individual acetanilide herbicides scaled according to the sensitivity of the ELISA method for each compound, based on product literature (Millipore Corporation, 1990).

3.4.3 FUMIGANTS

Among the pesticide classes examined, the fumigants were the only group for which county-based information on use was not available from Gianessi and Puffer (1990, 1992a,b). However, because fumigants are used primarily for nematode control, the distribution of areas treated with nematocides in 1987—based on the 1987 Census of Agriculture (U.S. Dept. of Commerce, 1990)—provides a reasonable indication of the areas where fumigants have been applied. As noted earlier for the insecticides, spatial correlations between fumigant detections (Figure 3.33) and use (Figure 3.34) appear to be strongest in coastal regions of the nation. Although many of the VOCs used as industrial solvents—such as methylene chloride, tetrachloroethylene and trichloroethylene—have also been used as fumigants (see Table 3.2), Figure 3.33 does not include data from sites where ground-water contamination by these compounds has occurred because of improper disposal or accidental releases in industrial areas.

3.4.4 GEOGRAPHIC RELATIONS BETWEEN OCCURRENCE AND USE FOR SPECIFIC PESTICIDES

The pesticides that have been the most widely used, sampled for, and detected in the ground waters of the United States are the herbicides (Table 3.6). Most prominent among these are the triazines (atrazine, simazine, cyanazine, propazine, and metribuzin), acetanilides

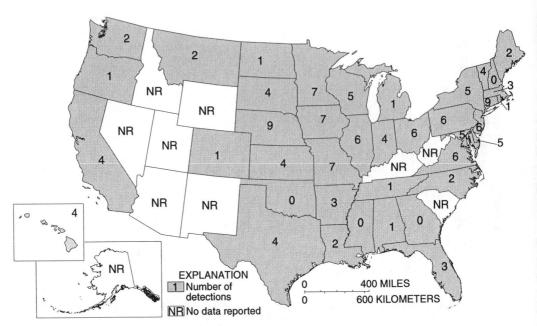

Figure 3.19. Number of triazine and acetanilide herbicides detected in one or more wells in each state, based on the Pesticides In Ground Water Database (data from U.S. Environmental Protection Agency, 1992b).

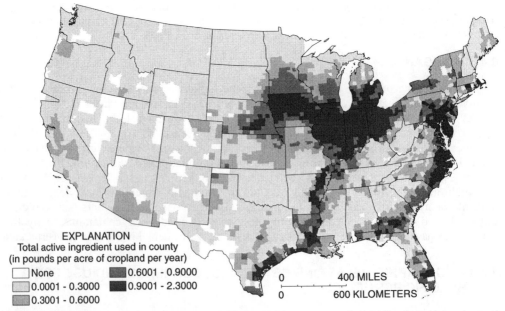

Figure 3.20. Annual use of triazine and acetanilide herbicides per acre of agricultural land in each county in the conterminous United States, based on the index years from 1987 to 1991 (data from Gianessi and Puffer, 1990).

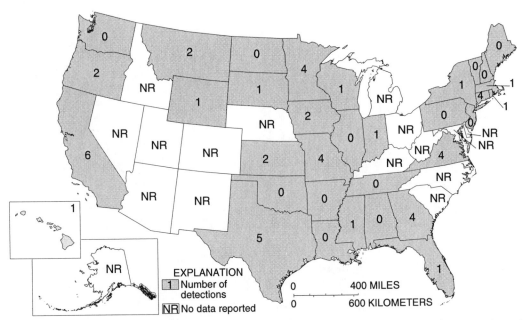

Figure 3.21. Number of chlorophenoxy acid and urea herbicides detected in one or more wells in each state, based on the Pesticides In Ground Water Database (data from U.S. Environmental Protection Agency, 1992b).

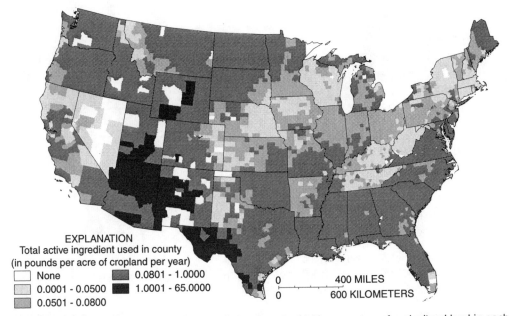

Figure 3.22. Annual use of chlorophenoxy acid and urea herbicides per acre of agricultural land in each county in the conterminous United States, based on the index years from 1987 to 1991 (data from Gianessi and Puffer, 1990).

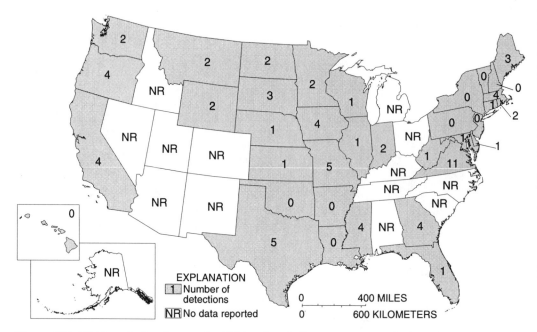

Figure 3.23. Number of miscellaneous herbicides detected in one or more wells in each state, based on the Pesticides In Ground Water Database (data from U.S. Environmental Protection Agency, 1992b).

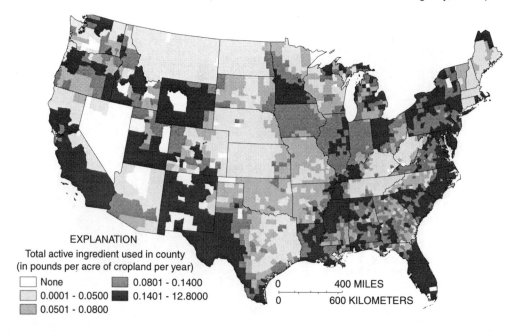

Figure 3.24. Annual use of miscellaneous herbicides per acre of agricultural land in each county in the conterminous United States, based on the index years from 1987 to 1991 (data from Gianessi and Puffer, 1990).

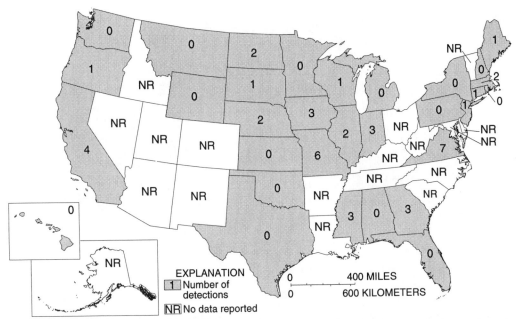

Figure 3.25. Number of organophosphorus insecticides detected in one or more wells in each state, based on the Pesticides In Ground Water Database (data from U.S. Environmental Protection Agency, 1992b).

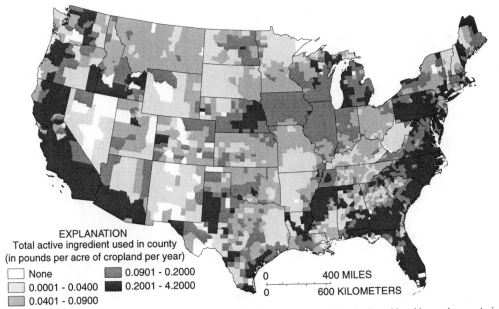

Figure 3.26. Annual use of organophosphorus insecticides per acre of agricultural land in each county in the conterminous United States, based on the index years from 1987 to 1991 (data from Gianessi and Puffer, 1992b)

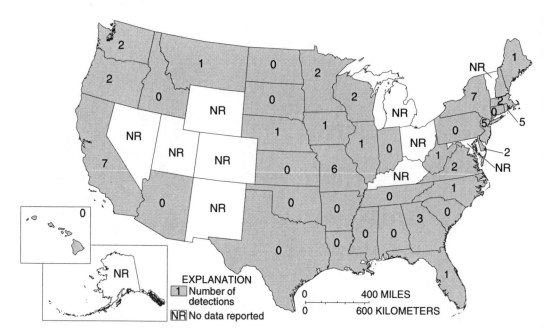

Figure 3.27. Number of carbamate and thiocarbamate insecticides detected in one or more wells in each state, based on the Pesticides In Ground Water Database (data from U.S. Environmental Protection Agency, 1992b).

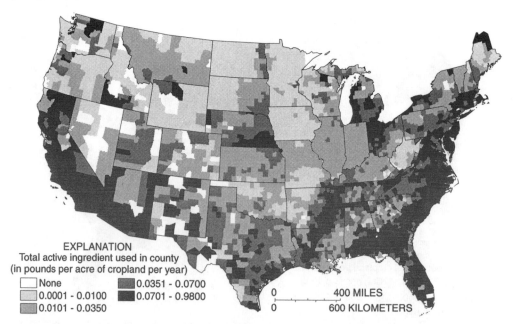

Figure 3.28. Annual carbamate and thiocarbamate insecticide use in each county in the conterminous United States, based on the index years from 1987 to 1991 (data from Gianessi and Puffer, 1992b).

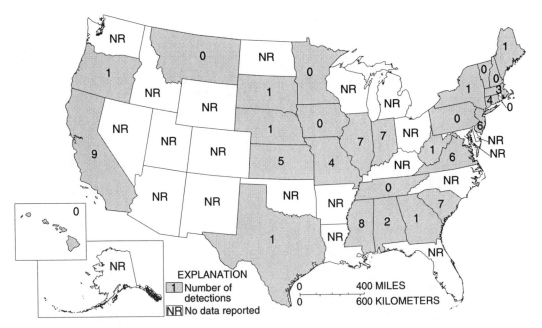

Figure 3.29. Number of organochlorine pesticides detected in one or more wells in each state, based on the Pesticides In Ground Water Database (data from U.S. Environmental Protection Agency, 1992b).

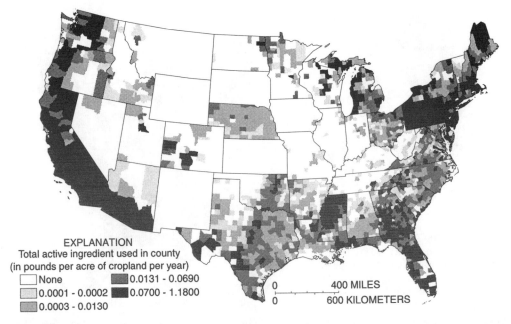

Figure 3.30. Annual use of organochlorine pesticides per acre of agricultural land in each county in the conterminous United States, based on the index years from 1987 to 1991 (data from Gianessi and Puffer, 1992b).

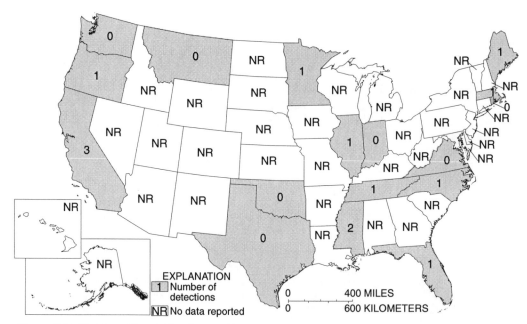

Figure 3.31. Number of fungicides detected in one or more wells in each state, based on the Pesticides In Ground Water Database (data from U.S. Environmental Protection Agency, 1992b).

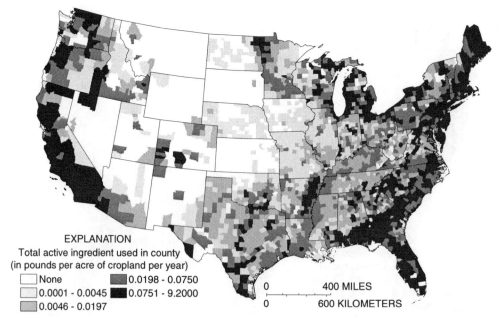

Figure 3.32. Annual use of fungicides per acre of agricultural land in each county in the conterminous United States, based on the index years from 1987 to 1991 (data from Gianessi and Puffer, 1992a).

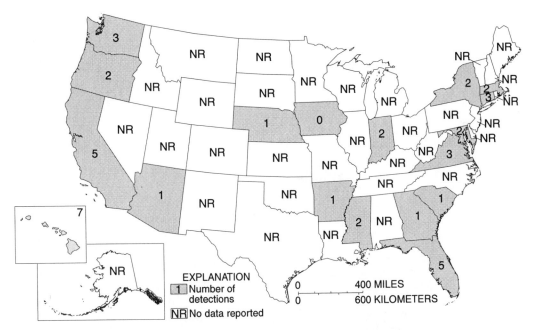

Figure 3.33. Number of fumigants detected in one or more wells in each state, based on the Pesticides In Ground Water Database (data from U.S. Environmental Protection Agency, 1992b).

(alachlor, metolachlor, and propachlor), trifluralin, dicamba, DCPA, and 2,4-D. Several of the most commonly used insecticides have also been found relatively frequently in ground waters, particularly aldicarb and its degrades and carbofuran (Figure 3.2). Frequent detections of other widely applied carbamate, organophosphorus, and organochlorine insecticides, such as carbaryl (U.S. Environmental Protection Agency, 1992b), terbufos, fonofos (Kelley, 1985; Kelley and Wnuk, 1986; Exner and Spalding, 1990), methamidophos, endosulfan (Neil and others, 1989), and lindane (U.S. Environmental Protection Agency, 1992b) have also been reported in some areas.

Some heavily used fumigants also have been found to be widespread in ground water within application areas. Chief among these compounds are 1,2-dibromo-3-chloropropane (DBCP), 1,2-dibromoethane (ethylene dibromide, or EDB), and 1,2-dichloropropane (Oki and Giambelluca, 1987; McConnell, 1988; Louis and Vowinkel, 1989; Choquette and Katz, 1989; Roaza and others, 1989; Domagalski and Dubrovsky, 1991; and Mayer and others, 1991). Because of concern about risks to human health associated with the presence of these three fumigants in ground water, their agricultural use has been cancelled in the United States (U.S. Environmental Protection Agency, 1990b; 1992b).

Several process and matrix-distribution studies have documented direct relations between rates of pesticide application and the concentrations or masses of the applied compounds measured in subsurface waters or soils (e.g., Lutz and others, 1973; Wehtje and others, 1981; Jernlås, 1985; Goodman, 1991; Huang and Ahrens, 1991; and Barnes and others, 1992), but such relations are not always evident from the results of the larger-scale studies. In some cases, pesticides that are used heavily within a given area have not been detected in ground water, while other studies have found that the compounds detected most frequently were

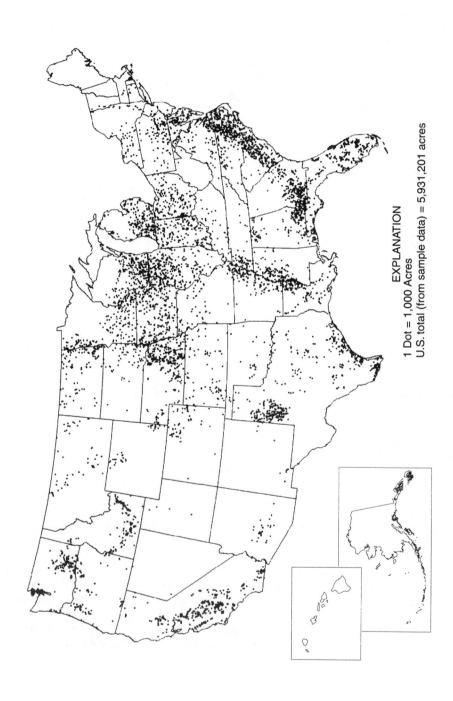

EXPLANATION

1 Dot = 1,000 Acres

U.S. total (from sample data) = 5,931,201 acres

Figure 3.34. Acres of crops treated for nematodes in the United States in 1987, based on the 1987 Census of Agriculture (modified from U.S. Department of Commerce, 1990).

not those applied most extensively (e.g., Hindall, 1978; Detroy and others, 1988; Klaseus and others, 1988; Hallberg, 1989; Neil and others, 1989; Miller and others, 1989; Exner and Spalding, 1990; Ando, 1992; Gomme and others, 1992; U.S. Environmental Protection Agency, 1992a; and Burkart and Kolpin, 1993). Pesticides have also been detected in ground water beneath areas for which there was no record of their use, or where their use was prohibited (Hallberg, 1989). Examples include the detection of sulprofos in Iowa (Kelley and Wnuk, 1986), trifluralin in California (Cardozo and others, 1988), cyanazine, metribuzin, and parathion in South Dakota (Goodman, 1991), and dinoseb in Illinois (Goetsch and others, 1992).

Given the relatively long residence times of shallow ground water (see Section 4.1), pesticide detections often reflect historic, rather than current application patterns. This explains, in part, the continued detection in ground water of persistent compounds whose use has been discontinued. Examples include the three fumigants mentioned earlier (DBCP, 1,2-dichloro-propane, and EDB), the chlorinated insecticides aldrin, dieldrin, endrin (McKenna and others, 1988), chlordane (Cohen and others, 1990), DDT and hexachlorobenzene, dinoseb, and the EBDC fungicides (U.S. Environmental Protection Agency, 1990b, 1992a,b). These observations provide additional evidence that current rates and locations of pesticide application are not always reliable predictors of pesticide occurrence in ground water.

Several studies have reported detections of pesticides that were not used within the immediate vicinity of the sampled wells, but were used on nearby fields. Examples include the detection of DBCP in Maryland (Pinto, 1980) and Hawaii (Oki and Giambelluca, 1987), EDB and 1,2,3-trichloropropane in Hawaii (Oki and Giambelluca, 1987), aldicarb in Massachusetts (Scarano, 1986), and atrazine and other pesticides in Arkansas (Cavalier and others, 1989), Iowa (Hallberg and others, 1992b), and Kansas (Steichen and others, 1988; Helgeson and Rutledge, 1989). These observations demonstrate the need to account for upgradient use in ascertaining the sources of pesticides detected in individual wells. Recent efforts, for example, have employed solute-transport models to identify source areas, or "contributing areas" of contamination in New York (Buxton and others, 1991), Oklahoma (Christenson and Rea, 1993), and Massachusetts (Barlow, 1994). Information on the size and configuration of contributing areas is of legal, as well as public-health significance in states such as California (Maes and others, 1991) and Massachusetts (Scarano, 1986), which sometimes restrict the handling and application of certain agrichemicals within specified areas surrounding water-supply wells.

The preceding discussion indicates that current rates and distributions of pesticide application may not, by themselves, be reliable predictors of pesticide contamination of ground water. The remainder of this section will examine the extent to which these relations are improved for individual compounds as the spatial scale of observation decreases.

Results from Data Compilations

When the proportions of wells with atrazine detections (based on the PGWDB) are compared with atrazine use in individual states (Figure 3.35), the observed pattern is similar to those noted earlier for all pesticides viewed as a group (Figure 3.13) and for the CPWTP herbicide data (Figures 3.17 and 3.18). Atrazine has been detected in a relatively small percentage of the wells sampled in those states with low atrazine use, while the range of detection rates is considerably broader in states with moderate or high atrazine use. Analogous patterns were noted among the PGWDB data for the three other high-use compounds examined in this manner, i.e., alachlor, carbofuran, and aldicarb (data not shown). Similar patterns were also observed when percentages of sampled wells with pesticide detections, based on statewide compilations of monitoring data in Nebraska (Spalding and others, 1989) and Indiana (Risch, 1994), were plotted against statewide use for individual compounds. Figure 3.36 shows this

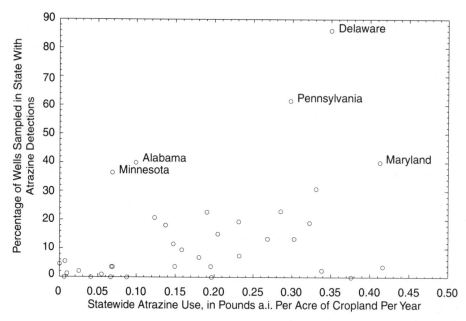

Figure 3.35. Percentages of sampled wells with atrazine detections in states with 10 or more wells sampled for atrazine from 1971 to 1991, based on the Pesticides In Ground Water Database (U.S. Environmental Protection Agency, 1992b), versus statewide use of atrazine, expressed as pounds of active ingredient (a.i.) per year (Gianessi and Puffer, 1990), per acre of cropland in state (U.S. Department of Commerce, 1990).

relation for the Indiana data. As observed earlier for several other studies (Section 3.1.2), the pesticide compounds detected most frequently in Indiana were transformation products.

Results from Multistate Monitoring Studies

Of the five multistate investigations reviewed (Table 2.4), three—the NPS, MCPS, and National Alachlor Well Water Survey (NAWWS)—examined the occurrence of more than one specific pesticide in ground water. However, data on the spatial distributions of detected compounds were available only for the MCPS and NAWWS.

Figure 3.37, reproduced from Burkart and Kolpin (1993), compares the locations of atrazine detections during the 1991 MCPS sampling with the distribution of atrazine use across the 12-state study area. Because atrazine dominated the herbicide detections, locations where atrazine was not detected generally correspond to those where no herbicides were detected in Figure 3.16. As noted by Burkart and Kolpin (1993), atrazine detections were generally more frequent in areas with heavier atrazine use, with the exception of comparatively infrequent detections through much of Indiana and Ohio, where use is also high.

The histogram in Figure 3.38 provides a statistical summary of atrazine detection rates as a function of countywide application rates within the MCPS area. Although there was a monotonic increase in detection rates with increasing use of the herbicide among the three lowest ranges of use, no statistically significant differences in detection rates were observed among the four use ranges ($\alpha = 0.05$). Of all the multistate studies reviewed, the MCPS provided the most detailed and systematic summary of pesticide use within the study area of interest. It also

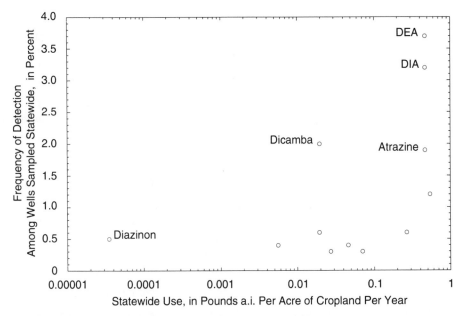

Figure 3.36. Percentages of sampled wells with detections of individual pesticides in Indiana (Risch, 1994) versus their statewide use, expressed as pounds of active ingredient (a.i.) per year (Risch, 1994; Gianessi and Puffer, 1990, 1992a,b), per acre of cropland (U.S. Department of Commerce, 1990).

employed the most rigorous criteria for the selection of wells for sampling (Kolpin and Burkart, 1991; Burkart and Kolpin, 1993). The fact that a clear relation between the rates of pesticide use and detection in ground waters was still not observed during this study demonstrates the difficulties associated with efforts to detect such relations across large, multistate areas, particularly at sampling densities as low as one well per county. Perhaps most significant among these difficulties is the fact that county-level use data—the smallest scale at which consistent information on pesticide use are currently available for such large areas—may not accurately reflect pesticide use near individual wells.

In agreement with the results from the CPWTP and MCPS, countywide percentages of sampled wells with atrazine detections during the NAWWS were generally found to be highest in high-use areas, such as the northern midcontinent and eastern Pennsylvania (Figure 3.39). Furthermore, atrazine was detected in northern Indiana and Ohio during the NAWWS at surprisingly low rates, given its extensive use in this region—an observation that is also consistent with the results discussed earlier for the 1991 MCPS. Generally, the counties where simazine, cyanazine, alachlor, and metolachlor were detected during the NAWWS represent subsets of those where atrazine was detected during the study (data not shown). The principal exceptions to this pattern included high countywide percentages (i.e., in the uppermost quartile) of wells with detections of simazine in central California, alachlor and metolachlor in northeastern Louisiana, and metolachlor in central Ohio.

Semi-logarithmic plots of herbicide detection frequencies in individual counties from the NAWWS versus countywide use show the same general appearance as analogous relations presented earlier (Figures 3.17, 3.18, 3.35 and 3.36). As illustrated for atrazine in Figure 3.40, low percentages of sampled wells exhibited detections in low-use counties, while broad ranges

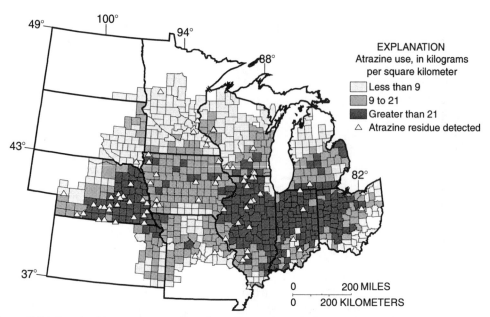

Figure 3.37. Countywide atrazine use (data from Gianessi and Puffer, 1990) and detections of atrazine or atrazine transformation products in ground water during the 1991 sampling of the Midcontinent Pesticide Study (reporting limit: 0.05 micrograms per liter). Redrawn with permission from Burkart and Kolpin (1993).

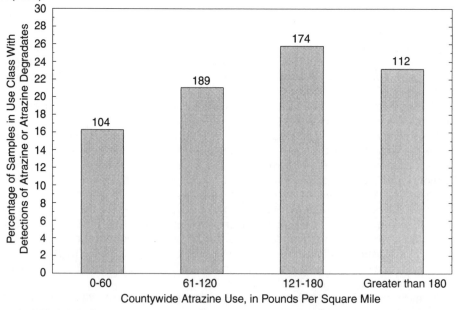

Figure 3.38. Relation of frequencies of detection of atrazine or atrazine transformation products in ground water (reporting limit: 0.05 micrograms per liter) in individual counties within the midcontinent in 1991 to estimated countywide atrazine use. Total number of samples in each category shown above bar. Redrawn from Kolpin and others (1994) and published with permission.

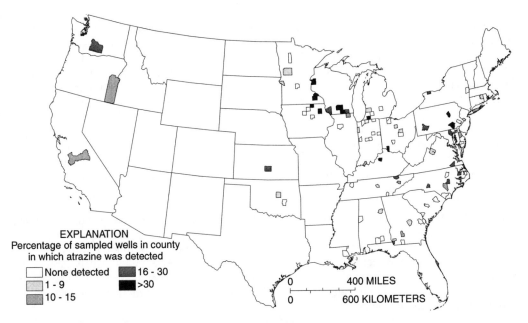

Figure 3.39. Frequencies of atrazine detection (detection limit: 0.03 micrograms per liter) in counties sampled for the National Alachlor Well Water Survey (data from Klein, 1993).

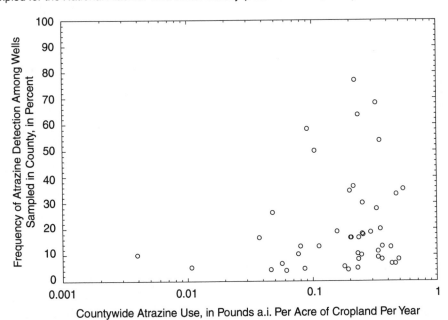

Figure 3.40. Frequency of atrazine detection among wells sampled in each county with 10 or more wells sampled during the National Alachlor Well Water Survey (Klein, 1993) versus countywide use, expressed as pounds of active ingredient (a.i.) per year (Gianessi and Puffer, 1990) per acre of cropland (U.S. Department of Commerce, 1990).

of detection rates were seen among high-use counties. Similar patterns were observed for simazine, alachlor, and metolachlor. Cyanazine was detected in only four counties.

An analogous trend was seen when the overall frequencies of detection for individual herbicides during the NAWWS (Table 3.3) were plotted against their respective use in all of the sampled counties (Figure 3.41), similar to what was done for the Indiana data (Figure 3.36). The wide range in herbicide detection frequencies observed at high use in Figure 3.41, however, was driven exclusively by the high detection rate for atrazine.

In discussing the NAWWS results, Holden and others (1992) compared countywide pesticide detection rates with use rates, but quantified the latter in terms of the probability of herbicide use within 0.5 mi of each sampled well, based on the opinions of local county experts. For each of the five herbicides examined, the wells near which herbicides were "probably used" exhibited significantly higher detection rates ($\alpha = 0.05$) than those from which the distance to "probable use" areas was greater than 0.5 mi (Holden and others, 1992). The authors did not state whether the herbicides detected in the wells were the same in each case as those presumed to have been used nearby. However, this result provides support for the hypothesis, noted earlier, that uncertainties regarding pesticide use within the vicinity of individual wells may be a significant reason why direct relations between pesticide occurrence and relatively large-scale estimates of use (i.e., countywide and larger) have generally not been observed. Other work, cited later in this section, implicates variations in well depth and construction as another major source of uncertainty in such relations.

Figure 3.42 compares the frequencies of pesticide detection reported during the NPS with rates of use for those compounds for which nationwide use information was available from

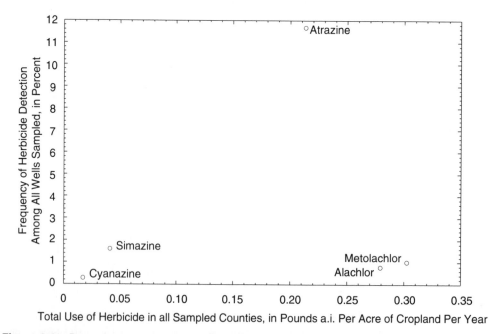

Figure 3.41. Percentage of sampled wells with detections of specific herbicides during the National Alachlor Well Water Survey (Holden and others, 1992) versus total use, expressed as pounds of active ingredient (a.i.) (Gianessi and Puffer, 1990), per acre of cropland (U.S. Department of Commerce, 1990) in all sampled counties (Klein, 1993).

Gianessi and Puffer (1990, 1992a,b). As seen for the NAWWS and Indiana data, the NPS results provide minimal evidence for a direct relation between use and frequencies of detection among different pesticides, either for the community water system or the rural domestic wells.

Of particular interest in Figure 3.42 are the high frequencies of detection noted in both types of wells for the products of DCPA hydrolysis (often referred to as "DCPA acid metabolites"). During the NPS, significant correlations ($\alpha = 0.05$) were observed between the frequencies of detection of DCPA hydrolysis products in community water system wells and either the rates of DCPA use by urban pesticide applicators or the rates of its use on golf courses. Extensive non-agricultural use also appears to explain the frequent detections of DCPA and prometon during the MCPS (Section 7.1). These examples indicate that data on agricultural use alone may not always be appropriate for explaining pesticide occurrence even in predominantly agricultural areas.

With the exception of the data for DCPA and its hydrolysis products, the NPS did not report any other significant relations between pesticide use in the vicinity of individual wells and the detection of those same compounds in the wells sampled. For both community water system and rural domestic wells, pesticides that were not reported to have been used nearby were detected more frequently than those that had been used within the general vicinity of the well. The authors of the NPS report suggested that the general absence of a detectable link between pesticide use and detections in ground water may have arisen in part from an inaccurate picture of actual pesticide use near the wells prior to the survey (U.S. Environmental Protection Agency, 1992a), a point also noted for the MCPS (Kolpin and others, 1994). A scientific advisory panel convened by the USEPA to comment on the NPS results suggested that the

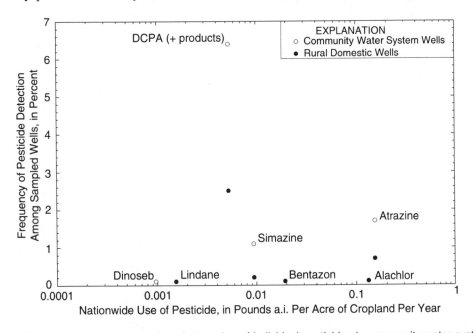

Figure 3.42. Nationwide frequencies of detection of individual pesticides in community water system wells (open circles) and rural domestic wells (filled circles) during the National Pesticide Survey (U.S. Environmental Protection Agency, 1990a), versus their annual nationwide use (Gianessi and Puffer, 1990, 1992a,b) per acre of cropland (U.S. Department of Commerce, 1990).

failure to control for such potentially confounding variables as well depth and construction may also have contributed to the apparent absence of significant associations between pesticide detections and a number of explanatory variables, including chemical use (Hallberg and others, 1992a).

Results from State and Local Monitoring Studies

Direct relations between the frequency of pesticide detection and pesticide use have been reported by a few state and local monitoring studies. Kross and others (1990) observed lower frequencies of atrazine detection in wells located on Iowa farms where herbicides had not been applied during the most recent growing season (4 percent), compared with farms where they had (10 percent). They also documented a direct relation between atrazine use and occurrence in ground waters beneath various areas of Wisconsin, based on data reported by LeMasters and Doyle (1989) for Grade A dairy farms across the state. A linear regression of atrazine detection rate versus the number of acres on which the herbicide was used in each Wisconsin Agricultural Statistics District was found by Kross and others (1990) to be statistically significant (Figure 3.43). Of all the monitoring studies reviewed, these results represent the strongest evidence for an increase in pesticide detection rate in ground waters with increasing pesticide use.

Two principal characteristics distinguish the occurrence-versus-use relation in Figure 3.43 from all others examined in this book: (1) the relative uniformity of well construction among the wells sampled; and (2) the use of treated acres, rather than the amounts applied, to quantify pesticide applications. Although, as noted earlier, the MCPS employed the most rigorous well-selection criteria of any of the reviewed studies, only the Wisconsin study sampled wells that met legally-based requirements for well construction. Because they were installed on Grade A dairy farms, the wells sampled during the Wisconsin study were required to meet state-mandated standards "for such things as minimum setback distances from septic tanks and barnyards, minimum casing depth and well depth, and proper plumbing" (LeMasters and Doyle, 1989). Constraints on the population of available wells made such requirements too restrictive for the MCPS. Because no other monitoring study has reported as strong a relation between pesticide occurrence and use as that observed in the Wisconsin study, differences in well construction may be one of the most important sources of uncontrolled variability in observed rates of pesticide detection. An alternative explanation for the strength of the relation shown in Figure 3.43 is that, although applied amounts and the number of treated acres are correlated (e.g., Figure 3.3), pesticide detection frequencies may be more closely related to the total area treated than to the total mass of compound applied (Kolpin, 1995a).

Pickett and others (1990, 1992) examined the relation between the rate of herbicide detection (simazine, bromacil, or diuron) in ground water and the intensity of citrus cultivation in Tulare County, California. The authors found the number of citrus growers per township (36 mi^2 or 94 km^2) responding to a mailing survey to be a statistically reliable index for the intensity of pesticide use. The number of respondents, in turn, exhibited a significant, linear correlation with the percentage of wells with herbicide detections in the townships of interest (Figure 3.44), but only when data were used from all twelve townships in question ($\alpha = 0.05$). When data were excluded for three townships where the mean depth to water was significantly greater (25-40 m) than that in the remaining nine (3-12 m), the correlation was still positive, but no longer statistically significant. Despite the variability, data in Figure 3.44 indicate a direct relationship between the intensity of pesticide use and the likelihood of ground-water contamination by pesticides within a single county. Data reported by two other monitoring studies in California provide additional opportunities to examine relations between pesticide

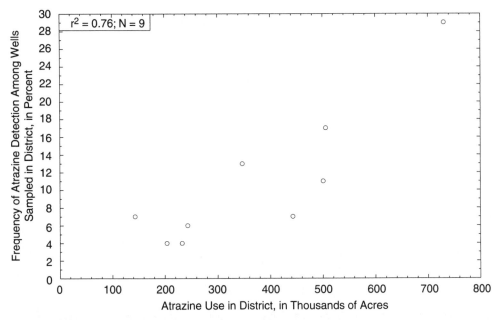

Figure 3.43. Frequencies of atrazine detection in Grade A dairy farm wells versus use in individual crop reporting districts in Wisconsin (data from LeMasters and Doyle, 1989). Relation found to be statistically significant ($\alpha = 0.05$).

detection rates and use among different counties, particularly for DBCP (Weaver and others, 1983) and DCPA (Ando, 1992). However, direct relations between occurrence and use were not observed for either compound (data not shown).

As noted earlier, the results from the NPS (Figure 3.42) suggest that relations between agricultural pesticide use and detections in ground water are most difficult to detect for compounds that are applied extensively for both agricultural and non-agricultural purposes. Studies conducted in the San Joaquin Valley of California and the Delmarva Peninsula east of Washington D.C. reinforce this point. As part of an analysis of historical data in the San Joaquin Valley, Domagalski and Dubrovsky (1991) compared the spatial distributions of application and detection in shallow ground water for four herbicides; atrazine, simazine, diuron, and bromacil. Although bromacil and diuron are applied primarily in orange orchards, atrazine and simazine are used for weed control along roadsides, as well as on crops. The distribution of bromacil detections in shallow ground water was highly correlated with application patterns. The spatial distribution of diuron also corresponded to areas of its application, but also varied with subsurface permeability. In contrast, the spatial distributions of atrazine and simazine detections did not correspond with use of these herbicides on crops, probably because of their widespread application along roadsides (Domagalski and Dubrovsky, 1991).

In a study of the quality of ground water beneath the Delmarva Peninsula, Koterba and others (1993) found that the pesticides detected in wells located near areas planted in corn, soybeans, or small grains were, with one exception, compounds that were commonly applied to those crops in that region. The single exception was hexazinone—an herbicide that, like the triazines examined by Domagalski and Dubrovsky (1991), is used to control brush and weeds in noncrop areas.

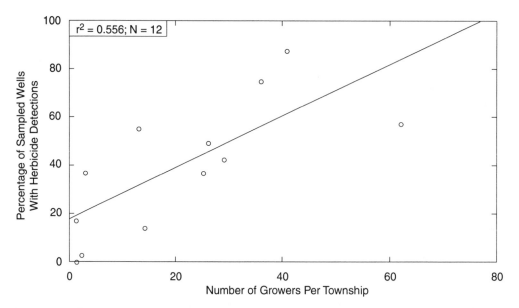

Figure 3.44. Percentage of wells with herbicide detections (simazine, bromacil, or diuron) in citrus-growing areas of Tulare County, California, versus use, as approximated by the number of growers responding to a survey of farming practices. Redrawn with permission from Pickett and others (1990). Relation found to be statistically significant ($\alpha = 0.05$).

Cohen and others (1990) observed a direct spatial relation between the intensity of application of pesticides and their detection in shallow ground water during a study of ground-water contamination by pesticides used on golf courses in Cape Cod, Massachusetts. Areas that received higher applications on the golf courses of interest were also those where pesticides were found more frequently and in greater concentrations. The establishment of a relatively close spatial connection between application and detection in this instance may have been aided by the fact that the areas of highest application—i.e., tees and greens—were highly localized, constituting only 3 percent of the total area of the golf courses of interest.

The data reported by Cohen and others (1990) also suggest that, in situations where the locations of pesticide application are well known and have been stable for decades, it may be possible to observe associations between historical use and pesticide occurrence—and perhaps even concentrations—in ground water. The golf courses had been in operation for at least 30 years when the study was initiated. Cohen and others (1990) found the turf pesticides that were registered for use at the time of the study to be present in the subsurface at concentrations that were considerably lower than the concentrations of those that had been banned several years earlier. This observation may reflect the general trend in recent years toward pesticides designed to be less persistent and applied at lower rates than those employed in the past. The higher concentrations of the older pesticides may also have arisen from their longer period of use and, consequently, their higher mass loading over time (Hallberg, 1995). The influence of pesticide use on ground-water quality beneath golf courses is discussed further in Section 7.1.

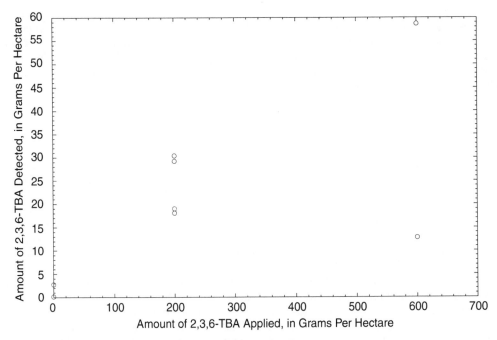

Figure 3.45. Amount of 2,3,6-trichlorobenzoic acid (2,3,6-TBA) leached from experimental plots in sandy soils above a tile drainage system in Björnstorp, Sweden, versus amount applied (redrawn from Jernlås, 1985).

Results from Process and Matrix-Distribution Studies

As noted earlier, studies conducted at the field scale have provided evidence for direct relations between pesticide concentrations or masses in subsurface waters and application rates. In Sweden, Jernlås (1985) measured larger amounts of 2,3,6-trichlorobenzoic acid draining through a sandy soil at higher application rates (Figure 3.45). Consistent with these results, Wehtje and others (1981) observed that the proportions of applied atrazine recovered from soil extractors beneath a corn field in Nebraska were essentially the same in two consecutive years (0.075 percent, versus 0.065 percent), despite a 50 percent reduction in the application rate for the second year. Similarly, Kördel and others (1991) detected no change in the percentage of applied ^{14}C-bentazone recovered in leachate from an outdoor lysimeter over a two-year period (0.03 percent), despite a threefold increase in application rate.

As with process and matrix-distribution studies in general, however (Tables 2.1 and 2.2), the majority of the field-scale studies examining relations between pesticide application rates and concentrations in the subsurface have involved the analysis of pesticides in soils, rather than in subsurface waters. In agreement with the studies of subsurface waters, cited above, direct relations between application rates and soil concentrations have been observed for picloram and 2,4,5-T in North Carolina (Lutz and others, 1973), bromacil in The Netherlands (Litwin and others, 1983), hexazinone in Alberta, Canada (Feng and others, 1989), alachlor in Connecticut (Huang and Ahrens, 1991), metolachlor in Arkansas (Barnes and others, 1992), and DDT in New Zealand (Boul and others, 1994).

A direct relation between pesticide applications and their subsequent concentrations in the subsurface is also supported by process and matrix-distribution studies that have reported increases in the maximum depth of pesticide migration—as well as overall increases in pesticide concentrations in the subsurface—with increasing application rates. This behavior has been observed for several compounds, including picloram, 2,4,5-T (Lutz and others, 1973), atrazine (Hall and Hartwig, 1978), bromacil (Litwin and others, 1983), and diflubenzuron (Sundaram and Nott, 1989).

3.4.5 SUMMARY

The pesticides for which the most extensive data on use are currently available are those applied in agricultural settings. Similarly, the majority of information on pesticide detections in ground water has been collected for agricultural pesticides within predominantly agricultural areas. The available data indicate that the pesticides detected most frequently in ground water are also those used most extensively in agriculture, namely, the triazine and acetanilide herbicides. Some of the more heavily used insecticides and fumigants, such as aldicarb, DBCP, and EDB, have also been among the pesticides detected most often in ground water.

The results from the studies discussed in this section provide varying degrees of support for the hypothesis that pesticide detections in ground water become more common with greater use. In general, direct relations between pesticide detection frequencies and use become more evident: (1) when data on pesticide use are aggregated over smaller areas; (2) when the analysis excludes pesticides that have limited use, those that are used extensively for non-agricultural purposes, or those that have been discontinued; and (3) when the study area includes smaller proportions of non-agricultural land. These patterns suggest that improvements in the ability to detect relations between pesticide occurrence and use are closely tied to improvements in the accuracy of data on pesticide use in the areas of interest. They also demonstrate the need for more reliable data on pesticide occurrence and use in non-agricultural settings.

When pesticide use is assessed at comparatively large spatial scales (i.e., on a countywide, statewide or nationwide basis), low use is typically associated with low frequencies of pesticide detection in ground water, while high use is associated with a broad range of detection frequencies. This pattern has been observed among different counties or states for the same pesticide (e.g., Figures 3.35 and 3.40), for the same set of pesticides (Figure 3.17 and 3.18), or for pesticides in general (Figure 3.13). It has also been observed among different pesticides within individual states (Figure 3.36) or across the nation (Figures 3.41 and 3.42). Thus, comparatively high pesticide use is a necessary but not sufficient condition for encountering high frequencies of pesticide detection in ground water.

Control over variations in well construction may also be important for detecting relations between pesticide occurrence and pesticide use. This is suggested by the fact that the clearest relation between occurrence and use was observed from the study in which constraints on well construction were the most restrictive (Figure 3.43), while substantially weaker relations were observed by investigations that exerted much less control over the nature of the sampled wells. Chapters 5 and 6 will examine the various influences of natural and anthropogenic factors on pesticide detections in ground water, including climate and agricultural practices (Chapter 5), as well as pesticide properties, soil characteristics, hydrogeology, well construction, and study design (Chapter 6).

3.5 TEMPORAL PATTERNS OF PESTICIDE DETECTION

3.5.1 LONG-TERM TRENDS

A comparison between the data compiled by the USEPA in the PGWDB (U.S. Environmental Protection Agency, 1992b) and those assembled 4 years earlier by the same agency during the 1988 Survey of State Lead Agencies (Parsons and Witt, 1989) indicates that the total number of states in which pesticides were detected in ground water, the total number of pesticide analytes detected (parents and degradates), and the number of analytes detected above their respective Health Advisory Levels, or HALs, had all increased (Table 2.4). The PGWDB reported the detection in United States ground waters of more pesticide compounds than did the 1988 Survey for nearly every chemical class examined by both compilations (Table 3.1).

These observations do not necessarily imply that contamination of ground waters in the United States became more widespread during the period from 1987 to 1991. Instead, they are more likely to reflect the expansion of sampling efforts across the nation over this period. As noted by Parsons and Witt (1989), "the principal criterion for whether pesticides had been detected in the groundwater in a state appears to be whether or not they have looked [sic]." Wells and Waldman (1991) reached a similar conclusion during their assessment of the occurrence and leaching potential of aldicarb across the nation, viz., "the likelihood of additional detections [of aldicarb in U.S. ground water] is more dependent on the extent of monitoring than any other single factor."

The expansion of monitoring efforts in a given area may lead to the detection of more pesticides in ground water, but does not necessarily result in increased frequencies of detection of individual compounds. Pesticide occurrence data collected from 1984 through 1990 in a heavily agricultural area in east-central South Dakota demonstrate this point. Although a substantial increase in sampling effort during this period was accompanied by an increase in the total number of pesticides detected in ground water each year, no significant increases in pesticide detection frequencies were observed in the area of interest (Goodman, 1991). In accord with these observations, a fourfold increase in the number of pesticides analyzed by the MCPS for the 1992 sampling (45), relative to 1991 (11), was accompanied by only a 3 percent increase in the number of wells with pesticide detections among the 100 wells sampled in both years (Kolpin and Thurman, 1995). (Relations between sampling effort and pesticide detection frequencies will be examined further in Section 6.6.6.)

Increases in the number of pesticides that have been detected in ground waters of the nation between 1987 and 1991 may also be related to improvements in analytical technology. Over the past three decades, there has been a steady decrease in analytical detection limits for pesticides and their transformation products. As noted by others (e.g., LeMasters and Doyle, 1989; Burkart and Kolpin, 1993), detection frequencies for a given solute in any environmental medium vary inversely with analytical detection limit. The influence of detection limits on pesticide detection frequencies is examined in further in Section 6.6.2.

Whether or not ground-water contamination has become more widespread across the nation—or within any particular area—over a specific time period can only be determined through monitoring programs whose design, spatial domain, and analytical scope have all remained stable over the time interval of interest. None of the multistate sampling programs carried out to date, however, have exhibited both the stability and the longevity required to address this issue in such a systematic manner.

The importance of study longevity to the assessment of long-term trends is illustrated by two different summaries of pesticide concentrations in the discharge from Big Spring in northeastern Iowa. The first of these assessments (Hallberg, 1986), based on 4 years of monitoring, from 1982 through 1985 (Figure 3.46a), noted that "the flow-weighted mean atrazine concentration has increased every year, in spite of major climatic pattern differences" and that "these Iowa observations suggest a trend of increasing pesticide residues in ground water in response to prolonged and widespread use." A subsequent assessment, however (Rowden, 1995), showed a pattern of fluctuating atrazine concentrations from 1982 through 1994 (Figure 3.46b). This example demonstrates the potential hazards associated with predicting long-term trends in pesticide concentrations from comparatively short periods of record.

Other monitoring efforts have also found little evidence for consistent shifts in pesticide occurrence in ground water over time. To date, the most systematic assessment of this issue on a large scale appears to have been the State-Wide Rural Well-Water Survey (SWRL) in Iowa (Kross and others, 1990; Libra and others, 1993; Rex and others, 1993). Repeat sampling of a "10 percent subset" of the original group of 686 SWRL wells on 4 occasions from 1988 to 1991 revealed considerable changes in pesticide detection frequencies over time. None of these trends were monotonic, however, regardless of whether the results were expressed in terms of the detection of any pesticide, any one of a fixed set of seven herbicides, atrazine plus its DEA and DIA degradates, or atrazine alone (Figure 3.47). Furthermore, the apparent increase in detection frequency of approximately 6 percent for the seven-herbicide set shown in the figure (alachlor, atrazine, butylate, cyanazine, metribuzin, metolachlor, and trifluralin) from 1988 to 1991 was not statistically significant ($\alpha = 0.05$). The observed shifts in pesticide occurrence over time shown in Figure 3.47 appear to have been controlled more by the prevailing or antecedent hydrologic conditions than by any long-term build-up or decline of pesticide residues in the subsurface.

If temporal changes in pesticide concentrations in ground water were controlled primarily by rates of application, then discontinuation of a particular pesticide within a given area might be expected to result in a gradual, monotonic decline in its concentrations in ground water over time. Indeed, with the exception of a few areas (e.g., Pacenka and others, 1987), the subsurface concentrations of aldicarb appear to have exhibited a general decrease in Long Island, New York, since its use was discontinued there in 1979 (Soren and Stelz, 1984; Pacenka and others, 1987; Suffolk County Department of Health Services, 1989). Aldicarb concentrations in ground water have also declined in Prince Edward Island, Canada since the early 1980's (Priddle and others, 1992). In addition, EDB concentrations exhibited substantial decreases in ground water drawn from the Upper Floridian aquifer in Seminole County, Georgia, from 1981 to 1987 (McConnell, 1988). However, temporal changes in the concentrations in ground waters of 1,2,3-trichloropropane, EDB, and DBCP in Oahu, Hawaii (Oki and Giambelluca, 1987), and DBCP in the Central Valley of California (Cohen, 1986; Schmidt, 1986a) exhibited conflicting trends following their discontinuation—increasing in some wells, and decreasing or remaining unchanged in others—depending on the history of use of the compound, the hydrologic conditions in the surrounding area, and the pumping history of the sampled well. Figures 3.48 and 3.49 illustrate these concentration fluctuations for DBCP over a 5-year period (1979-1984) in several water supply wells in the Central Valley of California, following the cancellation of its use in the state in 1977 (Cohen, 1986).

With the exception of a few compounds whose use was discontinued in the late 1970's, previous studies have thus provided little evidence of any obvious, monotonic trends in either the concentrations or frequencies of detection of pesticides in the ground waters of the nation. The apparent absence of any significant increases or decreases in pesticide occurrence,

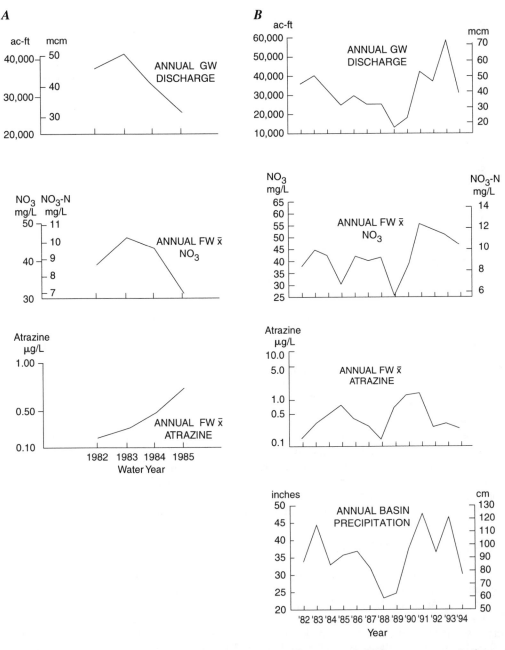

Figure 3.46. Annual ground-water (GW) discharge and flow-weighted (FW) mean concentrations of nitrate and atrazine in discharge from Big Spring, Clayton County, Iowa, for the periods from (*A*) 1982 to 1985 (redrawn from Libra and others, 1986, and reprinted by permission of the National Ground Water Association. Copyright 1986), and (*B*) 1982 to 1994 (redrawn with permission from Rowden, 1995). ac-ft, acre-feet; mcm, million cubic meters.

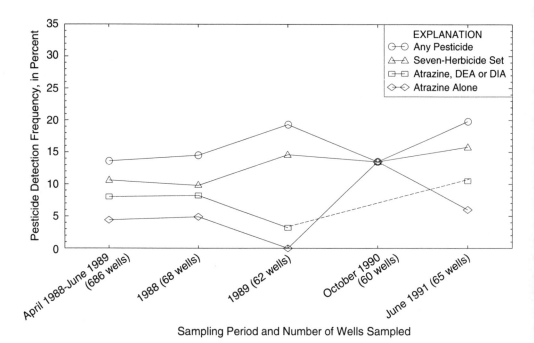

Figure 3.47. Frequencies of pesticide detection in Iowa ground waters from 1988 to 1991, based on the repeat sampling conducted for the Iowa State-Wide Rural Well-Water Survey (data from Libra and others, 1993). Apparent increase in detection frequency for the seven-herbicide set (alachlor, atrazine, butylate, cyanazine, metribuzin, metolachlor, and trifluralin) from 1988 to 1991 was not statistically significant ($\alpha = 0.05$). Analyses for DEA and DIA were not conducted for the October 1990 sampling, as indicated by the dashed line.

however, may result in part from a general paucity of long-term studies designed to reliably detect such trends, coupled with an insufficiently long period of record to date for investigations that are currently examining long-term trends of this type. Furthermore, because of the erratic nature of concentration fluctuations often observed in individual wells, reliable evaluations of long-term trends are more likely to arise from assessments of pesticide loadings to ground water across large areas—or entire aquifers—rather than in specific sampling locations.

3.5.2 SEASONAL VARIATIONS

Difficulties associated with observing long-term trends in pesticide concentrations or detection frequencies in ground water are exacerbated by the fact that both parameters vary during the course of the year. Typically, detection frequencies and median concentrations in shallow ground waters are lowest during winter, and reach peak values during late spring and early summer (Wehtje and others, 1983; Kelley, 1985; Schmidt, 1986a; Detroy and others, 1988; Hallberg and others, 1989; Neil and others, 1989; Pionke and Glotfelty, 1989; Goodman, 1991; Seigley and Hallberg, 1991; Kalkhoff and others, 1992; Blum and others, 1993; Rudolph and others, 1993; Risch, 1994; Barceló and others, 1995; and Kalkhoff and Schaap, 1995). Results from the MCPS reflect this pattern over a large scale; Burkart and Kolpin (1993) found the frequencies of herbicide detection in ground waters for the pre-planting period in March and

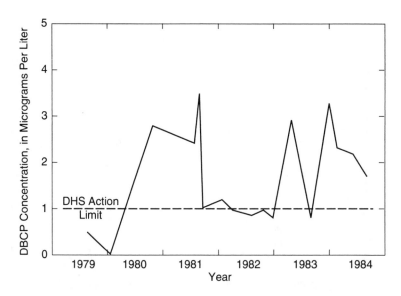

Figure 3.48. DBCP concentration fluctuations in a municipal well in Escalon, California, from 1979 to 1984. Use of DBCP was discontinued in California in 1977. Redrawn with permission from Cohen (1986). Copyright 1986, American Chemical Society. DHS, Department of Health Services.

April to be significantly lower (20.4 percent) in their 12-state study area ($\alpha = 0.05$) than those observed during the post-planting period (27.8 percent) in July and August of 1991 (Table 2.4).

The seasonal variability of pesticide detection frequencies in shallow ground water may be sufficiently large to mask long-term changes in detection frequencies over multi-year periods. Figure 3.50 illustrates the seasonal patterns of pesticide detection that have been observed in ground waters beneath the midcontinent. Whereas Libra and others (1993) observed an increase of approximately 6 percent in the detection frequencies for a set of seven herbicides in Iowa ground waters from 1988 to 1991 statewide (Figure 3.47), Figure 3.50 indicates that pesticide detection frequencies in shallow wells in a given area may vary by as much as 20 to 30 percent during a single year.

Results from a study of agrichemical contamination of rural ground waters in Iowa suggest that short-term variations in pesticide concentrations during the course of a day in a given well are probably less pronounced than these seasonal variations. Using ELISA methods, Glanville and others (1995) measured the concentrations of triazine and acetanilide herbicides repeatedly over an 8-hour period in a domestic supply well in Sac County, Iowa. For both herbicide groups, concentration fluctuations for nearly all of the samples were within the 20 percent precision specified for the ELISA methods used (Figure 3.51). According to the authors, these observations indicate that "day-to-day variability in water quality reflects true fluctuations in groundwater quality (or variability caused by laboratory analytical procedures) rather than differences caused by variations in the time of day that the sample was collected" (Glanville and others, 1995).

Seasonal variations in pesticide occurrence in ground waters are largely attributed to the application of pesticides during the spring, but these variations are also influenced by seasonal changes in temperature and precipitation, coupled with the timing of agricultural practices such

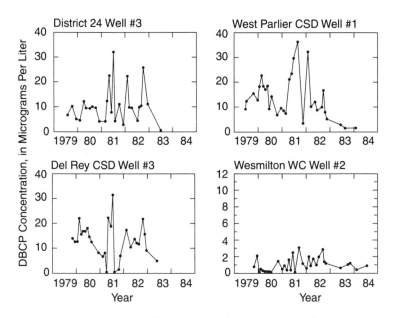

Figure 3.49. DBCP concentration fluctuations in four wells in Fresno County, California, from 1979 to 1984. Use of DBCP was discontinued in California in 1977. Redrawn with permission from Cohen (1986). Copyright 1986, American Chemical Society.

as irrigation and, perhaps, tillage. The extent to which the timing of variations in recharge, temperature, and agricultural management practices may be linked to variations in pesticide concentrations in the subsurface is examined in Chapter 5.

3.5.3 CHANGES IN TEMPORAL VARIABILITY WITH DEPTH

The temporal variability of pesticide concentrations in ground water generally decreases with increasing depth. This pattern has been reported for herbicides in the alluvial aquifers of the Platte River in Nebraska (Wehtje and others, 1983) and the Iowa River (Detroy, 1986; Detroy and Kuzniar, 1988) and Deer Creek (Kalkhoff and Schaap, 1995) in Iowa, in sand-and-gravel deposits in Minnesota (Klaseus and others, 1988), in the Central Sand Plain of Wisconsin (Fathulla and others, 1988), and in near-surface aquifers beneath twelve states of the midcontinent (Burkart and Kolpin, 1993). It was also demonstrated in the midcontinent during the 1993 floods, when the percentage of wells exhibiting detectable changes in total herbicide concentration decreased monotonically with increasing well depth (Kolpin and Thurman, 1995).

Such observations are consistent with general principles of solute transport through porous media (see Section 4.3). As pesticides (or any other surface-derived solutes) move downgradient, away from their source areas, temporal variations in their input concentrations— reflected in spatial variations along subsurface flowpaths—dampen out as a result of the "smoothing" effects of hydrodynamic dispersion (Valocchi and Roberts, 1983). This phenomenon has been observed during investigations of solute transport in the subsurface during artificial recharge (e.g., Roberts and others, 1980, 1982) and enhanced bioremediation (Roberts and others, 1990).

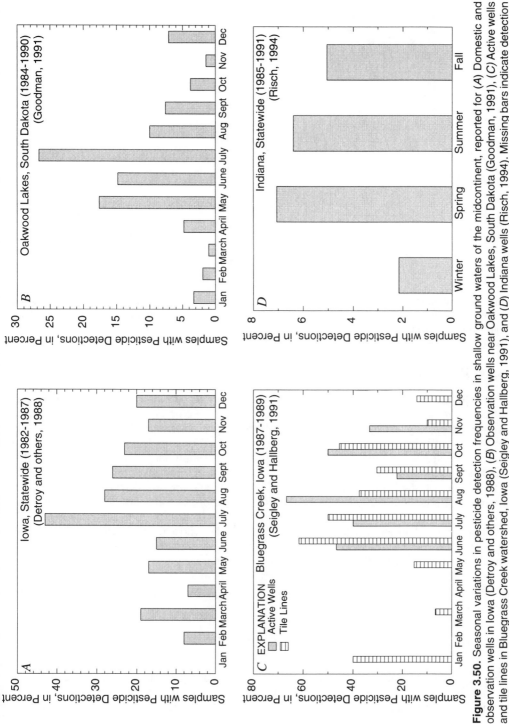

Figure 3.50. Seasonal variations in pesticide detection frequencies in shallow ground waters of the midcontinent, reported for (A) Domestic and observation wells in Iowa (Detroy and others, 1988), (B) Observation wells near Oakwood Lakes, South Dakota (Goodman, 1991), (C) Active wells and tile lines in Bluegrass Creek watershed, Iowa (Seigley and Hallberg, 1991), and (D) Indiana wells (Risch, 1994). Missing bars indicate detection frequencies of zero, rather than missing data.

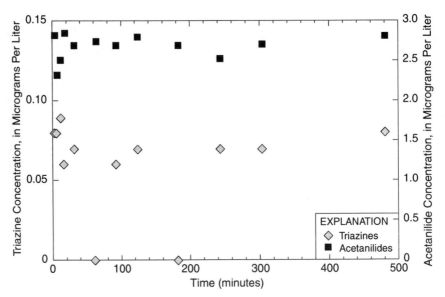

Figure 3.51. Variations in the concentrations of triazine and acetanilide herbicides (measured using immunoassay analyses) in a rural well in Sac County, Iowa, during eight hours of continuous pumping. Modified from Glanville and others (1995) and used with permission. Triazine concentrations of zero presumed to correspond to those below the detection limit of 0.05 micrograms per liter.

3.5.4 SUMMARY

The absence of long-term studies with stable design currently makes it impossible to evaluate whether the frequencies of pesticide detection in the ground waters of the United States have changed in any systematic fashion since large-scale monitoring efforts began in the early 1980's. Current evidence suggests that seasonal variations in pesticide detection frequencies in shallow ground water within agricultural areas are considerably larger and more reproducible than those observed over either longer or shorter time scales, and appear to be driven primarily by the timing of applications and recharge. The sparse systematic evidence for year-to-year variations in pesticide detection frequencies or concentrations over the past decade suggests that such variations, even if they are significant, have arisen mostly from changes in climatic conditions, rather than shifts in pesticide use. Work by Glanville and others (1995) indicates that, at least in rural agricultural areas, changes in pesticide concentrations in ground water over the course of a day are within the precision of current routine analytical methods. Finally, studies conducted in a variety of hydrogeological settings have demonstrated that, as might be expected, the temporal variability in pesticide concentrations in ground water diminishes with increasing depth.

3.6 SUMMARY

Despite the wealth of available data on the spatial distributions of pesticide detections in the wells of most states, no statistically representative summary of pesticide occurrence in ground water currently exists for the nation. Most of the monitoring studies that have been conducted—including all but one of the five multistate monitoring studies reviewed—have focused primarily on agricultural pesticides and agricultural areas. The principal exception, the National Pesticide Survey, provided a statistically representative summary of the occurrence of a large number of agricultural and non-agricultural pesticides in wells within both agricultural and largely non-agricultural areas, but did not account for the influence of well depth or construction (or many other factors) on pesticide occurrence in ground waters. Results from several studies, discussed in this chapter and elsewhere in this book, suggest that variations in well depth, construction, and integrity may represent some of the most significant sources of uncontrolled variability in pesticide occurrence data.

In general, pesticide use rates by themselves are not reliable predictors of pesticide detections in ground water. Rates of detection of pesticides and their transformation products in ground water—either among different states or counties for the same compound, or among different compounds within the same study area—are found to be low for areas or compounds with low use rates, but exhibit wide ranges at moderate to high use. Thus, high rates of pesticide application appear to be a necessary, but not sufficient, condition for high rates of pesticide detection in ground water.

Current limitations in the available data preclude an assessment of whether the frequencies of pesticide detection in ground waters of the United States have increased, decreased, or remained unchanged over the past decade. This is related to a paucity of long-term studies, as well as the fact that the seasonal variability in pesticide detection frequencies in shallow ground water is generally larger than the long-term changes in detection frequencies that have been reported to date.

Spatial and temporal patterns of pesticide detection in ground water are controlled by a variety of natural and anthropogenic factors. Much of the discussion in Chapters 5 and 6 will focus on the influence of these and other factors on the likelihood of detecting pesticides in ground water. Such an analysis, however, requires an understanding of the fundamental physical, chemical, and microbial processes that control the movement and fate of pesticides in the subsurface. Chapter 4 provides an overview of these phenomena.

Table 3.2. Pesticides and transformation products detected or not detected in ground waters of North America, based on either the National Pesticide Survey, the Pesticides In Ground Water Database, or the monitoring studies reviewed

[Pesticide or Transformation Product: Transformation products are indented. Data Source: MRL, Minimum Reporting Limit in micrograms per liter (µg/L), given in italics for all analytes for which a value was provided (U.S. Environmental Protection Agency, 1990a, Appendix E). A total of 1,349 wells were sampled in 50 states during the NPS (see Table 2.4). For Pesticides in Ground Water Database (PGWDB) entries, numbers in italics represent number of wells with detections/total number of wells sampled for the compound of interest. Detection limits not given for the PGWDB results because the data were drawn from a large number and variety of separate studies. National Pesticide Study (NPS) data were specifically excluded from the PGWDB (U.S. Environmental Protection Agency, 1992b), thereby eliminating any overlap between the two sources. PGWDB listing also excludes results from any multiresidue methods (e.g., triazine, carbamate, organochlorine or organophosphate "screens") or other grouping of analytes (e.g., "total aldicarb," "pyrethrins," "uracil/urea"). Monitoring Studies: Includes results from all multistate, state, and local monitoring studies reviewed (Tables 2.3 and 2.4), with the exception of the NPS. Data overlap with some of the PGWDB results. As with the PGWDB, variations in detection limits among the monitoring studies preclude designation of a specific detection limit for any single pesticide. ND, not detected; D, detected. Blank cells indicate no information applicable or available. Q, Compound may have been present, but "[could not] be quantified or reliably detected" (U.S. Environmental Protection Agency, 1990a)]

Pesticide or Transformation Product	Data Source			Pesticide or Transformation Product	Data Source		
	NPS (MRL, in µg/L)	PGWDB (Wells detected/wells sampled)	Monitoring Studies		NPS (MRL, in µg/L)	PGWDB (Wells detected/wells sampled)	Monitoring Studies
Triazine Herbicides				Cyprazine		ND (0/125)	ND
Ametryn	ND 0.13	D (2/1,146)	D	Hexazinone	ND 0.14	D (9/684)	D
Anilazine		ND (0/19)		Metribuzin	ND 0.18	D (232/5,452)	D
Atratone	ND 0.17	D (1/331)	ND	Metribuzin DA	ND 0.60	ND (0/121)	ND
Atrazine	D 0.12	D (1512/ 26,909)	D	Metribuzin DADK	Q	ND (0/120)	D
Deethyl atrazine (DEA)[1]	ND 2.2	D (27/689)	D	Metribuzin DK	Q	ND (0/120)	ND
Deisopropyl atrazine (DIA)[2]		D (24/689)	D	Prometon	D 0.15	D (48/2,215)	D
Didealkyl atrazine (D2A)		ND (0/125)		Prometryn	ND 0.10	D (1/1,656)	D
Cyanazine	ND 2.4	D (155/7,468)	D	Propazine	ND 0.10	D (15/1,428)	D
Cyanazine amide			D	Secbumeton		ND (0/210)	
Deethyl cyanazine			ND	Simazine	D 0.38	D (486/22,374)	D
Deethyl cyanazine amide			ND	Deethyl simazine			ND
				Simetone		ND (0/57)	ND

Table 3.2. Pesticides and transformation products detected or not detected in ground waters of North America, based on either the National Pesticide Survey, the Pesticides In Ground Water Database, or the monitoring studies reviewed—*Continued*

Pesticide or Transformation Product	Data Source			Pesticide or Transformation Product	Data Source		
	NPS (MRL, in μg/L)	PGWDB (Wells detected/wells sampled)	Monitoring Studies		NPS (MRL, in μg/L)	PGWDB (Wells detected/wells sampled)	Monitoring Studies
Simetryn	ND 0.050	ND (0/755)	ND	N-(1,1-dimethyl-acetonyl)-3,5-dichloro-benzamide	ND 3.1		
Terbuthylazine		ND (0/268)		3,5-dichloro-benzoic acid	ND 0.30	ND (0/87)	
Terbutryn	ND 0.15	ND (0/404)	ND	"Unspecified metabolite"		ND (0/126)	
Acetanilides				Propanil	ND 0.30	D (2/388)	D
Alachlor	D 0.50	D (467/25,993)	D	**Amines**			
Hydroxy-alachlor		D (1/297)	D	Amitrole (Amino-triazole)		ND (0/32)	ND
3-Hydroxy-alachlor			D	Benfluralin (Benefin)		ND (0/660)	D
2,6-Diethyl-aniline		ND (0/305)	D	Ethalfluralin		ND (0/188)	D
Alachlor ethane-sulfonic acid (ESA)			D	Fluchloralin		ND (0/533)	
Butachlor	ND 0.75	ND (0/6)	ND	Flumetralin		ND (0/126)	
Metolachlor	ND 0.75	D (213/22,255)	D	Glyphosate (Roundup)		D (7/247)	
Propachlor	ND 0.65	D (33/2,718)	D	Isopropalin		D (1/76)	
Other Amides				Naptalam		ND (0/40)	
Allidochlor (CDAA)			D	Pendimethalin		D (14/1,405)	D
Amitraz		ND (0/1)		Picloram	ND 0.50	D (74/2,934)	D
Chlordimeform		ND (0/287)		Profluralin		D (1/86)	
Diphenamid	ND 0.22	ND (0/812)	D	Trifluralin	ND 0.13	D (58/5,590)	D
Napropamide	ND 0.25	ND (0/391)	D	**Chlorophenoxy Acid Derivatives**			
Oryzalin		D (2/428)		Acifluorfen	Q	D (4/1,185)	D
Pronamide	Q	ND (0/432)	ND	2,4-D	ND 0.25	D (141/6,142)	D

Table 3.2. Pesticides and transformation products detected or not detected in ground waters of North America, based on either the National Pesticide Survey, the Pesticides In Ground Water Database, or the monitoring studies reviewed—*Continued*

Pesticide or Transformation Product	NPS (MRL, in μg/L)	PGWDB (Wells detected/wells sampled)	Monitoring Studies
2,4-DB	ND 1.0	D (1/928)	D
2,4-DP			D
Dichlorprop	ND 0.25	D (2/1,207)	D
Fenoprop			ND
MCPA		D (5/1,524)	D
MCPB		ND (0/197)	ND
Mecoprop (MCPP)		ND (0/460)	ND
2,4,5-T	ND 0.10	D (29/3,066)	D
2,4,5-TP (Silvex)	ND 0.10	D (28/3,876)	D
Chlorobenzoic Acids			
Chloramben	Q	D (12/1,895)	D
Dacthal (DCPA)	ND 0.060	D (5/2,033)	D
TPA (tetrachloro-terephthalic acid)			D
DCPA "acid metabolites" (unspecified)	D 0.10	D (59/118)	D
Dicamba	ND 0.10	D (72/3,172)	D
5-Hydroxy dicamba	ND 0.10	ND (0/87)	
2,4-Dichloro-benzoic acid		D (11/19)	
Urea and Sulfonylurea Herbicides			
Chlorbromuron			ND
Chloroxuron		ND (0/21)	
Chlorsulfuron		ND (0/8)	

Pesticide or Transformation Product	NPS (MRL, in μg/L)	PGWDB (Wells detected/wells sampled)	Monitoring Studies
Chlortoluron			ND
Difenoxuron			ND
Diuron	ND 0.32	D (160/17,865)	D
Fenuron		ND (0/158)	
Fluometuron	ND 0.48	D (68/530)	D
Linuron	ND 0.48	D (111/1,666)	D
Metabromuron			ND
Metoxuron			ND
Monolinuron			ND
Monuron		D (71/292)	ND
Neburon	ND 0.30	ND (0/325)	ND
Patoran			ND
Siduron		ND (0/182)	ND
Tebuthiuron	ND 0.23	D (3/233)	D
Uracil Herbicides			
Bromacil	ND 1.1	D (313/17,372)	D
Terbacil	ND 1.7	D (6/288)	ND
Miscellaneous Herbicides			
Bentazon	D 0.25	D (80/726)	D
Bromoxynil		ND (0/107)	ND
Dinoseb (DNBP)	D 1.3	D (28/2,030)	D
Diquat		ND (0/14)	
Endothall		ND (0/548)	
Fluazifop-butyl		D (7/12)	

Table 3.2. Pesticides and transformation products detected or not detected in ground waters of North America, based on either the National Pesticide Survey, the Pesticides In Ground Water Database, or the monitoring studies reviewed—*Continued*

Pesticide or Transformation Product	NPS (MRL, in µg/L)	PGWDB (Wells detected/wells sampled)	Monitoring Studies
Fluridone	ND 0.90	ND (0/4)	
Methazole		ND (0/40)	
Norflurazon	ND 0.18	ND (0/194)	
Octyl bicycloheptene dicarboximide		ND (0/6)	
Oxyfluorfen		ND (0/188)	
Paraquat		D (11/971)	D
Carbamates and Thiocarbamates			
Aldicarb	ND 0.71	D (3,002/ 43,786)	D
Aldicarb sulfone	ND 0.62	D (5,070/ 37,652)	D
Aldicarb sulfoxide	ND 0.85	D (4,991/ 37,593)	D
Aminocarb		ND (0/157)	
Barban	ND 1.9	ND (0/260)	
Baygon (Propoxur)	ND 0.95	D (5/21,405)	D
Bendiocarb		ND (0/4)	
Benomyl		D (1/1,043)	D
Bufencarb		ND (0/4)	
Butylate	ND 0.30	D (5/2,867)	D
Carbaryl (Sevin)	ND 0.60	D (106/25,712)	D
1-Naphthol		D (1/220)	ND

Pesticide or Transformation Product	NPS (MRL, in µg/L)	PGWDB (Wells detected/wells sampled)	Monitoring Studies
Carbofuran	ND 1.2	D (4,107/ 27,881)	D
3-Hydroxy carbofuran	ND 1.1	D (42/22,314)	D
Carbofuran phenol	ND 21	ND (0/126)	ND
3-Keto carbofuran		D (3/839)	D
3-Keto carbofuran phenol	ND 0.93		
CDEC		ND (0/6)	
Chlorpropham	ND 0.35	ND (0/561)	
Cycloate	ND 0.20	D (7/271)	D
Diallate		ND (0/649)	D
Dioxacarb		ND (0/4)	
EPTC (Eptam)	ND 0.15	D (2/1,752)	D
Methiocarb	ND 1.5	D (1/21,174)	ND
Methomyl	ND 0.60	D (154/23,250)	D
Mexacarbate (Zectran)		D (1/166)	
Molinate	ND 0.18	D (4/355)	ND
Molinate sulfoxide		D (1/196)	ND
Oxamyl	ND 0.95	D (904/23,305)	D
Pebulate	ND 0.19	ND (0/81)	ND
Propham	ND 5.5	D (1/1,060)	ND

Table 3.2. Pesticides and transformation products detected or not detected in ground waters of North America, based on either the National Pesticide Survey, the Pesticides In Ground Water Database, or the monitoring studies reviewed—*Continued*

Pesticide or Transformation Product	Data Source			Pesticide or Transformation Product	Data Source		
	NPS (MRL, in µg/L)	PGWDB (Wells detected/wells sampled)	Monitoring Studies		NPS (MRL, in µg/L)	PGWDB (Wells detected/wells sampled)	Monitoring Studies
Pirimicarb			ND	Chlorfenac		ND (0/4)	
Pirimicarb sulfone		ND (0/1)		Chlorfenson (Ovex)		ND (0/2)	
Promecarb		ND (0/4)		Chlorobenzilate	Q	ND (0/8)	
Swep	ND 0.15	ND (0/178)		Chloroneb	ND 0.70	ND (0/14)	
Thiobencarb		D (2/335)	ND	Chlorothalonil	ND 0.060	D (8/1,136)	D
Thiobencarb sulfoxide		ND (0/157)	ND	Chlorothalonil degradates (unspecified)			D
"Thiofenate"			ND	DDT	ND 0.15	D (108/3,115)	D
Thiophanate		ND (0/12)	ND	DDE	ND 0.060	D (34/2,918)	D
Thiophanate-methyl		ND (0/3)		DDD	ND 0.13	D (35/2,647)	D
Triallate			D	Dalapon	Q	ND (0/387)	
Vernolate	ND 0.19	ND (0/83)	ND	Dichlobenil		ND (0/1)	ND
Ziram		ND (0/319)		o-Dichlorobenzene		D (1/320)	D
Organochlorine Compounds				m-Dichlorobenzene			D
Aldrin	ND 0.060	D (62/3,044)	D	p-Dichlorobenzene		ND (0/97)	D
α-Chlordane	D 0.060	D (50/3,514) [isomer(s) not specified]	D(50/3,514) [isomer(s) not specified]	Dieldrin	ND 0.060	D (86/3,156)	D
γ-Chlordane	D 0.060			Endosulfan		D (32/2,410)	D
Oxy-chlordane		ND (0/165)		Endosulfan I (Thiodan I)	ND 0.060	D (1/906)	ND
trans-Nonachlor (impurity)		D (2/143)		Endosulfan II (Thiodan II)	ND 0.13	D (1/905)	ND
Nonachlor (impurity)		ND (0/22)		Endosulfan sulfate	ND 0.13	D (6/1,969)	D
Chlordecone (Kepone)		ND (0/2)					

Table 3.2. Pesticides and transformation products detected or not detected in ground waters of North America, based on either the National Pesticide Survey, the Pesticides In Ground Water Database, or the monitoring studies reviewed—*Continued*

Pesticide or Transformation Product	NPS (MRL, in µg/L)	PGWDB (Wells detected/wells sampled)	Moni- toring Studies	Pesticide or Transformation Product	NPS (MRL, in µg/L)	PGWDB (Wells detected/wells sampled)	Moni- toring Studies
		Data Source				Data Source	
Endrin	ND 0.13	D (39/4,084)	D	Toxaphene		D (9/4,273)	D
Endrin aldehyde	ND 0.13	ND (0/1,364)	D	Trichloroacetic acid		ND (0/97)	
γ-HCH (Lindane)	D 0.043	D (78/4,474)	D	1,2,4-Trichloro- benzene		ND (0/76)	
α-HCH	ND 0.060		D	Trichloronate		ND (0/172)	
β-HCH	D 0.060	D (26/1,848) [isomer(s) not specified]	D	Trichlorophenol		ND (0/178)	
δ-HCH	Q		ND	Triclopyr		D (5/379)	D
Heptachlor	ND 0.060	D (55/3,241)	D	**Organophosphorus Compounds**			
Heptachlor epoxide	ND 0.060	D (32/3,115)	D	Acephate		ND (0/1,019)	
Hexachloroben- zene	D 0.060	D (4/1,328)	ND	Azinphos-ethyl		ND (0/5)	
Kelthane (Dicofol)		ND (0/1,634)	ND	Azinphos-methyl		D (5/1,628)	D
Methoxychlor	ND 0.30	D (16/3,074)	D	Azinphos- methyl oxon			ND
Mirex		D (9/503)	D	Bensulide		ND (0/191)	
Nitrofen		ND (0/118)		Carbophenothion methyl		ND (0/347)	
PCNB		D (3/1,708)	D	Chlorpyrifos (Dursban)		D (32/5,398)	D
Pentachlorophenol	ND 0.10	D (213/526)	D	Chlorpyrifos- methyl		ND (0/237)	
Perthane (Ethylan)		ND (0/351)	D	Coumaphos		ND (0/173)	
Pyriclor		D (2/19)		Crufomate		ND (0/6)	
Ronalin			ND	Demeton		ND (0/1,386)	ND
TCBC (Tri- chlorobenzyl chloride)			D	Demeton-methyl (Metasystox)		ND (0/198)	ND
Tetradifon		ND (0/147)		Demeton-S		ND (0/8)	

Table 3.2. Pesticides and transformation products detected or not detected in ground waters of North America, based on either the National Pesticide Survey, the Pesticides In Ground Water Database, or the monitoring studies reviewed—*Continued*

Pesticide or Transformation Product	NPS (MRL, in µg/L)	PGWDB (Wells detected/wells sampled)	Monitoring Studies	Pesticide or Transformation Product	NPS (MRL, in µg/L)	PGWDB (Wells detected/wells sampled)	Monitoring Studies
Demeton-S sulfone		ND (0/188)		Fonofos		D (18/4,446)	D
Diazinon	Q	D (42/3,884)	D	Isofenphos		D (2/97)	ND
Dicapthon			ND	Malathion		D (12/3,252)	D
Dichlorvos (DDVP)	ND 0.12	ND (0/188)	ND	Malaoxon		ND (0/1)	
Dicrotophos		ND (0/14)		Merphos	Q	ND (0/434)	
Dimethoate		D (12/2,844)	D	Methamidophos		D (4/814)	D
Dioxathion		ND (0/22)		Methidathion		ND (0/262)	ND
Disulfoton	Q	D (20/2,468)	D	Methyl paraoxon	ND 0.15	ND (0/125)	
Disulfoton sulfone	Q	ND (0/308)		Methyl parathion		D (20/3,357)	D
Disulfoton sulfoxide	Q	ND (0/120)		4-Nitrophenol (also a fungicide)	Q	D (3/344)	
DMPA (Zytron)		ND (0/107)		Methyl trithion			ND
EPN		ND (0/272)		Mevinphos	ND 0.15	ND (0/665)	ND
Ethion		ND (0/1,613)	ND	Monocrotophos		D (1/152)	
Ethoprop	ND 0.060	ND (0/1,368)	D	Naled (Dibrom)		ND (0/247)	D
Ethyl parathion		D (3/3,529)	ND	Oxydemeton-methyl		ND (0/2)	ND
Fenamiphos	ND 0.15	(ND)[3] (0/1,236)	ND	Oxydisulfoton		ND (0/9)	
Fenamiphos sulfone	ND 29	(ND)[3] (0/180)	ND	Parathion			D
Fenamiphos sulfoxide	ND 4.7	(ND)[3] (0/180)	ND	Phorate		ND (0/3,341)	D
Fensulfothion		ND (0/290)	ND	Phorate sulfone		ND (0/12)	
Fenthion		ND (0/184)	ND	Phorate sulfoxide		ND (0/12)	

Table 3.2. Pesticides and transformation products detected or not detected in ground waters of North America, based on either the National Pesticide Survey, the Pesticides In Ground Water Database, or the monitoring studies reviewed—*Continued*

Pesticide or Transformation Product	Data Source			Pesticide or Transformation Product	Data Source		
	NPS (MRL, in µg/L)	PGWDB (Wells detected/wells sampled)	Moni-toring Studies		NPS (MRL, in µg/L)	PGWDB (Wells detected/wells sampled)	Moni-toring Studies
Phoratoxon		ND (0/9)		**Pyrethroid Insecticides**			
Phoratoxon sulfone		ND (0/9)		Cypermethrin		ND (0/311)	D
Phoratoxon sulfoxide		ND (0/9)		Deltamethrin			ND
Phosalone		ND (0/271)	ND	Fenvalerate		D (5/345)	D
Phosmet (Imidan)		D (1/307)	D	cis-Permethrin	ND 0.90	D (4/1097) [iso-mer(s) not specified]	D [iso-mer(s) not spec-ified]
"Phosmet oxygen analogue"		ND (0/3)		trans-Permethrin	ND 2.0		
Phosphamidon		ND (0/187)	D	Tralomethrin		ND (0/188)	
Profenofos		ND (0/188)		**Miscellaneous Insecticides**			
Prothiophos		ND (0/164)		Acenaphthene		ND (0/228)	
Reldan			ND	Arsenic[4]		D (106/271)	D
Ronnel		ND (0/647)	ND	Copper[4]		ND (0/6)	D
Sulprofos		D (1/230))	D	Fenbutatin oxide		ND (0/1)	
Terbufos	Q	D (11/4,224)	D	Isobornyl thiocyanoacetate		ND (0/1)	
Terbufos sulfone		ND (0/13)		MGK 264 (synergist)	ND 1.0		
Tetrachlorvinphos (Stirofos)	ND 0.18	ND (0/173)		Naphthalene		ND (0/82)	
Triazophos			ND	Propargite		ND (0/382)	ND
Tribufos (DEF)		ND (0/569)	ND	Rotenone		ND (0/12)	
Trichlorfon		D (12/459)		Rotenolone		ND (0/4)	
Trithion (Carbopheno-thion)		ND (0/682)	ND	"Other rotenone metabolite"		ND (0/8)	
				Sethoxydim		ND (0/65)	ND

Table 3.2. Pesticides and transformation products detected or not detected in ground waters of North America, based on either the National Pesticide Survey, the Pesticides In Ground Water Database, or the monitoring studies reviewed—*Continued*

Pesticide or Transformation Product	NPS (MRL, in µg/L)	PGWDB (Wells detected/wells sampled)	Monitoring Studies
Sodium bromide		ND (0/6)	
Xylene (o-, m-, and p-)		ND (0/3,160)	D
Fumigants			
Acrylonitrile		ND (0/99)	
Carbon disulfide		D (2/12)	ND
Carbon tetrachloride		ND (0/101)	D
Chloroform		D (7/193)	D
Chloropicrin		D (3/16,561)	ND
Dibromochloropropane (DBCP)	D 0.010	D (1829/20,545)	D
1,2-Dichloroethane		ND (0/305)	D
1,2-Dichloropropane (DCP)	ND 0.75	D (350/21,390)	D
cis-1,3-Dichloropropene	ND 0.010	D (6/21,270)	D [isomer(s) not specified]
trans-1,3-Dichloropropene	ND 0.10	D [isomer(s) not specified]	D [isomer(s) not specified]
Chloroallyl alcohol		ND (0/12)	
Ethylene dibromide (EDB)	D 0.010	D (2918/20,221)	D
Metam-Sodium			ND
Methyl bromide		D (2/20,429)	ND
Methylene chloride		D (18/292)	D
Methyl isothiocyanate		ND (0/14,864)	ND

Pesticide or Transformation Product	NPS (MRL, in µg/L)	PGWDB (Wells detected/wells sampled)	Monitoring Studies
Tetrachloroethylene		D (44/128)	D
Trichloroethylene		D (22/543)	D
Miscellaneous Fungicides			
Acrolein		ND (0/149)	
Captafol (Difolitan)		ND (0/265)	D
Captan		D (3/1,828)	D
Carbendazim		ND (0/208)	
Carboxin	ND 0.50	ND (0/265)	ND
p-Chloro-m-cresol		ND (0/40)	
p-Chloro-o-cresol		ND (0/1)	
Dinocap			ND
2,4-Dinitrophenol		ND (0/102)	
DNOC		ND (0/412)	
Etridiazole	ND 0.13	ND (0/6)	
ETU (EBDC degradate)[5]	D 4.5	D (1/183)	D
Fenarimol	ND 0.20	ND (0/6)	
Formaldehyde		ND (0/6)	
Iprodione		ND (0/15)	
Mancozeb		ND (0/60)	ND
Maneb (Manzate)		ND (0/437)	ND
Metalaxyl		D (17/352)	ND

Table 3.2. Pesticides and transformation products detected or not detected in ground waters of North America, based on either the National Pesticide Survey, the Pesticides In Ground Water Database, or the monitoring studies reviewed—*Continued*

Pesticide or Transformation Product	NPS (MRL, in μg/L)	PGWDB (Wells detected/wells sampled)	Monitoring Studies
4-Nitrophenol (also a methyl parathion degradate)	Q	D (3/344)	
PCNB (Quintozene)			D
Triadimefon	ND 0.16	ND (0/14)	
Tricyclazole	ND 0.60	ND (0/6)	
Zineb		ND (0.9)	ND

Pesticide or Transformation Product	NPS (MRL, in μg/L)	PGWDB (Wells detected/wells sampled)	Monitoring Studies
Miscellaneous Pesticides			
"Cyanide"		D (3/14)	
Dibutyl phthalate		ND (0/11)	
Dioctyl phthalate		ND (0/11)	
Ethyl alcohol		ND (0/1)	
Lead[4]			D
Mercury		D (3/26)	

[1]Also derived from propazine (Thurman and others, 1994).

[2]Also derived from cyanazine and simazine (Thurman and others, 1994).

[3]In an update on the PGWDB, Haaser and Waldman (1994) noted the reporting of detections of fenamiphos residues in Florida ground waters.

[4]Included in table due to the documented use of arsenic-, copper-, and lead-based pesticides.

[5]Ethylene bis-dithiocarbamate (EBDC) fungicides include Mancozeb, Maneb, Zineb, and Ziram.

Table 3.6. Estimates of pesticide use in the United States in agricultural and residential settings, and detections in ground and surface waters, rain, and air

[Compound: Other Uses [shown in brackets]: Fu, Fumigant; I, Insecticide; IS, Insecticide Synergist; Mi, Miticide; Mo, Molluscicide; R, Rodenticide; U, Use category not specified. Agricultural Use Sources: 1966: Eichers and others (1970); 1971: Andrilenas (1974); 1988: Gianessi and Puffer (1991, 1992a,b). Source for Home and Garden Use: Whitmore and others (1992). Sources for Detections in Environmental Media: GW (Ground Water), National Pesticide Survey (U.S. Environmental Protection Agency, 1990a) and Pesticides In Ground Water Database (U.S. Environmental Protection Agency, 1992b). SW (Surface Water), Wauchope (1978); Air and Rain, Majewski and Capel (1995). C, Detection reported in Canada; DMA, Dimethylamine. lb a.i., pounds of active ingredient]

Compound [Other Uses]	Agricultural Use (lb a.i. x 1000)			Home and Garden Use (1990)			Compound Detected In:			
	1966	1971	1988	Products (x 1000)	Outdoor Applications (x 1000)	Notes	GW	SW	Rain	Air
Organochlorine Insecticides										
Toxaphene	34,605	37,464		74			•	•	•	•
DDT	27,004	14,324		202			•	•	•	•
Aldrin	14,761	7,928					•		•	•
Methoxychlor	2,578	3,012	109	3,564	3,692		•	•	•	•
Chlordane	526	1,890		1,156	478		•		•	•
Endrin	571	1,427					•	•		•
Heptachlor	1,536	1,211		72	177		•			•
Endosulfan	791	882	1,992	111	561		•	•	•	•
HCH, γ- (Lindane)	704	650	66	1,638	1,355		•		•	•
Dieldrin	724	332					•	•	•	•
DDD	2,896	244					•			•
Strobane	2,016	216								
Dicofol		1,718		4,587	4,179					
p-Dichlorobenzene [Fu]				1,098	538					
Pentachlorophenol, total				576	89		•			
Others	347	293								
Total Organochlorine Insecticide Use	89,059	69,873	3,885	13,078	11,069					
Percent of Total Insecticide Use	62.75	43.03	3.59	4.02	1.82					
Organophosphorus Insecticides										
Chlorpyrifos			16,725	16,652	41,900		•		•	•
Methyl parathion	8,002	27,563	8,131				•	•	•	•
Terbufos			7,218				•			
Phorate	326	4,178	4,782							•
Fonofos			4,039				•	•	•	•
Malathion	5,218	3,602	3,188	9,551	16,597		•		•	•
Disulfoton	1,952	4,079	3,058	2,364	6,464		•			•
Acephate			2,965	4,940	19,167					
Dimethoate			2,960	301	132		•		•	
Parathion	8,452	9,481	2,848					•		•
Azinphos-methyl	1,474	2,654	2,477	37	348		•			
Diazinon	5,605	3,167	1,710	15,703	56,758		•	•	•	•
Ethoprop			1,636							

Table 3.6. Estimates of pesticide use in the United States in agricultural and residential settings, and detections in ground and surface waters, rain, and air—*Continued*

Compound [Other Uses]	Agricultural Use (lb a.i. x 1000)			Home and Garden Use (1990)			Compound Detected In:			
	1966	1971	1988	Products (x 1000)	Outdoor Applications (x 1000)	Notes	GW	SW	Rain	Air
Ethion	2,007	2,326	1,249	39						
Profenofos			1,224							
Methamidophos			1,135				•			
Phosmet			1,055	371	173		•			
Dicrotophos	1,857	807	963							
Sulprofos			874				•			
Fenamiphos			763							
Mevinphos			463							
Methidathion			402							•
Oxydemeton-methyl			370	1,032	670					
Naled			224	158						
Trichlorfon	1,060	617	36	41			•			
Dichlorvos	912	2,434		8,953	13,043					
Ronnel	391	479		38						
Tetrachlorvinphos				2,423						
Phosalone				118	93					
Chlorfenvinphos				105						
Isofenphos				100	43		•			
Crotoxyphos				40						
Others	2,710	9,319								
Total Organophosphorus Insecticide Use	39,966	70,706	70,495	62,966	155,388					
Percent of Total Insecticide Use	28.16	43.54	65.08	19.34	25.52					
Other Insecticides										
Carbaryl	12,392	17,838	7,622	18,437	31,735		•	•		•
Carbofuran		2,854	7,057				•	•	•	
Propargite			3,786							
Aldicarb			3,573				•			
Cryolite			2,970							
Methomyl		1,077	2,952	346	629		•			
Thiodicarb			1,714							
Permethrin, total			1,122	7,397	18,461		•			
Oxamyl			726				•			
Fenbutatin oxide			560	248	2,147					
Formetanate HCl			414							
Esfenvalerate			287							
Tefluthrin			197							
Cypermethrin			188	198						
Trimethacarb			131							
λ-Cyhalothrin			110							
Cyfluthrin			105	1,139	5,654					
Oxythioquinox			102							

Table 3.6. Estimates of pesticide use in the United States in agricultural and residential settings, and detections in ground and surface waters, rain, and air—*Continued*

Compound [Other Uses]	Agricultural Use (lb a.i. x 1000)			Home and Garden Use (1990)			Compound Detected In:			
	1966	1971	1988	Products (x 1000)	Outdoor Applications (x 1000)	Notes	GW	SW	Rain	Air
Amitraz			75							
Fenvalerate			73	3,192	2,937		•			
Metaldehyde [I/Mo]			44	5,144	27,094					
Tralomethrin			43							
Diflubenzuron			42							
Bifenthrin			28							
Abamectin			12							
Cyromazine			9							
Piperonyl butoxide [IS]				41,729	58,991					
Pyrethrins				34,609	39,289					
MGK-264 [IS]				27,558	13,249					
Diethyltoluamide				21,544	14,134					
Baygon (Propoxur)				21,484	53,594		•			
Allethrin, total				18,543	52,277					
Tetramethrin				12,962	31,464					
Resmethrin, total				12,506	34,576					
Sumithrin				8,089	31,856					
Rotenone				3,997	4,510					
Methoprene				2,709	1,999					
Hydramethylon				2,389	10,485					
Allethrin (isomer unspecified)				1,388	2,005					
Warfarin [R]				1,145	545					
Brodifacoum [R]				906	296					
Bendiocarb				697	1,778					
Tricosene				307	453					
Dienochlor [Mi]				265	1,591					
Methiocarb				237	386		•			
Diphacinone [R]				89						
Pindone [R]				81						
Bromadiolone [R]				70	177					
Fenoxycarb				54						
Cythioate				37						
Others	502	37								
Total Other Insecticide Use	12,894	21,806	33,941	249,496	442,312					
Percent of Total Insecticide Use	9.09	13.43	31.33	76.64	72.66					

Triazine and Acetanilide Herbicides										
Atrazine	23,521	57,445	64,236	134	488		•	•	•	•
Alachlor		14,754	55,187				•	•	•	•
Metolachlor			49,713				•	•		•
Cyanazine			22,894				•	•	•	

Table 3.6. Estimates of pesticide use in the United States in agricultural and residential settings, and detections in ground and surface waters, rain, and air—*Continued*

Compound [Other Uses]	Agricultural Use (lb a.i. x 1000)			Home and Garden Use (1990)			Compound Detected In:			
	1966	1971	1988	Products (x 1000)	Outdoor Applications (x 1000)	Notes	GW	SW	Rain	Air
Metribuzin			7,516				•	•	•	
Propazine	580	3,171	4,015				•	•		
Propachlor	2,269	23,732	3,989				•	•	•	
Simazine	193	1,738	3,964	172			•	•	•	•
Prometryn			1,807				•	•		
Terbutryn			1,113						•	
Prometon				1,244	1,281		•			
Others		450	1,022							
Total Triazine and Acetanilide Herbicide Use	26,563	101,290	215,456	1,550	1,769					
Percent of Total Herbicide Use	23.79	42.88	51.19	3.26	1.38					

Other Herbicides										
EPTC	3,138	4,409	37,191	125			•		•	
2,4-D	40,144	34,612	33,096	14,324	44,054	Total	•	•		•
Trifluralin	5,233	11,427	27,119	483	547		•	•	•	•
Butylate		5,915	19,107				•		•	
Pendimethalin			12,521	158	338		•		•	•
Glyphosate			11,595	8,110	25,618		•			
Dicamba	222	430	11,240	3,636	6,431	Total	•	•	C	
Bentazon			8,211				•			
Propanil	2,589	6,656	7,516				•			
MSMA			5,065	240	267			•		
Molinate			4,408				•	•		•
MCPA	1,669	3,299	4,338	119	121	DMA salt	•			
Ethalfluralin			3,518							
Triallate			3,509						C	C
Paraquat			3,025	201	79	Dichloride	•	•		
Chloramben (Amiben)	3,765	9,555	3,019				•			
Picloram			2,932				•	•		
Clomazone			2,715							
Bromoxynil			2,627						C	
Linuron	1,425	1,803	2,623				•	•		
Fluometuron		3,334	2,442				•	•		
Dacthal			2,219	733	368		•		•	•
Diuron	1,624	1,234	1,986				•	•		
Norflurazon			1,768							
DSMA			1,705	38	38					
Acifluorfen			1,475	1,845	7,081	Na salt	•			
Diclofop			1,452						C	

Table 3.6. Estimates of pesticide use in the United States in agricultural and residential settings, and detections in ground and surface waters, rain, and air—*Continued*

Compound [Other Uses]	Agricultural Use (lb a.i. x 1000)			Home and Garden Use (1990)			Compound Detected In:			
	1966	1971	1988	Products (x 1000)	Outdoor Applications (x 1000)	Notes	GW	SW	Rain	Air
Oryzalin			1,426	117	1,766		•			
2,4-DB			1,368				•			
Thiobencarb			1,359				•			•
Cycloate			1,175				•			
Benefin (Benfluralin)			1,167	79	81					
Bromacil			1,155				•			
Asulam			1,088							
Imazaquin			1,073							
Diphenamid			929					•		
Vernolate		3,739	855							
Sethoxydim			792							
Fluazifop-butyl			731	421	426		•			
Napropamide			699							
Naptalam (Alanap)	999	3,332	655							
Pebulate	150	1,062	653							
Bensulide			633	36	73					
Profluralin			621				•			
Tebuthiuron			608				•			
Oxyfluorfen			599	234	643					
Diethatyl Ethyl			502							
Dalapon	38	1,043	453							
Nitralin	14	2,706								
2,4,5-T	760	1,530		84			•	•	•	•
Flourodifen		1,330								
Norea	239	1,323								
MCPP, DMA salt				10,318	28,963					
Silvex, total				341	88		•			
Mecoprop				849	3,266					
Triclopyr, total				683	823		•			
Chlorflurenol, methyl ester				501	1,067					
Diquat dibromide				211	635					
Dichlobenil				167	124			•		
Amitrole				163	40			•		
Metam-Sodium				128	74					
Endothall, disodium salt				124	43					
Sodium thiocyanate				82						
Others	21,479	27,686	5,635							
Total Other Herbicide Use	85,114	134,903	205,407	44,425	123,054					
Percent of Total Herbicide use	76.21	57.12	48.81	93.53	95.96					

Table 3.6. Estimates of pesticide use in the United States in agricultural and residential settings, and detections in ground and surface waters, rain, and air—*Continued*

Compound [Other Uses]	Agricultural Use (lb a.i. x 1000)			Home and Garden Use (1990)			Compound Detected In:			
	1966	1971	1988	Products (x 1000)	Outdoor Applications (x 1000)	Notes	GW	SW	Rain	Air
Fungicides										
Chlorothalonil			9,932	1,399	2,602		•			
Mancozeb			8,661	113						
Captan	6,869	6,490	3,710	3,067	4,682		•			
Maneb	4,443	3,878	3,592	345	878					
Ziram			1,889	81						
Benomyl			1,344	684	3,704		•			
PCNB			800	64	64					
Iprodione			741							
Fosetyl-Al			689							
Metiram			641							
Metalaxyl			635	41						
Thiophanate-methyl			527							
Triphenyltin hydroxide			415							
Ferbam	2,945	1,398	337	82						
DCNA			286							
Dodine			275							
Propiconazole			274							
Thiram			238	397	249					
Triadimefon			149							
Anilazine			144	106	71					
Thiabendazole			139							
Myclobutanil			124							
Etridiazole			104							
Vinclozolin			103							
Streptomycin			88							
Triforine			81	3,150	14,286					
Fenarimol			58							
Oxytetracycline			37							
Carboxin			15							
Dinocap			14	703	796					
Dodine + Dinocap (combined totals)	1,143	1,191								
Zineb	6,903	1,969		684	1,413					
Folpet				3,314	4,347					
Dichlorophene				43						
Dichlone				40						
Limonene [U]				819						
Dextrin [U]				79						
Thymol [U]				37	166					
Others	3,334	10,814								
Total Fungicide Use	25,637	25,740	36,042	15,248	33,258					

CHAPTER 4

Processes that Govern Pesticide Concentrations in Ground Water

4.1 OCCURRENCE AND MOVEMENT OF WATER
WITHIN THE SUBSURFACE

Water is present at some depth beneath every point on the earth's surface (Figure 1.2). The water table is defined as the surface below which all interconnected voids are filled with water, and at which the total pressure, or hydraulic head, equals atmospheric pressure. (More precise definitions of most of the hydrogeologic terms used in this report can be found in Freeze and Cherry [1979] and Heath [1984].) Below the land surface, water occurs in one of two general regions, the vadose (or unsaturated) zone above, and the saturated zone below the water table. Ground water is defined as the water within the saturated zone. In the vadose zone, both aqueous and gaseous phases are present within interstitial voids (Figure 4.1). Immediately above the water table is a region, known as the capillary fringe, in which the voids are filled with water held at less than atmospheric pressure by capillary forces (Figure 4.2). Since capillary forces in the subsurface act primarily at the water-solid interface, the thickness of the capillary fringe increases as pore sizes decrease. This zone is therefore substantially thicker in clays, for example, than in well-sorted, sandy deposits. In this book, the term subsurface is used to denote the entire region from the land surface through the saturated zone.

An aquifer is defined as "a saturated permeable geologic unit that can transmit significant quantities of water under ordinary hydraulic gradients" (Freeze and Cherry, 1979). By contrast, an aquitard is any saturated geologic unit that is not sufficiently permeable to yield water at significant rates under ordinary gradients. If an aquifer is bounded above by an aquitard, it is said to be confined. Unconfined, phreatic or water-table aquifers are those for which the water table constitutes the upper boundary (Figure 4.3). From a practical standpoint, the depth of the water table at any given location is the depth to which water rises within a shallow well screened in an unconfined aquifer, as shown in Figure 4.2.

The level to which water will rise in a well screened in a confined aquifer delineates the potentiometric surface (or piezometric surface) for that aquifer at the screen location. Where the potentiometric surface for a given aquifer is above the land surface, the aquifer in question is said to be under artesian conditions. Springs occur where ground water discharges directly from the ground because either the water table or a potentiometric surface intersects the land surface. Most perennial surface waters occur where the water table is at or above the land surface (Figures 4.1 and 4.3).

In most areas, the topography of the water table generally follows that of the land surface, though usually in a more muted fashion. Exceptions to this pattern occur where significant pumping of ground water is taking place (Bond, 1987), or substantial lateral variations in soil

179

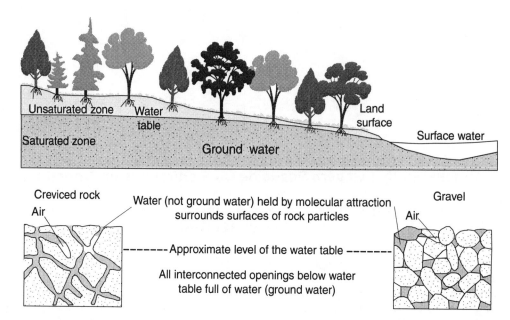

Figure 4.1. Where ground water occurs beneath the land surface (redrawn from Waller, 1991).

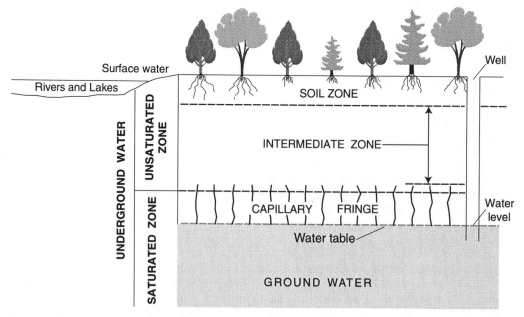

Figure 4.2. Location of the capillary fringe, relative to the saturated and unsaturated zones, within the subsurface (redrawn from Heath, 1984).

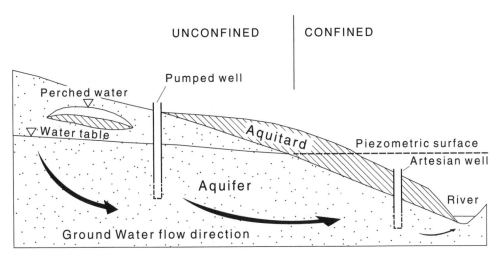

Figure 4.3. Confined and unconfined aquifers and perched water tables. Redrawn with permission from "Hydrogeology with respect to underground contamination" by C.J. Smith in *Environmental Fate of Pesticides*, D.H. Hutson and T.R. Roberts, eds. Copyright 1990 by John Wiley & Sons, Ltd.

permeability are evident (Phillips and Shedlock, 1993). Thus, the lateral direction of ground-water flow within a relatively unperturbed, unconfined aquifer is roughly similar to the direction water would flow along the land surface above it (Figure 4.4). Furthermore, the hydraulic gradient (essentially equivalent to the slope of the water table for shallow unconfined ground water) in most unconfined aquifers is sufficiently gentle that ground-water flow is typically quite slow in the absence of pumping, on the order of a meter per day to a meter per year.

The amount of time required for ground water to discharge, or re-emerge at the Earth's surface following recharge or infiltration, varies greatly with depth and hydrogeologic setting. Shallow ground water may discharge to the surface within days or weeks following recharge (Figure 4.4), the process often occurring most rapidly when the natural subsurface flow paths are short-circuited by tile drains or drainage ditches. Shallow ground waters, if contaminated, may thus represent a significant source of contaminants to nearby surface waters.

While the movement of ground water from areas of recharge to areas of discharge is predominantly horizontal, water within the unsaturated zone flows essentially downward, unless it encounters horizons of significantly lower permeability before reaching the water table. If the downward flux of water is sufficiently large, such horizons can give rise to a perched water table (Figure 4.3).

The rate and direction of water flow at any point within the subsurface are governed by the product of the hydraulic gradient and the hydraulic conductivity (Darcy's Law). Below the water table, the hydraulic conductivity of geologic materials is controlled primarily by the size, shape, and interconnectedness of the voids within those materials. Above the water table, however, the presence of a gas phase within the voids restricts the pathways through which water must pass as it percolates downward. Consequently, the hydraulic conductivity in the unsaturated zone is also controlled by the water content of the soil (typically expressed on a volumetric basis). Because water content varies considerably within the vadose zone—usually increasing with depth—the rates of water flow above the water table are more spatially variable, and hence less easily predicted than those within the saturated zone.

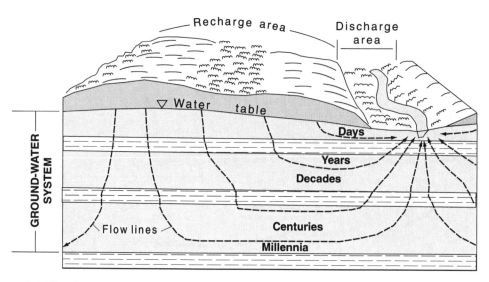

Figure 4.4. Directions and rates of ground-water movement (redrawn from Waller, 1991).

4.2 SOURCES OF CONTAMINATION BY PESTICIDES

Agriculture in the United States involves the handling and use of large quantities of pesticides, fertilizers, and other agrichemicals. These chemicals must be manufactured, distributed to dealerships, sold to users, transported to individual farms, placed in storage until they are needed, transferred to application equipment, and broadcast over large areas. The empty containers and excess formulation must then be discarded. Given all the stages that characterize the life cycle of most pesticides, it is not surprising that, in addition to their actual application to croplands and other targeted areas (e.g., lawns and rights-of-way), these compounds also may be inadvertently discharged to other nontargeted areas, as well. Figure 4.5 illustrates the major routes by which pesticides enter, move through, and exit from the subsurface environment in agricultural regions.

The most geographically widespread source of pesticides to ground waters is the atmosphere. Numerous pesticides have been detected in air, rain (Table 3.6), aerosols, snow, and fog. Majewski and Capel (1995) have reviewed the existing observations of pesticides in the atmosphere. Their review indicated that of the 63 pesticides that have been analyzed for in one or more atmospheric matrices, 45 have been detected, in addition to 18 transformation products. In general, the more volatile pesticides and those that are aerially applied have the greatest chance of entering the atmosphere. Pesticides can also enter the atmosphere during their manufacture and industrial use. Once in the atmosphere, pesticides can be transported by wind currents, undergo photochemical and hydrolytic transformation, and be deposited to aquatic and terrestrial surfaces.

The relative importance of atmospheric inputs of pesticides to ground water depends on the magnitude of other nearby sources. Atmospheric deposition of pesticides occurs globally, as demonstrated by observations of organochlorine insecticides in the Arctic (Majewski and Capel, 1995). If atmospheric deposition of pesticides to the land surface in active agricultural areas is

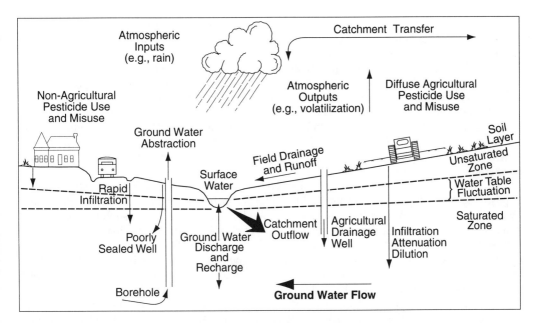

Figure 4.5. Routes by which pesticides enter, move through, and exit from the subsurface in agricultural regions. Modified from Cartwright and others (1991) and published with permission.

comparable to deposition rates in remote areas, the atmospheric contribution of currently used agricultural pesticides to the total loads reaching ground water may be overwhelmed by the amount coming directly from nearby application areas.

The reporting of detectable concentrations of pesticides in ground waters has frequently been accompanied by efforts to distinguish contamination derived from comparatively well-defined, localized areas, referred to as point sources, from that which results from the routine application of these compounds over extensive areas, referred to as a nonpoint source (Litwin and others, 1983; Cardozo and others, 1985; Exner and Spalding, 1985; Steichen and others, 1988; Turco and Konopka, 1988; Cavalier and others, 1989; Sitts, 1989; Voelker, 1989; Exner and Spalding, 1990; Fawcett, 1990; Kross and others, 1990; Walker and Porter, 1990; Little, 1994). Hallberg (1989) has provided the following definitions for the two types of sources:

> In general, the distinction between point and nonpoint sources is clear, and has legal definition, as well. Point sources would include problems associated with facilities that handle pesticides on a commercial basis, often with a large number of products, handled/stored/manufactured in concentrated form, and operating in a relatively small area. This would include manufacturing facilities, transport/transfer facilities, commercial storage, mixing, reformulating, rinsing and disposal sites, the latter group often being interrelated in rural areas. Nonpoint sources arise from the 'routine use' of pesticides in the process of their application to the soil or crops in the field or crops in storage. While this seems clear enough in theory, in practice it can be another story.

Attempts to draw distinctions between point and nonpoint sources for specific instances of ground-water contamination arise from the desire to distinguish between contamination that is largely preventable (point sources) from contamination that is viewed as an essentially unavoidable consequence of routine pesticide applications carried out in accordance with accepted agricultural practices (nonpoint sources). Whether point sources or nonpoint sources are responsible for most of the ground-water contamination by pesticides remains a topic of considerable debate, particularly in heavily agricultural areas such as Iowa and Nebraska (e.g., Hallberg and others, 1984; Kelley, 1985; Kelley and Wnuk, 1986; Mitchem and others, 1988; Turco and Konopka, 1988; Spalding, 1989; Fawcett, 1990, 1995; Little, 1994). The various approaches that have been adopted for distinguishing between point and nonpoint sources will be examined further in Section 8.9.

Hallberg (1986) used the term "quasi-point sources" to describe localized sources, such as pesticide dealerships and handling facilities, that are sufficiently numerous in areas dominated by agriculture that their aggregate impact on ground-water quality within such regions may be widespread. For example, there are over 1,500 agrichemical mixing-and-loading facilities in Illinois (Long, 1989) and 650 such facilities in Wisconsin (Habecker, 1989). Poorly constructed or abandoned wells, or those used to dispose of agricultural drainage, can also serve as quasi-point sources of ground-water contamination by pesticides. Improperly constructed wells may provide routes by which surface-derived contaminants can reach ground waters, most commonly through faulty seals at the well head or along the well annulus. Contamination through abandoned wells is a source of particular concern because the existence and locations of such wells are less well-known than those of operating wells.

Drainage wells, which are used to lower water tables beneath poorly drained agricultural fields (Seitz and others, 1977; Baker and others, 1985; Hoyer and Hallberg, 1991; Libra and Hallberg, 1993), or to dispose of runoff from urbanized areas (German, 1989; Wilson and others, 1990) or fluids from oil and gas drilling operations (Leidy and Taylor, 1992), are another type of quasi-point source. In areas where they are commonly used, drainage wells may be responsible for widespread contamination of ground waters by pesticides, nitrate, bacteria, and other pollutants (Whitehead, 1974; Seitz and others, 1977; Graham and others, 1977; Graham, 1979; Baker and others, 1985). The use of drainage wells can lead to ground-water contamination in regions where low-permeability strata close to the land surface substantially reduce the rate at which surface-derived contaminants can reach underlying aquifers (Leidy and Taylor, 1992). In Wright County, Iowa, for example, such wells provide routes by which agricultural contaminants may bypass thick glacial tills (>100 ft) overlying the bedrock aquifer (Libra and Hallberg, 1993).

Although the focus of this chapter is on the movement and fate of pesticides in the subsurface, pesticides applied to nearby areas may enter a well by routes other than migration through the subsurface. As discussed later (Section 8.8.1), pesticides applied in adjacent areas— in some cases within a land-use setting different from that surrounding the well of interest—may reach a well by spray drift or surface water "run-in," rather than subsurface transport (Frank and others, 1987b).

Following application, the concentration of a pesticide at the land surface will decline over time, because of advective and diffusive transport away from the site of application, as well as chemical or microbial transformation in situ. This decrease in concentration of the parent compound over time has often been referred to as "dissipation" (e.g., Nash, 1988). As noted in Section 2.10.2, this term represents some unknown mixture of the influences of these physical, chemical, and biological processes on the compound of interest. In the remainder of this chapter, each of these separate processes will be examined in some detail, to provide a more complete basis for understanding the various ways in which the physical, chemical, and biological characteristics of a given location, agricultural practices, and compound structure may affect the transport and fate of a pesticide in the subsurface.

4.3 PROCESSES GOVERNING SOLUTE DISTRIBUTIONS AND TRANSPORT IN THE SUBSURFACE

The movement of a dissolved solute through the subsurface is controlled by two principal phenomena, advection, and hydrodynamic dispersion, both of which may be affected, in turn, by the partitioning of the solute among the various liquid, solid, and gaseous phases present. Advection of a solute is "the process by which solutes are transported by the bulk motion of the flowing ground water" (Freeze and Cherry, 1979). Hydrodynamic dispersion causes the broadening of a contaminant pulse over time, and represents the combined result of mechanical dispersion and molecular diffusion (Figure 4.6).

Mechanical dispersion arises from variations in the directions and distances traveled by different solute molecules in a porous medium. As shown in Figure 4.6, this occurs through two principal mechanisms: (1) variations in advection velocity across individual pores, and (2) differences in the lateral distances traversed by different flow paths (Freeze and Cherry, 1979). These processes are largely governed by the properties of the porous medium, rather than by those of the solute. Figure 4.7 shows the effects of hydrodynamic dispersion on subsurface contaminant distributions on a macroscopic scale.

Solute movement related to molecular diffusion, also depicted in Figure 4.6, occurs in response to chemical gradients within the medium of interest. The random nature of Brownian motion causes molecules of any chemical species to migrate from areas of high concentration toward those of lower concentration. In contrast with mechanical dispersion, the rate of molecular diffusion is dependent on the physical and chemical properties of the solute molecule and the solvent itself (in this case, water), rather than on the physical properties of the porous medium. Estimates of molecular diffusion coefficients for a variety of neutral molecules in aqueous solution are available in the literature (e.g., Hayduk and Laudie, 1974), or may be computed using the methods of Tucker and Nelken (1990).

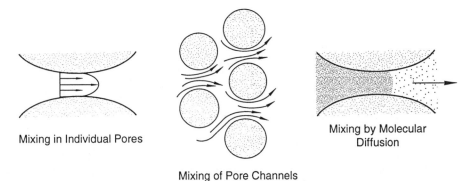

Mixing in Individual Pores

Mixing by Molecular Diffusion

Mixing of Pore Channels

Figure 4.6. Processes of dispersion on a microscopic scale. Freeze and Cherry, *Groundwater*, Copyright 1979, Fig. 2.30. Redrawn by permission of Prentice-Hall, Upper Saddle River, New Jersey.

4.3.1 SOLUTE PARTITIONING IN THE SUBSURFACE

Partitioning between the aqueous, solid, and vapor phases has three principal effects on the distribution and movement of solutes within the subsurface. First, it leads to a reduction in the initial solute concentration in the aqueous phase. A mass-balance model introduced by Katz (1993) illustrates the quantitative effects of such partitioning on the initial subsurface concentrations estimated for EDB in selected areas of Florida where the fumigant was applied.

Second, the tendency of a solute to associate with nonaqueous (or less-mobile aqueous) phases strongly influences the rate at which it moves through the subsurface. In general, the greater the degree of association between the solute and these other phases, the slower its migration. This phenomenon is also responsible for the chromatographic separations used widely in analytical chemistry.

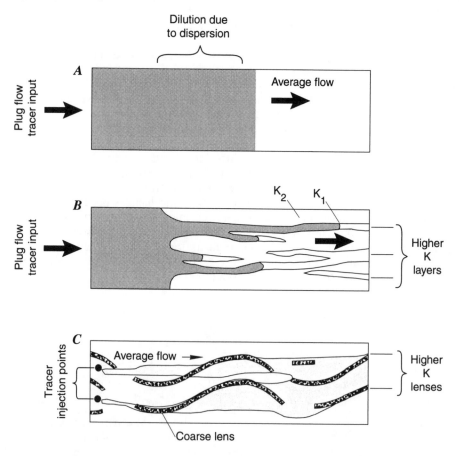

Figure 4.7. Hydrodynamic dispersion of a ground-water contaminant caused by spatial variations in hydraulic conductivity. (*A*) Spatially uniform hydraulic conductivity (idealized case); (*B*) Influence of high-conductivity layers; (*C*) Influence of high-conductivity lenses. Freeze and Cherry, *Groundwater,* Copyright 1979, Fig. 9.9. Redrawn by permission of Prentice-Hall, Upper Saddle River, New Jersey.

Finally, partitioning in the subsurface may cause the nonaqueous phases to serve as temporary repositories for a significant fraction of the solute of interest. Following application or discharge to the land, partitioning among solid, liquid, and gaseous media in the subsurface leads to a transient accumulation of some of the solute in one or more of the nonaqueous phases. Mass transfer of the contaminant from these phases back into the aqueous phase, once the initial contaminant pulse has passed, can then result in a broadening of the pulse to occupy an increasing volume of the subsurface over time, albeit at lower concentrations. Transfer between regions of "mobile" and relatively "immobile" waters in the subsurface can give rise to similar effects, as discussed later in this chapter (Section 4.3.3). This phenomenon constitutes the principal reason why, for example, aquifer remediation via "pump-and-treat" methods requires the removal of such large quantities of water—i.e., many pore volumes—from the subsurface.

Partitioning Between Solid and Aqueous Phases

Most pesticides are organic compounds (exceptions include arsenic-based pesticides, copper and elemental sulfur). Because their chemical structures are more similar to those of natural organic matter than to water, most pesticides exhibit some degree of affinity for organic matter in the subsurface. It is this affinity—the "like prefers like" rule of solution chemistry—that causes pesticides and other organic compounds to partition into soil organic matter from aqueous solution.

Compounds showing strong tendencies to associate with organic matter are said to be hydrophobic (or lipophilic), while those with a greater affinity for the aqueous phase are said to be hydrophilic. The best known examples of hydrophobic pesticides are the organochlorine insecticides such as DDT, toxaphene, aldrin, endrin, and dieldrin, which have been banned in the United States because of their tendency to bioaccumulate at high trophic levels within food webs and their toxicity to wildlife. Recent studies have also suggested a link between organochlorine compounds and increased risks of breast cancer in women (Messina, 1994) and endocrine disruption in wildlife and, perhaps, in humans (Colborn and others, 1993; Hileman, 1994; Raloff, 1994). Decades after their discontinuation, these compounds continue to be detected widely in sediments and soils within and downgradient from their original application areas, because of their strong affinities for natural organic matter, and their environmental persistence. The latter property has also facilitated their detection in polar areas and other remote regions of the planet where they were never applied. Among the pesticides that are currently registered for use in the United States, most have been designed to be more hydrophilic and less persistent than those that have been discontinued.

The equilibrium partitioning of a solute between an aqueous phase and subsurface solids is typically quantified using a soil-water partition coefficient, or K_p. As K_p increases, so does the proportion of the compound's mass that is associated with the solid, as opposed to the aqueous phase. This parameter is also sometimes referred to as a water-solids distribution coefficient, or K_d. Values of K_p have been observed to vary directly with the mass fraction of organic carbon in the soil (Lambert and others, 1965), or f_{oc}. From this relation, the partitioning of organic solutes between organic matter and water may be predicted, using a compound-specific, soil organic-carbon partition coefficient, or K_{oc}, where $K_{oc} = K_p/f_{oc}$. Conversely, this equation is often used to estimate the degree of partitioning between the solid and aqueous phases (K_p) at specific locations, based on the K_{oc} of the compound and the f_{oc} of the earth material of interest (Karickhoff and others, 1979).

Because of the relatively nonspecific nature of soil organic matter as a sorbent for most hydrophobic or weakly hydrophilic compounds (Lambert, 1967; Chiou and others, 1979; Karickhoff and others, 1979; Schwarzenbach and Westall, 1981), values of K_{oc} for individual compounds are widely applicable across a broad range of different types of soils and aquifer solids. Limitations on the use of partition coefficients for characterizing the sorption of pesticides and other organic solutes have been discussed elsewhere (e.g., Cheng and Koskinen, 1986; Curtis and others, 1986).

Not surprisingly, the partitioning of organic solutes between water and solid-phase organic matter closely mimics their partitioning between aqueous solution and biological tissues. For several decades, the partitioning of organic compounds between n-octanol and water has been used as a model for predicting their partitioning between aqueous and lipid phases in vivo, giving rise to an extensive body of data on octanol-water partition coefficients (K_{ow}) for a large number and variety of organic compounds (Hansch and Leo, 1979). Values of K_{ow} are highly correlated with K_{oc} (Schwarzenbach and Westall, 1981; Curtis and others, 1986). Through the use of such correlations, as well as from direct measurement, K_{oc} values are now available for a wide variety of pesticides and other synthetic organic compounds. The most recent and comprehensive summary of K_{oc} values and other physical and chemical properties of pesticides has been provided by Wauchope and others (1992).

The affinity of a compound for organic matter generally increases with decreasing water solubility (S_w). Studies have shown S_w to be inversely correlated with K_p (Chiou and others, 1979), K_{ow} (Curtis and others, 1986), and K_{oc} (Kenaga, 1980; Wauchope and others, 1992). Kenaga (1980) used a regression relation between K_{oc} and S_w to estimate K_{oc} values for 358 compounds, most of which are pesticides.

Association with Dissolved Organic Matter in Solution

Pesticides and other organic solutes may exhibit an affinity for organic matter in dissolved or colloidal form, as well as in the solid phase (e.g., Madhun and others, 1986; Kögel-Knabner and others, 1990; Pennington and others, 1991). The term dissolved organic matter (DOM) refers to organic materials that are not removed by filtration, and may include substances present in either dissolved or colloidal form. Association with DOM can increase the solubility of organic molecules in water, leading to a decrease in the apparent K_p, and a concomitant increase in subsurface mobility, a phenomenon sometimes referred to as facilitated transport (e.g., Cohen and others, 1990). As might be anticipated, this effect becomes more pronounced with increasing hydrophobicity. For example, although the removal of natural colloidal material—much of which consists of organic matter—exerted no discernible effect on the concentrations of atrazine (K_{oc} = 640 [Kenaga, 1980]) measured in ground waters obtained from an alluvial aquifer in Nebraska (Blum and others, 1993), the concentrations of chlordane (K_{oc} = 21,300 [Kenaga, 1980]) in ground water at a golf course in Cape Cod, Massachusetts, dropped below detection limits when the samples were filtered prior to analysis (Cohen and others, 1990). Consistent with the latter observation, the mobility of chlordane through a silt-loam soil was found by Johnson-Logan and others (1992) to be directly related to DOM concentrations. The mobility of viruses through soils is also enhanced in the presence of DOM (Scheuerman and others, 1979).

Partitioning Between Aqueous and Vapor Phases

Several pesticides, particularly those that are nonionic at neutral pH and of low molecular weight, have a tendency to volatilize under environmental conditions, a characteristic that is essential to the design of soil fumigants. Volatilization is of greatest importance in the unsaturated zone, where the gas phase can occupy a significant volume. The equilibrium partitioning of a solute between vapor and aqueous phases is described by its Henry's Law constant (H). A critical review of Henry's Law constants for a wide range of pesticides has been provided by Suntio and others (1988), while recent measurements of H for a substantial number of volatile organic compounds—including a number of previously and currently used fumigants—have been published by Gossett (1987) and Ashworth and others (1988). An earlier review of Henry's Law constants by Mackay and Shiu (1981) also included data for several fumigants and other pesticides. In general, volatilization is considered to be significant for compounds with H values of 10^{-5} atm-m^3/mole or greater at 20°C (Thomas, 1990).

Field Measurements of Solute Distributions Between Aqueous and Solid Phases

Complete mass balances of applied pesticides among subsurface media are rarely reported (e.g., Sophocleous and others, 1990; Poletika and others, 1995). However, many of the process and matrix-distribution studies estimated, or provided data that could be used to estimate, the proportions of applied pesticides occurring in different media. These results are summarized in Table 4.1 (at end of chapter). As is the case with pesticide occurrence data in general, the table indicates that matrix-distribution assessments involving pesticide transformation products are rare.

Although the atmosphere was not included among the various environmental media listed in Table 4.1, losses of applied pesticides into the air—caused by either drift from the target site during application or volatilization from the soil following application—often constitute a large proportion of the applied pesticide mass. Depending on the volatility of the compound, application method and soil moisture, the mass of applied pesticide lost to the atmosphere has been found to range from 2 percent to 90 percent (Jury and Ghodrati, 1989; Smith and others, 1991a; and Majewski and Capel, 1995).

Based on comparisons between available estimates of soil dissipation half-lives and the time elapsed between application and sampling, dissipation in situ probably introduced a pronounced negative bias to many of the proportions listed in Table 4.1. Indeed, time-series sampling over periods considerably longer than the estimated half-lives of the target compounds indicates substantial declines in these proportions during some of the studies.

The data summarized in Table 4.1 show a relatively consistent pattern with respect to the partitioning of pesticides among various environmental media. Most of the parent compound delivered to the soil surface resides in plant tissues and soils, rather than in the aqueous phases above or below the land surface. The proportions of applied pesticides measured in aqueous media—either in vadose zone waters, tile drainage, ground water, or surface waters—are generally at least an order of magnitude lower than those determined in the solid phases. Thus, although the fraction of applied pesticides found in soil and plant tissue is usually on the order of several tens of percent, the proportion transported away from the application site in either subsurface or surface waters is typically a few percent or less.

The data in Table 4.1 are in general agreement with those compiled by authors of previous summaries of pesticide loadings in ground waters (Hallberg, 1986; Boesten, 1987; Nash, 1988) and surface waters (Wauchope, 1978; Wauchope and others, 1990; Squillace and Thurman, 1992). They are also consistent with the moderately hydrophobic nature of most pesticides. However, although the percentages of applied pesticide reaching the receiving waters of interest are relatively small, the resulting concentrations may still be high enough to pose a significant concern for the health of humans and nearby aquatic ecosystems—a point also noted by Logan and others (1994). The small proportions of applied pesticides that enter ground and surface waters are also comparable to the amounts that typically reach target pests (Hallberg, 1986).

Data compiled in Table 4.1 can be used to examine the extent to which the proportions of applied pesticides measured in solid and aqueous environmental media are consistent with their equilibrium partitioning in the laboratory. These proportions (from Table 4.1) are plotted in Figure 4.8 as a function of K_{oc}. To reduce the potential negative bias arising from the dissipation of target compounds in situ, only data from samples taken within two soil half-lives from the time of application were selected for display.

Figure 4.8 shows the marked difference, noted earlier, between the proportions of individual pesticides residing in soils and those measured in natural waters. Regardless of the compound or the study, nearly all of the proportions measured in soils are larger than those in the aqueous phase (ground, surface, and vadose-zone waters). However, neither the proportions of pesticide residing in the aqueous phase nor those associated with soils were found to be significantly correlated with K_{oc} ($\alpha = 0.05$). The absence of significant correlations among these parameters may result from variations in environmental conditions (e.g., temperature, soil type, and f_{oc}) among the different studies.

4.3.2 INFLUENCE OF PARTITIONING ON SOLUTE TRANSPORT

As noted earlier, the affinity of a particular solute for solid and gaseous phases in the subsurface exerts an inverse influence over its rate of migration through the subsurface. Thus, the relative rates of movement of different pesticides in the subsurface have been found to increase with decreasing K_{oc} values (e.g., Jury and others, 1986b; Hall and others, 1989; Kladivko and others, 1991; Flury and others, 1995), although in part for reasons discussed in Section 4.3.3 below, such trends are not always observed (e.g., Ritter and others, 1989; Shirmohammadi and others, 1989; Clendening and others, 1990). Solutes that show negligible tendency to partition

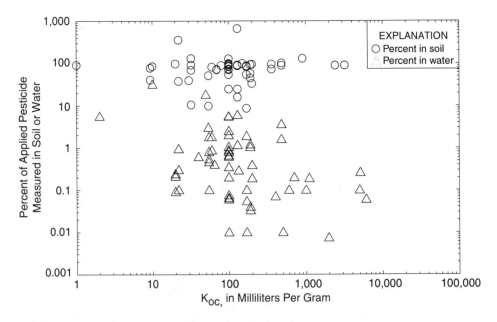

Figure 4.8. Percentage of applied pesticides measured in soil and aqueous phases (ground, vadose-zone, and surface waters) in the field versus K_{oc} values for compounds sampled within two dissipation half-lives of application. Partitioning data from Table 4.1. K_{oc} data from Wauchope and others (1992) or the original studies, with the exception of the value for EDB (Pignatello and others, 1990).

into nonaqueous phases—as well as negligible reactivity in situ—migrate at the same rate as water molecules, and are said to be conservative. Nonconservative solutes exhibit a significant affinity for solid or gas phases, and undergo retardation in relation to the movement of the water. Solutes commonly used as conservative tracers, typically during process and matrix-distribution studies, include bromide, chloride, tritium (^3H), and sulfur hexafluoride (SF_6).

The extent to which interactions with solid-phase organic matter retard the movement of a pesticide through the subsurface is directly related to the K_{oc} of the solute and the f_{oc} of the porous medium. Experience has shown that the retardation of organic solutes by sorption to natural organic carbon tends to dominate retardation by sorption to mineral surfaces for $f_{oc} > 0.001$, i.e., when the mass fraction of organic matter in the soil exceeds 0.1 percent (McCarty and others, 1981; Schwarzenbach and Westall, 1981).

4.3.3 PREFERENTIAL TRANSPORT

The subsurface migration of solutes at rates exceeding those predicted by the most commonly used transport models is often ascribed to movement along "preferred" flow paths, a phenomenon known as preferential transport. All solute-transport models in widespread use, however (see Chapter 9, as well as overviews by Matthies, 1987, and Pennell and others, 1990), are based on the traditional form of the advection-dispersion (or convection-dispersion) equation, which neglects this phenomenon. This equation assumes that the movement of fluid and solutes in the subsurface is governed by uniform transport, i.e., migration within a single, uniform flow regime (Figure 4.9a).

Preferential transport involves the movement of water and solutes through two different flow regimes within a porous medium (Figure 4.9b): (1) "dynamic regions" characterized by

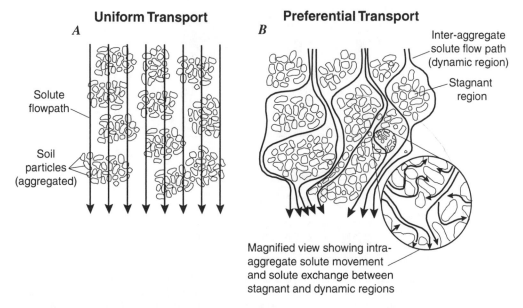

Figure 4.9. Contrasting views of solute movement through an aggregated porous medium: (*A*) Uniform transport, versus (*B*) Preferential transport.

rapid, advection-dominated transport along preferred flow paths, and (2) "stagnant regions" between preferred flow paths, in which solute migration is largely controlled by molecular diffusion. The aqueous phases in these two regions are often referred to as "mobile" and "immobile waters," respectively (Figure 4.10). Within a fractured clay, for example, mobile waters flow through the fractures, while immobile waters seep through the intact clay "matrix." Solute exchange between the two regions is largely controlled by molecular diffusion (van Genuchten and Wierenga, 1976; Rao and Jessup, 1983; Sudicky and others, 1985; Goetz and Roberts, 1986). Although this two-domain model has been shown by many studies to provide accurate predictions of solute movement through porous media (as discussed later in this section), dynamic and stagnant regions actually represent the end members at either extreme of a continuum of flow conditions encountered in situ.

The more rapid movement of solutes through a porous medium during preferential transport—relative to uniform transport—occurs for two principal reasons: (1) migration occurs along pathways with relatively high hydraulic conductivity, and (2) much of the sorptive capacity of the medium is bypassed. The macroscopic movement of solutes through a porous medium is accurately described by the uniform-transport model if preferred flow paths are absent. If the medium is traversed by one or more conduits of comparatively high hydraulic conductivity, however, the slow migration of solute through the porous matrix is short-circuited by the more rapid transport through these channels. As a result, some fraction of the initial solute pulse moves through the medium considerably faster than the rate predicted from the hydraulic conductivity of the matrix.

By postulating a single flow regime, the uniform-transport model assumes that all of the sorptive capacity of soils and other earth materials is in immediate and continuous contact with all fluids migrating through the subsurface (Figure 4.9a). During preferential transport, however,

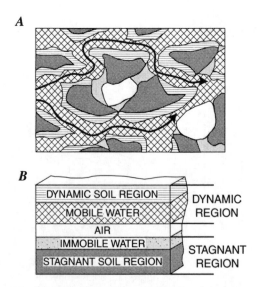

Figure 4.10. Schematic diagram depicting regions of "mobile" and "immobile" water within an unsaturated, aggregated porous medium. (*A*) Actual model, (*B*) Simplified model. The shading patterns in A and B represent the same regions. Modified with permission from van Genuchten and Wierenga (1976).

much of this capacity is bypassed by solutes moving along preferred flow paths (Figure 4.9b); in comparison with the uniform-transport model, this reduces the extent to which solute migration can be retarded by interactions with soil surfaces.

Laboratory experiments have indicated that the effects of changes in water flow velocity on preferential transport are markedly different under unsaturated and saturated conditions. Kung (1993) examined the "funneling" of flow along preferred pathways under unsaturated conditions in a laboratory sand tank containing a fine sand with an inclined layer of coarse sand imbedded at mid-depth. The extent of funneling was found to diminish when the water flux was increased above a certain critical value—approximately 2 percent of the saturated hydraulic conductivity of the media examined. This threshold was deemed to represent the flow rate above which capillary forces no longer prevented the entry of water from the smaller pores of the finer sand into larger, air-filled pores in the underlying coarse layer (Kung, 1990b). These results suggest that in the unsaturated zone, preferential transport may control the migration of pesticides and other surface-derived solutes at low flow rates, but becomes less pronounced when infiltration rates are high enough to overwhelm the ability of the porous medium to channel flow along preferred flow paths.

Under saturated conditions, however, laboratory studies suggest that preferential transport becomes more pronounced at higher flow velocities. In a fully saturated porous medium, flow is not diverted by air-water interfaces, and preferential transport is governed primarily by the exchange of solutes between zones of mobile and immobile water (Figure 4.9). Laboratory column studies by van Genuchten and others (1974) and Schwarzenbach and Westall (1981) indicate that departures from uniform transport become significant for velocities at or above 10^{-3} cm/s. At lower flow rates, more time is available for solute exchange between mobile and immobile regions. As flow rates increase, solutes have less time for such exchange, and departures from uniform transport become more evident.

Field Evidence for Preferential Transport

Over the past two decades, results from numerous investigations have challenged many of the assumptions behind the uniform-transport model, prompting Jury and others (1986a) to state, "a growing body of evidence suggests that the consensus laboratory scale model for solute movement, the convection-dispersion equation, may not correctly describe field-scale mobile chemical movement near the soil surface." Several lines of evidence indicate that preferential transport may affect the subsurface transport of solutes and even very small particulate bodies (e.g., colloids, bacteria, and viruses) within a broad range of hydrogeologic settings.

Many process and matrix-distribution studies have found that solutes entering the subsurface can migrate downward through the soil at rates that are substantially faster than those predicted on the basis of uniform transport (e.g., Willis and Hamilton, 1973; Rao and others, 1974; Jury and others, 1986b; Kelley and Wnuk, 1986; White and others, 1986; Richard and Steenhuis, 1988; Priebe and Blackmer, 1989; Blair and others, 1990; Clendening and others, 1990; Isensee and others, 1990; Sauer and others, 1990; Gish and others, 1991a; Goodman, 1991; Kladivko and others, 1991; Nash and others, 1991; Rice and others, 1991; Shaffer and Penner, 1991; Komor and Emerson, 1994; Czapar and others, 1994; and Traub-Eberhard and others, 1994). Movement along preferred flow paths has also been invoked to explain why viruses have been observed to travel substantially faster through the subsurface than nonsorbing solutes (e.g., McKay and others, 1993; Rossi and others, 1994).

One of the strongest lines of evidence for the widespread occurrence of preferential transport in subsurface environments has been the detection of hydrophobic pesticides in ground waters. Compounds with high K_{oc} values are typically assumed to be immobilized in soils by sorption to solid-phase organic matter. However, the detection in ground waters of strongly hydrophobic compounds such as toxaphene (LaFleur and others, 1973; Parsons and Witt, 1989; Maes and others, 1991), dieldrin, endrin, DDT, DDE, aldrin, endosulfan, lindane, heptachlor, mirex (Willis and Hamilton, 1973; Achari and others, 1975; Richard and others, 1975; Myott, 1980; Katz and Mallard, 1981; Weaver and others, 1987; Parsons and Witt, 1988; McKenna and others, 1988; Louis and Vowinkel, 1989; Exner and Spalding, 1990; Maes and others, 1991), trifluralin (Hallberg, 1989; Exner and Spalding, 1990), and chlordane (Steichen and others, 1988; Cohen and others, 1990; Puri and others, 1990) indicates that this assumption is not always valid.

Transport along preferred flow paths may also explain why pesticides with widely varying K_{oc} values have sometimes been observed to migrate at essentially identical rates through the subsurface (e.g., Bottcher and others, 1981; Everts and others, 1989; Kladivko and others, 1991; and Traub-Eberhard and others, 1994). As noted by Maes and others (1991), "in the presence of cracks, any pesticidal active ingredient, regardless of physicochemical characteristics could move to ground water."

Additional evidence for preferential transport comes from several studies suggesting that the partitioning of pesticides among solid and aqueous phases often does not reach equilibrium during the migration of these compounds through the subsurface. The advection-dispersion equation assumes that the mass transfer of solutes among aqueous, solid, and gas phases is instantaneous and, thus, all solutes are always at local equilibrium with respect to partitioning among these phases. However, Hall and Hartwig (1978) detected atrazine in lysimeter leachate approximately 0.5 m below the deepest point at which soil residues were detected in a silty clay-loam in Pennsylvania. Analogous behavior has been observed for picloram in aggregated Hawaiian soils (Rao and others, 1974), atrazine, deethyl atrazine (DEA), and metolachlor in clay-loam soils in Ontario, Canada (Frank and others, 1991a,b), ethoprop in sandy Florida soils (Norris and others, 1991), metribuzin in a loamy sand in Minnesota (Burgard and others, 1994), mecoprop and isoproturon in a clay soil in England (Harris and others, 1994), and atrazine, terbuthylazine, and triasulfuron in a loamy soil in Switzerland (Flury and others, 1995). These results are consistent with laboratory studies showing that organic solutes may take tens to hundreds of days to attain partitioning equilibrium in soil-water systems (e.g., Ball and Roberts, 1991a). They also demonstrate the potential hazards associated with the assumption (e.g., Shaffer and Penner, 1991) that pesticides will not leach below the depth at which their residues are no longer detectable in soil.

Field Evidence for Preferential Flow Paths in the Subsurface

Bouwer (1990) has noted that "for many years, preferential flow was thought to be associated with large pores, like cracks, root holes, wormholes, and macropores in structured clay soils." The formation of fractures is particularly favored in clayey or other fine-grained soils that are subject to desiccation or freeze-thaw cycles (e.g., Graham and others, 1992). Channels formed by plant roots, earthworms, and rodents are common in undisturbed or minimally tilled soils. A detailed set of dye tracer studies on sandy clay-loam soils in New York by Steenhuis and others (1990) indicated that earthworm burrows, if present, may constitute the most extensive and temporally stable conduits for the transport of solutes through the vadose zone.

Preferential transport can also occur in settings where cracks or other macropores are not visible. From dye tracer studies, preferential transport has been observed in irrigated agricultural soils exhibiting little or no evidence of fracturing (Bouwer, 1990; Kung, 1990a; Steenhuis and others, 1990; Czapar and others, 1994; Flury and others, 1994). Evidence of preferential solute transport through sandy glacial outwash deposits has been reported for pesticides in Cape Cod, Massachusetts (Cohen and others, 1990) and Rhodamine WT dye in the Central Sands area of

Wisconsin (Kung, 1990a), although visible cracks do not typically develop in such coarse-grained soils. Indeed, the frequency with which this phenomenon is observed prompted Flury and others (1994) to state that "the occurrence of preferential flow is the rule, rather than the exception."

Results from the dye tracer study in the Central Sands area of Wisconsin (Kung, 1990a) suggest that the volume of the subsurface occupied by preferential flow paths in the unsaturated zone diminishes with depth. Eighty days after the initial dye application, excavation of the site revealed that the proportion of the whole soil volume stained by the dye within three depth intervals (1.5-2.0 m, 3.0-3.5 m, and 5.6-6.6 m) decreased from approximately 50 percent to less than 10 percent, and less than 1 percent, respectively.

Hydrographs for the discharge from springs and tile lines have been used to infer the relative significance of "macropore flow" versus "matrix flow" in governing solute transport through the subsurface (e.g., Hallberg and others, 1986; Libra and others, 1987; Everts and Kanwar, 1988). Transient peaks in flow and solute loads in spring or tile-line effluents are presumed to correspond to transport within macropores (mobile waters), while the more constant, lower-flux component is assumed to be dominated by discharge from the surrounding matrix (immobile waters). This approach is derived from the hydrograph-separation techniques used to distinguish event-derived flow from that contributed by ground-water derived base flow in surface waters.

The exchange of solutes between dynamic and stagnant regions in the subsurface can only occur if the solute molecules are significantly smaller than the internal diameters of the pores themselves. However, pores in a typical soil (Figure 4.11) are sufficiently large—and ground-water contaminant molecules sufficiently small—that this constraint is likely to be significant only for the smallest pores ($<<1$ µm), or where solute diffusion is hindered by soil organic matter (e.g., Ball and Roberts, 1991b). Microscopic analysis of undisturbed soil cores taken from cultivated silt-loam soils in South Dakota indicated that the majority of soil pores with diameters greater than 100 µm were located within the upper meter of the soil profile (Goodman, 1991), an observation that is consistent with dye tracer studies using intact soil monoliths (Wildenschild and others, 1994).

Influence of Sorption

Exchange between mobile and immobile regions retards the migration of all solutes, regardless of whether they interact with solid surfaces, but the retardation of a solute pulse is enhanced if the molecule has significant interaction with solid surfaces (e.g., Starr and others, 1985). Whether these interactions involve the sorption of hydrophobic molecules to solid-phase organic matter, ion exchange at mineral surfaces, or both, the net result is that the residence time on the internal or external surfaces of an individual porous particle or aggregate will be extended for an interacting solute, relative to that for a noninteracting one. The internal surfaces of porous particles or aggregates typically exhibit larger total areas and are less accessible than the external surfaces. Consequently, solute interactions with the walls of internal pores generally result in a substantially greater degree of retardation than is caused by interactions with only the external surfaces of subsurface particles or aggregates.

In addition to anomalously high rates of transport for pesticides and other dissolved species, several other departures from the predictions of the uniform-transport model may also be explained by solute exchange between stagnant and dynamic regions in the subsurface. These include: (1) the continued detection of persistent, hydrophilic pesticides in subsurface waters, long after the initial "pulse" of contaminant has moved downgradient; (2) decreases in the amount of solute that can be desorbed from aquifer materials as the time allowed for uptake is

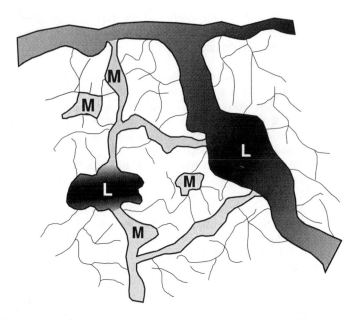

Figure 4.11. Sizes and configurations of pores in a typical soil. L, pores with diameters larger than 100 micrometers; M, pores with diameters of 10-100 micrometers; Single lines, pores with diameters of less than 10 micrometers. Modified and redrawn from Riley (1976) and published with permission.

increased; (3) the apparent existence of two different time scales over which pesticides and other solutes are taken up by aquifer solids; and (4) the observation of multiple peaks in pesticide concentration profiles in the subsurface, even following a single application.

Tailing of Solute Distributions in Porous Media

Several investigators have reported the continued detection of hydrophilic pesticides in subsurface waters long after the original pulse of applied compound has migrated downgradient, away from the point of measurement. An analogous effect is observed during laboratory column studies, in the form of a pronounced "tailing" in solute breakthrough curves, particularly at higher flow rates (e.g., van Genuchten and others, 1977; Hutzler and others, 1986). Among the more hydrophilic pesticides for which such behavior has been observed in the field are DBCP (e.g., Cohen, 1986; Schmidt, 1986a), atrazine, metolachlor (Pignatello and Huang, 1991), dicamba (Hall and Mumma, 1994), and simazine (Poletika and others, 1995).

Figure 4.12 shows the hypothetical distributions of a persistent, hydrophilic pesticide in the subsurface at different times following its application to a field. Spatial variations in hydraulic conductivity and other soil properties give rise to highly irregular distributions of the compound in the subsurface. Considerable small-scale variability in pesticide concentrations has been observed, for example, beneath fields treated with aldicarb in Wisconsin (Harkin and others, 1986). In addition, the distribution of horizontally averaged concentration with depth shows the pronounced tailing often seen upgradient from concentration maxima in contaminated subsurface environments. While Figure 4.12 focuses on solute distributions in the vadose zone, Figure 4.13 illustrates the tailing that is often observed for a contaminant pulse as it is transported downgradient in the saturated zone.

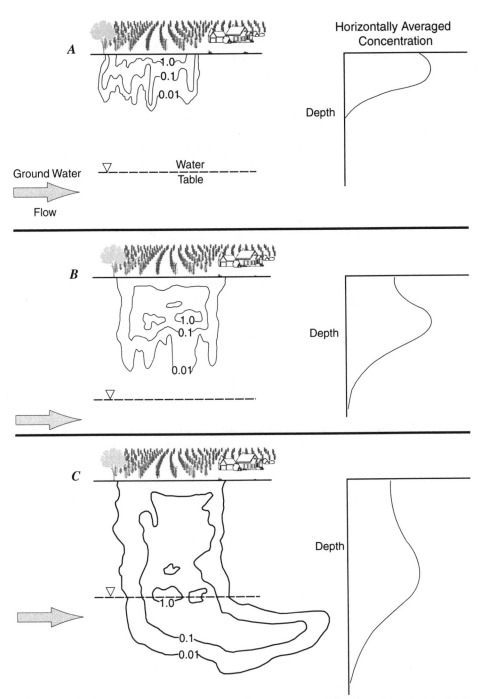

Figure 4.12. Idealized distribution of a hypothetical, persistent pesticide in the subsurface (*A*) 1, (*B*) 2, and (*C*) 3 years after application to a field. Contour values represent the ratio between the observed concentration of the pesticide and its initial concentration in the aqueous phase. Note "tailing" of concentrations at later times.

The role of solute exchange between regions of mobile and immobile water in giving rise to these tailing phenomena is illustrated by the movement of a solute pulse through a saturated porous medium traversed by macropores (Figure 4.14), such as an aggregated soil or a fractured clay. During their movement through a macropore, many of the solute molecules that are close to the walls diffuse into the surrounding matrix in response to concentration gradients (Figure 4.14a). Tailing of the solute concentration within the macropore occurs from the slow diffusion of the solute from the matrix back into the macropore (in response to comparatively small concentration gradients) after the peak concentration within the macropore has passed (Figure 4.14b). Solute molecules that diffuse into, and then out from the matrix thus take longer to traverse a given length of the macropore than those that remain within the macropore (Sudicky and others, 1985).

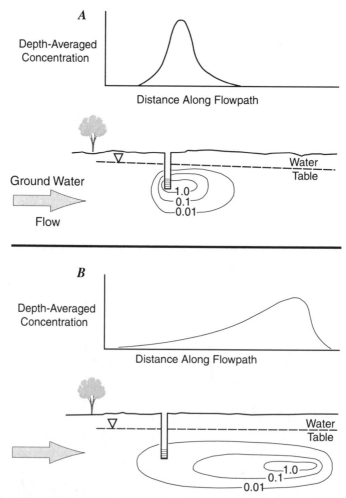

Figure 4.13. Idealized distribution of a hypothetical, persistent contaminant in the saturated zone (*A*) 1, and (*B*) 10 years after introduction through a well. Contour values represent the ratio between the observed concentration of the contaminant and its initial concentration in the aqueous phase. Note "tailing" of concentrations upgradient from center of mass of plume at later time.

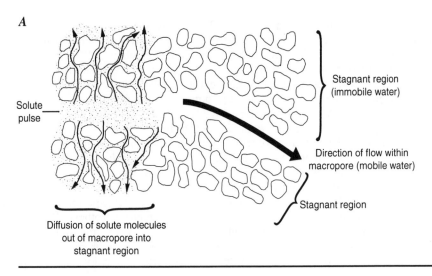

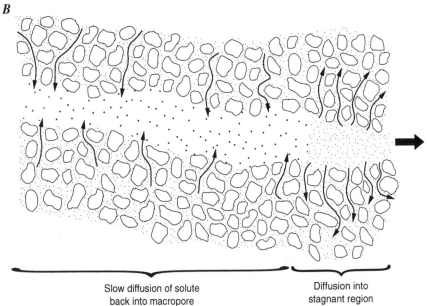

Figure 4.14. Diffusive exchange of solute molecules between mobile and immobile waters (*A*) before, and (*B*) after passage of a solute pulse through a macropore in a porous medium (e.g., a fractured clay).

With respect to solute transport, porous grains or aggregates of soil and other earth materials may be viewed as microcosms of the medium depicted in Figure 4.14. Voids between grains (inter-particle porosity) are regions of relatively mobile fluid, by comparison with the regions within grains (intra-particle porosity). Similarly, in an aggregated soil (e.g., Figure 4.9), mobile waters traverse the inter-aggregate spaces, while immobile waters reside within the aggregates themselves. Diffusion, and limited advective flow into and out of the pores within the particles or aggregates may thus retard solute transport through the more mobile regions between these entities. The effects of these internal pores on the velocity, broadening, and tailing of solute pulses can be significant, even in earth materials with relatively low intra-particle or intra-aggregate porosities (e.g., Crittenden and others, 1986; Ball and Roberts, 1991a,b).

Several investigations have demonstrated that the tailing and retardation of solute pulses observed in the field—as well as in breakthrough curves for laboratory columns—can best be explained by transport models that subdivide the porous medium of interest into regions of mobile and immobile water. Models of this type have been shown to provide better descriptions of transport behavior than uniform-transport models for the migration of chloride, iodide, carbon tetrachloride, trichloroethylene, tetrachloroethylene, and hexachloroethane in ground water during large-scale field experiments in a relatively homogeneous, glacial-outwash aquifer in Ontario, Canada (Goltz and Roberts, 1986; Thorbjarnarson and Mackay, 1994), and simazine and bromide in lysimeters installed in a loamy sand in Riverside, California (Poletika and others, 1995). Similar conclusions have been reached during laboratory model experiments for a variety of pesticides and other solutes, including 2,4,5-T (van Genuchten and others, 1977), atrazine (Gamerdinger and others, 1991; Steenhuis and others, 1994; Yiacoumi and Tien, 1994), simazine, cyanazine (Gamerdinger and others, 1991), carbofuran, ethoprophos (Dowling and others, 1994), fluometuron (Gaston and Locke, 1995), alachlor, dichlorophenol, quinoline (Brusseau and Rao, 1991), trichloroethylene, bromoform (Hutzler and others, 1986), tritium, calcium (Roberts and others, 1987), strontium (Starr and others, 1985), bromide (Hall and Webster, 1993), and chloride (Sudicky and others, 1985; Hutzler and others, 1986; Roberts and others, 1987; Hall and Webster, 1993). Intra-particle diffusion has also been invoked to account for the relatively slow uptake from aqueous solution that has been observed for tetrachloroethylene, 1,2,4,5-tetrachlorobenzene (Ball and Roberts, 1991b), several polychlorobenzenes (Yiacoumi and Tien, 1994), hexachloroethane (Curtis, 1991), and a variety of other organic molecules (Brusseau and Rao, 1989).

Decreases in Recovery of Sorbed Compounds Over Time

Evidence from a large number of field studies has indicated that the longer a pesticide (or any other solute) remains in contact with a soil, the smaller the amount of the compound that can be leached back into aqueous solution over a given period of time. To account for this observation, Maes and others (1991) hypothesized that, as the time allowed for uptake by the soil is increased, the compound becomes "more resistant...to leaching because the pesticide becomes more tightly bound to soil over time." Although this is a common assertion, its physicochemical basis is unclear.

Rather than becoming "more tightly bound to soil over time," the greater likelihood is that less of the applied compound becomes accessible to leaching over time—particularly over the relatively short time intervals that characterize most leaching experiments or transient leaching events—because of diffusion of the solute into zones of less mobile water or into soil organic matter. Thus, what appears to be the formation of "bound," "unextractable," or "resistant" residues (e.g., Bouchard and others, 1985; Uchrin and Katz, 1985; Nicholls, 1988; Freitag and Scheunert, 1990; Goodman, 1991; Loch, 1991; Obreza and Ontermaa, 1991; Winkelmann and Klaine, 1991; and Demon and others, 1994) may actually be a consequence of diffusion control over the movement of pesticide molecules out of the soil matrix into the mobile aqueous phase. A similar conclusion has been reached by other authors (e.g., Sawhney and others, 1988; Brusseau and Rao, 1989; and Pignatello, 1991). Furthermore, in the absence of data on transformation product concentrations (data that are seldom reported by these studies), the extent to which these observations are simply the result of the transformation of the parent compounds in situ cannot be assessed.

Multiple Apparent Time Scales for Sorption

The involvement of inter- and intra-particle diffusion may help to explain why the sorption of even moderately hydrophobic organics onto soil or aquifer solids often occurs over two different time scales; a rapid equilibration taking place within a few days and a much slower equilibration observed over a time scale of months to years (e.g., Newland and others, 1969; LaFleur, 1980; Barrett and Lavy, 1984; Hill and Schaalje, 1985; Karickhoff and Morris, 1985; Brusseau and Rao, 1989; Harmon and others, 1989; Ball and Roberts, 1991a; Gish and others, 1991a; Pavlostathis and Nathavan, 1992; Pignatello and others, 1993; Cooper and Zheng, 1994; Novak and others, 1994; and Gaston and Locke, 1995). The most common explanation for this phenomenon is that solute uptake arising from sorption to the external surfaces of soil particles and aggregates is comparatively rapid, whereas solute molecules require more time to diffuse through the narrow, tortuous channels leading to internal sorption sites within these entities. Experimental evidence suggests that the rate of the slow, diffusion-limited equilibration step decreases with increasing K_{oc} among different compounds, and with increasing f_{oc} among different soils (LaFleur, 1979; Brusseau and Rao, 1989; Poletika and others, 1995).

Multiple Peaks in Subsurface Concentrations

Preferential transport may also be responsible for the observation of double or multiple peaks in laboratory column breakthrough curves or in concentration profiles in soils and vadose-zone waters (e.g., Jury and others, 1986b; Czapar and others, 1994; Komor and Emerson, 1994; Wildenschild and others, 1994; and Poletika and others, 1995). According to Wildenschild and others (1994), "multiple peaks are interpreted to be a result of differences in pore size, tortuosity, and connectivity of the macropore system as well as redistribution of entrapped air" within the soil. Similarly, the early solute peak often observed during laboratory column experiments is usually ascribed to rapid transport along the internal walls of the column, a phenomenon known as the "wall effect" (Nose, 1984).

Other Contributing Effects

In addition to preferential transport, other mechanisms have been proposed to explain seemingly anomalous rates of solute transport through soils. Association of a relatively hydrophobic solute with high concentrations of dissolved organic matter in solution (discussed earlier in Section 4.3.1) can lead to a decrease in its apparent soil-water partition coefficient (K_p), and result in a migration rate more rapid than that predicted from its K_{oc} value and the soil f_{oc}. Alternatively, if a sorption isotherm is nonlinear, the K_p for the solute will be a function of its aqueous concentration. This situation may result in deviations from expected solute behavior with respect to both migration rate and the symmetry of the breakthrough curve. Pronounced hysteresis in sorption may also be responsible for deviations from expected solute transport behavior. A more extensive discussion of the influences of nonlinear sorption isotherms and sorption hysteresis on solute transport has been provided by Jury and Ghodrati (1989).

Observations of preferential transport in the subsurface may thus be accounted for through the simultaneous influences of one or more different physical or chemical phenomena. These include: (1) mobile-immobile water exchange; (2) mass-transfer controls on sorption arising from intra-particle or intra-aggregate diffusion; (3) increases in apparent aqueous solubility caused by the presence of colloids or dissolved organic matter; and (4) nonlinear or hysteretic sorption.

Significance to Prediction of Solute Behavior

Despite the evidence that preferential transport in the subsurface is a common occurrence, widespread confidence in the validity of the uniform-transport model for predicting solute transport continues to exist. This is demonstrated by statements such as "true leaching of simazine due to normal agricultural use appears to be a rare event, even in vulnerable areas of the country" (Roux and others, 1991b); and "[metolachlor] is generally considered non-leachable" (Turco and Konopka, 1988). This confidence, to some extent, also underpins the practice (examined in detail in Section 6.1) of classifying pesticides into "leachers" and "non-leachers" (or similar categories), based on specific cut-off values of parameters such as soil dissipation half-life, K_{oc}, or water solubility (Cohen and others, 1984; Bishop, 1985; Creeger, 1986; U.S. Environmental Protection Agency, 1986; Matthies, 1987; Pionke and others, 1988; Gustafson, 1989; Cohen, 1990b; Deubert, 1990; Ritter, 1990; Becker and others, 1991; Behl and Eiden, 1991; Domagalski and Dubrovsky, 1991; Tooby and Marsden, 1991; Yen and others, 1994).

Concern over the reliability of uniform-transport models in this regard was voiced by Hallberg (1989), who noted that "using PRZM [a commonly used simulation code based on the uniform-transport approach] to model pesticide leaching conditions in Iowa, suggests that none of the pesticides being detected in ground water, should leach to ground water, even using 'worst-case' situations." These difficulties are explored further in Section 9.1.2, as part of a discussion of the use of solute-transport models to predict the behavior of contaminants in the subsurface.

Although mobile zones typically occupy a relatively minor portion of the total volume of the subsurface in most settings, movement along preferred flow paths may still cause significant contamination of ground waters by pesticides or other surface-derived solutes. For example, although Everts and Kanwar (1988) found preferential flow to account for only 2 percent of the water reaching a tile drainage system beneath an Iowa silt loam planted with corn, preferential transport accounted for the movement of between 10 percent and 25 percent of the bromide and nitrate applied.

Hallberg (1989) has noted that the "preferential movement of solutes becomes an all-important mechanism when we must be concerned with the occurrence of ppm and ppb concentrations of toxic substances in ground water." Similarly, Utermann and others (1990) point out that, for nonconservative environmental contaminants, "it may be more important in some instances to know the time of arrival of the first 0.1 or 1 percent of the chemical at the groundwater, rather than the center of mass." Support for this perspective was provided by Steenhuis and others (1990), who estimated that a compound applied at 2 kilograms per hectare (kg/ha) could appear in ground water at concentrations exceeding 1 µg/L, even if only 0.1 percent of the applied mass reached the water table.

The need to account for the potential influence of preferential transport is underscored by the public's desire for complete protection of all ground water from contamination—regardless of whether such a goal is achievable in practice (Cohen, 1986). In California, the simple detection in ground water of a pesticide that has been registered for agricultural use automatically triggers an investigation by the Department of Pesticide Regulation into the possible reasons for the contamination and, where necessary, an exploration of remedial measures for its "mitigation" (Mackay and Smith, 1990; Maes and others, 1991). Although accounting for preferential transport adds considerable complexity to the prediction of pesticide transport in the subsurface (see Section 9.1), the failure to do so seriously compromises our ability to anticipate future contamination of ground-water supplies.

4.4 TRANSFORMATION

Prior to the 1960's, it was generally assumed that pesticides applied to the land would eventually undergo complete mineralization, which is the breakdown to simple chemical species such as H_2O, CO_2, N_2, NO_3^-, Cl^-, and PO_4^{3-}, primarily through the actions of soil micro-organisms. However, the widespread detection of organochlorine insecticides in polar ecosystems and other regions located far away from application areas (e.g., Carson, 1962; Puri and others, 1990; Majewski and Capel, 1995) has since demonstrated that these compounds are much more resistent to transformation than originally expected. Many of the more recently introduced pesticides are designed to transform to less harmful products relatively soon after reaching their target pests.

Within the present context, the terms transformation products, or degradates, are used to denote the compounds produced by the transformation of pesticides in situ. Somasundaram and Coats (1991) listed 19 different terms that have been used for this purpose. The word parent is used here to denote the original compound from which the degradate of interest was derived.

4.4.1 REACTION PATHWAYS

As might be expected, there are a wide variety of reaction pathways by which pesticides undergo transformation in the subsurface. This section provides only a brief overview of the principal mechanisms by which these reactions take place. For more detail, the reader is referred to more extensive reviews of the subject, such as those by Alexander (1981), Kuhn and Suflita (1989), and Coats (1991).

The transformation of a pesticide or any other synthetic organic compound in natural waters occurs as a result of biochemical, photochemical, or nonphotochemical processes, or some combination of these pathways. Phototransformation reactions will receive only minor attention in this book, as they are likely to be significant only during the period immediately following application, when pesticides temporarily reside on soil and plant surfaces prior to being carried away in surface runoff or into the ground in recharge waters. Comprehensive reviews of the phototransformations of pesticides and other organic compounds in water have been provided by other authors (e.g., Mill and Mabey, 1985; Mill, 1989; and Harris, 1990).

Table 4.2 lists the principal types of nonphotochemical pathways by which pesticides may be transformed in the subsurface. For each reaction, one or more examples are provided, each of which depicts the parent compound, as well as the product(s) of interest. The various mechanisms by which pesticides undergo transformation within the subsurface may be classified according to the manner in which the overall oxidation state of the molecule is altered in the process, if at all. The oxidation of a pesticide involves an increase, while reduction results in a decrease in its overall oxidation state. Neutral reactions leave the oxidation state unchanged. All three types of transformation may occur through either biochemical or abiotic pathways. The oxidation of a pesticide typically leads to an increase in its water solubility, while reduction usually gives rise to a product that is less polar, and hence less water-soluble than the parent compound (Coats, 1991). Neutral reactions may increase or decrease water solubility, depending on the manner in which the chemical structure is modified.

Direct assessments of the significance of transformation reactions in relation to advection, dilution, and sorption in situ have been carried out through the use of transformation product-to-parent concentration ratios, although such investigations are uncommon. Indeed, field investigations of product-to-parent ratios for any organic contaminants in ground water are rare.

Table 4.2. Principal pathways by which pesticides may be transformed in subsurface environments

[Taken from Coats (1991), unless otherwise specified. Blank cells indicate no information applicable or available]

Reaction Type	Example(s)	Reference
	Parent → Principal Product(s)	
Oxidations		
N-dealkylation[1]	Ametryn → Deethyl ametryn	Miles and others, 1990
	Atrazine → Deethyl atrazine (DEA) + Deisopropyl atrazine (DIA)	Adams and Thurman, 1991
	Cyanazine → Deisopropyl cyanazine (=DIA)[2]	See Table 3.4
	Cyprazine → Decyclopropyl cyprazine (=DEA)[2]	See Table 3.4
S-dealkylation[1]	Prothiophos → Despropyl prothiophos	
O-dealkylation[1]	Methoxychlor → Desmethyl methoxychlor	
	2,4-D → 2,4-Dichlorophenol	Kuhn and Suflita, 1989
	2,4,5-T → 2,4,5-Trichlorophenol	Kuhn and Suflita, 1989
Epoxidation	Aldrin → Dieldrin	
	Heptachlor → Heptachlor epoxide	See Table 3.4
Sulfoxidation	Aldicarb → Aldicarb sulfoxide → Aldicarb sulfone	
	Molinate → Molinate sulfoxide	See Table 3.4
	Terbufos→ Terbufos sulfoxide → Terbufos sulfone	
Oxidative desulfuration	Fonofos → Fonofoxon	
Side-chain oxidation	Diazinon → Hydroxy diazinon	
Amine oxidation	Dichloroaniline → Dichloronitrobenzene + Dichlorophenol	
Ring hydroxylation	Carbaryl → 4-Hydroxy carbaryl	
	Carbofuran → 3-Hydroxy carbofuran	See Table 3.4
Reductions		
Reductive dehalogenation	DDT → DDD	Kuhn and Suflita, 1989
	2,4-D → 2-Chlorophenoxyacetate + 4-Chlorophenoxyacetate	Kuhn and Suflita, 1989
	1,2-Dichlorobenzene → Chlorobenzene	Watts and Brown, 1985
	2,4,5-T → 2,4-Dichlorophenoxyacetate + 2,5-Dichlorophenoxyacetate	Kuhn and Suflita, 1989
	EDB → Bromoethane	Watts and Brown, 1985
	Heptachlor → Chlordene	Kuhn and Suflita, 1989
	Leptophos → Desbromoleptophos	
	Pentachlorophenol → 2,3,4,5-Tetrachlorophenol + 2,3,4,6-Tetrachlorophenol + other polychlorophenols	Kuhn and Suflita, 1989
Dihalo-elimination	Tralomethrin → Deltamethrin	
Sulfoxide reduction	Phorate sulfoxide → Phorate	
Nitro group reduction	Parathion → Aminoparathion	
	Pentachloronitrobenzene —> Pentachloroaniline	Kuhn and Suflita, 1989
Neutral Reactions		
Nucleophilic Substitution - By H$_2$O (Hydrolysis)	Aldicarb → Aldicarb oxime	
	Aldicarb sulfoxide → Aldicarb sulfoxide oxime	

Table 4.2. Principal pathways by which pesticides may be transformed in subsurface environments—
Continued

Reaction Type	Example(s) Parent → Principal Product(s)	Reference
	Atrazine → Hydroxyatrazine	Armstrong and others, 1967; Harris, 1967
	Heptachlor → 1-Hydroxychlordene	Kuhn and Suflita, 1989; Puri and others, 1990
	Cyanazine → Hydroxycyanazine	Beynon and others, 1972
	Daminozide (Alar) → UDMH + 1,2-dicarboxyethane	
	DCPA → Monomethyl tetrachloroterephthalate (MTP) → Tetrachloroterephthalic acid (TPA)	Ando, 1992
	1,3-Dichloropropylene → 3-Chloroallyl alcohol	Roberts and Stoydin, 1976; Leistra and others, 1991
	Diflubenzuron → 2,6-diflurobenzoic acid + p-chlorophenylurea	
	EDB → Ethylene glycol	Weintraub and Moye, 1986; Haag and Mill, 1988b
	Malathion → Carboxy malathion	
	Parathion → Diethyl thiophosphate + p-nitrophenol	
	Propazine → Hydroxypropazine	Harris, 1967
	Simazine —> Hydroxysimazine	Harris, 1965; 1967
- By bisulfide ion (HS⁻)	EDB → 1,2-Dithioethane 1,2-Dichloroethane → 1,2-Dithioethane	Barbash and Reinhard, 1989a,b; 1992a
Dehydrohalogenation	DDT → DDE	Kuhn and Suflita, 1989
	Hexachlorocyclohexane (HCH) → Pentachlorocyclohexene → Trichlorobenzene	Miller and Pedit, 1992; Ngabe and others,1993
	EDB → Bromoethylene (vinyl bromide) 1,2-Dichloroethane → Chloroethylene (vinyl chloride)	Barbash and Reinhard, 1992b
Hydration	Cyanazine → Cyanazine amide	Sirons and others, 1973
Rearrangements		
Isomerization	D-fenvalerate ↔ L-fenvalerate	
	γ-HCH (Lindane)→α-HCH + δ-HCH	Newland and others, 1969
Cyclization	Maneb → Ethylene thiourea (ETU)	Neil and others, 1989
	Dieldrin → Photodieldrin	
Migration	Diclofop methyl → 2,5-Dichloro-3-hydroxy diclofop methyl	
Coupled Processes		
Dehydrohalogenation/ Hydrolysis	DBCP → Bromochloropropene + Dibromopropene → Bromoallyl alcohol	Burlinson and others, 1982
Oxidation/Hydrolysis/ Dehydration	Aldicarb → Aldicarb sulfoxide → Aldicarb sulfoxide oxime → Aldicarb sulfoxide nitrile	Bank and Tyrrell, 1984; Moye and Miles, 1988

Table 4.2. Principal pathways by which pesticides may be transformed in subsurface environments— *Continued*

Reaction Type	Example(s)	Reference
	Parent → Principal Product(s)	
Deamination/Oxidation	Metribuzin → Metribuzin DA + Metribuzin DK → Metribuzin DADK[3]	Harper and others, 1990
Ring hydroxylation/ Oxidation	Carbofuran → 7-Hydroxycarbofuran → 7-Ketocarbofuran	

[1]Reactions involving the replacement of an alkyl group attached to an oxygen, nitrogen, or sulfur atom with a hydrogen atom (dealkylation) are referred to as oxidations (Coats, 1991), despite the fact that this does not change the formal oxidation state of the molecule. The fact that the departing alkyl groups are oxidized during the process of their removal from the parent compound (Kaufman and Kearney, 1970) is presumed to be the reason why these reactions are classified as oxidations.

[2]The structure of deisopropyl cyanazine is identical to that of deisopropyl atrazine (DIA), while that of decyclopropyl cyprazine is identical to that of deethyl atrazine (DEA).

[3]Abbreviations refer to deaminated metribuzin (DA), diketometribuzin (DK), and deaminated diketometribuzin (DADK) (Harper and others, 1990).

Most of the reviewed studies that employed this technique for ground water focused on the deethyl atrazine-to-atrazine concentration ratio (DAR). Jayachandran and others (1994) examined the utility of the deisopropyl atrazine-to-atrazine ratio in this regard, but found it to be of substantially less diagnostic value than the DAR. As part of their study of triazine concentrations in midwestern rivers, Thurman and others (1994) used the ratio of deisopropylatrazine (DIA) to DEA (D2R) to deduce that at least 25 percent of the DIA measured in their study area was produced from cyanazine, rather than atrazine transformation. Results reported by Kolpin and others (1996) suggest a similar pattern in ground water.

The DAR has been examined for a variety of purposes, including: (1) monitoring atrazine transformation over time at individual locations in the subsurface (Frank and others, 1979; Isensee and others, 1990; Adams and Thurman, 1991; Jayachandran and others, 1994); (2) accounting for atrazine losses with increasing distance downgradient in the subsurface (e.g., Denver and Sandstrom, 1991); (3) comparing ground-water residence times in different hydrogeologic settings (Burkart and Kolpin, 1993) or different fields (Jayachandran and others, 1994); and (4) distinguishing waters—either below or above the ground surface—that have been in extensive contact with soil microorganisms from those that have not (Adams and Thurman, 1991; Pereira and Hostettler, 1993; Schottler and others, 1994). Many of these results are examined in later chapters.

4.4.2 INFLUENCE OF TRANSFORMATION ON MOBILITY IN SUBSURFACE WATERS

Although the transformation of a pesticide may lead to either an increase, a decrease, or no change in its water solubility, the transformation products detected in subsurface waters— particularly those generated under aerobic conditions—are usually more water-soluble, and hence more mobile than their respective parent compounds. For example, Mills and Thurman (1994) found that one of the products of atrazine transformation, deethyl atrazine (DEA), sorbs less strongly to soils than does its parent compound, atrazine. Two additional observations by these authors are in agreement with this finding. First, the deethyl atrazine-to-atrazine concentration ratio (DAR) was found to be consistently higher in pore waters than in the solid phase of a silt-loam soil treated with the herbicide. Second, although the DAR increased monotonically over time in both phases, the disparity between the DAR values in the solid phase and the pore water also increased monotonically over time (Mills and Thurman, 1994b). Muir and Baker (1976) and Frank and others (1991a) have also found DEA to be more mobile in soils than atrazine.

Increases in subsurface mobility following transformation have been observed for several other pesticides, as well. Examples include cyanazine amide, produced from the transformation of cyanazine (Muir and Baker, 1976) and the sulfoxide and sulfone products of phorate oxidation (Weaver and others, 1988b). Similarly, decreases in soil-water partition coefficients accompany the oxidation of aldicarb to its sulfoxide and sulfone products (Moye and Miles, 1988), as well as the hydrolysis of heptachlor to form 1-hydroxychlordene (Kuhn and Suflita, 1989; Puri and others, 1990).

Increases in water solubility arising from chemical transformations do not always result in greater mobility through the subsurface, however. For example, although the atrazine products that have been detected most frequently in ground waters are DEA and DIA, the principal pathway of atrazine transformation involves hydrolysis to form hydroxyatrazine, as observed both in the laboratory (Armstrong and others, 1967; Skipper and others, 1967; Skipper and Volk, 1972; Mandelbaum and others, 1993; Demon and others, 1994) and in situ (Sophocleous and others, 1990; Loch, 1991). Hydroxyatrazine is more water-soluble than atrazine, but studies involving batch equilibration in laboratory flasks (Armstrong and Chesters, 1968) and solute transport in the field (Schiavon, 1988; Loch, 1991; Demon and others, 1994) have shown that it exhibits a substantially greater tendency to sorb to soils—particularly to the clay fraction—than either atrazine, DEA, or DIA. This difference in sorption behavior has been attributed to the fact that the replacement of the chlorine on the atrazine ring by a hydroxyl group raises the acid-base dissociation constant (pK_a) of the molecule (Armstrong and others, 1967), increasing the proportion of its protonated form, and thus leading to more extensive association with the surfaces of clays and other negatively charged soil components. The transformation of atrazine within the subsurface thus gives rise to products that are more water-soluble, but may be either more (DEA, DIA) or less mobile (hydroxyatrazine) than the parent compound. Studies of the environmental transport and fate of hydroxyatrazine have been limited, however, partly because of difficulties associated with its detection in environmental samples.

4.4.3 INFLUENCE OF TRANSFORMATION ON TOXICITY

The transformation of a pesticide in situ can increase, decrease, or have a negligible effect on its toxicity to nontarget organisms. In general, acute toxicity tends to decrease with increasing water solubility. Indeed, one of the principal strategies employed by living organisms for detoxifying xenobiotic compounds is to increase their water solubility—through hydrolysis or oxidation—to facilitate elimination of these compounds from the body (Coats, 1991).

Many exceptions to this general pattern exist, however, because of specific chemical interactions of pesticides with biological systems. For example, some transformations may generate carcinogenic products from noncarcinogenic parent compounds, as is the case with the formation of ethylene thiourea (ETU) from the ethylene bis-dithiocarbamate (EBDC) fungicides (Nowell and Resek, 1994), or the hydrolysis of daminozide (Alar) to form unsymmetrical dimethylhydrazine (UDMH). For some compounds, known as propesticides, it is the trans-formation product itself—rather than the applied chemical—that serves as the active ingredient whose pesticidal properties are exploited. Examples include the proinsecticide carbosulfan, which hydrolyzes to form carbofuran; tralomethrin, which undergoes dihalo-elimination to deltamethrin (Coats, 1991); and Vel-4207, which hydrolyzes to release dicamba (Hotzman and Mitchell, 1977).

In general, the majority of research on the biological effects of pesticides and their transformation products (as well as other synthetic organic compounds) has focused on acute toxicity, mutagenesis, tumor formation, and carcinogenicity. Comparatively minor attention has been devoted to the investigation of other chronic effects.

4.4.4 REACTION RATES

The rate at which a pesticide undergoes a specific transformation depends on the physical, chemical, and biological circumstances under which the reaction takes place. Most fundamental among these influences is the dependence of reaction rate on temperature; all reactions, whether or not they are biologically mediated, will be accelerated with increasing temperature, according to the Arrhenius equation, viz.,

$$\ln(k) = \ln(A) - \frac{E_a}{R T}$$

where k is the rate constant for the rate-limiting step of the reaction of interest, A is the Arrhenius pre-exponential factor, E_a is the activation energy, R is the universal gas constant, and T is the absolute temperature of the reaction, in Kelvin. This relation does not apply above a temperature at which the decomposition of the reactants or, for biochemical reactions, the inhibition of biological functions occurs. Other attributes of the surrounding environment may also influence the rates and pathways by which a pesticide undergoes transformation in aqueous solution. These include pH, ionic strength, and the concentrations of other reactive species that may be present, such as nucleophiles, oxidants, reductants, catalysts, and surface-active compounds.

Influence of pH

Changes in pH can influence the rates and pathways of pesticide transformation either directly, through changes in reactant concentrations, or indirectly, through the effects of pH on organisms or surfaces involved in the reaction of interest. For compounds whose aqueous concentrations are pH-independent (as is the case for most pesticides), rates of attack by a pH-sensitive reactant exhibit the same pH dependence as does the concentration of the attacking species. This effect has been demonstrated not only for the attack of hydronium, H_3O^+, and hydroxide, OH^- (e.g., Mabey and Mill, 1978), but for reactions with other Brønsted acids and bases as well, including bisulfide, HS^- (Haag and Mill, 1988a), and anthraquinone, a model for quinone-like moieties in organic matter (Curtis, 1991; Curtis and Reinhard, 1994).

Among the chemical reactions influenced by changes in pH, that which has received the most attention is hydrolysis (e.g., Mabey and Mill, 1978; Ellington and others, 1986, 1987; and Jeffers and others, 1989, 1994; Washington, 1995), in which a substituent is replaced by a hydroxyl group (—OH). "Acid-catalyzed" hydrolysis, promoted by attack of hydronium, proceeds more rapidly with increasing acidity below a particular, compound-specific pH, known as the I_{AN} (Figure 4.15). Similarly, "base-catalyzed" hydrolysis, promoted by the attack of hydroxide, is accelerated in aqueous solution when the pH is greater than I_{NB}, which is also compound specific. Within the range $I_{AN} < pH < I_{NB}$ for a given compound, the "neutral" process predominates, the water molecule is the principal attacking species, and the reaction rate is independent of pH. If the neutral process is not significant, I_{AB} is the pH below which acid catalysis predominates, and above which base catalysis predominates (Mabey and Mill, 1978).

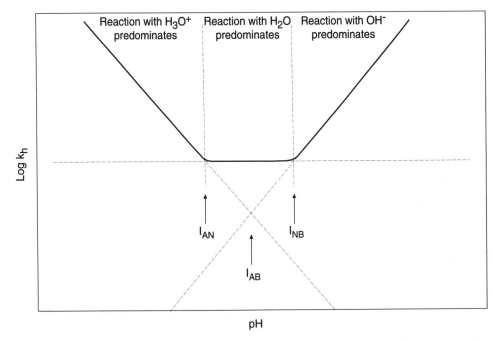

Figure 4.15. Dependence of hydrolysis rate (k_h) on pH for acid-catalyzed, neutral, and base-catalyzed processes (adapted from Mabey and Mill, 1978).

Some pesticides undergo hydrolysis via all three mechanisms, resulting in a rate that increases as the solution becomes either more acidic or more basic, relative to the rate observed at neutral pH. Rates of hydrolysis reactions involving atrazine (Armstrong and others, 1967), diazinon (Konrad and others, 1967), prometryn (Kaufman and Kearney, 1970), methyl isothiocyanate (Geddes and others, 1994), chlorsulfuron, metsulfuron-methyl (Hemmamda and others, 1994), and most epoxides, esters, amides, and carbamates (Mabey and Mill, 1978) exhibit this type of dependence on pH. Figure 4.16 shows this pattern for the disappearance of aldicarb in water at 90°C (Bank and Tyrrell, 1984), a trend that is consistent with the finding that the field persistence of aldicarb increases as pH decreases from 7.5 to 4.5 (Harkin and others, 1986). Data reported by Hansen and Spiegel (1983) indicate the I_{NB} values for the hydrolysis of aldicarb, aldicarb sulfoxide, and aldicarb sulfone are all between 7.0 and 7.5 at typical ground water temperatures (5°-25°C), and that I_{NB} for these reactions generally increases with increasing temperature. For ciodrin (Konrad and Chesters, 1969), malathion (Konrad and others, 1969), and most other organophosphorus pesticides (Mabey and Mill, 1978), hydrolysis occurs primarily through reaction with hydroxide, rather than with water or hydronium ion, whereas the hydrolysis reactions of many of the sulfonylurea herbicides are acid catalyzed (Smith and Aubin, 1993). As a general rule, the hydrolytic displacement of halide ions from haloalkanes—many of which are fumigants—does not appear to be subject to acid catalysis (Mabey and Mill, 1978). For example, the rate of EDB hydrolysis at 63°C is independent of pH below pH 9 (Weintraub and Moye, 1986).

Dehydrohalogenation involves the loss of a proton and a halide ion (HX) from adjacent carbons on the molecule of interest. Under the relatively mild reaction conditions typical of

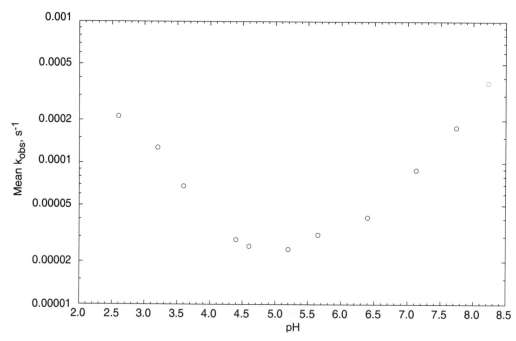

Figure 4.16. Observed first-order rate constant for aldicarb disappearance in water (k_{obs}), versus pH at 90°C (data from Bank and Tyrrell, 1984).

ground waters, HX will only be lost from adjacent carbons connected by a single bond (rather than a double bond), forming the corresponding alkene. The reaction is initiated by the attack of hydroxide, or some other Brønsted base (at the most acidic hydrogen atom on the molecule), so reaction rates increase with increasing pH. The dehydrohalogenation of DBCP in water, for example (Table 4.2), was shown by Burlinson and others (1982) to be subject to base catalysis above an I_{NB} of ~5 at 85°C, but not to acid catalysis (compare Figure 4.17 with Figure 4.15). This has also been observed for the dehydrochlorination of several other halogenated fumigants (Jeffers and others, 1989), as well as the α-, β- (Ngabe and others, 1993) and γ-hexachlorocyclohexane insecticides (Miller and Pedit, 1992). The most well-known example of this reaction for pesticides is the dehydrochlorination of DDT to DDE (Table 4.2).

Most of the available data on the pH-dependence of transformation rates for pesticides and other organic compounds are derived from laboratory, rather than field studies. It is therefore important to recognize that pH buffers, which are typically used to stabilize pH during laboratory studies, may accelerate reactions or shift the relative rates of different pathways of transformation in aqueous solution, as has been observed for the fumigants EDB, 1,2-dichloroethane (Barbash and Reinhard, 1989a; 1992b), and DBCP (Deeley and others, 1991). Other concerns regarding the use of laboratory-derived kinetic data to predict hydrolysis rates in natural waters have been discussed by Mabey and Mill (1978).

Influence of Ionic Strength

Transition-state theory predicts that the rates of reactions involving charged chemical species will be influenced by changes in ionic strength (Lasaga, 1981). However, increases in ionic strength up to 2.0 M have been observed to exert a relatively minor influence over the rates

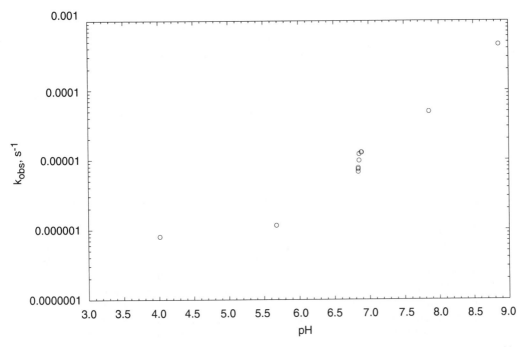

Figure 4.17. Observed first-order rate constant for disappearance of DBCP in water (k_{obs}), versus pH at 85°C (data from Burlinson and others, 1982).

at which a variety of organic compounds undergo hydrolysis (Mabey and Mill, 1978; Barbash and Reinhard, 1992b; Jeffers and others, 1994; and Liquiang and others, 1994). For comparison, the ionic strength of seawater is approximately 0.73 M (Liquiang and others, 1994). Thus, ionic-strength effects on reactivity are likely to be unimportant in most natural waters, particularly those that might be used as a source of drinking water.

Influence of Redox-Active Species in Solution

Despite considerable effort spanning several decades, a reliable system for characterizing the oxidation-reduction (or "redox") status of natural waters has not yet been devised. Electrode potentials, measured in the field (e.g., Mansell and others, 1977; Leistra and others, 1991) or in the laboratory (Smelt and others, 1983; Klecka and Gonsior, 1984; Wolfe and others, 1986), have been the parameters used most widely to date for this purpose, but they are poor predictors of the concentrations of redox-active species in solution (Champ and others, 1979; Lindberg and Runnels, 1984; Thorstenson, 1984; Barcelona and others, 1989; Wolfe and Macalady, 1992; and Grundl, 1994). As a result, more recent attempts to characterize the redox status of natural waters have focused instead on the concentrations of individual redox-active species themselves, such as oxygen, iron (III), iron (II), manganese (II), nitrate, nitrite, nitrous oxide (N_2O), ammonium, sulfate, sulfide, hydrogen (H_2) and methane (Baedecker and Back 1979; Champ and others, 1979; Berner, 1981; Bourg and Richard-Raymond, 1994; Lovley and others, 1994; Bjerg and others, 1995; and Chapelle and others, 1995). Other efforts have involved the measurement of the oxidation-reduction capacities of aquifer solids and subsurface waters (Barcelona and Holm, 1991).

Laboratory investigations have established that the rates of electron-transfer reactions increase with increasing concentrations of any oxidant(s) or reductant(s) participating in the rate-limiting step for the reaction of interest. This has been demonstrated for both nonphotochemical (e.g., Kray and Castro, 1964; Wade and Castro, 1973; Sevcik and Khir, 1980; Klecka and Gonsior, 1984) and indirect photochemical electron-transfer reactions in aqueous solution (e.g., Mill, 1980).

Although the rates and pathways of redox transformations have been studied extensively for many pesticides in laboratory systems (Kuhn and Suflita, 1989; Somasundaram and Coats, 1991), few field observations are available for comparison with laboratory results. The detection of bromoethane in hypoxic ground waters (i.e., containing no detectable dissolved oxygen) suspected to have been contaminated by EDB in Florida (Watts and Brown, 1985; see Table 4.2) suggests that reductive dehalogenation of this fumigant, which has been reported in the laboratory (Castro and Belser, 1968), may also occur in situ. Most of the field observations that can be used to examine relations between redox conditions and pesticide occurrence, however, focus on atrazine.

As has been seen for other s-triazines (Kaufman and Kearney, 1970), alachlor (Pothuluri and others, 1990; Isensee, 1991), dicamba (Hall and Mumma, 1994), and other pesticides, the transformation of atrazine has been found during laboratory investigations to be considerably slower under hypoxic conditions than in oxic environments (Kaufman and Kearney, 1970; Goswami and Green, 1971; and Nair and Schnoor, 1992). Evidence that atrazine shows similar behavior in ground water was provided by Druliner (1989), who observed an inverse correlation between atrazine and dissolved oxygen concentrations (Spearman's $\rho = -0.80$) in the High Plains aquifer of Nebraska (Figure 4.18). Consistent with this, both atrazine and simazine have been

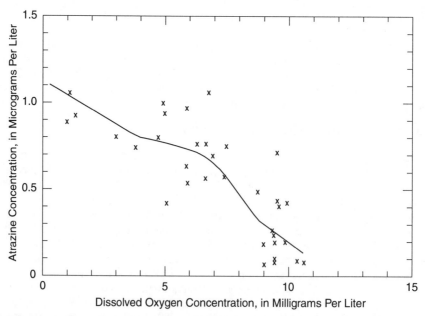

Figure 4.18. Inverse relation between atrazine and dissolved oxygen concentrations observed in the High Plains aquifer of Nebraska (redrawn from Druliner, 1989).

found to be relatively persistent in ground waters beneath sanitary landfills (Spillmann, 1989), where hypoxic conditions are commonly encountered. Similarly, Detroy (1986) found detectable concentrations of atrazine to persist in shallow ground water under denitrifying conditions in the Iowa River alluvial aquifer in Iowa County, Iowa.

Observational evidence on the stability of atrazine under sulfate-reducing conditions, however, is inconclusive. Although the herbicide was detected in 9.6 percent of the wells sampled statewide during the Iowa State-Wide Rural Well-Water Survey (SWRL), it was not detected in any ground waters that smelled of hydrogen sulfide (Hallberg and others, 1992b). In contrast, the two wells with the highest atrazine concentrations measured in an agricultural area of Pennsylvania (Pionke and others, 1988) also contained hydrogen sulfide, suggesting that atrazine is stable under sulfate-reducing conditions. No studies appear to be available, however, that have specifically examined the reactivity of atrazine under sulfate-reducing conditions, either in the laboratory or in situ.

In apparent contrast with the trend displayed for atrazine in Figure 4.18, Kolpin and others (1994) observed a significant, direct relation ($\alpha = 0.05$) between herbicide detection frequencies and dissolved oxygen concentrations in ground waters of the midcontinent. However, the authors hypothesized that this correlation arose from simultaneous losses of herbicides and dissolved oxygen from solution as recharge waters moved downgradient, rather than from any direct, causal relation between oxygen and herbicide concentrations. Further work—incorporating analyses of transformation products—is clearly needed to determine the extent to which the concentrations of triazines and other herbicides are influenced by redox conditions in ground water.

Influence of Other Reactive Species in Solution

Reduced sulfur anions are some of the most powerful chemical reactants commonly encountered in natural waters. As nucleophiles, they are capable of displacing halide from haloalkanes at significant rates under environmentally relevant conditions. Although this is particularly true for brominated species, the nucleophilic displacement of chloride may also take place, albeit at substantially lower rates (Barbash and Reinhard, 1989a,b; 1992a; Roberts and others, 1992). Field evidence that such reactions may affect pesticides in the subsurface was provided by the detection of a variety of reduced-sulfur products in sulfide-rich ground waters suspected to have been contaminated by EDB in Florida (Watts and Brown, 1985).

The bisulfide ion is also a powerful reductant, and has been observed to participate in the reduction of polyhaloalkanes in aqueous solution (Kriegman-King and Reinhard, 1992). As with the nucleophilic displacement of halide, such reactions could be responsible for the transformation of halogenated fumigants and other pesticides in situ.

Several studies have demonstrated that cupric ion (Cu^{2+}) accelerates the ester hydrolysis of several organophosphorus pesticides in aqueous solution—apparently by forming a complex with the compound and, through electron withdrawal, rendering the ester more susceptible to nucleophilic attack (Smolen and Stone, 1994). The pesticides for which this effect has been documented include diazinon (Mortland and Raman, 1967; Smolen and Stone, 1994), dursban, ronnel, zytron (Mortland and Raman, 1967), chlorpyrifos, chlorpyrifos-methyl (Meikle and Youngson, 1978; Blanchet and St.-George, 1982), methyl parathion (Lee and Macalady, 1990), azinphos-methyl, phosalone, and zinophos (Smolen and Stone, 1994). No such effect, however, was observed for methidathion or phosmet (Smolen and Stone, 1994). These findings are particularly significant because copper has been detected in ground waters, possibly as a result of being used as a pesticide in some areas (Table 3.2).

Influence of Surface-Active Substances

Surface-active substances ("surfactants") may also influence the rates and pathways of transformation of pesticides in subsurface aqueous systems. The significance of this to pesticide behavior and fate in ground water arises from the fact that surfactants are commonly employed as "adjuvants" in pesticide formulations (see Sections 5.3.3 and 9.2.4). Furthermore, natural organic matter (NOM) exhibits surface-active properties because it typically possesses both hydrophobic and hydrophilic regions within its overall structure (e.g., Thurman, 1985). As a result, the sequestration of pesticides within more hydrophobic regions of NOM may influence the accessibility of these molecules to other attacking chemical species.

Because most of the functional groups within NOM are negatively charged, the association of a hydrophobic pesticide with the more hydrophobic regions of NOM may reduce the rate of attack of the molecule by anionic reactants, but accelerate reactions with cationic species (Barbash, 1987). This may explain why the rates of base-catalyzed hydrolysis of chlorpyrifos (Macalady and Wolfe, 1985), the octyl ester of 2,4-D (Perdue and Wolfe, 1982), other organic esters (Liquiang and others, 1994), and other organic solutes (Perdue and Wolfe, 1982) are slowed in the presence of humic substances, while the rates of the corresponding neutral processes (i.e., attack by H_2O) remain unaffected. Similarly, the rate of lindane dehydrochlorination is significantly reduced in the presence of solid-phase NOM (Miller and Pedit, 1992). The fact that the rates of dehydrochlorination of DDT (Otero and Rodenas, 1986) and DDD (Nome and others, 1982; Rezende and others, 1983) are accelerated by cationic surfactants suggests that these reactions, too, would be slowed in the presence of NOM, particularly given the relatively hydrophobic nature of these two organochlorine compounds.

Consistent with this model, the acid-catalyzed hydrolysis of atrazine—a reaction that, as noted earlier, is facilitated by the attack of hydronium ion—is accelerated in the presence of increasing concentrations of fulvic acid (Khan, 1978) and other dissolved humic substances (Li and Felbeck, 1972). Similar results have been reported for propazine as well (Nearpass, 1972). For these reactions, the observed acceleration may arise from the attraction of the hydronium ion to the anionic functional groups of the dissolved humic or fulvic acids with which the triazine molecule is associated in solution. A review of the various ways in which surface-active materials—either natural or anthropogenic—may influence the rates and pathways of transformation of pesticides and other organic compounds in aqueous solution has been provided by Barbash (1987). Macalady and Wolfe (1987) reviewed the effects of aquatic humic substances on hydrolysis rates.

4.4.5 BIOCHEMICAL TRANSFORMATIONS

Most of the reactions listed in Table 4.2 are mediated by microorganisms. In addition to the abiotic factors discussed above, the rate at which a given pesticide will undergo biotransformation in the subsurface also depends on the availability of one or more essential nutrients or electron acceptors, soil moisture (e.g., Felsot and others, 1982; Nash, 1988; Nicholls, 1988; Goetz and others, 1990; Puri and others, 1990; and Rowden and others, 1993), soil organic matter (Morrill and others, 1982, cited by Troiano and others, 1990), the size, health, and species composition of the microbial population present, and the chemical structure and concentration of the pesticide itself (Alexander, 1981; Smith and Aubin, 1994). During biotransformation, the compound of interest may be consumed as a primary food source (primary substrate) or as an inadvertent consequence of the consumption of some other primary substrate (secondary utilization, or co-metabolism). The dependence of microbial transformation rates on substrate concentrations and the size of the microbial population has been described quantitatively through the use of Monod kinetics (e.g., Rittmann and others, 1980). Unlike most abiotic reactions, biotransformations often exhibit a "lag period," or "acclimation period," before substrate concentrations begin to decline (Aelion and others, 1987, 1989). When it occurs, this delay usually represents the time interval required for either the population of the appropriate microorganisms or the concentrations of the necessary enzymes to reach high enough levels to effect measurable reductions in substrate concentrations.

In general, biotransformation rates decrease with increasing depth in the soil (e.g., Rao and others, 1986; Pothuluri and others, 1990; McMahon and others, 1992; Johnson and Lavy, 1994; and Yen and others, 1994), in part because both microorganisms (e.g., Wilson and others, 1983) and organic matter (e.g., Loague and others, 1994) are much less abundant with increasing depth in the soil and below the water table. The resulting decrease in microbial activity below the water table may explain why, for example, atrazine, mecoprop (Wehtje and others, 1983; Agertved and others, 1992), aldicarb (Jones and others, 1988), and methomyl (Jones and others, 1989) have all been found to undergo transformation at considerably faster rates in surficial soils than in the unsaturated or saturated zones.

4.4.6 INFLUENCE OF THE SOLID-WATER INTERFACE ON TRANSFORMATIONS

The subsurface environment is dominated by solid-water interfaces. Consequently, any manner in which the solid-water interface affects reactivity will have a significant impact on the rates and pathways of reaction in subsurface waters. Previous research has indicated that the nature and extent of these influences vary considerably among different reactions and settings.

The effects of subsurface solids on hydrolysis reactions provide a valuable illustration. The presence of aquifer materials exerts negligible effects on the rates of hydrolysis of parathion (Graetz and others, 1970), chlorpyrifos, diazinon, Ronnel (Macalady and Wolfe, 1985), EDB (Haag and Mill, 1988b), 1,1,1-trichloroethane (Haag and Mill, 1988b; Jeffers and others, 1994), and DBCP (Deeley and others, 1991), relative to the rates measured in aqueous solution alone at neutral pH. By contrast, the introduction of soil leads to pronounced increases in the rates at which atrazine (Armstrong and others, 1967; Harris, 1967), diazinon (Konrad and others, 1967), ciodrin (Konrad and Chesters, 1969), malathion (Konrad and others, 1969), and methyl isothiocyanate (Geddes and others, 1994) hydrolyze in aqueous solution. For atrazine, the reaction is catalyzed by protonation of the molecule on clay surfaces (Armstrong and Chesters, 1968; Russell and others, 1968). This is consistent with the observation by Roeth and others (1969) that atrazine disappearance rates increase in the presence of soils of higher clay content.

The solid-water interface may also modify the rates of electron-transfer reactions in aqueous solution. Rates of tetrachloromethane reduction in the presence of bisulfide anion in aqueous solution increase substantially upon the addition of either biotite, vermiculite, or amorphous silica (Kriegman-King and Reinhard, 1992). Furthermore, as noted earlier, Wolfe and Macalady (1992) observed that the rates at which halogenated organic compounds undergo reduction in natural soils and aquifer sediments are correlated with the organic carbon content of the solid phase.

The influence of sorption on the rates of biotransformation of organic compounds continues to be a source of some debate (Alvarez-Cohen and others, 1993). The majority of research on this subject, however, suggests that sorption to earth materials reduces the rate of biotransformation of organic compounds in the subsurface. This observation has been accounted for by the hypothesis that only compounds in aqueous solution can be utilized by microorganisms, and hence, that sorbed compounds are unavailable for biotransformation (e.g.,

Ogram and others, 1985; Loch, 1991; and Miller and Alexander, 1991). Indeed, the close agreement between experimental results and simulation models that presume sorbed compounds to be unavailable for biotransformation (e.g., Miller and Alexander, 1991; Alvarez-Cohen and others, 1993) provides considerable support for this hypothesis.

Many of the uncertainties associated with resolving this question appear to derive from the practical difficulties involved with determining whether the disappearance of the substrate of interest from aqueous solution results from: (1) reaction; (2) genuine sorption to the solid phase; (3) sequestration into zones of immobile water within the solids fraction; or (4) some combination of these three factors. These uncertainties are perpetuated by the fact that complete mass balances—which must account for all major transformation products, as well as the parent compound—are difficult to obtain and have rarely been reported by studies of contaminant transformation in soils.

4.4.7 LIMITATIONS OF AVAILABLE DATA ON TRANSFORMATION RATES IN SITU

In addition to the difficulties mentioned earlier (Section 2.10.2), literature data on rates of pesticide transformation in situ, typically presented in terms of soil dissipation half-lives (e.g., Nash, 1988; Wauchope and others, 1992), seldom include ancillary information on the chemical, physical, or biological circumstances under which the measurements are made. This limits the extent to which these data may be generalized to other environmental conditions. Comparisons among soil dissipation half-lives are likely to be more reliable if the results are accompanied by data on the temperature, pH (Jones, 1986; Moye and Miles, 1988), f_{oc}, percent clay, and water content of the soil (Moye and Miles, 1988). For example, Moye and Miles (1988) showed that a surprising degree of uniformity is observed among aldicarb transformation rate constants from different studies when the rates of individual transformation pathways are isolated and the same temperature is used for comparisons. As a rule, however, specific pathways of pesticide transformation are rarely reported by field dissipation studies.

For compounds that are subject to electron-transfer reactions, useful comparisons among rates in different settings also require a characterization of the redox status of the surrounding environment. As noted earlier, the aqueous concentrations of redox-active species (e.g., oxygen, iron, manganese, ammonium, nitrate, nitrite, sulfate, sulfide, hydrogen gas, and methane) provide a more reliable description of redox status than do measured redox potentials. Studies that have examined the influence of redox conditions on pesticide transformation rates in situ, however, are rare (e.g., Mansell and others, 1977; Druliner, 1989; Isensee, 1991).

4.5 SUMMARY

Pesticides enter the subsurface from both point and nonpoint sources. The movement of these compounds below the land surface following application is governed primarily by the flow of water through the unsaturated and saturated zones. As with all other solutes, the subsurface migration of pesticides occurs by both advection and hydrodynamic dispersion. In addition, the widespread occurrence of preferential transport adds considerable complexity to the movement of pesticides, as well as other solutes, beneath the land surface, causing migration rates to be both faster and slower than those predicted by uniform transport. The rates at which most pesticides move through the subsurface are slowed by interactions with the surfaces of earth materials (primarily organic matter), and by solute exchange between regions of mobile and immobile water.

Many pesticides also undergo chemical transformation in the subsurface, either with or without microbial involvement. The rates and pathways of these reactions are influenced by a variety of physical, chemical, and biological factors, including temperature, pH, the presence of other reactive species in solution, nutrient concentrations, and the nature of the resident microbial population. Transformation reactions—either abiotic or microbial—are also influenced by the presence and characteristics of the surfaces of earth materials, the nature of these influences being dependent, in turn, on the chemical structure of the pesticide of interest.

This chapter has provided an overview of the various physical, chemical, and microbial processes that control the transport and fate of pesticides in the subsurface. In Chapters 5 and 6, the information summarized here is used to help account for many of the ways in which climate, agricultural practices, hydrogeology, and other environmental factors have been found to govern the distribution and fate of pesticides in the subsurface.

Table 4.1. Pesticide distributions among environmental media at and below the land surface, based on process and matrix-distribution studies

[Site locations placed within brackets indicate probable sites, since exact location was not given. Compound Abbreviations for Pesticides: DEA, deethyl atrazine; DIA, deisopropyl atrazine; DAA, diamino atrazine; HOA, hydroxy atrazine; decycl. cyprazine, decyclopropyl cyprazine (same structure as DEA); TTR, total toxic residue (sum of concentrations of parent and specified product[s]). Soil dissipation half-life given in days (unless specified otherwise) for pesticide of interest. Data taken either from study in question (or literature cited therein) or from Wauchope and others (1992), the latter values denoted by the superscript "(w)". NV, no value given if data not provided, either by Wauchope and others (1992) or by the study in question. Note that soil dissipation half-life values provided by different studies may display considerable variation for the same compound (cf. atrazine, bromacil, hexazinone). Time Elapsed Since Application: Time elapsed between pesticide application and sampling for pesticide-distribution measurements. While the amounts given for aqueous compartments were usually cumulative measurements, those given for solid-phase compartments typically represented values obtained at specific times, and hence were usually seen to decrease over time. Abbreviations: min, minutes; hr, hours; d, days; wk, weeks; mo, months; yr, years; ST, results obtained for a shallow-tilled soil; DT, deep-tilled soil; DTL, deep-tilled soil, mixed to a depth of 1.05 m with 2.24 mt/ha/yr dolomitic limestone; NTM, no-till management; CT, conventional tillage; CP, chisel plow; MP, moldboard plow; RT, ridge-till. Soil: Data entered for soil are percentages of applied pesticide detected throughout sampled interval within soil columns. Depth of sampling varied considerably among different studies. Ranges pertain to varying recoveries among soil depths, replicates, or application rates. Pesticide concentrations measured in soils were frequently found to be dependent upon soil type, as well as upon timing of sampling, tillage practice and intensity of irrigation. Vadose Zone Water: Includes measurements of pesticide recoveries in lysimeter leachate and tile drainage waters. ND, not detected; NS, not significant; NG, not given. Blank cells indicate no information applicable or available. <, less than; ≤, less than or equal to; m, meters]

Reference	Study Location(s)	Pesticide(s)	Soil Dissipation Half-Life (days)	Time Elapsed Since Application	Percentage of Applied Pesticide Measured In				
					Plant Biomass	Soil	Vadose Zone Water	Ground Water	Surface Water
Edwards and Glass, 1971	Ohio—Coshocton	Methoxychlor	120(w)	14 mo			ND[1]		0.004
		2,4,5-T	24(w)	14 mo			NS[2]		0.05
Glass and Edwards, 1974	Ohio—Coshocton	Picloram	90(w)	0–24 mo			0.2		0.007
Liu, 1974	Puerto Rico—Río Piedras	Diuron	90(w)	0–22 wk			3.6		
		Fluometuron	85(w)	0–22 wk			5.5		
Muir & Baker, 1976	Canada—Quebec	Atrazine	60(w)	9 mo			0.074		
		DEA					[3]0.094		
		DIA					[3]0.0080		
		Cyanazine	14(w)				0.0006		
		Cyanazine amide					[3]0.011		
		Cyprazine	NV				0.11		
		Decyc. cyprazine[4]					[3,4]0.13		
		Metribuzin	40(w)				0.013		

Table 4.1. Pesticide distributions among environmental media at and below the land surface, based on process and matrix-distribution studies—*Continued*

Reference	Study Location(s)	Pesticide(s)	Soil Dissipation Half-Life (days)	Time Elapsed Since Application	Percentage of Applied Pesticide Measured In				
					Plant Biomass	Soil	Vadose Zone Water	Ground Water	Surface Water
Mansell and others, 1977	Florida—Fort Pierce	2,4-D	3-14	14 d / ST / DT / DTL			[5,6]0.21 / 0.09 / 0.24		
		Terbacil	120-225	ST / DT / DTL			1.8 / 0.54 / 0.46		
Hall and Hartwig, 1978	Pennsylvania—Centre County	Atrazine	60[w]	52 d / 128 d		20-71 / 0-24			
Hebb and Wheeler, 1978	Florida—NG [Marianna]	Bromacil	150-180	156 d / 770 d		10.6 / 2.2			
Fryer and others, 1979	England—Yarnton, Oxford	Picloram	90[w]	1 yr		2-6			
Mansell and others, 1980	Florida—Fort Pierce	Terbacil	120[w]	2 wk / DT / ST			0.1 / 0.80		
Spalding and others, 1980	Nebraska—Central Platte Valley	Atrazine	60[w]	<6 mo				1	
Zandvoort and others, 1980	The Netherlands—Maarn	Bromacil	180	1-4 yr		12-130			
Bottcher and others, 1981	Indiana—Woodburn	Alachlor / Carbofuran	15[w] / 50[w]	20 d / 20 d			[7]0.1 / [7]0.1		
Wehtje and others, 1981, 1984	Nebraska—Central Platte Valley	Atrazine	60[w]	1-16 wk (1979) / 1-16 wk (1980)			0.075 / 0.065		
Barrett and Lavy, 1984	Arkansas—Stuttgart (greenhouse)	Oxadiazon	60[w]	45 d		[7]45-90			
Leistra and others, 1984a	The Netherlands (greenhouses)	Methomyl	3-14	NG			<1 (drainage)		

Table 4.1. Pesticide distributions among environmental media at and below the land surface, based on process and matrix-distribution studies—Continued

Reference	Study Location(s)	Pesticide(s)	Soil Dissipation Half-Life (days)	Time Elapsed Since Application	Percentage of Applied Pesticide Measured In				
					Plant Biomass	Soil	Vadose Zone Water	Ground Water	Surface Water
Bouchard and others, 1985	Arkansas—Fleming Creek	Hexazinone	77	42 d	<0.10 (litter)	10			2-3
Wyman and others, 1985	Wisconsin—Cameron	Aldicarb + aldicarb sulfoxide + aldicarb sulfone (TTR)	NV	4 d		0.46-0.93			
				29 d		1.8-2.1			
				110 d		0.61-1.1			
				188 d		0.034-0.11			
	Hancock			12 d		0.37-1.8			
				45 d		0.85-1.9			
				113 d		0.004-0.19			
Welling and others, 1986	California—Tulare County / Ventura County	Simazine	37-234	1 mo		0.8-16			
		Bromacil	90-349	5 mo		56-69		ND	
		Diuron	328			68-90		ND	
		Simazine	37-234			25			
Basham and others, 1987	Arkansas—Fayetteville (greenhouse)	Imazaquin	60[w]	0 d		> 97			
Albanis and others, 1988	Greece—Ioannina	Methyl parathion	14-20	21-182 d			0.15-0.26		
		Lindane	400[w]	21-247 d			0.15-0.19		
		Atrazine	60[w]	21-247 d			0.47-0.66		
Helling and others, 1988	Maryland—Beltsville	Alachlor	39	[8]<7 h	[8]76	[8]55.2			
				<7 h		8.7			
				14 d		2.0			
				40 d		0.1			
				42 d		1.9			
				124 d		2.8			
				139 d		0.2			
				218 d		1.9			

Table 4.1. Pesticide distributions among environmental media at and below the land surface, based on process and matrix-distribution studies—Continued

Reference	Study Location(s)	Pesticide(s)	Soil Dissipation Half-Life (days)	Time Elapsed Since Application	Percentage of Applied Pesticide Measured In				
					Plant Biomass	Soil	Vadose Zone Water	Ground Water	Surface Water
Helling and others, 1988—Continued	Maryland—Beltsville—Continued	Atrazine	60[w]	<7 h	75	74.2			
				<7 h		85.0			
				14 d		60.4			
				40 d		17.9			
				42 d		31.1			
				124 d		28.3			
				139 d		3.1			
				218 d		7.3			
		Cyanazine	31	<7 h	79	92.9			
				<7 h		45.5			
				14 d		19.2			
				40 d		1.4			
				42 d		2.9			
				124 d		0.8			
				139 d		<0.1			
				218 d		0.7			
Klaine and others, 1988	Tennessee—Shelby County	Atrazine	21.5	0 d		94.4			
				28 d		5			1.5
				31 d					
				238 d		1.88			
Kubiak and others, 1988	Germany—Merzenhausen	Metamitron	NV	41-53 d		1.7			
		Desamino metamitron	NV	41-53 d		1.9-2.9			
		14C-metamitron + products[9]	NV	160 d	<1	50	0.13		
		14C-methabenz-thiazuron + products[9]	NV	127-133 d	<1	37-59	<0.1		

Table 4.1. Pesticide distributions among environmental media at and below the land surface, based on process and matrix-distribution studies—*Continued*

Reference	Study Location(s)	Pesticide(s)	Soil Dissipation Half-Life (days)	Time Elapsed Since Application	Percentage of Applied Pesticide Measured In				
					Plant Biomass	Soil	Vadose Zone Water	Ground Water	Surface Water
Leonard and others, 1988	Georgia—Tifton	Aldicarb + aldicarb sulfone Atrazine Butylate EDB Fenamiphos	NV 20-100 30 4.1 yr[10] 2 d	0-30 d				2.5 0-0.2 0-0.07 0.29 0-0.01	
Mitchem and others, 1988	Iowa—Big Spring Basin	Atrazine	60[w]	Variable (3-year study)				5	
Schiavon, 1988	France—[Nancy]	14C-atrazine 14C-DEA 14C-DIA 14C-DAA 14C-HOA	60[w]	1 wk-13 mo			5.6 9.6 3.9 3.0 0.06		
Bowman, 1989	Canada—Ontario	Atrazine Metolachlor Terbuthylazine	60[w] 90[w] NV	0-21 wk 0-21 wk 0-21 wk			<0.8 ND ND		
Foy and Hiranpradit, 1989	Virginia—Blacksburg	Atrazine	60[w]	NG (one irrigation) NT CT			0.47		1.7 [11]0.13 (ss)

Table 4.1. Pesticide distributions among environmental media at and below the land surface, based on process and matrix-distribution studies—*Continued*

Reference	Study Location(s)	Pesticide(s)	Soil Dissipation Half-Life (days)	Time Elapsed Since Application	Percentage of Applied Pesticide Measured In				
					Plant Biomass	Soil	Vadose Zone Water	Ground Water	Surface Water
Hall and others, 1989	Pennsylvania—Centre County			**1984** soils: 3 mo vadose-zone: 0-7 mo					
		Atrazine	60[W]	CT		14.5	<0.01		
				NT		15.8	0.60		
		Cyanazine	14[W]	CT		1.3	<0.01		
				NT		1.5	0.13		
		Simazine	60[W]	CT		14.2	<0.01		
				NT		10.1	0.66		
		Metolachlor	90[W]	CT		5.4	<0.01		
				NT		6.9	0.17		
				1985 soils: 2 mo vadose-zone: 0-7 mo					
		Atrazine	60[W]	CT		38.7	0.75-0.85		
				NT		17.1	0.21-9.60		
		Cyanazine	14[W]	CT		4.7	0.32-0.56		
				NT		0.9	<0.10-4.73		
		Simazine	60[W]	CT		27.0	1.50-1.63		
				NT		9.1	0.18-8.36		
		Metolachlor	90[W]	CT		27.6	0.25-0.61		
				NT		2.8	<0.10-4.19		0.6
Johnson and Morton, 1989	Arizona—Prescott, Flagstaff (5 sites)	Tebuthiuron	360[W]	1 yr		19-89			
Lavy and others, 1989	West Virginia—Parsons	Hexazinone	90[W]	1 mo	0.2	51.3			4.7
				5 mo		8.6			
				24 mo					
Reiml and others, 1989	Germany—Neuherberg	Buturon products[12]	NV	1 yr			0.2		
				12 yr			1.7-2.1		

Table 4.1. Pesticide distributions among environmental media at and below the land surface, based on process and matrix-distribution studies—*Continued*

Reference	Study Location(s)	Pesticide(s)	Soil Dissipation Half-Life (days)	Time Elapsed Since Application	Percentage of Applied Pesticide Measured In				
					Plant Biomass	Soil	Vadose Zone Water	Ground Water	Surface Water
Bergström, 1990	Sweden—Kjettslinge Bulstofta	Chlorsulfuron Metsulfuron-methyl	50 40	1 wk-7 mo			0.02-0.61 0.02-0.06		
Bush and others, 1990	South Carolina—Barnwell	Hexazinone	30-360	1 mo 2 yr				ND ND	
Clark, 1990	Montana—Havre (2 sites)	Atrazine	60[w]	3 wk 24-25 d		25 40			
	Ronan (2 sites)	Aldicarb Aldicarb sulfoxide	30[w]	8 wk 8 wk		40 15		1	
Clendening and others, 1990	California—[Riverside]	Atrazine	94	<24 h 8 d 15 d		99.7 98.9 89.2			
		Bromacil	34-269	<24 h 8 d 15 d		91.9 88.1 76.3			
		EPTC	23	<24 h 8 d 15 d		33.4 20.3 11.3			
		Prometon	57	<24 h 8 d 15 d		99.8 91.4 81.6			
		Triallate	31	<24 h 8 d 15 d		90.0 76.5 66.5			

Table 4.1. Pesticide distributions among environmental media at and below the land surface, based on process and matrix-distribution studies—*Continued*

Reference	Study Location(s)	Pesticide(s)	Soil Dissipation Half-Life (days)	Time Elapsed Since Application	Percentage of Applied Pesticide Measured In				
					Plant Biomass	Soil	Vadose Zone Water	Ground Water	Surface Water
Goetz and others, 1990	Arkansas—Fayetteville Rohwer	Imazethapyr	80-320	4 wk 52 wk		66-86 13-34			
Lee and others, 1990	Delaware—Greenhouse	^{14}C-monocrotophos	4	"zero time" 30 d 120 d		92 23 ND			
Smith and others, 1990	Georgia—Tifton	Atrazine	60[w]	187 d				5	
Frank and others, 1991a	Canada—Ottawa, Ontario	Atrazine + DEA	42-327	7.5 mo 9 mo 7 mo 4 mo 684 d 5 mo 3 mo 12 mo		77.8 17.3 29.6 	 1.9 0.2 0.09	 0.3 0.6 	 0.003 0.004 0.006
Frank and others, 1991b	Canada—Ottawa, Ontario	Metolachlor	51-142	11 mo 13 mo 12 mo		18.1 12.9 23.7	0.003 0.008 0.01	0.19 0.06 0.07	0.008 0.001 0.006
Gish and others, 1991a	Maryland—Beltsville	Atrazine	19-103	2 hr 7 hr 6 d 13 d 14 d 20 d 30 d 6 mo		83 84 69 65 71 62 42 9			
		Cyanazine	11-15	2 hr 7 hr		58 46			

Table 4.1. Pesticide distributions among environmental media at and below the land surface, based on process and matrix-distribution studies—*Continued*

Reference	Study Location(s)	Pesticide(s)	Soil Dissipation Half-Life (days)	Time Elapsed Since Application	Percentage of Applied Pesticide Measured In				
					Plant Biomass	Soil	Vadose Zone Water	Ground Water	Surface Water
Hall and others, 1991	Pennsylvania—Centre County	Atrazine	30-100	2-7 mo[7,13] CT			[13]0.12		[13]<0.01
				NT			0.88		<0.01
		Cyanazine	≤30	CT			0.02		<0.01
				NT			1.04		<0.01
		Simazine	30-100	CT			0.12		<0.01
				NT			1.16		<0.01
		Metolachlor	90	CT			0.11		<0.01
				NT			0.39		<0.01
Kladivko and others, 1991	Indiana—Butlerville (Jennings County)	Alachlor	18	1 wk -1 yr			0.00-0.01		
		Atrazine	64				0.01-0.06		
		Carbofuran	40				0.05-0.94		
		Chlorpyrifos	30[w]				<0.06		
		Cyanazine	14				0.00-0.04		
		Nitrapyrin	NV				ND		
		Terbufos	5[w]				<0.01		
Kördel and others, 1991	Germany—Schmallenberg	Cloethocarb	30-60	1 yr		<1	ND		
		Cloethocarb prods.[14]	NV				ND		
		Bentazone	30-60	2 yr		<0.1	0.03		
		Bentazone products[14]	NV		0.8	40.9-59.7	3.6-3.9		
Leake, 1991	England—Essex	Benazolin-ethyl	NV	6 mo		0.6-7.7			
		Benazolin[15]	NV			1.3-27.4			
		Benzothiazolinone[15]	NV			4.2-47.6			
		Benazolin-ethyl + benazolin + benzothiazolinone (TTR)	NV	18 mo		20.9-49.6	<0.1-1.0		

Table 4.1. Pesticide distributions among environmental media at and below the land surface, based on process and matrix-distribution studies—*Continued*

Reference	Study Location(s)	Pesticide(s)	Soil Dissipation Half-Life (days)	Time Elapsed Since Application	Percentage of Applied Pesticide Measured In				
					Plant Biomass	Soil	Vadose Zone Water	Ground Water	Surface Water
Neal and others, 1991	California—Riverside	Thiobencarb Simazine Carbofuran	8-160 10-300 7-60	70 d		4-130 220-660 130-360			
Norris and others, 1991	Florida—Polk County Manatee County	Ethoprop	13-40	<1 d <1 d 35 d 91 d		70 72 21		ND ND	
Obreza and Ontermaa, 1991	Florida—Collier County	1,3-dichloropropene	3-37	3-167 d				ND	ND
Rose and others, 1991	England—Brimstone Farm (South Central)	Isoproturon Mecoprop	[16]11-21 21(w)	1-6 mo[7]			1 ND		
Wietersen and others, 1993	Wisconsin—Greenhouse (simulated rainfall)	^{14}C-atrazine Alachlor Metolachlor	60(w) 15(w) 90(w)	1-156 d 1-156 d 1-156 d			1.4 0.010 0.12		
Cooper and Zheng, 1994	Martinique (French West Indies)—Saint Anne Bochet France—Montpelier	Metolachlor	21-31	1 hr 1 hr 1 hr		>96 >96 >96		1	
Czapar and others, 1994	Iowa—Kanawha	Alachlor Cyanazine	15(w) 14(w)	1-3 d (1988) 1-2 d (1989) 1-3 d (1988) 1-2 d (1988)			0.005-0.2 0.01-1.9 0.01-0.2 0.02-1.2		
Gish and others, 1994	Maryland—Upper Marlboro	Alachlor Atrazine	4-41 36-110	30 min 30 min	57-88 20-55				

Table 4.1. Pesticide distributions among environmental media at and below the land surface, based on process and matrix-distribution studies—*Continued*

Reference	Study Location(s)	Pesticide(s)	Soil Dissipation Half-Life (days)	Time Elapsed Since Application	Percentage of Applied Pesticide Measured In				
					Plant Biomass	Soil	Vadose Zone Water	Ground Water	Surface Water
Hall and Mumma, 1994	Pennsylvania—Centre County	Dicamba	25-58	1-3 mo NTM CT 1-6 mo NTM CT			0.39-5.56 <0.01-0.58		<0.01-0.12 <0.01-0.81
Harris and others, 1994	England—Brimstone Farm, Oxfordshire	Isoproturon Mecoprop	55 d 4.5 d	2-3 mo 112 d 2-3 mo 56 d		35 <1	0.05-1.0 ND		
Hassink and others, 1994	Germany—Schmallenberg	Diuron Methabenzthiazuron Simazine	29-102 47-198 32-62	<1 d <1 d - 150 d <1 d <1 d - 150 d <1 d <1 d - 150 d	93.9-102.2 95.7-102.8 77.9-87.7	0.06-1.6 0.04-0.9 1.1-6.0			
Isensee and Sadeghi, 1994	Maryland—Beltsville	Atrazine	60[w]	<1 d		85-92		<2	
Jayachandran and others, 1994	Iowa—Ames	Atrazine	60[w]	110 d			0.50-1.05		
Kanwar and others, 1994	Iowa—Nashua	Alachlor (continuous corn)	15[w]	6 mo[7] CP MP RT NTM			[6]0.014 0.0027 0.014 0.015		

Table 4.1. Pesticide distributions among environmental media at and below the land surface, based on process and matrix-distribution studies—Continued

Reference	Study Location(s)	Pesticide(s)	Soil Dissipation Half-Life (days)	Time Elapsed Since Application	Percentage of Applied Pesticide Measured In				
					Plant Biomass	Soil	Vadose Zone Water	Ground Water	Surface Water
Kanwar and others, 1994—Continued	Iowa—Nashua—Continued	Alachlor—Continued (corn/soybean)		CP			0.0064		
				MP			0.028		
				RT			0.054		
				NTM			0.022		
		(soybean/corn)		CP			0.0027		
				MP			0.0027		
				RT			0.0095		
				NTM			0.0064		
		Atrazine (continuous corn)	60[w]	CP			0.28		
				MP			0.078		
				RT			0.41		
				NTM			0.62		
		Cyanazine (corn/soybean)	14[w]	CP			0.0082		
				MP			0.0089		
				RT			0.033		
				NTM			0.012		
		Metribuzin (soybean/corn)	40[w]	CP			0.29		
				MP			0.38		
				RT			1.8		
				NTM			1.0		
Lee and others, 1994	Korea—Kakyung-dong, Cheong Ju	[14]Carbofuran	50[w]	1-47 wk	16.7	37.89			
				1-85 wk		21.32			
				1-167 wk					
Logan and others, 1994	Ohio—Wood County	Atrazine	60[w]	1-1,460 d			<0.1-2.5, 0.30		<0.2
Mills and Thurman, 1994a	Kansas—Topeka	Atrazine	60[w]	58 d					1-5.5

Table 4.1. Pesticide distributions among environmental media at and below the land surface, based on process and matrix-distribution studies—Continued

Reference	Study Location(s)	Pesticide(s)	Soil Dissipation Half-Life (days)	Time Elapsed Since Application	Percentage of Applied Pesticide Measured In				
					Plant Biomass	Soil	Vadose Zone Water	Ground Water	Surface Water
Odanaka and others, 1994	Japan—Ibaraki prefecture (putting greens)	Trichlorfon	10(W)	37 d			31.8		
		Metalaxyl	70(W)				18.2		
		Isoprothiolane					9.4		
		Bensulide	120(W)				<0.1		
		Isofenphos	150(W)				<0.1		
		Fenitrothion					<0.1		
		Fenarimol	360(W)				<0.1		
		Iprodione	14(W)				0.2		
		Triflumizole					<0.1		
		Tetrachlorvinphos					<0.1		
		Flutolanil					3.7		
		Isoxathion					<0.1		
		Prothiofos					<0.1		
		Tolclofos-methyl					<0.1		
		Pendimethalin	90(W)				<0.1		
		Pyridaphenthion					<0.1		
		Oxine-copper					<0.1		
Traub-Eberhard and others, 1994	Germany—Welver-Borgeln; Bad Sassendorf-Bettinghausen	Isoproturon	11-21	1 d-5 mo			0.09-0.4		
		Pendimethalin	90-120				<0.001		
Flury and others, 1995	Switzerland—Obfelden	Atrazine	60(W)	1 d		128			
		Terbuthylazine	NV			96			
		Triasulfuron	NV			78			
	Les Barges	Atrazine	60(W)	1 d		95			
		Terbuthylazine	NV			73			
		Triasulfuron	NV			41			

Table 4.1. Pesticide distributions among environmental media at and below the land surface, based on process and matrix-distribution studies—
Continued

Reference	Study Location(s)	Pesticide(s)	Soil Dissipation Half-Life (days)	Time Elapsed Since Application	Percentage of Applied Pesticide Measured In				
					Plant Biomass	Soil	Vadose Zone Water	Ground Water	Surface Water
Poletika and others, 1995	California—Etiwanda	Simazine	60[16]	10 d		87-92			
Weed and others, 1995	Iowa—Nashua	Alachlor	36	48 d		84	0.0002-0.10		
		Atrazine	55			70	0.02-0.35		
		Metribuzin	32			82	0.14-0.87		

[1]Concentration of analyte found to have been below method detection limit.

[2]Recovery of compound deemed "not significant" by study authors.

[3]Recoveries of transformation products expressed as a proportion of the applied parent compound, after adjusting for molecular weight differences between parent and transformation product.

[4]Recovery of decyclopropyl cyprazine, which is structurally identical to deethyl atrazine, represents a maximum estimate, since the compound may have appeared as a result of the transformation of atrazine—carried over from atrazine applications during previous years—rather than cyprazine transformation.

[5]Data given are those for the only experiment (1974, drain outlet submerged) from which data from all three treatments (ST, DT and DTL) were reported for both compounds.

[6]All proportions calculated from data reported in original study.

[7]Approximated.

[8]Data given are those reported over the course of three growing seasons, hence the use of the same sampling time more than once. Authors did not explain why the sums of the proportions in soil and plant biomass, when reported, greatly exceeded 100 percent.

[9]Products of metamitron transformation include desamino metamitron and other, unspecified compounds. Products of methabenzthiazuron transformation include 1-methyl-1-(2-benzthiazolyl)-urea and other, unspecified compounds.

[10]Half-life at 25 °C in a sterile aqueous solution in contact with a low-carbon soil (Haag and Mill, 1988b).

[11]Percentage of applied pesticide in suspended sediment. Data for CT only, since no runoff was observed under no-till management.

[12]Analyses conducted for three buturon transformation products in lysimeter leachate: 4-chloroaniline, methyl N-(4-chlorophenyl)carbamate and "conjugated 4-chloroaniline." Liquid scintillation counting used to detect reaction products from ^{14}C-labeled buturon parent.

[13]For ease of display, only data from the most recent year of sampling are shown (1988).

[14]Unspecified products of cloethocarb and bentazone transformation, although one bentazone transformation product was identified as chlorobentazone.

[15]Transformation product of compound given above (i.e., reaction sequence: benazolin-ethyl—>benazolin—>benzothiazolinone).

[16]Soil dissipation half-life obtained from Traub-Eberhard and others (1994).

CHAPTER 5

Influence of Climate and Agricultural Practices

The discussion in Chapter 3 showed that much of the spatial variability in pesticide detection frequencies in ground water cannot be accounted for solely by differences in the amounts of pesticide applied—particularly in areas with moderate to high use rates. As a basis for examining some of the reasons for this unexplained variability, Chapter 4 provided an overview of the physical, chemical, and biological processes that control the entry, transport, and fate of pesticides and their transformation products in the subsurface. The present chapter examines the extent to which these fundamental principles can account for the various ways in which climate and agricultural practices affect the behavior of pesticides in situ. These two groups of effects are examined together because agricultural practices, particularly the timing of planting, pesticide applications, and irrigation, are closely linked to climate. Furthermore, in many cases, the effects of climatic variations on the transport and fate of pesticides in the subsurface are sufficiently intertwined with anthropogenic factors that the two influences cannot be easily distinguished. Chapter 6 focuses on the influence of pesticide properties, hydrogeology, well characteristics, and study design on pesticide detection frequencies.

Because of the inherent complexity of the subsurface environment, conclusions regarding the relative importance of different factors in governing pesticide behavior in ground water must be based on evidence from studies conducted over a variety of spatial scales, environmental settings, and agricultural management regimes. Thus, results from laboratory and process and matrix-distribution studies, as well as local- to multistate-scale monitoring investigations, will be drawn upon to address the subjects examined. Multiscale approaches of this nature have been used with considerable success in the past to understand, for example, the environmental and anthropogenic controls on the spatial distributions of agrichemicals in the subsurface beneath the Central Valley of California (Domagalski and Dubrovsky, 1991), Oklahoma City (Christenson and Rea, 1993) and the Delmarva Peninsula (Koterba and others, 1993), as well as selenium in the Central Valley (Gilliom, 1989; Dubrovsky and others, 1993).

Although the emphasis of this report is on pesticide behavior in subsurface waters, the majority of the process and matrix-distribution studies reviewed have focused on their measurement in soils, as noted earlier (Table 2.5). Data from investigations of pesticide residues in soils are, therefore, incorporated into this discussion, under the dual assumptions that: (1) the transport and fate of pesticides in soils is largely controlled by the movement and distribution of subsurface waters, and (2) the presence of pesticides at detectable concentrations in soils provides the potential for these compounds to enter ground water.

5.1 INFLUENCE OF RECHARGE ON PESTICIDE MOVEMENT AND FATE

It is well established that pesticide transport is enhanced by increases in water flux through the subsurface (e.g., Junk and others, 1980; Wyman and others, 1985; Schiavon, 1988; van Biljon and others, 1988; Troiano and others, 1990; Isensee and others, 1990; Gish and others, 1991a,b; Kladivko and others, 1991; Maes and others, 1991; and Flury and others, 1995). Studies of solute transport in the subsurface often exploit this connection by irrigating experimental plots for the specific purpose of inducing solute migration (e.g., Mansell and others, 1977; Bergström, 1990; and Smith and others, 1991a,b), or irrigating at minimal rates to decrease the rate of solute movement (Clark, 1990; Schneider and others, 1990). The potential impact of recharge rate—particularly during periods of pesticide application—has achieved sufficient recognition that, to avoid periods of heavy rainfall, the state of California has considered imposing restrictions on the time of year during which aldicarb may be applied (Mackay and Smith, 1990).

5.1.1 RELATIONS BETWEEN RECHARGE VOLUME AND PESTICIDE CONCENTRATIONS

Studies examining relations between recharge volume and pesticide concentrations or detection frequencies in the subsurface have produced a diversity of results. Most of these studies have reported transient increases in the concentrations of pesticides and other contaminants in subsurface waters in response to individual recharge events (e.g., Hebb and Wheeler, 1978; Kelley and Wnuk, 1986; Boesten, 1987; McKenna and others, 1988; O'Neill and others, 1989; Isensee and others, 1990; Goodman, 1991; Kladivko and others, 1991; Seigley and Hallberg, 1991; Smith and others, 1991b; Priddle and others, 1992; Rowden and others, 1993; and Harris and others, 1994).

Direct relations between pesticide concentrations in ground water and precipitation recharge have also been observed with respect to longer term climatic patterns. On the island of Oahu, Hawaii, Oki and Giambelluca (1987) observed fumigant concentrations in ground water to be higher beneath fields receiving larger amounts of precipitation than beneath drier areas. Similarly, quarterly sampling in Floyd and Mitchell Counties, Iowa, during a period of steadily decreasing precipitation from 1986 to 1989 documented concurrent declines in pesticide detection frequencies, pesticide concentrations, and the number of pesticides detected in the underlying ground waters as the drought progressed (Kross and others, 1990). Furthermore, during the Iowa State-Wide Rural Well-Water Survey, or SWRL (Kross and others, 1990), and the well surveys conducted by the Minnesota Departments of Agriculture and Health (Klaseus and others, 1988), pesticides were generally detected more frequently in areas that experienced larger amounts of precipitation during the year preceding sampling.

Several studies, however, have reported data indicating an inverse relation between pesticide concentrations in subsurface waters and recharge volume (e.g., Feng and others, 1989; Pionke and Glotfelty, 1989; Mayer and others, 1991; U.S. Environmental Protection Agency, 1992a). For example, Mayer and others (1991) subjected a 27-month set of time-series data on air temperature, precipitation volume, and EDB concentrations in ground waters in Whatcom County, Washington (Figure 5.1) to auto- and cross-correlation analyses. A highly significant, inverse correlation ($\alpha = 0.05$) was observed between the concentrations of the fumigant in a domestic well located 15 m from a commercial raspberry farm and antecedent precipitation during the three months prior to each sampling. Similarly, the concentrations of hexazinone in soil leachates collected from beneath a recently clear-cut forest near Grande Prairie, Alberta were observed by Feng and others (1989) to vary inversely with leachate volume. At Big Spring, in

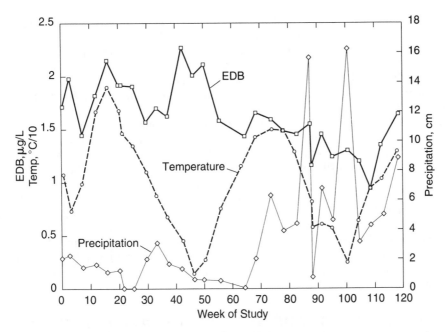

Figure 5.1. EDB (ethylene dibromide) concentrations in a well in an unconfined aquifer, air temperatures, and precipitation volumes over a 27-month period in Whatcom County, Washington. Study began in March 1988. Redrawn with permission from Mayer and others (1991).

Iowa, what appears to have been an inverse relation between atrazine concentrations and spring discharge between 1982 and 1992 (see Figure 3.46) may have been the result of a time lag between the movement of peaks in discharge and atrazine concentrations through the subsurface. During the National Pesticide Survey (NPS), pesticide detection frequencies in community water system wells were inversely correlated ($\alpha = 0.05$) with the amount of precipitation in the previous 2 years (U.S. Environmental Protection Agency, 1992a). Kimball and Goodman (1989) did not observe any significant correlations between annual pesticide detection frequencies and total annual precipitation in central South Dakota, based on monthly ground-water monitoring data from 1984 to 1989.

   The apparent inconsistencies in relations between recharge and pesticide concentrations in ground water among different studies are likely to arise from the varying influences of dilution and lag periods on pesticide levels in individual wells. As discussed in the next section, however, relations between recharge and overall pesticide flux through the subsurface are more consistent among different investigations.

5.1.2 RELATIONS BETWEEN RECHARGE VOLUME AND PESTICIDE FLUX

   In contrast with the variable relations observed between recharge volume and pesticide concentrations, pesticide fluxes appear to be directly related to the amount of recharge entering the subsurface. For example, during the period from 1971 through 1974, Muir and Baker (1976) observed the recoveries of atrazine, cyprazine, cyanazine, and metribuzin—as well as their principal transformation products (Table 3.4)—in tile drainage from sandy loams in Quebec, Canada, to increase with increasing precipitation. Similarly, Hall and co-workers found the proportions of atrazine, simazine (Hall and others, 1989), and dicamba (Hall and Mumma, 1994) recovered in pan-lysimeter percolates following spring applications to a Pennsylvania silty clay loam to be directly related to the number of leaching events following application. Consistent with these findings, Isensee and Sadeghi (1994) observed the residual amounts of atrazine to be lower in Maryland silt loam soils receiving higher amounts of recharge.

It is reasonable that relations between pesticide fluxes and recharge are more consistent than those between pesticide concentrations and recharge. As discussed in the previous chapter, the aqueous concentration of any solute at a specific time and location in the subsurface is a complex function of the different time scales of water flow, partitioning, molecular diffusion, and transformation in situ. By contrast, the overall fluxes of water or solute mass through the subsurface—parameters that display more consistent relations with recharge—represent an integration of these various phenomena over time and space, and are therefore less sensitive to the particular history of the ground water sampled at a specific location and time. Within the context of ground-water vulnerability assessments, Carsel and others (1985) reached an analogous conclusion, stating that "experience to date suggests that mass fluxes rather than concentrations are the most appropriate values for extrapolation and interpretation relative to potential ground water threat." Similarly, Hallberg and others (1986) noted that mass fluxes of nitrate through the unsaturated zone are more strongly related to water flux than are nitrate concentrations.

5.1.3 INFLUENCE OF TIMING OF RECHARGE IN RELATION TO PESTICIDE APPLICATIONS AND PERSISTENCE

Although previous work has established that pesticides are, in most cases, more likely to be detected in shallow ground water following recharge events than at other times, the likelihood of detection beneath a treated area is also dependent on the timing of application and recharge, relative to the rate of transformation of the compound of interest. Pesticides are likely to be detected in shallow ground waters in response to recharge events following application if: (1) the transformation half-life of the parent compound is long relative to the time intervals between successive recharge events (e.g., Hebb and Wheeler, 1978; Bottcher and others, 1981; Boesten, 1987; McKenna and others, 1988; Isensee and others, 1990; Katz and others, 1990; Kladivko and others, 1991; Rowden and others, 1993; and Harris and others, 1994); or if (2) comparatively stable transformation products are monitored in conjunction with the parent, as is typically done for aldicarb (e.g., Priddle and others, 1992).

The tendency for recharge events to induce increases in pesticide concentrations in shallow ground waters—independent of the timing of applications—has been demonstrated most clearly by studies of the long-term behavior of persistent pesticides whose use was discontinued for several years prior to sampling. In such situations, there is no influence of application timing on pesticide distributions in the subsurface. For example, McKenna and others (1988) demonstrated connections between individual recharge events and transient increases in endrin concentrations in ground water in Mason County, Illinois, from 1985 to 1987 (Figure 5.2), despite the fact that endrin had not been used in the area since the early 1970's.

The repeated appearance of transient peaks in pesticide concentrations in ground waters following significant recharge events, or upon the resumption of pumping from previously inactive wells (e.g., Schmidt, 1986a), long after the compounds are applied is likely to be a result of the transport of pesticide residues stored in regions of immobile water in the subsurface. On a microscopic scale, pesticides that had diffused into zones of less mobile water soon after application will gradually migrate back into zones of more mobile water, after the initial

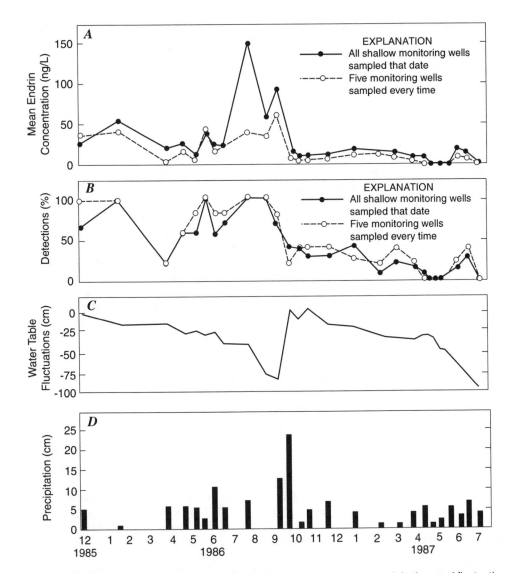

Figure 5.2. Relations between endrin occurrence in shallow ground water, precipitation, and fluctuations in water-table depth over time in Mason County, Illinois: (*A*) mean concentration of endrin in ground water; (*B*) frequency of endrin detection (detection limit not given, but inferred to be 0.004 micrograms per liter); (*C*) water-table fluctuations (reference level: December 1985); and (*D*) total precipitation in the 2-week period prior to each sampling. Figures (*A*) and (*B*) reflect results for all of the wells sampled on any day (filled circles: N = 7-14 wells) and a subset of 5 wells sampled every time (open circles). Redrawn from McKenna and others (1988) by permission of the National Ground Water Association. Copyright 1988.

contaminant pulse has passed (Figure 4.14), to be swept downgradient with the next significant flush of recharge water, or upon the resumption of pumping. This process, described in detail in Section 4.3.3, will repeat itself until the mass of compound stored in stagnant regions is depleted. Experience with discontinued persistent pesticides and with industrial solvents in relation to ground-water remediation efforts has shown that the depletion of stored residues to this point may take several years to decades.

If, by contrast, transformation is rapid compared to the intervals between recharge events, the timing of application relative to sampling can determine whether or not the parent compound is detected (e.g., Boesten, 1987; Hall and others, 1989; Keim and others, 1989; Troiano and Garretson's unpublished data cited by Maes and others, 1991; and Harris and others, 1994). This will also be the case for relatively persistent pesticides if the interval between successive recharge events is long enough to allow significant uptake of the compounds into stagnant regions in the subsurface (Figures 4.10 and 4.14).

Several studies provided evidence of decreases over time in the amount of applied pesticide available for leaching following application (e.g., King and McCarty, 1966; Willis and Hamilton, 1973; Hebb and Wheeler, 1978; Bottcher and others, 1981; Zalkin and others, 1984; Wyman and others, 1985; McKenna and others, 1988; Sawhney and others, 1988; Druliner, 1989; Hall and others, 1989; Keim and others, 1989; Goodman, 1991; Kladivko and others, 1991; Pignatello and Huang, 1991; Priddle and others, 1992; Troiano and Garretson's unpublished data cited by Maes and others, 1991; Harris and others, 1994; and Ritter and others, 1994). At their Pennsylvania study site, Hall and co-workers found that the proportions of atrazine, simazine (Hall and others, 1989), and dicamba (Hall and Mumma, 1994) recovered in the unsaturated zone were directly related not only to the number of leaching events, but to their temporal proximity to the date of pesticide application, as well. Hall and others (1989) concluded that recharge events were likely to carry detectable amounts of atrazine and simazine to subsurface waters if they occurred in the first 60-70 days following application. This interval is bracketed by the range in soil half-lives of 60-75 days compiled by other authors for these two compounds (Jury and others, 1984c; Wauchope and others, 1992).

Similarly, Keim and others (1989) observed that the highest concentrations and frequencies of detection of alachlor, atrazine, and cyanazine occurred in tile lines draining a sandy loam in Ohio during the first 6 to 8 weeks following application. Willis and Hamilton (1973) found that the concentration of endrin measured in ground water beneath a Louisiana silty clay loam decreased as the time between application and significant rainfall events increased. The concentrations of bromacil measured by Hebb and Wheeler (1978) in ground water over a period of 2 years following a single application to a sandy Florida soil also exhibited this pattern (Figure 5.3). Finally, Bottcher and others (1981) observed the concentrations of carbofuran and alachlor in tile drainage emerging from a silty clay soil in Indiana to decrease with successive recharge events (Figure 5.4).

The interplay between pesticide persistence and the timing of recharge and application has also been noted in settings where the water is supplied primarily by irrigation, rather than by precipitation. Troiano and Garretson (unpublished data cited by Maes and others, 1991) investigated whether delaying the onset of irrigation following pesticide application would influence either the persistence or the downward leaching of simazine and bromacil. Following application of each herbicide, weekly irrigation of the soil was commenced after delays of 1, 7 and 14 days during separate treatments. For simazine, the longer the interval between pesticide application and irrigation, the smaller the amount of parent compound that was available within the upper meter of the soil to leach to greater depths (Figure 5.5). However, variations in the

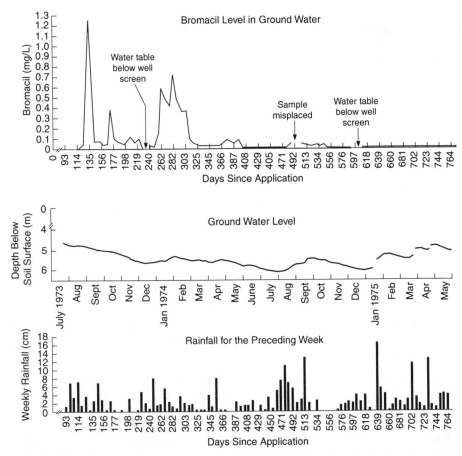

Figure 5.3. Bromacil concentrations, ground-water levels, and rainfall during the 2 years following bromacil application to a sandy soil in Florida. Modified from Hebb and Wheeler (1978) and published with permission.

delay between application and irrigation did not appear to influence the amount of bromacil detected within the upper meter of the soil column (Figure 5.6). The contrasting behavior of the two herbicides may have resulted from the greater persistence of bromacil, for which a soil dissipation half-life has been estimated at 350 days, compared with 75 days for simazine (Jury and others, 1984c).

None of the reviewed studies that examined this issue analyzed for individual transformation products in parallel with the parent compounds of interest (Willis and Hamilton, 1973; Hebb and Wheeler, 1978; Zalkin and others, 1984; Sawhney and others, 1988; Druliner, 1989; Hall and others, 1989; Keim and others, 1989; Kladivko and others, 1991; Pignatello and Huang, 1991; Troiano and Garretson's unpublished data cited by Maes and others, 1991; and Priddle and others, 1992). Consequently, the extent to which the decreases in pesticide concentrations arose from the in situ transformation of the compounds of interest could not be determined. However, the pronounced stability of endrin under aerobic conditions, as well as its strong affinity for soil

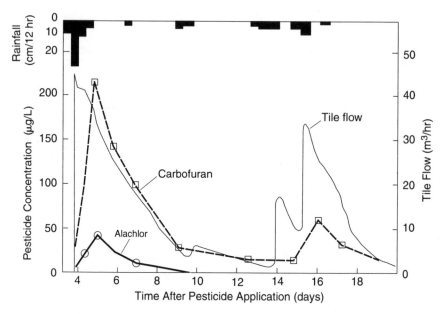

Figure 5.4. Concentrations of carbofuran (squares) and alachlor (circles) in tile outflow from a silty clay soil near Woodburn, Indiana. Tile flow and rainfall also shown. Modified with permission from Bottcher and others (1981). m³/hr, cubic meters per hour.

organic matter, indicate that the observations by Willis and Hamilton (1973), discussed earlier, arose from solute exchange between stagnant and dynamic regions within the subsurface, rather than from transformation.

The findings of the studies cited in this section illustrate the complex interplay between transformation rates and the timing of application and recharge in controlling the concentrations of pesticides that reach ground water. In settings where recharge is relatively continuous during the year, the timing of application appears to largely determine when pesticide residues are likely to enter the subsurface in the greatest quantities, typically in response to the first major recharge events after application. By contrast, where recharge is intermittent during the year because of seasonal variations in either precipitation or irrigation, the time intervals between application and recharge are likely to influence the amount of pesticide that enters ground water, this influence becoming less pronounced for increasingly persistent pesticides. Mass transfer into zones of less mobile water also might generate a similar effect, regardless of the persistence of the compound.

Given that the transformation of pesticides in water typically gives rise to products that are more hydrophilic than their parent compound (Section 4.4.2), the relative proportions of parent and product compounds that reach ground water are also likely to be influenced by the length of time elapsed between application and significant recharge (e.g., Mills and Thurman, 1994b). The proportion of the applied compound that enters subsurface waters in the form of transformation products would thus be expected to increase over time for most compounds, especially in aerobic environments, where the formation of products more soluble than their parent compound is more common. This hypothesis, however, can only be tested through studies that incorporate analyses for transformation products, in parallel with those for the parent compounds.

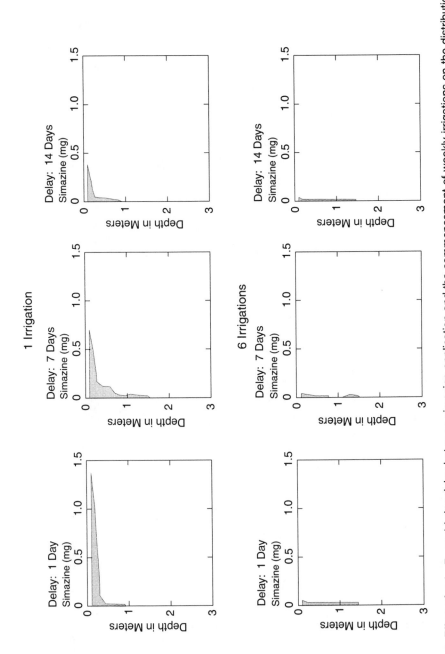

Figure 5.5. Effect of a 1-, 7-, or 14-day delay between simazine application and the commencement of weekly irrigations on the distribution of simazine in the upper 3 meters of a loamy sand in Fresno, California. Unpublished data from Troiano and Garretson, redrawn from Maes and others (1991) and published with permission.

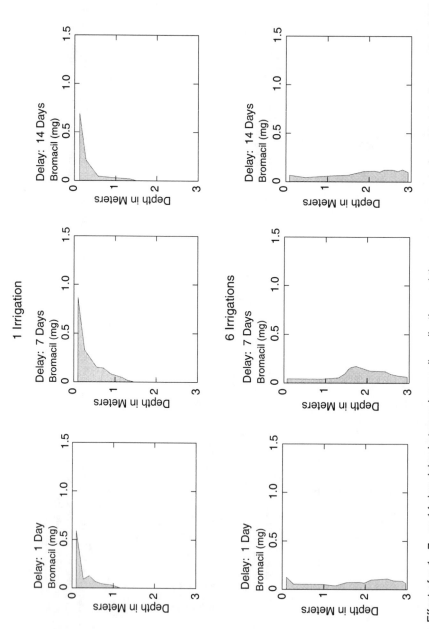

Figure 5.6. Effect of a 1-, 7-, or 14-day delay between bromacil application and the commencement of weekly irrigations on the distribution of bromacil in the upper 3 meters of a loamy sand in Fresno, California. Unpublished data from Troiano and Garretson, redrawn from Maes and others (1991) and published with permission.

5.1.4 EFFECTS OF IRRIGATION PRACTICES

As might be expected from the previous discussion of recharge effects, irrigation generally increases the rates with which pesticides and other solutes move through the unsaturated zone to the water table. However, only a relatively small number of studies have systematically examined this issue.

Effects of Different Irrigation Rates and Methods

One of the most thorough and systematic investigations of the effects of irrigation on pesticide transport was carried out by Troiano and others (1990) on a loamy sand in Fresno, California. They investigated the effects of irrigation on the migration rates of atrazine and two conservative tracers (chloride and bromide) through the unsaturated zone. Four irrigation methods (basin, drip, furrow, and sprinkler) were examined at three water application rates, selected to represent 75, 125, and 175 percent of the locally measured rate of evapotranspiration. For each of the three solutes, migration rates were found to be directly related to the net flux of water through the soil, regardless of whether the different irrigation methods were compared using the same irrigation rate, or different irrigation rates were compared for a given method. Figure 5.7 displays this relation for atrazine relative to the sprinkler and basin methods.

Figure 5.8 shows the differences in the vertical distributions of atrazine in the unsaturated zone arising from the use of the sprinkler, basin, and furrow methods of irrigation, each under the three irrigation rates listed earlier. Among the three methods, the sprinkler method produced the smallest downward fluxes of water and atrazine because it is most subject to evaporative losses, while the furrow method gave rise to the largest fluxes because it involves only half of the water surface area exposed during basin irrigation. Solute movement under drip irrigation was

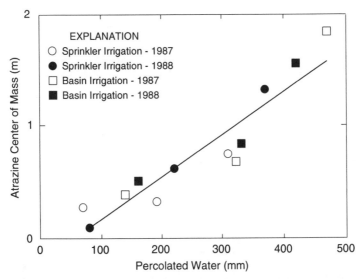

Figure 5.7. Location of the center of atrazine mass in relation to the amount of percolating water produced by sprinkler and basin irrigations. Data from Troiano and others (1990), redrawn from Maes and others (1991) and published with permission.

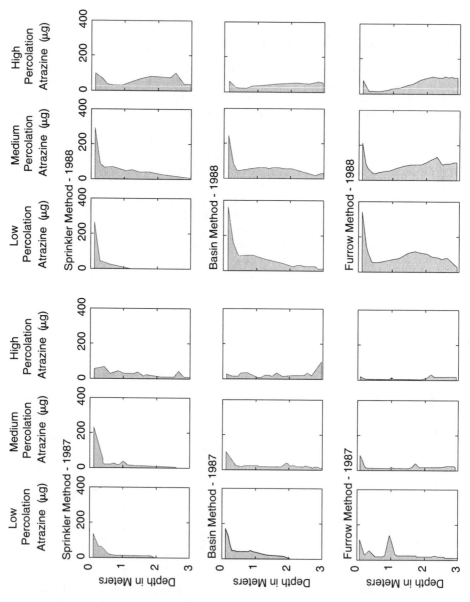

Figure 5.8. Distribution of atrazine in the unsaturated zone following the application of irrigation water using three different methods (sprinkler, basin, and furrow) and three different rates ("low," "medium," and "high") of irrigation. Data from Troiano and others (1990), redrawn from Maes and others (1991) and published with permission.

negligible (Troiano and others, 1990). Viewed in a more general context, these results also indicate the likely effects of storms of different sizes on pesticide migration through the subsurface.

During a survey of agricultural practices associated with citrus production in Tulare County, California, Pickett and others (1990) examined the influence of three different irrigation methods on the frequencies of herbicide detection in ground water. Significant, positive correlations ($\alpha = 0.05$) were observed between herbicide detection frequencies and the number of growers employing either furrow or low-volume (micro, drip, or mister) irrigation in individual townships. There was no significant correlation for growers using the drag-line method ("one or more sprinklerheads attached to a hose line pulled through the orchard"), probably because of the lower volumes of water involved, and the comparatively deep water tables that characterized the areas where this method was employed (Pickett and others, 1990). At least five other studies found increases in irrigation rates to enhance pesticide dissipation in soils. During these investigations, higher rates of irrigation were found to give rise to either reductions in the concentrations of pesticide residues in soils and plants (Brewer and others, 1982; Barrett and Lavy, 1983, 1984; Wyman and others, 1985), decreased pesticidal control, or diminished crop yields (Cooke, 1966; Barrett and Lavy, 1983; Wyman and others, 1985).

As might have been anticipated, the influence of irrigation rate on solute transport appears to become more pronounced with more rapid infiltration. Using soil columns subjected to simulated irrigation outdoors, Dejonckheere and others (1983) found that increases in irrigation rate depleted aldicarb and thiofanox residue concentrations in the soil and enhanced their transport to the greatest extent in what appeared to be the most permeable of the three soils investigated (the authors provided no quantitative estimates of soil permeabilities).

Although none of the studies discussed in this section analyzed for specific transformation products, in some cases, the dissipation of the parent compounds may have been caused largely by their transformation in situ, rather than their downward transport alone. The oxidations of atrazine, aldicarb, and thiofanox are favored under the aerobic conditions examined by Troiano and others (1990) and Dejonckheere and others (1983). By contrast, reduction may have been responsible for much of the dissipation that has been observed under long-term flood irrigation for trifluralin, fluchloralin (Brewer and others, 1982), and pendimethalin (Barrett and Lavy, 1983), since hypoxic conditions often develop in waterlogged soils.

Influence of Proximity and Intensity of Nearby Irrigation

Large-scale monitoring studies also suggest that irrigation influences pesticide contamination of shallow ground water. Such contamination appears to become more likely with increasing proximity to irrigated areas. During the Midcontinent Pesticide Study (MCPS), Burkart and Kolpin (1993) found that the percentage of wells containing detectable levels of herbicides was nearly twice as high for wells located within 3.2 km of irrigated crops (35 percent) than it was for wells located beyond 3.2 km from such areas (19 percent). Further work on this study (Kolpin, 1995a) has indicated that total herbicide concentrations in shallow wells (≤ 50 ft) screened in unconsolidated aquifers are significantly correlated with the amount of irrigated land within a 2-km radius ($\alpha = 0.05$), a correlation that was not significant for the deeper wells (>50 ft).

Results from other monitoring studies indicate that pesticide concentrations in ground water may increase with more intense irrigation. In studies of ground-water quality in the High Plains aquifer beneath 12 counties in Nebraska, Chen and Druliner (1987) and Druliner (1989) observed positive relations between median atrazine concentrations in ground water and either: (1) the spatial density of irrigation wells, or (2) the number of irrigated acres within the 4 mi^2

surrounding each of the sampled wells (Figure 5.9). The scatter in Figure 5.9 is considerable, but higher atrazine concentrations were clearly associated with areas of more intense irrigation.

The occurrence of herbicides in ground water was also found to be related to the intensity of nearby irrigation during the study by Pickett and others (1990) in Tulare County, California. A significant, positive correlation ($\alpha = 0.05$) was observed between the number of growers employing irrigation to protect their crops from frost damage in a given township and the frequency of herbicide detection (simazine, bromacil, or diuron) in ground water within the township (Figure 5.10). A significant, positive correlation was also observed among growers who used wind machines in conjunction with irrigation for frost protection, but not among growers using wind machines alone for this purpose (Pickett and others, 1990).

The proportion of the landscape under irrigation may have an indicative, as well as a causative, relation with pesticide occurrence in ground water (e.g., Burkhart and Kolpin, 1993). Areas where irrigation is practiced in the midwest generally have soils that are more permeable and contain lower amounts of organic matter than unirrigated areas. Indeed, an analysis of land use within 2 km of 100 of the MCPS wells screened in unconsolidated aquifers found the area of irrigated land to be significantly correlated with soil infiltration rate ($\alpha = 0.05$). Ground waters beneath irrigated areas in the midwest may, therefore, be more vulnerable to pesticide contamination than those beneath unirrigated areas (Kolpin, 1995a; Kolpin and others, 1996).

Confounding Effects of Precipitation

The effects of irrigation on pesticide occurrence in ground water are more difficult to discern in agricultural areas that receive significant amounts of rainfall during the growing season (e.g., Helling and others, 1988; Kolberg and others, 1989; Isensee and others, 1990; and Gish and others, 1991b). McKenna and others (1988) found no significant difference in pesticide detection frequencies between wells located immediately downgradient from irrigated areas and

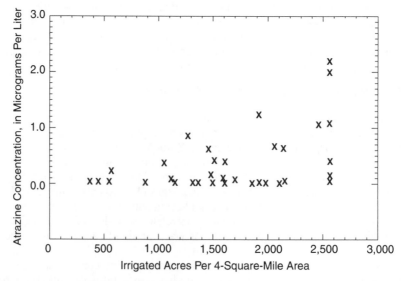

Figure 5.9. Relation between median atrazine concentration and the number of irrigated acres per 4-square mile area surrounding each sampled well in the High Plains aquifer of Nebraska. Modified from Druliner (1989).

those located downgradient from nonirrigated areas in Mason County, Illinois. The marked response of water-table elevations to precipitation, however (Figure 5.2), indicates that the relatively abundant rainfall in the study area during the 19-month sampling period may have obscured any detectable effects of irrigation on solute transport to ground water.

The varying influence of precipitation in different regions of the United States may also partially explain why no significant associations were observed between irrigation practices and pesticide detections in community water system wells during the NPS (U.S. Environmental Protection Agency, 1992a), although, once again, this may have been the result of a failure to control for other confounding influences, such as well depth and construction (Hallberg and others, 1992a). In contrast, the effects of irrigation practices on ground-water contamination could be observed by Pickett and others (1990) and Troiano and others (1990) with minimal interference from natural recharge, because precipitation in the Central Valley of California is negligible during the growing season.

5.2 INFLUENCE OF TEMPERATURE

In addition to variations in recharge, seasonal fluctuations in temperature may also affect the likelihood of detecting pesticides in the subsurface. Because the rates of chemical reactions decrease with decreasing temperature (Section 4.4.4), pesticides are more likely to persist long enough to reach ground waters if they are applied at colder times of the year. Among the studies examined for this book, soil half-lives have been observed to increase with decreasing temperature at the time of application for atrazine (Buchanan and Hiltbold, 1973; Frank and others, 1991a), metriflufen (Rogers and Talbert, 1981), metolachlor (Chesters and others, 1989),

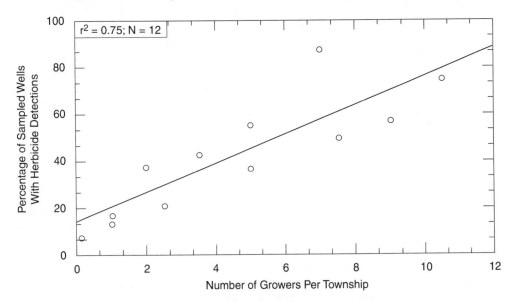

Figure 5.10. Relation between the percentage of wells contaminated by herbicides (simazine, bromacil, or diuron) and the number of growers employing irrigation (alone) for frost protection within a given township in citrus-growing areas of Tulare County, California. Redrawn with permission from Pickett and others (1990). Relation found to be statistically significant ($\alpha = 0.05$).

and aldicarb (Wyman and others, 1985; Lorber and Offutt, 1986). Consistent with these findings, a compilation assembled by Jones (1986), and expanded upon by Moye and Miles (1988) indicates that the minimum estimated soil dissipation half-life for aldicarb (0.3 months) was measured in both Arizona and South Carolina, while the maximum value (3.3 months) was determined in Maine. Similarly, under identical rates of application to either turfgrass or fallow soil, Racke and others (1993) reported chlorpyrifos to be more persistent in Indiana soils than in Florida soils. It has also been reported that when atrazine is applied to corn in the midwest, soybean crops planted in the same field the following year are more likely to fail if the intervening fall is cold than if it is warm (Chesters, 1995).

These observations indicate a greater potential for subsurface contamination by pesticides applied during cooler times of the year, but the influence of temperature may be reinforced by the effects of recharge—colder times of the year often coincide with periods of higher precipitation. Several authors have concluded that application at warmer times of year can reduce the amount of applied pesticide that reaches ground water (e.g., Wyman and others, 1985; Jones and others, 1986c; Priddle and others, 1989, 1992; Pickett and others, 1990; Porter and others, 1990; and Traub-Eberhard and others, 1994). However, none of the reviewed studies provided direct evidence of significant increases in either pesticide concentrations or detection frequencies in ground water with decreasing temperatures at application, while controlling for variations in recharge. Chemical analyses for specific transformation products in the subsurface, in parallel with the parent compounds, may have helped to distinguish the separate influences of temperature and recharge on the dissipation of the applied compounds during these investigations, but none of the reviewed studies on the subject provided such data.

5.3 INFLUENCE OF AGRICULTURAL PRACTICES OTHER THAN IRRIGATION

Among process and matrix-distribution studies of pesticide behavior in the subsurface, the influence of agricultural management practices is one of the topics that has been examined most extensively (Tables 2.5 and 2.6). In contrast, relatively little attention has been devoted to this topic by monitoring studies, regardless of their spatial scale. Most of the discussion on this subject is therefore drawn from process and matrix-distribution studies. A valuable guide to "best management practices" and their effectiveness in controlling the movement of pesticides and other agricultural contaminants to ground and surface waters has been provided by Brach (1989).

5.3.1 "ROUTINE USE" VERSUS "STANDARD OPERATING PROCEDURES"

In most agricultural settings, procedures employed for the transfer, mixing, application, and disposal of pesticides are considered "routine use." However, Hallberg (1989) has pointed out that the manner in which these activities are actually carried out, referred to as "standard operating procedures," may be substantially different from what is officially recognized as routine use, usually in ways that are likely to pose a greater threat to ground-water quality. While it is likely that some of the standard operating procedures that have posed the most severe threats to ground-water quality are no longer employed by most farmers (Hallberg, 1989), this is not always the case. For example, the disposal of leftover pesticide formulations on treated crops, into drainage ditches, or on the road returning from the field, or the rinsing and burning of empty pesticide containers in the field, is still practiced by many farmers in Iowa (Kross and others, 1990), and the disposal of excess formulation and equipment rinsewater on treated fields continues to be a recommended practice (Brennan, 1987). A summary of methods used by commercial applicators, dealers, and farmers for disposing of unwanted pesticides and used pesticide containers has been provided by Litwin and others (1983).

5.3.2 EFFECTS OF TILLAGE PRACTICES

Heightened concern over soil erosion and the input of pesticides, nutrients, and soil to surface waters in agricultural areas has led to an exploration of the use of agricultural practices in which the mechanical manipulation of the soil is either diminished (reduced-till management) or eliminated entirely (no-till management). Conventional tillage typically involves the use of a moldboard plow and one or more subsequent tillage operations before planting, while reduced tillage has been defined as "any tillage and planting system that maintains at least 30 percent of the soil surface covered by residue after planting" (Dick and Daniel, 1987). Of particular concern with respect to ground-water quality is the possibility that reductions of agrichemical inputs to surface runoff accomplished through reduced- or no-till management may be accompanied by commensurate increases in pesticide and nutrient infiltration into the subsurface (e.g., Brach, 1989).

Attempts to discern the influence of tillage practices on pesticide movement to ground water are beset by a number of complicating factors. First, the effects of tillage on infiltration capacity are seasonal. Second, both the placement of pesticides during application (i.e., surface applied or incorporated) and the magnitude of individual recharge events may influence the effects of tillage on pesticide transport. Furthermore, although a variety of different agricultural management practices—such as the amount of crop residue remaining and the timing, magnitude, and placement of pesticide applications—can be (and often are) adjusted simultaneously to reduce the amounts of pesticide that enter the subsurface, most studies of tillage effects on agrichemical behavior have examined individual management practices in isolation from, rather than in combination with, one another. These and other confounding influences are examined further below.

Influence on Water Flux Through the Subsurface

The observed effect of reduced tillage on water infiltration rates depends, in part, on the time scale of observation. Conventional tillage leads to transient increases in soil permeability, relative to an untilled soil. Viewed over the course of an entire growing season, however, long-term infiltration rates tend to be higher under reduced tillage than under conventional tillage (Baker, 1987).

Several studies have reported higher infiltration capacities in recently tilled soils than in nearby, untilled soils. Rates of infiltration measured by Steenhuis and others (1990) in the spring were an order of magnitude higher in conventionally tilled plots on a sandy clay loam in New York than in adjacent no-till areas. The authors attributed this observation to a decrease in the bulk density of the soil, and thus a concomitant increase in infiltration capacity, relative to an undisturbed soil. This may also explain why enhanced degrees of tillage often lead to increases in both the rapidity and the magnitude with which the water table rises in response to recharge, a phenomenon that has been documented in New York by Steenhuis and others (1990), as well as for a sandy soil in Florida (Mansell and others, 1977; 1980) and a silt loam in Virginia (Foy and Hiranpradit, 1989). While the increased permeability caused by tillage may be transient, it can have a pronounced effect on pesticide transport, because it occurs soon before pesticides are applied and, in many areas, during periods of significant recharge.

Over the course of a growing season, however, the total amounts of water that pass through the soil generally increase with decreasing degrees of tillage (Baker, 1987; Hall and others, 1989, 1991; Hall and Mumma, 1994). In the absence of tillage, Steenhuis and others (1990) observed the water table to be closer to the land surface throughout the growing season than beneath adjacent plots subjected to conventional tillage. In addition, tile drain flow was observed in direct response to individual storm events throughout the year under no-till conditions, but only during periods of greatest precipitation (in the fall) under conventional tillage.

There appear to be a number of reasons why reduced tillage increases long-term infiltration rates. First, reductions in tillage lead to less disruption of soil macropores that provide conduits for the passage of water into the subsurface. Second, reduced tillage typically leaves substantial amounts of crop residue on the ground, reducing the compaction caused by the direct impact of raindrops on the soil surface. Finally, reductions in tillage involve less of the soil compaction arising from equipment traffic.

Hall and others (1989) described the effects of tillage on soil structure in the following manner:

> The nature of a well-drained, untilled soil, a more undisturbed environment, compared with tilled soil, seemingly provides more continuous topsoil macropores that contact the cracks and voids between the larger structural ped faces of the subsoil. These characteristics collectively provide a matrix for rapid water transmission through the rooting zone. On the other hand, yearly tillage would initially destroy the macropore continuity between soil layers, particularly those created by decayed roots or macrofauna activity.

Similarly, Isensee and others (1990) state that "conventional tillage disrupts the macropores in the surface 15 to 20 cm, whereas macropores appear to remain largely intact under no-tillage." In areas where earthworms are encountered, their burrowing activities appear to constitute the single most significant mechanism by which soil macropores are created (Baker, 1987; Steenhuis and others, 1990).

In addition to leaving much of the soil structure intact, reduced tillage typically involves an increase in the amount of plant residue left on the soil, relative to conventional tillage. Among the studies of tillage effects reviewed by Berryhill and others (1989), infiltration rates were observed to increase with the amount of plant residue left on the land surface. Indeed, at a study site in Georgia where different tillage treatments had been maintained for 10 years prior to their experiment, Radcliffe and others (1988) found infiltration rates to be significantly related to the amount of residue, but not to the degree of tillage. By absorbing most of the kinetic energy of falling raindrops, the plant residues shield the soil from compaction (or "sealing") during rain storms. In addition, during more intense rainstorms, intact plants may guide the flow of rain water down to the plant base, where permeability is higher than in the surrounding soil (Goodman, 1991).

Another type of mechanical manipulation of the soil that, by design, leads to substantial increases in soil permeability is the "ripping" of low-conductivity, "hardpan" horizons to improve drainage. This practice is commonly employed in the eastern San Joaquin Valley, California (Pickett and others, 1990; Domagalski and Dubrovsky, 1991) and other agricultural areas underlain by hardpan soils.

Influence on Pesticide Concentrations in the Subsurface

Although numerous studies have examined the effect of tillage practices on pesticide concentrations in the subsurface, the results are not always consistent among different investigations, or even among different treatments for the same investigation. Much of this inconsistency arises from the complex interplay among several different factors controlling pesticide concentrations, particularly the timing and extent of tillage, pesticide application, and recharge.

In general, reduced tillage gives rise to pesticide distributions in the subsurface that are markedly different from those observed under conventional tillage. While pesticide concentrations are typically higher in surficial soils under conventional tillage than under reduced tillage, the reverse is often observed at greater depths in the soil (Hall and others, 1989; Isensee and Sadeghi, 1994; Ritter and others, 1994; Sadeghi and Isensee, 1994; and Weed and others, 1995). In addition, pesticides are usually detected more frequently and at higher concentrations in subsurface waters beneath no-till and reduced-till areas than beneath conventionally tilled fields (Mansell and others, 1980; Steenhuis and others, 1990; Gish and others, 1991b; Hall and others, 1991; Hall and Mumma, 1994; and Ritter and others, 1994).

These results may be explained in part by the effects of tillage on infiltration, discussed earlier. By increasing the permeability and reducing the bulk density of the soil, tillage allows applied pesticides to become more uniformly distributed throughout the tilled layer, providing more opportunity for the compounds to sorb to soil surfaces or diffuse into immobile waters, resulting in higher concentrations in surficial soils than in untilled areas. In the absence of tillage, much of the surficial soil is bypassed by infiltrating waters moving through subsurface macropores, carrying a larger proportion of the applied compounds to greater depths. It is also for these reasons that when pesticides are incorporated during application, their concentrations in subsurface waters are higher under conventional tillage than under reduced tillage, while the opposite pattern is observed following surface application (Hallberg, 1995).

In several cases, however, pesticide concentrations have displayed inconsistent relations with tillage practices in subsurface waters. Pesticide concentrations in subsurface waters beneath conventionally tilled areas have been found to be either higher or lower than those in reduced tillage areas, depending on the compound or the year examined (Hall and others, 1989; Shirmohammadi and others, 1989; Hall and Mumma, 1994; and Kanwar and Baker, 1994). Although the reasons for the differences between compounds in the same year are unclear, different trends observed in different years for the same compound may arise from variations in several key parameters related to tillage, pesticide application, and recharge.

For example, a heavy rainstorm following surface application may lead to high concentrations in subsurface waters beneath an untilled area, as the applied compounds are carried rapidly downward through intact macropores, while in an adjacent tilled area, most of the applied pesticide is carried offsite in runoff, leaving little behind to enter the subsurface. Following a moderate rainstorm during which runoff is not initiated, however, infiltrating waters have more time to dissolve pesticides in a tilled surficial soil than in an untilled soil, resulting in higher pesticide concentrations in recharge waters beneath the tilled area. This example is just one of many different potential scenarios, but it illustrates the complex—and largely unknown— interplay among some of the factors that govern the influence of tillage practices on pesticide concentrations in the subsurface.

Influence on Pesticide Flux Through the Subsurface

The available data comparing the fluxes of pesticides through conventionally tilled and no-till soils are more consistent than those on pesticide concentrations. This contrast is analogous to that discussed earlier with regard to relations between recharge volumes and pesticide flux, as opposed to pesticide concentration, in subsurface waters (Section 5.1).

Numerous studies have documented increased fluxes of pesticides through the subsurface in response to reductions in tillage of relatively permeable soils (see Table 4.1). Mansell and others (1977, 1980) showed that the quantities of terbacil, 2,4-D, and nutrients (nitrate, phosphate, and potassium) discharged from tile lines draining an acid, sandy spodosol beneath a citrus grove in Florida were all substantially higher beneath shallow-tilled than beneath deep-tilled soils (plowed to 15 cm and 105 cm depths, respectively). Similarly, Kanwar and others (1994) observed the proportions of applied herbicides (alachlor, atrazine, cyanazine, and metribuzin) entering tile drainage water from a fine loam in Iowa to be generally higher under ridge-till and no-till regimes than when the soil was worked with a chisel plow or a moldboard plow. Hall and co-workers reported increased fluxes of several pesticides through a silty clay loam in Pennsylvania under no-till, relative to conventional tillage regimes (Hall and others, 1989, 1991; Hall and Mumma, 1994). The proportions of applied herbicides recovered in pan-lysimeter percolates were three to eight times higher beneath the no-till than beneath the conventionally tilled areas for atrazine, simazine, cyanazine, and metolachlor (Hall and others, 1991). Differences in herbicide fluxes between the two tillage regimes were found to be even more pronounced for dicamba (Hall and Mumma, 1994).

Work performed to date suggests that—all other factors being equal—the reduction or elimination of tillage, by reducing runoff in general, leads to an increase in the proportion of applied pesticide mass that is carried to ground water, rather than to surface water (Berryhill and others, 1989). Regardless of the season or the timing of application relative to rainfall events, the proportions of applied herbicides carried away in runoff under conventional tillage were found by Hall and others (1991) to be insignificant in comparison with those leached to ground water under no-till conditions. Similarly, under conventional tillage, nearly all of the atrazine loss from an application site on a silt loam in Virginia during a single irrigation event was observed by Foy and Hiranpradit (1989) to occur through surficial runoff, while under no-till management, most of the applied atrazine entered the subsurface in recharge waters. Differences in tillage practice may have less impact on pesticide transport through low-permeability soils, compared to more permeable soils. Logan and others (1994) observed no discernible difference between the losses of herbicides (atrazine, alachlor, metolachlor, and metribuzin) in tile drainage from tilled and untilled plots on a poorly-drained, silty clay in Wood County, Ohio. Despite the importance of this issue, however, studies that have systematically examined the effects of tillage on the relative losses of applied pesticides in ground and surface waters are rare.

Graphic demonstrations of the increased propensity for solutes to migrate downward through untilled soils, relative to conventionally tilled soils, were provided by dye tracer studies conducted by Steenhuis and others (1990) on sandy clay-loam soils near Lake Champlain in New York. Tillage of the upper 2-3 cm led to substantial decreases in the tendency of surface-applied solutes to reach areas beneath the root zone, an observation noted for both strongly and weakly adsorbed dyes, i.e., methylene blue and Rhodamine WT, respectively. Differences in the subsurface migration rates of these two solutes have been demonstrated during field experiments showing the mobility of methylene blue in soil to be quite limited (Timlin and others, 1994), but the mobility of Rhodamine WT in the unsaturated zone to be comparable to that of alachlor and cyanazine (Czapar and others, 1994).

Despite the greater variability in the effects of tillage on pesticide concentrations compared to fluxes, the majority of the available evidence suggests that, all other factors being equal, reduced-tillage agriculture increases the mass loading of pesticides to ground water. This tendency is of particular concern because the reduction or elimination of tillage are often accompanied by elevated rates of pesticide application, particularly for herbicides (Wauchope, 1987; Berryhill and others, 1989). Recent advances in agricultural management, however, may reduce the need for such increases in pesticide application rates (Dick and Daniel, 1987; Chesters and others, 1989b). Indeed, most studies of the effects of tillage practices on pesticide transport in the subsurface have, in the interest of experimental design, employed approaches that are not necessarily representative of the manner in which conservation tillage is often carried out. In actual practice, the methods and timing of pesticide application and (limited) tillage are all tailored to minimize pesticide losses either to surface or ground waters (Hallberg, 1995). Uncontrolled variations in the timing and amount of recharge can also influence the effects of reduced tillage on pesticide movement to ground water. In any event, the information that is currently available from published studies does not provide a clear answer to the question of whether reductions in tillage lead to overall reductions in the proportions of applied pesticides lost to the hydrosphere.

5.3.3 EFFECTS OF PESTICIDE APPLICATION METHODS

Several studies have examined the effects of different pesticide application strategies on pesticide residue levels and, in some cases, persistence in the subsurface. The influences of application rate and the timing of application relative to recharge were discussed earlier (Sections 3.4.4 and 5.1.3, respectively). Additional factors that have been examined include: (1) pre- versus post-emergence application; (2) the use of single versus split applications; (3) placement; (4) formulation; and (5) the method of delivery (chemigation versus conventional spraying). For several studies, however, the specific influence of each of these individual factors was difficult to discern, because of the study design or experimental uncertainty.

Pre- Versus Post-Emergence Application

A number of studies have investigated the effect of pre-emergence applications compared to post-emergence applications on the likelihood or severity of subsurface contamination by pesticides, but results are unclear. During a field study in Ontario, Canada, Sirons and others (1973) measured lower concentrations of triazine herbicides in soils following pre-emergence treatment than following post-emergence treatments. In contrast, higher recoveries of aldicarb residues were reported by Jones and others (1986a) from Maine soils in potato fields treated with the insecticide at planting, relative to those observed following emergence treatment. No consistent differences in the recoveries of aldicarb residues from soils were evident, however, between application at planting and at emergence during studies in upstate New York (Porter and others, 1990) or in Wisconsin (Wyman and others, 1985). Similarly, no consistent differences in atrazine concentrations were evident between pre-planting and pre-emergent applications to a fine sandy loam in central Alabama (Buchanan and Hiltbold, 1973). All of these observations were based on pesticide concentrations measured in soils, and thus do not provide any direct indication of the effects of pre- versus post-emergence applications on pesticide concentrations in ground water. Furthermore, in only two of the studies cited (Buchanan and Hiltbold, 1973; Sirons and others, 1973) were the same application rates used for the pre- and post-emergence treatments. As is often the case in practice, the post-emergence application rates for the other three studies were lower than the pre-emergence application rates (Wyman and others, 1985; Jones and others, 1986a; and Porter and others, 1990).

The results from these studies were caused by a combination of the influences of application rate, application method, temperature, and recharge on pesticide mobility and fate in the subsurface. Because of the uncontrolled variability among these parameters, conclusions about the effects of pre- versus post-emergence applications on subsurface contamination still await resolution.

Split Versus Single Application

Subsurface concentrations of a pesticide may be affected by dividing applications among two or more different periods, rather than applying it all at once. However, none of the reviewed studies appear to have examined this issue in the field. Results from soil-column experiments by Dejonckheere and others (1983) indicated that for aldicarb and thiofanox, split application (four treatments, spaced at 3-week intervals) led to generally lower concentrations of total toxic residues in the uppermost soil horizon over a 92-day sampling period, compared to one application of the same total mass. This indicates that, as might be expected, pesticide concentrations in the subsurface are likely to be higher if application is carried out all at once, rather than split between two or more occasions, but field verification of this conclusion is needed.

Data reported by Hall and Hartwig (1978) suggest that increasing the time elapsed between multiple applications leads to a decrease in pesticide concentrations in shallow subsurface waters. The authors compared the concentrations of atrazine in lysimeter leachates collected from beneath a clay loam soil in Pennsylvania under two application regimes following disking and cultimulching: (1) fall surface application followed by spring pre-emergence application, and (2) spring surface application followed by spring pre-emergence application. Atrazine was applied at 2.2 kg a.i./ha on all occasions. The concentrations of atrazine in leachates recovered from the shallowest depth sampled (15 cm) during the period from May through September were found to be slightly but consistently higher when the surface application was carried out in the spring than in the previous fall.

The observations of Hall and Hartwig (1978) are expected because pesticide concentrations in the subsurface tend to decrease with increasing time elapsed since application. However, the absence of data on transformation products makes it impossible to determine the extent to which the depletion of atrazine during their investigation occurred as a result of transformation, rather than transport away from the sampling locations. Furthermore, the effects that splitting pesticide applications may exert on subsequent pesticide behavior in situ may, in practice, be influenced by the timing of rainfall in relation to applications.

Placement and Incorporation

Limiting the area of the land surface to which pesticides are applied—when it involves concomitant reductions in the total mass applied—appears to reduce pesticide concentrations and depths of migration in the subsurface. In a study of fields cropped to continuous corn in Iowa, Kanwar and Baker (1994) found concentrations of atrazine in tile drainage to be generally lower following banded, rather than broadcast application of the herbicide. In this instance, the use of banding reduced the mass of applied pesticide to one-third of that employed during broadcast application.

Clay and others (1992) compared the effect of applying alachlor along ridges versus along troughs at the same rate in a ridge-tillage system established on a sandy loam in Minnesota.

Under ridge application, alachlor concentrations were highest at the soil surface and decreased with depth, while under the trough application, the opposite pattern was observed. The authors hypothesized that the herbicide was carried deeper into the soil under the troughs because of the ponding of water during recharge events, and concluded that banding of pesticides along ridge tops in ridge-tillage systems reduces the transport of the applied chemicals to the subsurface.

Similarly, Wyman and others (1985) showed that the rate of downward migration of aldicarb and its oxidation products in sandy Wisconsin soils could be reduced by applying the compound directly above the seed piece, rather than on either side, while simultaneously applying the parent compound at a lower rate and at crop emergence rather than at planting. The fact that the placement, rate, and timing of application were all changed simultaneously, however (rather than one parameter at a time), makes it difficult to discern the importance of placement, relative to the rate and timing of application, in limiting the migration of aldicarb residues during this investigation.

The depth to which pesticides are incorporated into the soil may also affect their persistence in situ. Placement of oxadiazon at increasing depths within the soil was found by Barrett and Lavy (1984) to reduce its rate of disappearance, relative to surface application, in greenhouse experiments. This observation was attributed by the authors to different rates of oxadiazon uptake by plants as a function of soil depth. Buchanan and Hiltbold (1973) observed no difference in the persistence of atrazine between incorporation (prior to planting) and surface application (pre-emergence) to a fine sandy loam in central Alabama, although the two treatments were carried out at different times of the year. The diverse conclusions reached by these two studies indicate that the influence of application depth on pesticide persistence, like the effect of the timing of application, may be related to the reactivity of the compound of interest.

Formulation

Variations in pesticide formulations are developed by manufacturers to facilitate the handling and application of the active ingredient and, in many cases, to influence its solubility or transport behavior. Pesticide leaching may be restricted by the use of various types of "controlled-release formulations" in which the active ingredient is associated with, or encapsulated within, a solid matrix from which it gradually diffuses into the soil over time. The addition of cationic surfactants to pesticide formulations (Bayer, 1966) has also been shown to retard the release of active ingredients to aqueous solution from soil.

Several studies have demonstrated that the incorporation of pesticides into controlled-release formulations diminishes the rate at which the active ingredient enters the soil solution, resulting in slower depletion of the compound from surficial soils (Wilkins, 1989; Hill and others, 1991; Huang and Ahrens, 1991; Gish and others, 1994; and Mills and Thurman, 1994b), lower overall losses to subsurface waters (Alva and Singh, 1991b; Wietersen and others, 1993; Gish and others, 1994; and Mills and Thurman, 1994b), and shallower depths of penetration within the soil column per unit time (Koncal and others, 1981; Sundaram and Nott, 1989; Bowman, 1991). Compounds for which such behavior has been demonstrated include alachlor (Koncal and others, 1981; Huang and Ahrens, 1991; Buhler and others, 1994; and Gish and others, 1994), atrazine (Buhler and others, 1994; Gish and others, 1994; and Mills and Thurman, 1994b), deltamethrin (Hill and others, 1991), diflubenzuron (Sundaram and Nott, 1989), EPTC (Koncal and others, 1981), isazofos (Bowman, 1991), metolachlor (Koncal and others, 1981; Wietersen and others, 1993; Buhler and others, 1994), propachlor (Wilkins, 1989), and several substituted ureas (Bayer, 1966). Relatively little information is available, however, on the long-term effects of controlled-release formulation use on pesticide persistence and transport in the subsurface (Hallberg, 1995).

Technical pesticide formulations often contain surfactants for the specific purpose of increasing the aqueous solubility of the active ingredient. Using a technical formulation of chlordane containing 8 percent anionic/cationic surfactants for a series of soil-column experiments, Puri and others (1990) found that the surfactants increased the migration rate for all chlordane constituents. This influence became less pronounced with increasing degree of chlorination. Other studies cited by Puri and others (1990) suggest that surfactants can enhance the mobility of other hydrophobic solutes through soils, as well.

Most studies indicate that different formulations influence the rates at which pesticide active ingredients are released to soils and subsurface waters. However, data are not yet sufficient to evaluate distinctions between different formulations for different compounds and to predict results.

Chemigation Versus Conventional Spraying

Barnes and others (1992) compared the subsurface migration and persistence of metolachlor applied by chemigation (application by addition of the compound to irrigation water prior to discharge) with that observed following conventional application and spray irrigation. No significant differences were detected between the two application methods, either in the disappearance rates or the subsurface distributions of the herbicide. This is expected, because the volumes and chemical compositions of the applied solutions were identical for the two treatments.

Effect of Application Rate on Persistence

Process and matrix-distribution studies have generally not found any obvious effects of application rates on pesticide persistence. Increases in the mass of active ingredient applied to soil appear to have no consistent effect on the rates of disappearance of a number of different pesticides, including atrazine (Buchanan and Hiltbold, 1973), pendimethalin (Barrett and Lavy, 1983), simazine, ethofumesate, picloram, linuron (Hurle and Walker, 1980), tebuthiuron (Johnsen and Morton, 1989), and chlorpyrifos (Cink and Coats, 1993). The lack of influence of application rate is not unexpected. Except for advection, most of the processes that govern the rates at which solute concentrations decrease in soil—diffusion away from the point of measurement and most pathways of transformation, assuming minimal interference from lag periods—are generally first-order with respect to the solute concentration. (Rates of advection are presumed to be independent of concentration.) The half-lives for such processes should therefore be independent of the initial concentration of the solute of interest, within certain limits. Upper limits on biotransformation rates are typically governed by the levels above which toxic effects occur (e.g., Hurle and Walker, 1980). Toxicity to soil microflora may help to explain the increased persistence sometimes observed at higher application rates for several pesticides in soils, including methyl parathion (Butler and others, 1981), chlorpyrifos (Cink and Coats, 1993; Racke and others, 1993), and metolachlor (Zheng and others, 1993). For diffusion, upper limits on first-order reductions in concentration may be governed by solubility constraints or the saturation of exchange sites on the sorbents of interest.

5.4 PESTICIDE OCCURRENCE IN RELATION TO PROXIMITY AND INTENSITY OF NEARBY AGRICULTURAL ACTIVITIES

The studies examined for this book have employed one of two general approaches to investigate relations between pesticide occurrence in ground water and the proximity of agricultural activities. One approach has been to compare pesticide detection frequencies or concentrations with the distances between the sampled wells and the nearest cropped areas. The second approach compares pesticide occurrence data with the proportions of the land surrounding the sampled wells on which agricultural activities—and, by inference, pesticide applications—are taking place. Each approach is examined below.

5.4.1 PROXIMITY TO CROPPED AREAS

Pesticide concentrations in ground water are generally higher for wells located within or immediately adjacent to cropped areas than for those located farther away (e.g., Zaki and others, 1982; Wehtje and others, 1983; Scarano, 1986; Pacenka and others, 1987; Porter and others, 1990; and Baker and others, 1994). With increasing distance from potato farms in Suffolk County, New York, Zaki and others (1982) found fewer wells to contain high aldicarb concentrations (Figure 5.11). Beyond 2,000 ft from the farms, aldicarb concentrations in all of the sampled wells fell within the lowest concentration range (<8 µg/L). Similar data from Massachusetts (Figure 5.12) also showed that wells with higher aldicarb concentrations were located closer to application sites; based on these findings, the state of Massachusetts prohibited the use of aldicarb within 1,000 ft of any well (Scarano, 1986).

Data reported from Nebraska and North Dakota indicate similar relations between pesticide concentrations and proximity to cropland and other pesticide use areas. Wehtje and others (1983) observed atrazine concentrations in wells along the Platte River in Buffalo and Hall Counties, Nebraska, to decrease with increasing distance from irrigated corn fields (Figure 5.13).

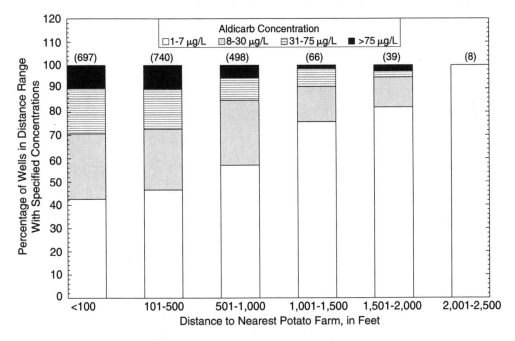

Figure 5.11. Percentage of wells with different concentrations of aldicarb in Suffolk County, New York (among wells with aldicarb detections), as a function of their distance from the nearest potato farm (data from Zaki and others, 1982). Numbers in parentheses denote the number of wells in each distance range.

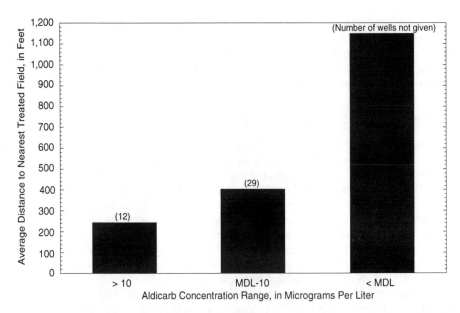

Figure 5.12. Average distances of wells with different aldicarb concentrations from nearest treated field in Massachusetts. Method detection limit (MDL) not given (data from Scarano, 1986). Numbers in parentheses denote the number of wells in each concentration range.

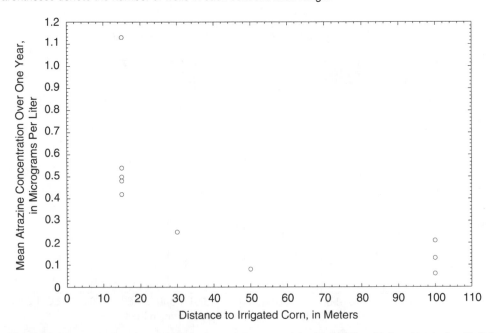

Figure 5.13. Mean atrazine concentration in ground waters of the Platte River Valley, Nebraska (Buffalo and Hall Counties; April 1980 - April 1981), versus distance of well from nearest irrigated corn field (data from Wehtje and others, 1983).

Studies of pesticide occurrence in rural wells in Nebraska, cited by Exner and Spalding (1990), reported higher frequencies of pesticide detection with increasing proximity to row-cropped areas, and Lym and Messersmith (1988) found that all wells contaminated with picloram during a monitoring investigation in North Dakota were located within 1.5 km of an area where the herbicide was used.

The most recently published results from the Cooperative Well Water Testing Program (Baker and others, 1994) indicate that the pesticide occurrence-versus-distance relations noted previously on a county scale (Zaki and others, 1982; Wehtje and others, 1983), and on a statewide scale (Scarano, 1986; Exner and Spalding, 1990), may also be observed on a regional scale. Baker and others (1994) found that frequencies of occurrence of triazine and acetanilide herbicides (and their degradates) decrease with increasing distance from the nearest cropland (Figures 5.14 and 5.15, respectively). Such relations may be difficult to discern, however, in predominantly agricultural areas, such as many parts of the midcontinent, where farmland is nearly ubiquitous (Hallberg, 1995).

5.4.2 INTENSITY OF NEARBY AGRICULTURAL ACTIVITIES

The results from the studies discussed above provide strong evidence that the occurrence of pesticides in shallow ground water becomes more likely with increasing proximity to cropped areas. Some investigators, however, have approached this issue by comparing the occurrence of pesticides in ground water with a variety of measures of the intensity of agricultural activity—and, by inference, pesticide use—within the area surrounding each sampled well. Table 5.1 summarizes the principal features of these studies. The specific results of the statistical analyses were not included in the table due to the considerable variability in the nature of the statistical tests employed. Instead, these findings are discussed below, first among studies that did not specify a particular spatial scale to characterize nearby land use, followed by those that did.

Influence of Nearby Agriculture

Several investigators have found pesticide contamination of ground waters to be more common in agricultural areas where the compounds are used than in areas where they are not. Koterba and others (1993) found the proportions of wells containing detectable pesticide residues on the Delmarva Peninsula to be significantly higher when pesticides were used on nearby land under cultivation for corn, soybeans, or small grains than when the nearby land was used to grow other crops or for non-agricultural purposes. Higher pesticide detection frequencies also were observed in agricultural areas, relative to non-agricultural or mixed-use areas, when these classifications were based on major-ion chemistry, rather than on the nature of the activities visible from the wells (Koterba and others, 1993). Similarly, Grady (1989) found the frequencies of atrazine and 1,2-dichloropropane detection in the surficial, stratified-drift aquifers of Connecticut to be significantly higher beneath agricultural areas than beneath undeveloped, residential, or industrial/commercial areas. In the Conestoga River Basin in Pennsylvania, Fishel and Lietman (1986) detected atrazine, simazine, alachlor, and metolachlor in ground water at significantly higher rates in agricultural areas than in residential neighborhoods or small towns. McMahon and others (1994) found atrazine concentrations in ground water to be higher beneath agricultural areas than beneath forested, rangeland, or urban areas in the South Platte River Basin of Colorado.

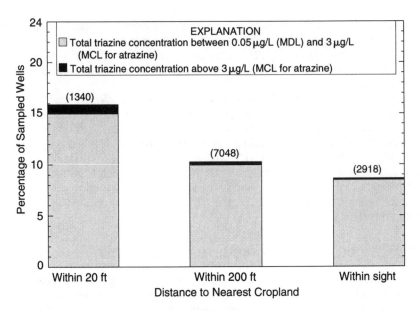

Figure 5.14. Relations between triazine herbicide concentrations in wells (based on immunoassay analyses) and proximity to cropland for the Cooperative Private Well Testing Program. Number of wells in each category given in parentheses (data from Baker and others, 1994). MDL, Method Detection Limit; MCL, Maximum Contaminant Level.

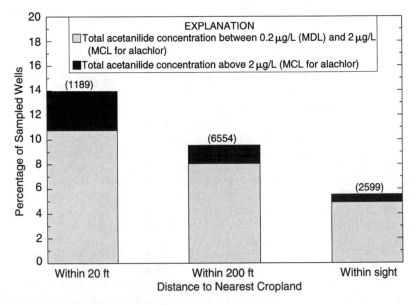

Figure 5.15. Relations between acetanilide herbicide concentrations in wells (based on immunoassay analyses) and proximity to cropland for the Cooperative Private Well Testing Program. Number of wells in each category given in parentheses (data from Baker and others, 1994). MDL, Method Detection Limit; MCL, Maximum Contaminant Level.

Table 5.1. Principal features of studies relating pesticide occurrence in ground water to land use in agricultural, versus non-agricultural settings

[MCPS, Midcontinent Pesticide Study; NAWWS, National Alachlor Well-Water Survey; SWRL, Iowa State-Wide Rural Well-Water Survey. Land Use: R, radius of zone of influence, in miles. Statistical Tests Employed, See text for details. mi, mile; ppb, parts per billion]

Reference	Study Location	No. Wells	Pesticides Examined	Parameters Used to Characterize		Number of Land-Use Categories Examined	Statistical Tests Employed
				Pesticide Occurrence	Land Use		
Greenberg and others, 1982	New Jersey (Statewide)	40	Organochlorine insecticides	Percent of wells containing 1-5 ppb pesticides	Percent of area in land-use category (R = 1)	17	Comparisons of land use among "contaminated" wells
Fishel and Lietman, 1986	Pennsylvania (Upper Conestoga River)	42	Atrazine, alachlor, metolachlor, simazine, and several unidentified herbicides	Detection frequency	Predominant land use	2	None
Barton and others, 1987	New Jersey (Central)	65	Organochlorine, organophosphorus, and triazine pesticides	Detection frequency	Predominant land use Presence/absence (R = 0.25)	3	Kruskal-Wallis
Rutledge, 1987	Florida (Orlando area)	32	Organochlorine, triazine, organophosphorus, phenoxy-acid pesticides, and EDB	Detection frequency	Predominant land use	4	None
Pionke and others, 1988; Pionke and Glotfelty 1989	Pennsylvania (Mahantango Creek)	21	Atrazine, simazine, alachlor, metolachlor, 2,4-D, dicamba, chlorpyrifos, fonofos, and terbufos	Concentration (atrazine only)	Corn Production Intensity (see text)	Not applicable	None
Eckhardt and others, 1989a	New York (Long Island)	903	Aldicarb, carbofuran, DDT, heptachlor epoxide, and chlordane	Detection frequency	Predominant land use (R = 0.5)	10	Parametric tests, non-parametric tests and contingency-table analyses

Table 5.1. Principal features of studies relating pesticide occurrence in ground water to land use in agricultural, versus non-agricultural settings—Continued

Reference	Study Location	No. Wells	Pesticides Examined	Parameters Used to Characterize		Number of Land-Use Categories Examined	Statistical Tests Employed
				Pesticide Occurrence	Land Use		
Eckhardt and others, 1989b	New York (Long Island)	90	Carbamate, organochlorine, and organophosphorus insecticides; triazines and chlorophenoxy acids	Detection frequency	Predominant land use	5	Nonparametric tests, contingency-table analyses
Grady, 1989	Connecticut (Drift aquifers)	83	Atrazine, 1,2-dichloropropane, and at least 5 others	Detection frequency	Predominant land use	4	Analysis of variance, contingency-table analyses
Helgeson and Rutledge, 1989	Kansas (South-central)	82	Atrazine, 2,4-D	Concentration	Predominant land use	2	Wilcoxon-Mann-Whitney
Kross and others, 1990 (SWRL); Hallberg and others, 1992b	Iowa (Statewide)	686	27 pesticides (atrazine most common)	Detection frequency (atrazine; any pesticide)	Presence of area in land-use category ($R = 0.5$)	8	Comparisons of 95% confidence intervals among land-use categories
Holden and others, 1992 (NAWWS)	Nationwide (89 counties)	1430	Alachlor, metolachlor, atrazine, cyanazine, and simazine	Detection frequency	Percent of area growing row crops ($R = 0.5$)	Not applicable	One- and two-tailed comparisons among means
Koterba and others, 1993	Delmarva Peninsula	100	36 pesticides and 4 degradates	Detection frequency	Predominant crops	2	Mann-Whitney, Kruskal-Wallis
Kolpin and others, 1994 (MCPS)	Midcontinent (12 States)	303	11 herbicides and 2 atrazine degradates	Detection frequency	Percent of area in land-use category ($R = 0.02, 0.25, 2$)	9	Spearman's rank correlation, Mann-Whitney, Kruskal-Wallis
McMahon and others, 1994	South Platte River (Colorado, Nebraska, and Wyoming)	24 (mini-piezometers)	Atrazine and its transformation products (via immunoassay)	Concentration	Predominant land use ($R = 0.6$)	4	Kruskal-Wallis
Szabo and others, 1994	New Jersey (Southern Coastal Plain)	36	Organochlorine insecticides, triazines, acetanilides, and carbamates	Detection frequency	Percent of surrounding land in agriculture ($R = 0.5$)	Not applicable	Discriminant analysis

Effects of Agricultural Activity Within Specified Areas of Influence

Many investigators who have examined relations between ground-water quality and land use have computed the areal proportions occupied by different land-use settings within a "zone of influence" of specified radius surrounding each well of interest. The reasoning behind the particular radius selected for this purpose, however, is seldom provided. Among studies that did provide this reasoning, the selection of the radius has usually been driven by the desire to minimize spatial autocorrelation among adjacent wells, either with respect to ground-water chemistry or nearby land use.

Of the studies reviewed, only two examined zones of influence whose size was derived explicitly from hydrogeologic considerations. Eckhardt and others (1989a) chose a radius of influence of 0.5 mi for assessing land use near wells sampled in Long Island, New York, based on an estimate of the lateral distance traversed by ground water in the study area during the 7-year period of sample collection. Similarly, the sizes of the areas used by Helgeson and Rutledge (1989) to characterize land use surrounding individual wells in south-central Kansas (3-10 mi^2) were based on local estimates of ground-water travel times. Migration rates within the subsurface have also been incorporated into regulations governing atrazine applications in Germany, where the herbicide is "not to be applied within a distance from the wells where the groundwater flow to a well takes less than 50 days" (Milde and others, 1988).

A radius of 0.5 mi appears to have been employed most commonly in investigations of this nature (e.g., Eckhardt and others, 1989a; Hardy and others, 1989; Kross and others, 1990; Vowinkel, 1991; Holden and others, 1992; and Szabo and others, 1994), but the results from several studies conducted in New Jersey provide empirical support for the use of radii between 0.5 and 1 mi surrounding individual wells (Greenberg and others, 1982; Vowinkel and Battaglin, 1989; Hay and Battaglin, 1990; and Vowinkel, 1991). None of the latter studies firmly established a single optimal size to be used for this purpose, however, because the radii at which the strongest correlations were observed between land use and ground-water quality were found to vary among different chemical constituents (Hay and Battaglin, 1990; Vowinkel, 1991), study areas (Vowinkel and Battaglin, 1989; Vowinkel, 1991) and land-use categories (Greenberg and others, 1982). Additional discussions of the various difficulties associated with the use of zones of influence for examining relations between land use and ground-water chemistry have been provided by Barringer and others (1990) and Hay and Battaglin (1990).

Most of the studies that have employed zones of influence to characterize land use surrounding individual wells (Table 5.1) have observed positive correlations between the frequencies of pesticide occurrence (or pesticide concentrations) in ground water and the intensity of nearby agricultural activity. McMahon and others (1994) observed detectable concentrations of atrazine in ground waters of the South Platte River Basin in areas where irrigated agriculture was the predominant land use within 0.6 mi, but not in areas where the predominant land use was forest, rangeland, or urban development. Kolpin and others (1994) observed a highly significant increase ($\alpha = 0.05$) in the frequency of corn-and-soybean herbicide detections from the MCPS as the percentage of forest land (where the compounds were not used) decreased within 2 mi of the sampled wells. Additional analysis of the MCPS data (Kolpin, 1995a), based on high-resolution assessments of land use near each of the sampled wells, indicates that, as might be anticipated, correlations between pesticide concentrations and nearby land use are stronger for shallow wells (less than 50 ft deep) than for deeper wells (more than 50 ft deep). Data from the National Alachlor Well-Water Survey (NAWWS) showed an increase in the percentage of sampled wells containing detectable levels of alachlor, metolachlor, and atrazine as a greater percentage of the area within 0.5 mi of individual wells was devoted to row crops (Figure 5.16). For all five of the herbicides investigated during the NAWWS, the

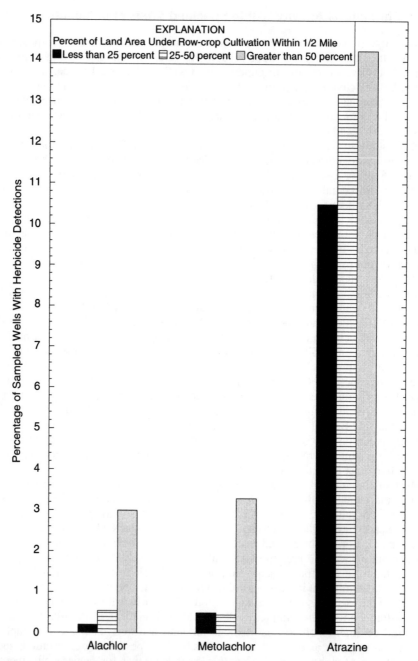

Figure 5.16. Frequencies of alachlor, metolachlor, and atrazine detection relative to the proportions of land under row-crop cultivation within 1/2 mile of the wells sampled for the National Alachlor Well Water Survey (adapted from Holden and others, 1992).

percentages of wells with detections were also higher for wells near probable use areas ("based on opinions reported by county experts"), relative to wells near areas where the herbicides were "probably not used" within 0.5 mi (Holden and others, 1992).

Studies of both existing (Eckhardt and others, 1989a) and newly acquired data (Eckhardt and others, 1989b) on ground-water quality in Long Island, New York, also documented strong relations between the frequencies of pesticide detection in wells and the presence of agricultural activities nearby. For their analysis of existing data, Eckhardt and others (1989a) classified the area within a 0.5-mi radius surrounding each of 903 wells into one of 10 different categories on the basis of the predominant land use during the period of interest (1978-1984). Consistent with its use as an agricultural insecticide, carbofuran was detected with high frequency (42 percent) among wells in predominantly agricultural areas, but in only one well in another land-use category (recreation). Surprisingly, aldicarb was not detected in any of the 43 wells located in agricultural areas, despite the extensive aldicarb contamination of Long Island ground waters documented in numerous previous studies (e.g., Baier and Robbins, 1982b; Zaki and others, 1982; Zaki, 1986) at levels considerably higher than the detection limit of 2 µg/L employed by Eckhardt and others (1989a).

Eckhardt and others (1989b) observed analogous relations following their own sampling of 90 shallow wells on Long Island for pesticides in five different land-use areas. The pesticides most frequently detected were the carbamate and organochlorine insecticides. As with the earlier data (Eckhardt and others, 1989a), carbamates were detected primarily within agricultural areas. Organochlorine insecticides—primarily chlordane and dieldrin—were detected in ground waters beneath both residential and agricultural areas (Eckhardt and others, 1989b). The latter observation was consistent with the extensive agricultural use of chlordane prior to its discontinuation in 1987 (Puri and others, 1990), and the detection of dieldrin in ground waters beneath agricultural areas in Illinois (McKenna and others, 1988) and Nebraska (Exner and Spalding, 1990). The detection of organochlorine compounds in ground water in relation to their use in residential settings will be examined in Section 7.2.1.

Several studies carried out in New Jersey also detected significant relations between the occurrence of pesticides in ground water and the presence of agricultural lands nearby. Using a 0.25-mi radius for characterizing land use in central New Jersey, Barton and others (1987) found the frequencies of pesticide detection to be higher in ground water beneath agricultural areas than beneath either urban or undeveloped areas, based on the sum of all detections of organochlorine, organophosphorus, and triazine pesticides. Consistent with these observations, Szabo and others (1994) detected a significant, positive relation ($\alpha = 0.05$) between the proportion of agricultural land within 0.5 mi of individual wells and the frequencies of pesticide detection within the southern Coastal Plain of New Jersey. Greenberg and others (1982) found that wells containing 1-5 µg/L of organochlorine insecticides in New Jersey were surrounded by a disproportionately high percentage of agricultural land within a 1-mi radius, relative to residential, commercial, industrial, or other urban land uses.

Pionke and Glotfelty (1989) employed a different measure of the intensity of agricultural activity for their study of ground-water contamination by pesticides applied to corn near Mahantango Creek in central Pennsylvania. This parameter, referred to as the Corn Production Intensity (CPI), was defined by the authors as "the average percentage of the near well area [a semi-circle of radius 100 m, extending upslope from the well] planted to corn over the crop rotation cycle." The CPI therefore accounts for changes in cropping patterns over time, and thus for potential changes in the pesticides present in ground water as a function of location along a presumed flow path. (Among the other studies reviewed, only the work of Koterba and others [1993] examined the pesticides used throughout entire crop-rotation cycles to account for those detected in ground water beneath agricultural areas.) Consistent with expectation, Figure 5.17, taken from an earlier report on the Mahantango Creek watershed (Pionke and others, 1988), shows a greater prevalence of elevated atrazine concentrations at greater CPI values.

Among the reviewed studies that examined this issue, two failed to observe significantly higher frequencies of pesticide occurrence in ground waters beneath agricultural lands, relative to those beneath non-agricultural areas (Helgeson and Rutledge, 1989; Kross and others, 1990). During the SWRL study in Iowa (Kross and others, 1990), the frequencies of pesticide occurrence—for either atrazine alone or any pesticide in general—showed no obvious correlations with nearby pesticide use, either within a 0.5-mi radius around individual wells or within the immediate vicinity (Table 5.2).

Actual variations in pesticide detection frequencies among different land-use settings during the SWRL study were likely to have been masked by the considerable degree of redundancy inherent in the land-use data employed. The design of the questionnaire used to gather the land-use data (Hallberg and others, 1990) was such that an individual well was assigned to every land-use category observed nearby. Specification of a single, predominant land-use category for each well, a method employed by many other studies reviewed on this subject (Table 5.1), would have avoided this overlap. As noted earlier, however (Section 5.4.1), the fact that agricultural land is nearly ubiquitous in rural areas of Iowa probably interferes with the detection of relations between pesticide occurrence and non-agricultural land use in the state. Among the 686 wells sampled for the SWRL study, only two did not have farmland present within 0.5 mi (Hallberg, 1995).

An analysis of data on ground-water quality in the High Plains aquifer beneath south-central Kansas (Helgeson and Rutledge, 1989) failed to detect significant differences in pesticide concentrations between water drawn from small-yield wells located in irrigated cropland and

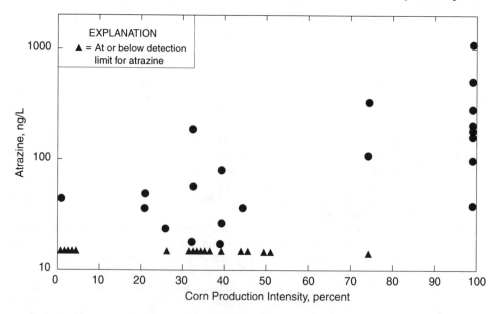

Figure 5.17. Atrazine concentration versus Corn Production Intensity (CPI) in the Mahantango Creek watershed in central Pennsylvania. Redrawn with permission from Pionke and others (1988).

water obtained from similar wells in nearby, noncropped, rangeland areas. However, the apparent similarity in atrazine detection frequencies observed by Helgeson and Rutledge (1989) between the different settings may have been a consequence of a relatively small sample size; atrazine was detected in only 7 of 30 wells in irrigated cropland and 1 of 22 wells in nonirrigated rangeland.

The proportion of agricultural land within the vicinity of a well may also influence its susceptibility to changes in pesticide concentrations in response to flooding. Following the 1993 floods in the midcontinent, Kolpin and Thurman (1995) examined changes in the total concentrations of all herbicides detected in wells screened in near-surface, unconsolidated aquifers in the Upper Mississippi River Basin, relative to the concentrations measured in the wells during the 1991 and 1992 sampling for the MCPS. Increases of 20 percent or more in total herbicide concentrations were found to be more common in wells that had higher proportions of the land (within 400 m) in corn and soybean production. Conversely, as the proportion of nearby land in corn and soybean production increased, the percentage of wells with no herbicide detections, or with total herbicide concentration changes smaller than 20 percent, decreased.

Viewed in aggregate, the studies discussed above provide strong support for the hypothesis that pesticide detection frequencies and concentrations in ground water are likely to increase with increasing intensity of agricultural activity—and, by extension, pesticide use—within roughly 1 mi of the sampled well. While some of the observations reported by Kross and others (1990) and Helgeson and Rutledge (1989) appear to be at variance with this conclusion, significant relations of this type, even if present, would have been difficult to detect during these

Table 5.2. Relations between proximity to nonpoint sources of pesticides and the frequency with which either any pesticide or atrazine was detected in Iowa wells, based on the State-Wide Rural Well-Water Survey (SWRL)

[Data excerpted from Kross and others (1990). Percentages of wells with detections are percentages among all wells sampled that contained detectable levels of either any pesticide or atrazine, and that exhibited the characteristic indicated in the first column. "+" and "-" indicate values that are between 1 and 5 percent above or below the statewide 95 percent confidence intervals, respectively; "++" denotes a value that is more than 5 percent above this confidence interval. ft, feet; mi, mile]

Nearby Land Use	Percentage of Wells With Detections	
	Any Pesticide	Atrazine
• Statewide percentage from total SWRL sample	13.6	8.0
• Land use within 0.5 mi of well[1]		
Feedlot	15	10
Farmland	13	8
Row Crops	13	8
Pasture	15	8
Forest	14	10
Nonfarm, "suburban" houses	17	16 ++
• Land use immediately adjacent to well (within 100-200 ft)[1]		
Feedlot	15	11 +
Farmland	14	9
Row Crops	15	9
Pasture	11	6 -
Forest	7 -	6
Nonfarm, "suburban" houses	13	11

[1]Percentages of surrounding land assigned to specific land-use categories not given.

two studies, for the reasons stated earlier. In particular, the results from the studies discussed in this section illustrate the central importance of sampling relatively large numbers of wells and a broad diversity of land-use settings when investigating relations between ground-water quality and land use.

5.5 SUMMARY

Extensive research over the past three decades has established that variations in climate and agricultural practice can influence the transport and fate of pesticides, and hence the likelihood of their detection, in the subsurface. With increasing recharge, derived from either precipitation or irrigation, the proportion of applied pesticide that enters the subsurface generally increases. For less persistent compounds, this proportion decreases with increasing time elapsed since application. Consistent with these findings, pesticide detections in ground water become more likely with increasing proximity to irrigated areas or greater intensity of irrigation. Similarly, pesticide concentrations in the subsurface have been found to increase in response to recharge, although this relation is less consistent than that involving overall pesticide fluxes. Pesticides are more persistent in soils—and thus more likely to reach ground water—when they are applied at colder times of the year, but direct evidence of the effects of temperature on detections in ground water appear to be lacking.

Reductions in tillage, employed to reduce the movement of agrichemicals and soil to surface waters, usually lead to increased rates of pesticide entry into the subsurface. Increases in pesticide concentrations in the subsurface have also been found to be associated with reduced tillage, although, as with the effects of recharge, this relation is less clear than that involving pesticide fluxes. The effects of different formulations on pesticide persistence and transport in the subsurface are not well established, but pesticide detections tend to be more likely following single applications than following split applications of the same total amount. Confining applications to more limited areas of the land surface—such as along ridge tops in ridge tillage systems, rather than broadcast over the whole field—appears to reduce the amount of applied compound that enters the subsurface, largely by reducing the total mass applied. As expected, pesticide persistence does not appear to be affected by application rate or the method of delivery (chemigation versus conventional spraying). Finally, the likelihood of detecting pesticides in ground water increases with increasing proximity to cropped areas, increasing intensity of nearby agricultural activity, and increasing proportions of agricultural land within the surrounding area. While the present chapter has focused on the effects of climate and the nature and proximity of agricultural practices on pesticide occurrence in ground water, Chapter 6 examines the extent to which pesticide detections are related to pesticide properties, environmental setting, and study design.

CHAPTER 6

Influence of Pesticide Properties, Environmental Setting, and Study Design on Pesticide Detections

Among the factors presumed to control the transport and fate of pesticides in the subsurface, those most often cited include the chemical properties of the compounds of interest and the various edaphic and hydrogeologic characteristics of the site. Another major, but infrequently acknowledged influence on the conclusions reached by investigations of pesticide occurrence in ground water is the design of the studies themselves.

6.1 INFLUENCE OF PESTICIDE PROPERTIES ON OCCURRENCE IN GROUND WATER

It is generally assumed that pesticides that are more resistant to transformation, or that exhibit a weaker affinity for soil organic carbon are more likely to be detected in ground water than those that are more reactive or more hydrophobic. Although a few process and matrix-distribution studies (e.g., Kladivko and others, 1991; Hassink and others, 1994; and Logan and others, 1994), and numerous laboratory investigations have supplied justification for this hypothesis, patterns of pesticide detection and nondetection derived from ground-water monitoring investigations provide only marginal supporting evidence.

6.1.1 PESTICIDE DETECTIONS IN RELATION TO HYDROPHOBICITY AND REACTIVITY

The effects of persistence and hydrophobicity on the occurrence of pesticides in ground water can be examined through relations between soil organic-carbon partition coefficient (K_{oc}) and soil dissipation half-life values for pesticides that have and have not been detected in ground water during individual monitoring investigations—particularly those studies that have analyzed for a relatively large number and variety of pesticides. According to expectation (see Figure 6.1), the compounds most often detected in ground water should be those with comparatively long dissipation half-lives and low K_{oc} values, while those not detected should exhibit shorter dissipation half-lives and higher K_{oc} values, all other factors, such as use, hydrogeologic setting, and detection limits, being equal. The validity of this assumption is of particular importance because published values of soil dissipation half-lives and K_{oc} are widely used to predict which compounds will, and will not, leach to ground water (Cohen and others, 1984; Bishop, 1985;

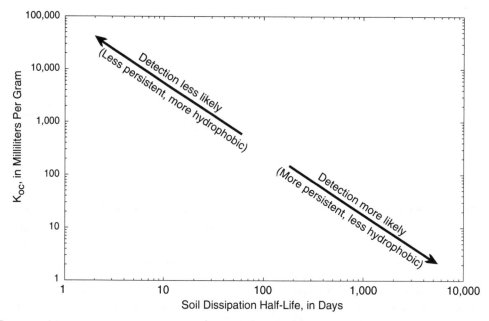

Figure 6.1. Expected trends in detection of pesticides in ground water, relative to their respective soil dissipation half-life and K_{oc} values.

Creeger, 1986; U.S. Environmental Protection Agency, 1986; Matthies, 1987; Pionke and others, 1988; Berryhill and others, 1989; Gustafson, 1989; Cohen, 1990b; Deubert, 1990; Ritter, 1990; Becker and others, 1991; Behl and Eiden, 1991; Domagalski and Dubrovsky, 1991; Tooby and Marsden, 1991; and Yen and others, 1994).

Figure 6.2 displays a plot similar to Figure 6.1 that summarizes the results from the seven ground-water monitoring studies that analyzed for the largest numbers of pesticides and transformation products. These include the National Pesticide Survey, or NPS (U.S. Environmental Protection Agency, 1990a); the Midcontinent Pesticide Study, or MCPS (Kolpin and others, 1993); and statewide well-water surveys conducted in Minnesota (Klaseus and others, 1988), Wisconsin (LeMasters and Doyle, 1989), Iowa (Kross and others, 1990), Illinois (Goetsch and others, 1992), and the Delmarva Peninsula east of Washington, D.C. (Koterba and others, 1993). Each data point in the figure represents an individual compound detected (filled symbols), or not detected (open symbols), during the study in question. The soil dissipation half-life and K_{oc} values used were those assembled from the literature by Wauchope and others (1992).

The results displayed in Figure 6.2 indicate that literature values for soil dissipation half-lives and K_{oc} provide unreliable predictions of the pesticides that are likely to be detected, or not detected, in ground water. Although there appears to be some segregation of compounds according to the trends indicated in Figure 6.1 for most of the studies, the overlap between detected and nondetected compounds is substantial in every case. Indeed, for the two largest-scale investigations—the NPS and the MCPS—the distributions of detected and nondetected pesticides among dissipation half-lives and K_{oc} values are essentially random. Consistent with these observations, logistic regressions carried out during the NPS (U.S. Environmental

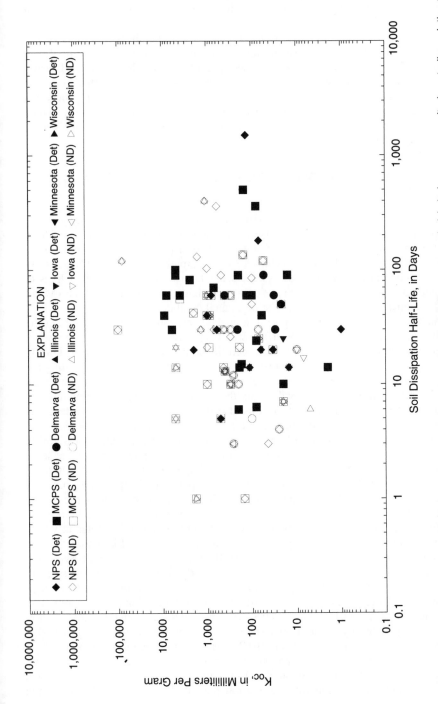

Figure 6.2. Pesticide analytes detected (Det - filled symbols) and not detected (ND - open symbols) during seven monitoring studies relative to their respective soil dissipation half-life and K_{oc} values. Pesticide occurrence data from U.S. Environmental Protection Agency, 1990a (National Pesticide Survey, or NPS); Kolpin and others, 1993 (Midcontinent Pesticide Study, or MCPS); Koterba and others, 1993 (Delmarva Peninsula); Goetsch and others, 1992 (Illinois); Kross and others, 1990 (Iowa State-Wide Rural Well-Water Survey); Klaseus and others, 1989 (Minnesota); and LeMasters and Doyle, 1989 (Wisconsin).   Data on pesticide properties from compilation by Wauchope and others (1992).

Protection Agency, 1992a) indicated that the likelihood of detecting individual pesticides in the sampled wells was not significantly correlated with K_{oc}, and only weakly correlated with dissipation half-life ($\alpha = 0.05$).

Diagrams analogous to Figure 6.2 have been generated by other authors for ground-water monitoring surveys conducted in a diversity of locations (Boesten, 1987), including Iowa, Minnesota, Wisconsin, West Germany (Gustafson, 1989), and England (Hollis, 1991). Despite claims to the contrary, however (Gustafson, 1989), these analyses also indicate that threshold values for soil dissipation half-lives and K_{oc} cannot be used to distinguish reliably between pesticides that have and have not been detected in ground water during a given study. Thus, although these parameters may provide a general indication of which compounds are likely to be detected in ground water, they are not diagnostic.

6.1.2 POTENTIAL REASONS FOR DISPARITIES BETWEEN PREDICTION AND OBSERVATION

The minimal utility of soil dissipation half-lives and K_{oc} values for predicting ground-water contamination by pesticides is likely to arise from a number of different factors. These include: (1) the potentially significant influence of preferential transport on pesticide movement; (2) the substantial variability among literature values for soil dissipation half-life and K_{oc} for any given compound; (3) inherent uncertainties associated with the use of dissipation half-lives to predict pesticide fate; and (4) spatial variability of initial pesticide concentrations in soil water following application during a given study, in relation to analytical detection limits.

As noted in Section 4.3.3, the use of partition coefficients (such as K_{oc}) to predict the relative rates of advection of pesticides in subsurface waters assumes that all solutes are always at local equilibrium with respect to partitioning between the aqueous and solid phases. During preferential transport, however, most of this sorptive capacity is bypassed, severely limiting the value of partition coefficients as predictors of relative rates of transport among different contaminants. The limited ability of K_{oc} values to distinguish detected from nondetected pesticides in ground water thus suggests that, as was also noted in Section 4.3.3, preferential transport of pesticides may be substantially more common than is generally acknowledged.

Literature values for soil dissipation half-lives and K_{oc} exhibit pronounced variability among different studies, in some cases spanning two or more orders of magnitude for the same pesticide (e.g., Moye and Miles, 1988; Nash, 1988; and Wauchope and others, 1992). Although some of this variability is probably a result of differences in experimental approaches, much of it derives from the inherent variability in the physical, hydrologic, and chemical properties of the different soils and environmental settings in which these measurements were carried out, even after accounting for variations in organic carbon content. The environmental settings in which ground waters were sampled during the surveys examined in Figure 6.2 were likely to have been substantially different not only from those used to measure the K_{oc} values and soil half-lives used in the figures, but also from one site to another for the same study. This variability gives rise to a considerable degree of uncertainty regarding the placement of individual compounds in a K_{oc}-versus-soil dissipation half-life plot, and hence in the confidence with which these parameters can be used to distinguish pesticides likely to be detected in ground water from those that are not.

The variability exhibited by literature-derived soil dissipation half-life values may also arise from the fact that, as noted earlier (Section 2.10.2), the rate of dissipation of a pesticide at a given location is the net result of the simultaneous influence of transformation, volatilization, advection, and diffusion in situ. (Jury and others [1987] have used the term "surface zone residence time" as an alternative name for "dissipation half-life," to distinguish the simple reduction in concentration of a compound from its actual transformation within the soil.) Additional uncertainty is introduced when a single dissipation half-life is used for more than one compound simultaneously, as has often been the practice for aldicarb in conjunction with its sulfoxide and sulfone transformation products (e.g., Dejonckheere and others, 1983; Jones, 1986). Other difficulties associated with using dissipation half-life data to predict pesticide fate have been discussed elsewhere (e.g., Lavy and others, 1985; Cheng and Koskinen, 1986; and Jury and others, 1987).

The use of partition coefficients and soil dissipation half-lives to predict the compounds that are likely to be detected in ground water implicitly assumes that the ratio between the initial concentration (C_o) in subsurface waters immediately following application and the method detection limit (C_o/MDL) is the same for all compounds examined during a given study. Although the method detection limit (MDL) is fixed by the analytical method used for a given compound during a particular investigation, the magnitude of C_o is likely to be highly variable throughout the study area, as it is controlled by spatial and temporal variations in application rates (e.g., Logan and others, 1994) and environmental conditions. Compounds with large C_o/MDL ratios in a given setting would be more likely to be detected in ground water than those that exhibit comparable persistence and hydrophobicity, but have substantially lower C_o/MDL values. None of the studies examined on this subject confronted this difficulty. However, spatial variations in C_o/MDL ratios among different pesticides probably contributed to the inability of either K_{oc} or dissipation half-life values to predict patterns of pesticide detection and nondetection among the studies examined in Figure 6.2.

6.1.3 SIGNIFICANCE OF WATER SOLUBILITY

Water solubility is another factor that has often been used as a guide for anticipating the degree to which different pesticides will leach to ground water (e.g., Rodgers, 1968; Croll, 1972; García and others, 1984; Bishop and Lawyer, 1987; Albanis and others, 1988; Perry and others, 1988; Pionke and Glotfelty, 1989; Deubert, 1990; Isensee and others, 1990; Maes and others, 1991; Koterba and others, 1993; Logan and others, 1994; and Odanaka and others, 1994). However, as noted by Nicholls (1988), "despite being widely invoked in the literature, water solubility is rarely an important factor influencing deep leaching of pesticides. The concentration of pesticides in soil water, under conditions when leaching occurs, rarely approaches the water solubility of the pesticide," a point also noted by other authors (Gustafson, 1989; Logan and others, 1994). While the water solubilities of most pesticides fall within the range of 10^{-1} - 10^5 mg/L (Wauchope and others, 1992), the aqueous concentrations measured in subsurface waters are usually from 10^{-5} - 10^{-2} mg/L, either below the water table (e.g., Hallberg and others, 1984; Hallberg, 1989) or in the unsaturated zone (e.g., Glass and Edwards, 1979; Hall and others, 1989; Keim and others, 1989; and Kolberg and others, 1989).

The water solubility of a pesticide gives little information regarding its tendency to associate with the solid phase in the subsurface, where sorptive interactions are likely to be significant. For this reason, partition coefficients, whether keyed to a specific soil (K_p) or normalized to soil organic carbon (K_{oc}), provide a more valuable indication of pesticide transport properties in the subsurface (Section 4.3) than water solubility.

The limited value of water solubility as a predictor of ground-water contamination by pesticides is illustrated by the results from several process and matrix-distribution studies. Copin and others (1985) measured the concentrations in soil of three pesticides following their application to beet fields. Although the three compounds were chosen because they spanned a wide range of aqueous solubilities (100, 500, and 1,800 mg/L for ethofumesate, carbofuran, and metamitron, respectively), none were detected below the uppermost soil horizon sampled during either of two field seasons. In their study of herbicide movement through a clay loam, Hall and others (1989) found the maximum concentrations of simazine in pan-lysimeter percolates to equal or exceed those of cyanazine and metolachlor, even though: (1) the aqueous solubility of simazine is lower than those of the other two herbicides by two orders of magnitude, and (2) the three herbicides were applied at similar rates. Furthermore, the direct relations that have been noted between application rates and subsurface concentrations of pesticides (Section 3.4.4) would not have been observed if water solubility represented a significant control on pesticide behavior in the subsurface following applications at normal rates.

Whereas water solubility, by itself, does not appear to be a reliable predictor of relative mobilities among different pesticides, a considerable amount of effort has been invested in developing pesticide formulations that either increase or decrease the rate at which the active ingredient dissolves following application. These modifications influence the amount of pesticide available in solution at any given time. As noted earlier (Section 5.3.3), several investigations have examined the extent to which differences in pesticide formulations influence the concentrations and behavior of the active ingredients in the subsurface—albeit with variable results.

The foregoing discussion assumes that application rates are sufficiently low that dissolution of a pesticide at its site of deposition is complete within hours to days, rather than months to years. Upper limits on solubility and dissolution rates in water will, however, become more important following spills or other accidental releases of large quantities of these compounds onto the land surface or down wells. For example, a study by Butler (1981) demonstrated the substantially greater persistence of a microencapsulated formulation of methyl parathion, relative to an emulsifiable concentrate, in an undissolved state following a simulated spill.

6.2 INFLUENCE OF SOIL PROPERTIES

Soil permeability and organic-carbon content (f_{oc}) are of primary importance in controlling the degree to which pesticide residues pass through the rhizosphere to underlying ground waters. However, these two factors are so intimately correlated that their separate influences over pesticide transport and fate within the subsurface are not easily discerned (e.g., Nash, 1988). The organic-carbon content of soils tends to increase with increasing clay content (Nicholls, 1988), largely because of the strong tendency for natural organic solutes to form surface complexes with clays (Sposito, 1984). Thus, low-permeability soils typically exhibit higher f_{oc} values than more highly permeable soils.

These two factors influence pesticide behavior in the subsurface in a concerted fashion. Whereas lower permeability restricts pesticide movement by slowing the downward migration of the bulk soil solution (e.g., Helling and Gish, 1986), the presence of larger amounts of organic matter reduces the mass of pesticide reaching greater depths through the combined effects of enhanced microbial activity (Hurle and Walker, 1980; Felsot and others, 1982; Nash, 1988; Goetz and others, 1990; Harper and others, 1990; and Norris and others, 1991) and hydrophobic sorption. Indeed, it may be largely because of the difficulty of discerning the separate influences of soil permeability and organic-matter content that systematic attempts to determine the effect of one parameter on pesticide behavior in situ, while explicitly controlling for the other, appear to be absent from the published literature.

The covariation of f_{oc} with clay content also influences the results of laboratory studies. The observation that atrazine concentrations in soil slurries diminish more rapidly with increasing f_{oc} of the soil (Armstrong and others, 1967) may have arisen from the influence of increasing clay content (Section 4.4.6), rather than organic matter per se.

Studies examining pesticide contamination of ground waters in the San Joaquin Valley, California (Welling and others, 1986; Domagalski and Dubrovsky, 1991), Minnesota (Klaseus and others, 1988), Wisconsin (Krill and Sonzogni, 1986; LeMasters and Doyle, 1989), and the Delmarva Peninsula (Koterba and others, 1993) all observed higher frequencies of pesticide detection beneath coarse-grained (and usually low f_{oc}) soils than beneath finer grained (and usually high f_{oc}) soils. Similarly, Maes and others (1991) noted spatial correlations between areas in California exhibiting ground-water contamination by pesticides and those with soils containing comparatively low proportions of organic carbon. Consistent with these observations, spatial correlations have been reported between soil organic-carbon content and pesticide partition coefficients (K_p) at a variety of locations in the United States for atrazine (Moorman and others, 1994), aldicarb, metolachlor, and diuron (Rao and others, 1986).

Other large-scale monitoring investigations have not observed clear associations between soil properties and pesticide detections in ground water. Results obtained from wells in 17 states during the ongoing Cooperative Private Well Testing Program (CPWTP) indicated monotonic decreases in the frequencies of either detection or exceedance of 2.0 µg/L for acetanilides with decreasing soil permeability (i.e., sandy > "in-between" > clay), but no clear relation of this sort for the triazine herbicides (Baker and others, 1994). In addition, neither a survey of private wells in Ontario, Canada, conducted by the Ontario Ministry of the Environment, or OMOE (Ontario Ministry of the Environment, 1987a), nor a study of alachlor occurrence in 242 wells in 15 United States counties with substantial alachlor use (Holden and others, 1992) observed any significant associations between pesticide detections in ground water and soil permeability or organic matter.

One explanation for the absence of clear relations between soil texture and pesticide occurrence in ground water during these three investigations (Ontario Ministry of the Environment, 1987a; Holden and others, 1992; and Baker and others, 1994) is that the soil texture in the immediate vicinity of a well is not necessarily the same as that in the recharge area for the water drawn from the well (Hallberg, 1995). Furthermore, for the OMOE and alachlor studies, the effects of variations in soil properties on pesticide movement to ground water may have been overshadowed by uncontrolled variations in well construction or other factors. The Ontario study sampled a variety of different well types—including drilled, dug, bored, and sandpoint wells—and, indeed, documented significant differences in pesticide detection frequencies among these different well types (Ontario Ministry of the Environment, 1987a), as will be discussed in Section 6.5. No details on well construction were given for the alachlor study, but uncontrolled variations in construction were inferred from the design of the substantially more extensive follow-up investigation, the National Alachlor Well-Water Survey (NAWWS), during which well type was not reported to have been a criterion for well selection (Monsanto Agricultural Company, 1990; Holden and others, 1992).

In Iowa, a statewide effort to assess ground-water susceptibility to agrichemical contamination explicitly excluded consideration of varying soil types, in all areas except for those underlain by alluvial aquifers. According to a recent report by a committee of the National Research Council (NRC), "the [Iowa] Department of Natural Resources staff suggested that since the soil cover in most of the state is such a small part of the overall aquifer or well cover, processes that take place in those first few inches are relatively similar and, therefore, insignificant in terms of relative susceptibilities to ground-water contamination" (National Research Council, 1993). This assumption was also based on the relatively narrow range of soil properties encountered across much of Iowa (Hallberg, 1995).

Investigations of pesticide contamination in the subsurface over smaller scales, beneath parts of Fresno, Merced, and Tulare Counties in California (Teso and others, 1988; Wilkerson and others, 1985; and Zalkin and others, 1984), Portage and Waushara Counties in Wisconsin (Harkin and others, 1986), the Mahantango Creek Watershed in Pennsylvania (Pionke and others, 1988; Pionke and Glotfelty, 1989), and a 500-km^2 glacial outwash area in northern Germany (Milde and others, 1988), have provided either marginal or negligible evidence for any significant associations between soil texture or f_{oc} and pesticide contamination in the subsurface. Although a large number of studies have documented extensive pesticide contamination of ground waters in areas with highly permeable soils in Wisconsin (Harkin and others, 1986), Nebraska (Wehtje and others, 1983), Illinois (McKenna and others, 1988), Long Island, N.Y. (e.g., Zaki and others, 1982), and California (Cohen, 1986), these studies generally did not sample ground waters in nearby areas with less permeable soils for comparison. During the Pennsylvania studies (Pionke and others, 1988; Pionke and Glotfelty, 1989), the effects of soil properties on pesticide behavior in the subsurface may have been obscured by uncontrolled variations in the intensity of agricultural activity—and hence pesticide use—in the area investigated.

At the field scale, most process and matrix-distribution studies have also failed to detect consistent relations between soil properties and pesticide occurrence, concentrations, or persistence (Hurle and Walker, 1980; Nash, 1988; and Lagas and others, 1989). Keim and others (1989) found that atrazine migrated to greater depths and was detected at greater frequencies and higher concentrations beneath a sandy soil than beneath a clayey soil in northwestern Ohio. The atrazine was applied at a higher rate to the more permeable soil, however, making it difficult to distinguish between the effects of application rate and soil texture on atrazine behavior during the study.

A study of 20 agrichemical mixing-and-loading facilities in Wisconsin by Habecker (1989) provided extensive comparisons between pesticide occurrence in ground waters and in overlying soils. The detection of pesticides in surficial soils at these facilities was found to be an unreliable predictor of pesticide contamination in either water-supply or observation wells. At some sites, pesticides were detected in the ground water, but not in the soil. Some of these apparent inconsistencies may have arisen from the fact that the measurement of pesticide concentrations in soils often involves higher detection limits and greater variability than measurements in water (Hallberg, 1995). Although a variety of data were collected on soil texture, permeability, and organic-matter content, as well as on the types and depths of bedrock beneath the 20 facilities, no obvious correlations were evident between any of these characteristics and the levels of pesticide contamination in the ground water.

The fact that pesticide concentrations in ground waters beneath the Wisconsin agrichemical facilities were not correlated with the organic-carbon content of the soils suggests that high pesticide concentrations may overwhelm the capacity of surficial or sub-surface soils to retard the downward migration of spilled chemicals. Closer examination of the Habecker (1989) data, however, suggests that variations in hydrophobicity among different spilled compounds may have influenced their subsequent migration to underlying ground water. The pesticides detected in soils, but not in the ground waters at the Wisconsin facilities (chlorpyrifos, fonofos, glyphosate, parathion, pendimethalin, and trifluralin) all exhibited log(K_{oc}) values of 2.9 or higher (Figure 6.3). Conversely, every one of the compounds with log(K_{oc}) values below 2.9 was detected in the ground water beneath at least one of the sites and, in some cases, in the soils as well.

The results discussed above suggest that pesticide contamination of ground waters may be detected more commonly beneath coarse, highly permeable, organic-carbon-poor soils than

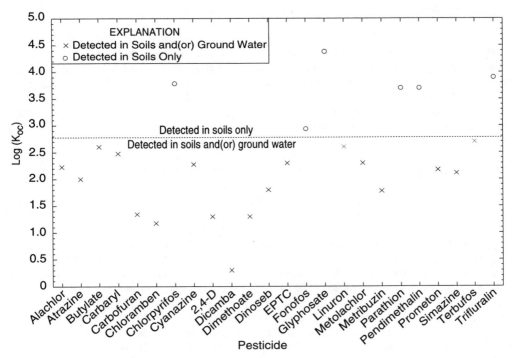

Figure 6.3. Pesticides detected in soils alone, compared with those detected in soils and(or) ground water at agrichemical mixing-and-loading facilities in Wisconsin (data from Habecker, 1989), versus their log(K_{oc}) values. K_{oc} data from Wauchope and others (1992).

beneath fine-textured, less permeable, organic-rich soils. However, such effects may only be discernible among areas that exhibit substantial contrasts in these properties, or where differences in soil properties are not overwhelmed by variations in other factors, such as well construction (Ontario Ministry of the Environment, 1987a; Holden and others, 1992), intensity of agricultural activity (Pionke and others, 1988; Pionke and Glotfelty, 1989), pesticide application rates (Keim and others, 1989), or pesticide concentrations in soil (Habecker, 1989).

6.3 SIGNIFICANCE OF HYDROGEOLOGIC SETTING

Despite the importance of the issue, few studies have systematically examined the influence of variations in hydrogeologic setting on pesticide occurrence in ground water. Furthermore, conclusions regarding this influence may be adversely affected by the fact that hydrogeologic setting often correlates with well type; dug, bored, or driven wells, for example, are only installed in unconsolidated materials or heavily weathered rock (Glanville and others, 1995; Hallberg, 1995). As noted elsewhere (Sections 3.4.4 and 6.5), few studies controlled for variations in well construction. These concerns notwithstanding, however, most of the research indicates that, consistent with expectation, shallow, unconfined, and unconsolidated aquifers and mature karst areas are the hydrogeologic environments most susceptible to pesticide contamination.

6.3.1 UNCONSOLIDATED VERSUS BEDROCK FORMATIONS

Investigations in Iowa (Kelley, 1985; Kelley and Wnuk, 1986) and in the 12 states sampled during the MCPS (Kolpin and others, 1993), found that frequencies of pesticide detection are significantly higher in unconsolidated formations than in bedrock. A similar pattern was also seen with respect to the occurrence of high nitrate concentrations in Ohio (Baker and others, 1989) and the MCPS study area (Kolpin and others, 1993).

Four lines of evidence indicate that, among the near-surface hydrogeologic settings sampled during the MCPS, the bedrock aquifers are more isolated from contamination by surface-applied pesticides than the unconsolidated, sand-and-gravel aquifers (Kolpin and others, 1993, 1994). First, as noted above, the frequencies of herbicide detection in the bedrock wells were significantly lower ($\alpha = 0.05$) than those in the wells screened in unconsolidated deposits (17 percent versus 28 percent, respectively). Second, although Kolpin and others (1994) observed a significant difference in herbicide detection frequencies between the pre- and post-planting samplings of ground waters taken from the unconsolidated, sand-and-gravel surficial aquifers, this difference was not significant among samples obtained from bedrock aquifers. The more muted oscillations in pesticide concentrations in the bedrock aquifers suggest longer travel distances from the contaminant sources (see Section 3.5.3). Third, the median deethyl atrazine-to-atrazine concentration ratio (DAR) in the unconsolidated aquifers (0.57) was less than half of that in the bedrock aquifers (1.25). The authors concluded that "the much smaller DAR for unconsolidated aquifers suggests either [more] rapid recharge rates or [shorter] distances from recharge areas" relative to the bedrock aquifers (Burkart and Kolpin, 1993). Potential limitations of the use of the DAR and similar ratios for understanding pesticide behavior in the subsurface are discussed in Section 8.9.5. Finally, the proportion of wells drawing water recharged before 1953, based on tritium analyses (Section 9.2.2), was significantly higher for the near-surface bedrock aquifers (50.0 percent) than for the near-surface unconsolidated aquifers (9.1 percent) as noted by Kolpin and others (1995).

In apparent contradiction of these conclusions, however, Neil and others (1989) detected pesticides more frequently, and found nitrate concentrations to be higher, in wells screened in bedrock than in those screened in either till or sand-and-gravel formations in northern and western Maine. Furthermore, all four of the wells containing detectable concentrations of more than one pesticide were drilled in bedrock. These results suggest that the rates of solute migration through fractures in the bedrock were more rapid than in the unconsolidated deposits of the study areas, leading to more frequent pesticide detections and less diminution of nitrate concentrations. This is consistent with the fact that the velocity of ground-water movement through fissures in a fractured rock is substantially higher than that through an unfractured granular medium exhibiting the same bulk hydraulic conductivity, when both media are subjected to the same hydraulic gradient (Freeze and Cherry, 1979).

Results from the Minnesota Department of Health study (Klaseus and others, 1988) indicate that, in some cases, pesticide contamination of bedrock wells may be controlled by transport rates through overlying deposits, rather than through the bedrock itself. Among the public wells that were screened in bedrock (sedimentary, igneous, or metamorphic) overlain by unconsolidated deposits, the frequencies of pesticide detection decreased with increasing depth to bedrock. Most of these wells were finished in the first bedrock formation encountered during drilling, or in formations in direct hydraulic connection with the overlying unconsolidated materials (Klaseus and others, 1988). Consequently, these observations appear to be a direct reflection of the inverse relation observed by the authors between the frequencies of pesticide occurrence and the depths of the well screens in the unconsolidated aquifers.

In contrast with the other studies discussed in this section, Christenson and Rea (1993) did not observe significant differences between the frequencies of pesticide occurrence in water-supply wells (mostly domestic) screened in unconsolidated alluvial deposits and those completed in bedrock (sandstones, siltstones, and mudstones) in an urban area of Oklahoma City. The authors suggested that mixing within the annular seals of the sampled wells, which were typically gravel-packed from the bottom of the well to within 3 m of the ground surface, may have obscured variations in pesticide occurrence among the different subsurface formations (Christenson and Rea, 1993).

6.3.2 INFLUENCE OF LOW-PERMEABILITY MATERIALS NEAR THE LAND SURFACE

The presence of low-permeability earth materials at or below the land surface sometimes provides a measure of "protection" of shallow aquifers from contamination by surface-applied pesticides. However, this has only been documented systematically for a few areas, most of which have been in the northern midcontinent. In Iowa, the surficial deposits that retard the downward migration of pesticides are primarily glacial tills (Libra and others, 1984; Detroy and others, 1988; and Hallberg, 1989). In Minnesota, lacustrine clays and tills serve this function (Klaseus and others, 1988). Both the NAWWS (Holden and others, 1992), and the NPS (U.S. Environmental Protection Agency, 1992a) found pesticide detection frequencies to be significantly higher in wells tapping unconfined aquifers than in those screened in confined formations.

Ground-water monitoring studies conducted in Floyd and Mitchell Counties, Iowa, have provided one of the clearest demonstrations of the influence of low-permeability surficial deposits on the occurrence of pesticides in the subsurface. Two separate investigations in these counties both showed monotonic declines in the frequencies of pesticide detection in wells screened in carbonate bedrock as the thickness of the overlying, low-permeability glacial till deposits increased from 1.5 m to greater than 15 m (Libra and others, 1984; Hallberg, 1989). Similarly, Glanville and others (1995) observed significantly lower triazine herbicide concentrations ($\alpha = 0.05$) in Iowa wells screened in aquifers overlain by more than 50 ft of till than where the till thickness was less than 50 ft.

These direct relations between near-surface permeability and the likelihood of detecting pesticides in ground water reflect the connection between pesticide occurrence and ground-water age. All other factors being equal, the lower the near-surface permeability, the older the underlying ground water, and the lower the likelihood of detecting pesticides.

Patterns of pesticide detection in Illinois provide indirect evidence that low-permeability materials at or near the land surface may reduce the incidence of pesticide contamination in underlying ground waters. Pesticide detection frequencies in Illinois ground waters seem remarkably low, given the extensive use of pesticides in the state, but studies are limited (Hallberg, 1989). The most recent results from the CPWTP (Baker and others, 1994) show a lower frequency of triazine herbicide occurrence (at concentrations above 3 µg/L) in Illinois than in Indiana and Ohio, the two other high-use states in the midcontinent from which most of the data from this program have been reported to date. The first year of sampling during the MCPS (1991) also detected this pattern, particularly in the east-central portion of Illinois (Burkart and Kolpin, 1993). According to McKenna (1990), such low frequencies of pesticide detection are likely to arise from the presence of thick deposits of fine-grained materials overlying the near-surface aquifers beneath this portion of the state (McKenna and Keefer, 1991).

Other studies have shown that pesticides are more likely to be encountered in ground waters beneath areas of greater subsurface permeability. Richards (1992) found that frequencies of detection and concentrations of triazine herbicides in ground water were substantially higher in an area underlain by a porous limestone reef than in the less permeable rocks surrounding the reef in Hancock, Seneca, and Wyandot Counties, Ohio. An analogous, though less consistent, pattern was observed for the acetanilide herbicides. Similarly, Chen and Druliner (1987) found triazine herbicide concentrations in the High Plains Aquifer beneath Nebraska to exhibit a significant, positive correlation with the hydraulic conductivity of the formations sampled ($\alpha = 0.05$).

6.3.3 KARST AREAS

Karst areas seem to be as susceptible to pesticide contamination as surficial, unconsolidated formations. This is important because, as noted by Hallberg (1987), such areas are considerably more common than is indicated by the limited attention they have received. Carbonate aquifers provide a substantial proportion of the ground-water supplies in 15 states (White, 1989), while 13.5 percent of Europe and 4 percent of the Earth's terrestrial area are underlain by bedrock subject to karstification (Simmleit and Herrmann, 1987a). Results from the long-term investigation of the Big Spring Basin, in northeastern Iowa (Rowden and others, 1993) illustrate the rapid response of pesticide concentrations in spring discharge to individual recharge events that may be observed in agricultural areas underlain by karst formations (Figure 6.4).

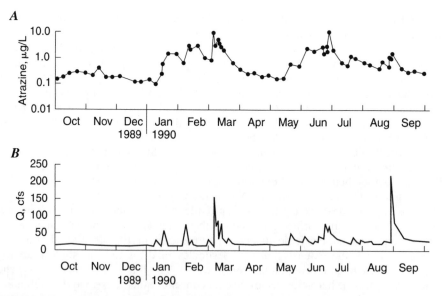

Figure 6.4. Atrazine concentrations (*A*) and ground-water discharge (*B*) at Big Spring, in northeastern Iowa, from October 1989 to September 1990. Redrawn with permission from Rowden and others (1993). cfs, cubic feet per second.

Susceptibility to Pesticide Contamination

Three characteristics of karst aquifers appear to be largely responsible for their high vulnerability to surface-derived pesticide contamination. First, sinkholes, fractures, "swallow holes," and other open conduits provide routes for the direct entry of surface-derived contaminants into the subsurface. Second, the rapid transport of solutes through solution channels in karst formations is analogous to the preferential transport of solutes through soil macropores, discussed earlier in Sections 4.3.3 and 5.3.2. Finally, the relatively low amounts of organic carbon in the earth materials within such systems limits their capacity to attenuate contaminant concentrations through sorption and transformation.

Surface-water hydrograph separation techniques have been adapted to distinguish the proportions of agrichemicals in spring discharge attributable to the more rapid "conduit flow," or surface-derived "run-in," from those contributed by infiltration, or "diffuse flow," through karst systems. From these methods, most of the pesticide (and other solute) mass discharged from Big Spring has been found to be derived from infiltration recharge, rather than conduit flow (Hallberg and others, 1983, 1984; Libra and others, 1986; and Mitchem and others, 1988). This observation may help explain why monitoring studies in Iowa have not found the presence of one or more sinkholes near wells to be associated with significantly higher frequencies of pesticide detection (Kross and others, 1990) or herbicide concentrations (Glanville and others, 1995), relative to wells that are not located near sinkholes.

Although the entry of surface-derived solutes through surface openings may only account for less than half of their total discharge from karst areas, such conduits still represent potential sources of short-lived, but severe contamination of such systems (e.g., Hallberg and others, 1983), much as individual precipitation events can contribute significant loads of contaminants to nearby surface waters over time scales of a few hours (e.g., Richards and Baker, 1992; Kuivila, 1993; Cessna and others, 1994; and Meyer and Thurman, 1995). Mitchem and others (1988), for example, related an instance in which "whey dumped from a cheese factory in 1963 ran into a sinkhole and into the aquifer, killing all the fish at a trout hatchery supplied by a spring more than 16 km (10 mi) away." Direct entry of contaminated surface waters into karst systems through surface openings is a concern in heavily agricultural areas because during major recharge events, pesticide concentrations in spring effluent are essentially the same as those in surface waters (e.g., Hallberg and others, 1984; Libra and others, 1986) which, in turn, often carry pesticides at concentrations substantially higher than those in percolating ground waters beneath such areas (e.g., Hallberg, 1987; Thompson, 1990; Squillace and Thurman, 1992; and Blum and others, 1993). In addition, sinkholes have frequently been used to dispose of empty pesticide containers, as well as farm machinery and other solid wastes.

Despite the relatively high susceptibility of karst ground waters to surface-derived contamination, the total proportions of applied pesticides discharged from springs draining such basins are comparable to those measured in other hydrogeologic environments. While the proportions of applied pesticides that reach subsurface waters in various hydrogeologic settings have ranged from 0.003 percent to 9.6 percent (Table 4.1), estimates of the mass of atrazine and other pesticides discharged from Big Spring range from less than 0.1 percent (Libra and others, 1986) to approximately 5 percent (Mitchem and others, 1988) of the total mass applied to the basin. In Germany, the proportions of α-HCH and γ-HCH (lindane) entering a karst system in Upper Franconia that were estimated to have been discharged in spring flow from February to August, 1984, were of similar magnitude, ranging from 0.3 percent to 0.8 percent (Simmleit and Herrmann, 1987b).

Comparisons with Other Hydrogeologic Settings

Few investigations have provided systematic comparisons of pesticide detection frequencies or concentrations between karst and other hydrogeologic settings. Studies conducted in Floyd and Mitchell Counties, Iowa (discussed earlier), detected pesticides with greater frequency in karst formations than in nonkarst bedrock overlain by low-permeability glacial tills varying in thickness from 1.5 m to greater than 15 m (Hallberg, 1989). In Minnesota, the highest frequencies of pesticide detection in ground water were encountered in either karst or shallow sand-and-gravel formations (Klaseus and others, 1988). During the MCPS, wells in the karst regions of northeastern Iowa and southwestern Wisconsin were found to be more susceptible to pesticide contamination than those completed in other near-surface bedrock aquifers (Kolpin, 1995a).

Atrazine detections in ground water during NAWWS were most common in areas underlain by karst in southeast Minnesota, northeast Iowa and west-central and southern Wisconsin. Although frequencies of atrazine detection were greater than 20 percent in each of the nine counties sampled in these areas, atrazine was detected in less than 5 percent of the wells sampled in the rest of the midcontinent corn belt (U.S. Environmental Protection Agency, 1995). Given the widespread occurrence of unconsolidated aquifers beneath the midcontinent (Burkart and Kolpin, 1993), this suggests that karst formations are even more susceptible to pesticide contamination than unconsolidated materials, a pattern that has also been noted in Germany on the basis of atrazine occurrence (Milde and others, 1988).

During their study of the influence of agricultural activities on ground-water quality in the Upper Conestoga River basin in southeastern Pennsylvania, Fishel and Lietman (1986) found nearly all of the pesticide detections to occur in areas underlain by carbonate, rather than conglomerate, shale, sandstone, or diabase bedrock. While it was not mentioned by the authors, these carbonate aquifers are probably karstified, since they are located in one of the major karst areas of the United States (Kurtz and Parizek, 1986; Hallberg, 1987; and White, 1989).

6.3.4 INFLUENCE OF AQUIFER DEPTH

It is reasonable to suppose that, other factors being equal, the greater the depth of an aquifer below the land surface, the older the ground water, and thus the less likely it is to be contaminated by surface-derived pesticides. However, only a few studies have examined this issue in a systematic manner. During the 1991 sampling of the MCPS (Burkart and Kolpin, 1993), a highly significant, inverse relation was observed ($\alpha = 0.05$) between herbicide detection frequencies and the depth below the land surface of the uppermost surface of the sampled aquifer—regardless of whether the earth materials were saturated or unsaturated at that depth. This was the strongest relation that the authors observed between herbicide detection frequencies and any of the "depth factors" examined (i.e., depth of the well, open interval, water level, or aquifer). The observation of lower frequencies of pesticide detection with increasing depths to bedrock in Floyd and Mitchell Counties, Iowa (Section 6.3.2), is consistent with these results.

Three other studies that examined this issue, all conducted in Illinois (Goetsch and others, 1992, 1993; Mehnert and others, 1995), provided further support for this trend. During their statewide survey of rural, private water-supply wells, Goetsch and others (1992) found the frequencies of pesticide detection to be more than twice as high where the uppermost surface of the sampled aquifers was within 20 ft of the land surface than where it was more than 20 ft deep (14.1 percent versus 6.8 percent, respectively), although this difference was not statistically significant ($\alpha = 0.05$). Similarly, from a sampling of 240 private wells in rural Illinois, Mehnert and others (1995) observed atrazine detection frequencies to increase with decreasing depth of the uppermost aquifer. The third Illinois study, discussed further in Chapter 8, examined pesticide occurrence in the subsurface beneath 52 agrichemical handling facilities in Illinois (Goetsch and others, 1993). While the statistical significance of the relation was not reported, the data suggested a general decrease in pesticide detection frequencies with increasing aquifer depth (Goetsch and others, 1993). The detection of similar trends in all three of the Illinois studies provided support for a ground-water vulnerability assessment system devised for the state on the basis of aquifer depth (McKenna and Keefer, 1991).

6.3.5 DEPTH OF THE WATER TABLE

As the thickness of the unsaturated zone increases, so does the time required for surface-derived solutes to reach ground water. This results in more time for transformation, sorption, hydrodynamic dispersion, and volatilization to reduce the amount of the parent compound that ultimately reaches the water table. Several process and matrix-distribution studies (e.g., Harkin and others, 1986; Troiano and others, 1990; Smith and others, 1991a; and Mills and Thurman, 1994b) have tracked the downward movement of pesticide pulses through the unsaturated zone over time. Results suggest that inverse relations between water-table depth and either pesticide concentrations or detection frequencies are more likely to be observed as the transit time to the water table increases. Repeated sampling of ground water in several potato-growing areas of Long Island, New York, by Pacenka and others (1987) indicated that the greater the water-table depth beneath a field to which aldicarb had been applied, the longer the time interval required before the peak concentration was observed, and the lower the peak concentration in nearby wells. Similarly, Priddle and others (1992) found depth-to-water to be one of the two strongest predictors of aldicarb residue concentrations in ground waters beneath potato fields in Prince Edward Island, Canada. The other parameter, noted earlier (Section 5.1.3), was the time elapsed since application.

Most of the reviewed studies, however, did not detect significant relations between pesticide occurrence and water-table depth. Concentrations of atrazine in ground water beneath the Mahantango Creek watershed, in Pennsylvania, were found to be more strongly controlled by the intensity of corn cultivation than by the depth to water (Pionke and others, 1988; Pionke and Glotfelty, 1989). During a study of pesticides in wells located immediately adjacent to corn and soybean fields along the Illinois River in Mason County, Illinois, McKenna and others (1988) found no discernible correlations between water-table depth and either pesticide concentrations or detection frequencies. Similarly, Koterba and others (1993) found the influence of water-table depth on the frequencies of pesticide detection in the ground waters beneath the Delmarva Peninsula to be nonsignificant ($\alpha = 0.05$). In addition, neither the MCPS (Burkart and Kolpin, 1993) nor the NAWWS (Holden and others, 1992) showed significant correlations between herbicide detection frequencies and water depth in the sampled wells.

During the Illinois and Delmarva studies, the absence of significant relations between water-table depth and pesticide occurrence in ground water was probably related to the shallow water tables and highly permeable soils and subsurface materials—and hence the rapid solute transport rates through the unsaturated zone—at these locations. By contrast, the relations observed between pesticide concentrations and water-table depths during the aldicarb studies may have resulted from relatively long solute-transit times through the unsaturated zone beneath these sites, caused either by a comparatively deep water table (Pacenka and others, 1987) or by the presence of low-permeability materials at the land surface (Priddle and others, 1992). The reasons why neither of the multistate studies that examined this question detected significant effects of water-table depth on pesticide detection frequencies remain unclear, although both investigations cited difficulties with the reliability of the water-level data as potential causes (Holden and others, 1992; Kolpin and others, 1994). Other studies have attempted to account for variations in pesticide detection frequencies through regional patterns of water-table depth (e.g., Chen and Druliner, 1987; Detroy and others, 1988; Exner and Spalding, 1990; Kross and others, 1990; Domagalski and Dubrovsky, 1991), but none reported on the statistical significance of such relations.

6.3.6 DEPTH OF THE WELL SCREEN BELOW THE WATER TABLE

Only two of the investigations reviewed here specifically examined the influence of the depth of the well screen below the water table (McKenna and others, 1988; Koterba and others, 1993). An Illinois study (McKenna and others, 1988) showed that neither the concentrations nor the frequencies of detection of pesticides exhibited any discernible relation with the depths of monitoring well screens beneath the water table. Koterba and others (1993), however, found the distribution of wells with pesticide detections in the Delmarva Peninsula to be skewed toward shallower screen depths below the water table, while the wells without pesticide detections were skewed toward greater screen depths. The difference between the two distributions was statistically significant ($\alpha = 0.05$), in marked contrast to the absence of any detectable relation between pesticide detections and water-table depth (Koterba and others, 1993), discussed earlier.

The difference between the conclusions reached in the Illinois and Delmarva investigations may have arisen because the latter study examined a substantially larger range of well-screen depths below the water table. Whereas information provided by McKenna and others (1988) suggests that the Illinois wells were screened between 2 and 11 m below the water table (more specific data were not provided), the Delmarva study sampled wells screened between 1.8 and 40 m below the water table (Koterba and others, 1993). Differences in subsurface permeabilities were unlikely to have been responsible for the contrasting conclusions, as both studies were conducted in highly permeable unconsolidated deposits.

The significant effect of well-screen depth on pesticide detection frequencies during the Delmarva study may also have been a consequence of historical changes in pesticide use in the study area (Koterba and others, 1993). Estimated ages of the ground waters obtained from some of the wells used during the study—based on chlorofluorocarbon concentrations—ranged from less than 1 year to greater than 50 years (Dunkle and others, 1993). Atrazine, the triazine herbicide used earliest and most frequently within the study area, did not come into widespread use in the area until the mid-1970's. Consistent with this, most of the detections of atrazine and other commonly used pesticides occurred in ground waters that had recharged after 1970 (Koterba and others, 1993).

Among the various hydrogeologic settings from which shallow ground waters are withdrawn, karst aquifers and unconsolidated, unconfined aquifers are likely to be the most susceptible to contamination by surface-applied pesticides. With the exception of the study conducted in Maine by Neil and others (1989), most studies indicate that bedrock aquifers (other than those in karst areas) are less likely to be contaminated by pesticides than unconsolidated deposits. Both direct and indirect evidence from several monitoring studies shows that the detection of pesticides in ground water becomes less likely with: (1) decreasing hydraulic conductivity of the aquifer materials surrounding the well screen; (2) increasing thickness of low-permeability materials at or close to the land surface; (3) increasing depth of the top of the sampled aquifer; (4) increasing depth of the water table; and (5) increasing depth of the well screen below the water table.

6.4 SIGNIFICANCE OF INTERACTIONS BETWEEN GROUND WATER AND SURFACE WATER

Rivers and streams interact with ground water in their associated aquifers. As a result, contaminant concentrations in aquifers near streams are influenced by contaminants in stream water in losing reaches, and the opposite occurs in gaining reaches. The nature and degree of these influences is determined by the rate at which water flows between a river and an aquifer, and the direction of flow, both of which vary over time and space.

6.4.1 EFFECTS OF SURFACE-WATER RECHARGE ON PESTICIDE CONCENTRATIONS IN SHALLOW GROUND WATER

Although it is not universally true (e.g., McMahon and others, 1994), pesticide concentrations are generally greater in surface waters than in ground waters in regions dominated by agriculture (Hallberg, 1987; Thompson, 1990; Squillace and Thurman, 1992; Blum and others, 1993; and Kalkhoff and Schaap, 1995). This may explain why, during periods of runoff, when the entry of surface waters into the subsurface is most likely, substantial increases in pesticide concentrations have been observed in ground waters drawn from alluvial aquifers in agricultural areas. This influence on ground-water quality is most likely in the spring, when pesticide concentrations in rivers (Goolsby and others, 1991; Thurman and others, 1991; Squillace and Thurman, 1992; Baker, 1993; Blum and others, 1993; Pereira and Hostettler, 1993; Richards and Baker, 1993; Schottler and others, 1994; and Kalkhoff and Schaap, 1995) and runoff volume typically exhibit their annual maxima.

Investigations of atrazine exchange between the Cedar River and its alluvial aquifer in Iowa illustrate the dynamic nature of these interactions in an agricultural region (Liszewski and Squillace, 1991; Squillace and others, 1993; and Wang and Squillace, 1994). In February, when river flow is supplied primarily by ground-water discharge ("base-flow" periods), atrazine concentrations in the river are low and similar to those in ground water (Figure 6.5). By March, when streamflow is sufficiently high to cause recharge of the underlying aquifer, the atrazine concentrations in the ground water close to the river bed increase markedly, reflecting the substantially higher levels in the river itself. The elevated atrazine concentrations in the aquifer upgradient from the river in March, however, indicate that surface-water recharge is not the only mechanism by which pesticides may enter alluvial aquifers—they may also reach ground water by direct infiltration from the land surface. By April, flow in the river is once again dominated by ground-water discharge and the atrazine concentrations in the river return to lower concentrations (Liszewski and Squillace, 1991; Squillace and others, 1993).

Consistent with these detailed observations of river-ground-water interaction, Burkart and Kolpin (1993) found that the frequency of herbicide detection in ground waters beneath the corn belt of the midcontinental United States was more than twice as high in wells located within 30 m of a stream (48 percent) than in wells located a greater distance from streams (22 percent). Similarly, Liszewski and Squillace (1991) found atrazine levels in the alluvial aquifer of the Cedar River to rise in response to concentration increases in the river water for wells located within 30 m of the river (compare Figures 6.6 and 6.7), but no such correlations were detected in wells located farther away.

Municipal water-supply wellfields are often installed adjacent to major rivers in the United States and Europe to ensure a relatively reliable source of supply, under the assumption that "bank filtration" by the earth materials in the alluvial aquifer can remove contaminants present in the river water. Blum and others (1993), however, showed that transient increases in atrazine concentrations in the Platte River near Ashland, Nebraska, led to similar but smaller

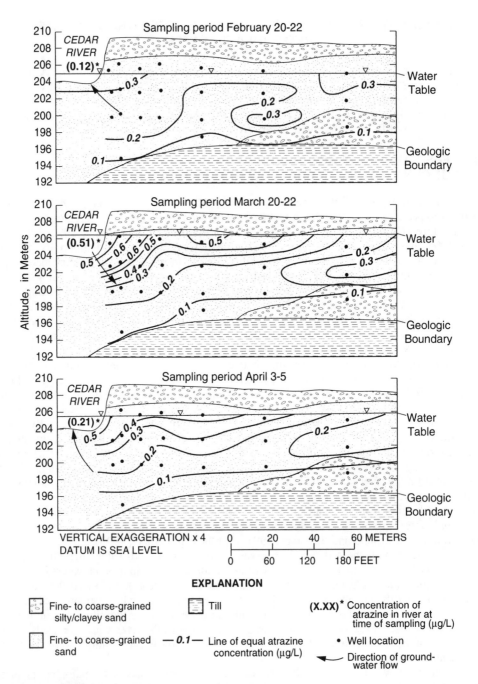

Figure 6.5. Cross-sectional distribution of atrazine in the Cedar River, Iowa, and the adjacent alluvium prior to (February), during (March), and following (April) the spring herbicide flush. Redrawn from Squillace and others, 1993. *Water Resources Research*, Volume 29, pages 1719-1729. Copyright 1993 American Geophysical Union.

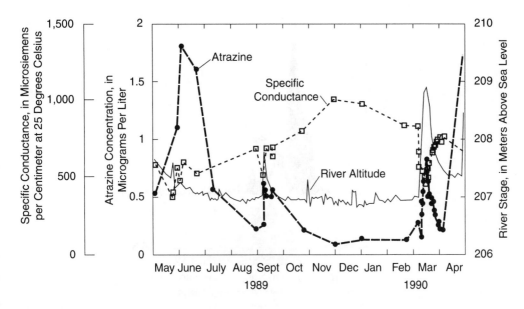

Figure 6.6. Atrazine concentration and specific conductance in the Cedar River, Iowa, compared to river stage, from May 1989 through April 1990 (Liszewski and Squillace, 1991).

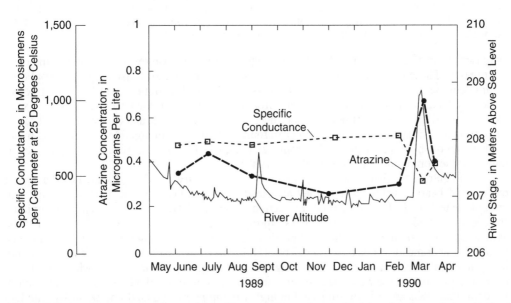

Figure 6.7. Atrazine concentration and specific conductance in an alluvial well screened at a depth of 6 meters and located 10 meters from the Cedar River, Iowa, compared to river stage, from May 1989 through April 1990 (Liszewski and Squillace, 1991).

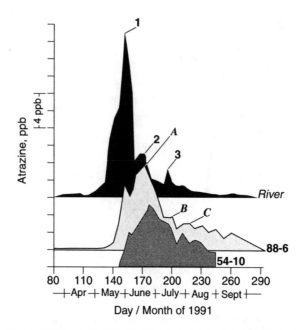

Figure 6.8. Atrazine concentrations in the Platte River and wells located 10 feet (monitoring well 88-6) and 200 feet (production well 54-10) from the Platte River near Ashland, Nebraska, between March and October 1991. Peaks identified as 1, 2, and 3 in the river were presumed to have given rise to peaks labeled A, B, and C, respectively, in monitoring well 88-6. Redrawn from Blum and others (1993) by permission of *Ground Water Monitoring and Remediation.* Copyright 1993.

increases in atrazine levels in both municipal-supply and monitoring wells adjacent to the river (Figure 6.8). These results demonstrate that, as observed for low molecular weight, hydrophobic organic compounds (Schwarzenbach and others, 1983), bank filtration may be an ineffective strategy for removing hydrophilic pesticides from contaminated surface waters. Consistent with these observations, water treatment methods designed to remove taste and odor problems through adsorption to activated carbon have also been found to be largely ineffective for removing many of the common, hydrophilic pesticides from water, although such processes may be optimized to remove these compounds, as well (Hallberg, 1987).

In contrast to other studies, the National Pesticide Survey (NPS) provided evidence that recharge of nearby surface waters may also lead to reductions in pesticide concentrations in ground water (U.S. Environmental Protection Agency, 1992a). The presence of unlined surface-water bodies (rivers, canals, bays, springs, or ponds) within 300 ft was found to be associated with a reduced likelihood of pesticide detections in community water system wells, though not in domestic wells ($\alpha = 0.05$). The NPS involved extensive sampling of wells not only in agricultural areas, but also in non-agricultural regions (Figure 2.4), where pesticide concentrations in surface waters are considerably lower (Gilliom and others, 1985). Community water system wells are usually subjected to higher pumping rates than domestic wells. The fact that this effect was significant in community water system wells during the NPS, but not in domestic wells, suggests that it may have been a result of induced recharge from nearby—and, on a nationwide basis, generally less contaminated—surface waters.

6.4.2 EFFECTS OF GROUND-WATER DISCHARGE ON PESTICIDE CONCENTRATIONS IN SURFACE WATER

Pesticides in alluvial aquifers also may contaminate adjacent surface waters. However, in contrast to the acute, transient pesticide loads that can be contributed to ground water by surface waters (Figure 6.8), the discharge of ground water to surface water appears to be responsible for maintaining the chronic, albeit lower concentrations of the more persistent pesticides observed in rivers during periods dominated by base flow (Hallberg, 1987). For example, Kalkhoff and Schaap (1995) reported significant, inverse correlations ($\alpha = 0.05$) between the percentage of ground-water inflow to Deer Creek, in Clayton County, Iowa, and the concentrations of atrazine and metolachlor in the creek.

Liszewski and Squillace (1991) found that several months may be required following even a moderate-sized recharge event before pesticide concentrations in a river and its associated aquifer approach similar levels (compare Figures 6.6 and 6.7). This appears to represent the amount of time required for the bank storage accumulated during the recharge event to return to the river (Squillace and others, 1993). It also implies that caution must be exercised when using pesticide concentrations in surface waters during base-flow periods as indicators of contaminant levels in adjacent shallow aquifers (e.g., Hallberg and others, 1984; Hallberg, 1986). The latter point is reinforced by the work of McMahon and others (1994), who found that atrazine concentrations in the South Platte River during base flow (July 27-August 7, 1992) differed from those in the ground water 30 cm below the streambed, in some locations by as much as an order of magnitude.

Squillace and others (1993) also demonstrated that shallow ground waters, rather than tributary inputs, constitute the principal source of the pesticide load carried by the Cedar River during base-flow periods. Although this may not be universally true for all rivers, it is likely to be a common pattern for large rivers in intensively farmed areas with relatively low relief. The entry of pesticides from ground water into surface waters has also been documented in areas of more pronounced topography (Neary and others, 1983; Gomme and others, 1992; Kalkhoff and Schaap, 1995).

6.5 INFLUENCE OF WELL CONSTRUCTION AND USE

6.5.1 IMPROPERLY CONSTRUCTED WELLS

A number of features of well construction influence the likelihood with which pesticides in runoff or shallow ground water may move downward through a well, either along the well annulus or through the well bore itself (e.g., Litwin and others, 1983). In Iowa, three of these features—improper seals around the well annulus, unsealed or poorly-sealed casings, and the placement of the wellhead in an unsealed frost pit—were found to be associated with a higher frequency of pesticide contamination (Kross and others, 1990; Hallberg and others, 1992b). Surveys of domestic wells in Nebraska, cited by Exner and Spalding (1990), reported lower frequencies of pesticide detection in domestic wells with proper (i.e., modern) annular seals than in those without them. Similarly, during the Iowa State-Wide Rural Well-Water Survey (SWRL), wells that were "grouted" (those for which the annular space between the well casing and the drilled borehole was sealed with cement or bentonite) had significantly lower frequencies of pesticide detection ($\alpha = 0.05$) than those that were not grouted (Hallberg and others, 1992b).

In addition, a survey of public water supplies in Iowa showed that all of the bedrock wells without a casing contained detectable concentrations of synthetic organic compounds, compared to only 32 percent of the bedrock wells that had casings (Kelley, 1985). Two of the reviewed studies, however, did not observe lower frequencies of pesticide detection in sealed wells, relative to unsealed wells. No relation was observed between pesticide detections and the grouting of wells sampled in Iowa by Glanville and others (1995), perhaps because of variations in the extent and effectiveness of grouting among the sampled wells. Similarly, data reported by Segawa and others (1986) exhibited no apparent difference in triazine herbicide detection frequencies between sealed and unsealed wells in Glenn County, California.

The importance of well construction features, relative to normal agricultural activities, in facilitating pesticide contamination of ground waters in predominantly agricultural regions continues to be a source of considerable debate, and definitive conclusions remain elusive. Furthermore, many older wells whose construction might be considered "faulty" because of their susceptibility to surface-derived pesticide contamination were installed according to generally accepted practices at a time before pesticides were in widespread use (Hallberg, 1995). Such practices are thus considered "improper" only in retrospect.

Some authors maintain that improper well construction and placement (e.g., downgradient from septic systems, barnyards, or other contaminant sources) are the primary reasons why pesticides and high nitrate concentrations are found in wells in agricultural regions (Exner and Spalding, 1985; Baker and others, 1989). During a study of rural well contamination in southeastern Nebraska, atrazine and alachlor were detected only in water from wells lacking a proper annular seal or located in pump pits (Exner and Spalding, 1985). The sampling for this work, however, focused on wells with high nitrate contamination, thus providing an unrepresentative picture of relations between well characteristics and pesticide (or nitrate) occurrence in rural wells across the study area.

Other authors have concluded that improper well construction alone does not account for most of the agrichemical contamination of shallow wells in agricultural areas. Kross and others (1990) found that the frequencies of pesticide detection in properly constructed wells in Iowa were comparable to those in all wells sampled statewide. In Wisconsin, the survey of Grade A dairy farm wells found all wells in which pesticides were detected—and for which the original construction records were available—to have been of high quality and proper construction (LeMasters and Doyle, 1989). Similarly, in seven California counties, all wells with confirmed pesticide detections were found to be in proper physical condition (Ando, 1992). These results indicate that, as stated by Kross and others (1990), "ground-water contamination is far from being a problem of well construction [alone]."

Several publications have provided guidelines for the design, placement, and construction of new wells, remediation of existing wells, or the sealing of abandoned wells to minimize the potential for contamination of ground water by surface-derived solutes. These include discussions by Briggs and Fiedler (1975), the U.S. Environmental Protection Agency (1975), Litwin and others (1983), the Minnesota Pollution Control Agency (Brach, 1989), and Waller (1991).

6.5.2 RELATIONS BETWEEN WELL DEPTH AND PESTICIDE CONTAMINATION

Well depth was the factor most commonly found to be correlated with pesticide detections in ground water. Inverse relations between well depths and pesticide detection frequencies or concentrations have been documented in many locations, including California (Domagalski and Dubrovsky, 1991); Illinois (McKenna and others, 1988); Iowa (Kelley, 1985;

Kelley and Wnuk, 1986; Thompson and others, 1986; Detroy and others, 1988; Cherryholmes and others, 1989; Kross and others, 1990; Libra and others, 1993; Glanville and others, 1995; and Kalkhoff and Schaap, 1995); Minnesota (Klaseus and others, 1988); Nebraska (Chen and Druliner, 1987; Exner and Spalding, 1990); New Jersey (Szabo and others, 1994); South Dakota (Goodman, 1991); the states sampled during the NPS (U.S. Environmental Protection Agency, 1992a), the MCPS (Burkart and Kolpin, 1993), and the CPWTP (Baker and others, 1994); Ontario, Canada (Rudolph and others, 1993); and northern Italy (Capodaglio and others, 1988). Indeed, it is because of this widely known relation that screens for water-supply wells are often installed at considerable depths below the water table, particularly where the shallow ground water is known to be contaminated (Clark, 1990; Koterba and others, 1993). Figure 6.9 shows this relation for both groups of herbicides that were examined during the CPWTP.

The inverse relation frequently observed between well depth and pesticide detection frequencies in ground water arises from the combined influence of two factors: (1) the depth of the water table below the land surface, and (2) the depth of the screened interval below the water table. These two parameters control the amount of time required for surface-derived contaminants to pass through the unsaturated zone to the water table, and to reach the well screen from the water table, respectively. Increases in either distance provide more time for pesticide concentrations to diminish through hydrodynamic dispersion, sorption, transformation, and (above the water table) volatilization. However, the separate influences of water-table depth and screen depth below the water table are not easily discerned from existing studies, because such information is not readily available (particularly in comparison to well depth). As a result, monitoring studies seldom relate pesticide occurrence to either parameter.

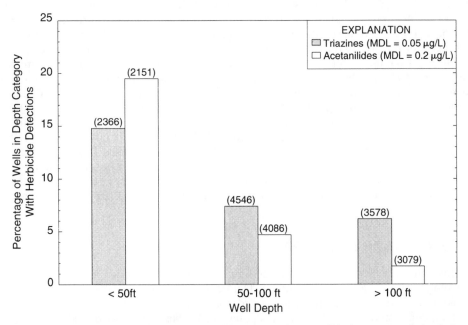

Figure 6.9. Frequencies of herbicide detection (triazines and acetanilides) versus well depth from the Cooperative Private Well Testing Program. Number of wells sampled in each category shown in parentheses (data from Baker and others, 1994). MDL, Method Detection Limit.

The importance of travel time in governing relations between well depth and contaminant occurrence is reinforced by one of the few instances where a relation of this type was investigated, but not observed—in this instance, for nitrate. During the SWRL study, inverse relations between nitrate concentration and well depth were documented in all but one of the six principal hydrogeologic regions of Iowa (Kross and others, 1990). The one region where such a relation was not observed was in the karstified northeast, where dissolution of the Paleozoic carbonate bedrock has led to deep dissection of the landscape, and extensive exposure of the permeable bedrock at the land surface. In this type of setting (see Section 6.3.3), channels in the solution-weathered rock may carry surface-derived contaminants rapidly to the screens of deeper wells, confounding the usual relation between well depth and contaminant occurrence that arises when the contaminants must move through the soil and rock matrices as they migrate in the subsurface. This explanation is consistent with the fact that the weakest relation observed between nitrate concentration and well depth among the five other regions of Iowa was seen in the eastern part of the state, where Paleozoic carbonates are also the predominant bedrock formation, but they are not as heavily weathered as in the northeast (Kross and others, 1990).

6.5.3 INFLUENCE OF PUMPING RATE

When high rates of ground-water withdrawal are sustained over sufficiently long periods, the directions of subsurface flow can be dramatically shifted, or even reversed (e.g., Bond, 1987). As a result, high pumping rates in agricultural areas have been observed to accelerate the movement of surface-derived agrichemicals toward the screens of high-volume wells. In eastern Long Island, New York, the downward movement of 1,2-dichloropropane was accelerated because of the pumping of large-capacity agricultural wells and numerous private wells (Baier and Robbins, 1982a). Similarly, the pumping of production wells in southwestern New Jersey has enhanced the downward migration of pesticides in the subsurface (Szabo and others, 1994).

Results from the NPS (U.S. Environmental Protection Agency, 1992a) are in agreement with these observations. During the NPS, the presence of another operating well within 500 ft was associated with a significantly increased likelihood of detecting pesticides in community water system wells ($\alpha = 0.05$), but not in domestic wells. This is consistent with the higher pumping rates employed for most community water-supply wells. During the SWRL study in Iowa, pesticide detection frequencies were significantly higher in rural wells ($\alpha = 0.05$), relative to the statewide average, when one or more operable wells were located less than 50 ft away (Hallberg and others, 1992b). Barton and others (1987) examined the influence of pumping rate on ground-water quality in agricultural areas of central New Jersey, but were unable to investigate this issue for pesticides, because of uncontrolled variations in analytical detection limits.

6.5.4 INFLUENCE OF WELL TYPE

Specific types of wells have a number of features in common with respect to construction, hydrogeologic setting, and pumping patterns. Domestic and monitoring (observation) wells tend to be shallow and are usually pumped at low rates for short intervals. Because of their manner of installation, dug, bored, or driven (e.g., sandpoint) wells, like springs, also tend to be shallow; furthermore, as noted earlier (Section 6.3), such wells are installed only in unconsolidated materials or highly weathered bedrock. By contrast, municipal, irrigation, and industrial water-supply wells are typically deeper and subjected to higher rates of withdrawal for extended periods (Cohen, 1986; Barton and others, 1987; Klaseus and others, 1988; and Koterba and others, 1993).

Many of the monitoring studies examined relations between well type and the frequencies of pesticide detection. Table 6.1 summarizes the contrasts reported in detection frequencies among different types of wells, based on construction and use. For studies that provided detection frequencies for individual compounds, but not for "any pesticide" (e.g., U.S. Environmental Protection Agency, 1990a; Seigley and Hallberg, 1991), the data for the most commonly detected compound were given to provide the most robust statistics possible. In addition to the work cited in the table, Helgeson and Rutledge (1989) and Exner and Spalding (1990) also discussed contrasts in ground-water quality among different well types, but neither study reported data on pesticide residues.

Well Construction

The relative frequencies of pesticide detection among the various types of well construction listed in Table 6.1 are in general agreement with their anticipated susceptibilities to contamination. With one exception among the studies listed (triazines in driven wells sampled for the CPWTP), springs and dug, bored, or driven wells show consistently higher frequencies of pesticide detection than do drilled wells (Figure 6.10), which are usually deeper and more effectively protected against contamination from surficial sources (e.g., McKenna and Keefer, 1991; Glanville and others, 1995). Consistent with the results from the CPWTP, Baker and others (1989) found that among the six water systems in Ohio observed to exhibit pesticide concentrations in excess of health advisory levels, two derived their water from dug wells, while three were supplied by springs. Glanville and others (1995), however, observed no significant difference in triazine herbicide concentrations between drilled and bored wells in Iowa ($\alpha = 0.05$)

Deficiencies in well construction may obscure the effects of other factors controlling pesticide detection frequencies in ground waters. During their second Province-wide survey in Ontario, Canada, Rudolph and others (1993) observed a significant, inverse correlation between well depth and the summed concentrations of atrazine and deethyl atrazine for drilled wells ($\alpha = 0.05$), but not for dug or bored wells.

Well Use

In contrast with the relatively consistent relations observed with regard to well construction, pesticide detection frequencies show substantially less clearly defined relations to well use. Among wells employed for different purposes, monitoring wells usually exhibit the highest detection frequencies, since they are typically shallow, and often screened close to the water table. In addition, monitoring wells are often installed to examine contamination in known or suspected problem areas, although this was not verified for any of the studies cited in Table 6.1. The rural well surveys conducted in Ontario in the winter of 1991/1992 and the summer of 1992 (Rudolph and others, 1992, 1993) provide the only exceptions to this pattern among the results compiled in Table 6.1. Although each of the small-volume monitoring wells installed for these two studies was screened at the same depth as the domestic well against which it was being compared, the domestic wells drew water from a broader depth range, and thus may have captured larger proportions of shallower, more contaminated ground waters.

During a study of contaminant occurrence in large-diameter (0.9 m) "seepage" wells in Audubon County, Iowa, Seigley and Hallberg (1991) found that pesticide detections are more frequent in inactive than in active wells (based on pesticide detection frequencies among samples, rather than among wells). Of the various explanations offered by the authors for this

Table 6.1. Frequencies of pesticide detection relative to well construction and use

[Study Location: Reference given in parentheses and italics. SWRL, Iowa State-Wide Rural Well-Water Survey; OMOE, Ontario Ministry of the Environment; CPWTP, Cooperative Private Well Water Testing Program; USEPA, U.S. Environmental Protection Agency; NPS, National Pesticide Survey; PGWDB, Pesticides In Ground Water Database. Blank cells indicate no information applicable or available. NG, not given; ≥, greater than or equal to]

Study Location (Reference)	Percent of Total in Each Class With Detectable Pesticide Concentrations (upper number) *Number of wells in class given in parentheses and italics, unless otherwise indicated (lower number)*								
	Well Construction				Well Use				
	Drilled	Springs	Dug or Bored	Driven	Monitoring	Domestic Supply	Public Supply	Irriga-tion	Inactive
Supply-Well Surveys									
Illinois *(Goetsch and others, 1992)*	9.5 *(190)*		22.7 *(141)*						
Iowa *(SWRL; Kross and others, 1990)*	10 *(425)*		19 *(233)*	23 *(27)*					
Ohio *(Baker and others, 1989)*	1.7 *(422)*	5.4 *(56)*	4.6 *(132)*						
Ontario, Canada *(Frank and others, 1987b)*	9.7 *(62)*	21 *(29)*							
Ontario, Canada *(OMOE, 1987a)*	41 *(71)*	61 *(175)*							
Ontario, Canada *(Rudolph and others, 1992)*[1]	3 *(701)*		15 *(419)*	8 *(64)*	4.5 *(133)*	9.8 *(133)*			
Ontario, Canada *(Rudolph and others, 1993)*[1]	6 *(687)*		22 *(406)*	14 *(71)*	6.7 *(120)*	17 *(120)*			
Illinois *(Mehnert and others, 1995)*	[2]0 *(48)*		46 *(48)*						
California *(Cohen, 1986)*						33.6 *(NG)*	14.6 *(NG)*		
Long Island, New York *(Zaki and others, 1982)*						26.8 *(8,051)*	14.8 *(68)*		
Illinois *(McKenna and others, 1988)*					100 *(18)*	37 *(19)*		18 *(11)*	
New Jersey *(Louis and Vowinkel, 1989)*						28 *(25)*	([3]) *(17)*	43 *(37)*	

Table 6.1. Frequencies of pesticide detection relative to well construction and use—*Continued*

Study Location (Reference)	Percent of Total in Each Class With Detectable Pesticide Concentrations (upper number) *Number of wells in class given in parentheses and italics, unless otherwise indicated (lower number)*								
	Well Construction				Well Use				
	Drilled	Springs	Dug or Bored	Driven	Monitoring	Domestic Supply	Public Supply	Irriga-tion	Inactive
Iowa *(Seigley and Hallberg, 1991)*						25 *(73 samples)*			75 *(72 samples)*
CPWTP *(≥8 States; Baker and others, 1994)* Triazines	6.8 *(10,172)*	32.4 *(451)*	30.2 *(756)*	5.3 *(1,328)*					
Acetanilides	4.1 *(9,077)*	10.2 *(294)*	18.5 *(617)*	19.5 *(1,260)*					
Nationwide *(NPS; USEPA, 1990a)*						2.5 [4]*(264,000)*	6.4 [4]*(6,010)*		
Nationwide *(PGWDB; USEPA, 1992b)*					42 *(664)*	18 *(37,714)*			
Agrichemical Facility Studies									
Illinois *(Goetsch and others, 1993)*	13 *(16)*		75 *(4)*	100 *(1)*					
Wisconsin *(Habecker, 1989)*					100 *(9)*		94 *(18)*		

[1]Percentages estimated from histograms presented in original report.
[2]Includes both drilled and driven wells.
[3]Wells sampled, but data not provided.
[4]Estimated numbers of wells nationwide.

observation, the most persuasive is that "the inactive wells may retain the water delivered during recharge events, and maintain the relatively high herbicide concentrations, suggesting that there is limited groundwater flow into and out of these wells under nonpumping conditions" (Seigley and Hallberg, 1991). This hypothesis is consistent with the low permeability of the surficial materials in which the wells were installed, as well as the comparatively high pesticide concentrations that are often observed in recharge waters during the spring when most pesticides are applied and recharge events are most common.

Patterns of pesticide detection among domestic, public-supply, and irrigation wells varied considerably among different studies. For example, although the NPS encountered higher frequencies of pesticide detection in community water system wells than in domestic wells (U.S. Environmental Protection Agency, 1990), Klaseus and others (1988) observed the opposite pattern in Minnesota (data appropriate for Table 6.1, however, were not provided by the latter

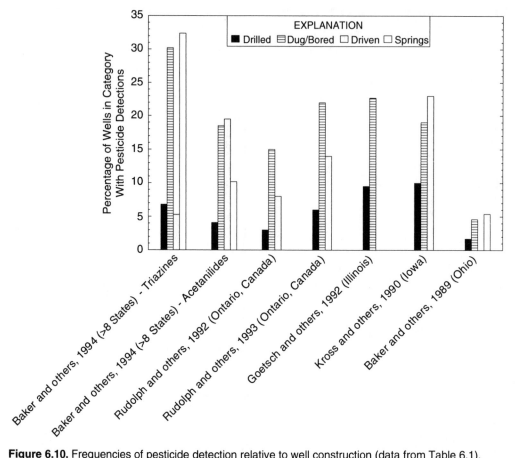

Figure 6.10. Frequencies of pesticide detection relative to well construction (data from Table 6.1).

study). Finally, Kolpin and others (1994) randomly selected 30 public-supply wells, 30 domestic wells, and 30 wells of other types (e.g., stock, irrigation and industrial) from the set of wells sampled during the MCPS and compared the frequencies of herbicide detection among the three categories. No significant difference in herbicide occurrence was observed among the three groups ($\alpha = 0.05$).

The apparent inconsistencies among wells used for different purposes may reflect the competing influences of well depth and pumping rate. Because irrigation and public-supply wells are pumped more heavily than domestic wells, they draw their water from larger areas. In agricultural areas with extensive nonpoint inputs of pesticides, this may increase the likelihood of detecting pesticides, relative to the lower-volume domestic wells. Alternatively, increased pumping rates may bring in larger proportions of uncontaminated ground water, particularly from greater depths, thus reducing pesticide concentrations by dilution. Irrigation and public-supply wells are generally deeper than domestic wells; as noted earlier, deeper wells exhibit lower frequencies of contamination than shallow ones. Furthermore, private domestic wells tend to be located closer to agricultural fields, where pesticides are most commonly applied (Klaseus and others, 1988). Given the multiplicity of factors that contribute to the susceptibility of an individual well to agrichemical contamination—some of which may act in opposition to one

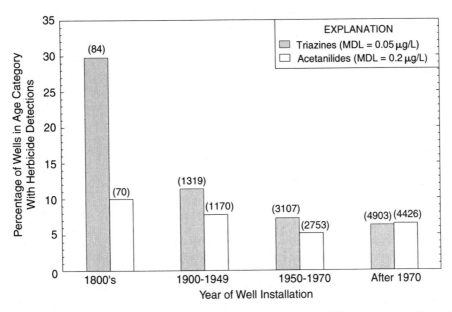

Figure 6.11. Frequencies of herbicide detection (triazines and acetanilides) versus well age for the Cooperative Private Well Testing Program. Number of wells sampled in each category given in parentheses (data from Baker and others, 1994). MDL, Method Detection Limit.

another—it is not surprising that consistent differences in pesticide detection frequencies are not observed among domestic, public supply, irrigation, and monitoring wells.

Well Age

As they have been developed over time, well construction practices designed to minimize ground-water contamination, both during drilling and in the completed well, have been gradually incorporated into installation operations. Consequently, well age is sometimes used as an approximate surrogate for well integrity. Direct correlations between frequencies of pesticide detection in ground water and well age have been documented during the NPS (U.S. Environmental Protection Agency, 1992a) and the CPWTP (Baker and others, 1994). Relations observed from the latter study are shown in Figure 6.11. In addition, Glanville and others (1995) observed significant, direct relations ($\alpha = 0.05$) between well age and the concentrations of either triazine or acetanilide herbicides in rural wells in Iowa. Several investigations have noted similar correlations for nitrate (Steichen and others, 1988; Rudolph and others, 1992; and Baker and others, 1994).

Studies conducted in Iowa (Hallberg and others, 1992b) and Glenn County, California (Segawa and others, 1986), however, did not observe clear relations between well age and pesticide detection frequencies. Thus, well age by itself does not appear to be a reliable indicator of pesticide contamination. Any direct relations that are detected between well age and pesticide occurrence—such as those described earlier—are likely to arise from the fact that older wells tend to be shallower, of larger diameter, and less effectively isolated from surface-derived contamination than newer wells (Hallberg and others, 1992b; Glanville and others, 1995). Consequently, well depth and construction are likely to be more reliable predictors of pesticide contamination than well age.

6.6 EFFECTS OF STUDY DESIGN ON FREQUENCIES OF
PESTICIDE DETECTION IN GROUND WATER

The frequencies and spatial distributions of pesticide detections reported from ground-water monitoring studies are strongly dependent on the designs of the investigations themselves. This section examines several of the ways in which variations in design may influence the conclusions reached by such studies. Additional discussion of this subject has been provided by Hallberg (1989).

6.6.1 RANGES OF COMPOUNDS EXAMINED

The set of compounds for which chemical analyses are conducted during any monitoring study is limited by the number of pesticides and transformation products that can be examined within time and budgetary constraints. As a result, analytes that are eliminated from consideration may include not only those of marginal or negligible interest, but also compounds whose analysis is prohibitively costly, despite appreciable interest. The complexity of the laboratory procedures required to detect glyphosate (Newton and others, 1994) and hydroxyatrazine (Armstrong and others, 1967; Kolpin, 1995a), for example, has restricted the attention devoted to these important compounds to only a handful of studies.

As of 1989, glyphosate (Roundup) was the tenth most heavily used herbicide in the United States, and was applied to 44 percent of the corn and soybeans grown in the nation (Gianessi and Puffer, 1990). However, despite the widespread use of glyphosate, the Pesticides In Ground Water Database (PGWDB) contains monitoring information from only four states for the herbicide (U.S. Environmental Protection Agency, 1992b). Process and matrix-distribution studies of glyphosate are also rare; of the four reviewed studies that examined its environmental behavior (Subramaniam and Hoggard, 1988; Feng and Thompson, 1990; Schowanek and Verstraete, 1991; Newton and others, 1994), only two were field investigations (Feng and Thompson, 1990; Newton and others, 1994). Similarly, although hydroxyatrazine has been detected in field soils (Sophocleous and others, 1990; Loch, 1991), none of the reviewed studies analyzed for it in ground water, despite the fact that it is the principal transformation product (Section 4.4.2) for the most widely used herbicide in the United States (Gianessi and Puffer, 1990). Difficulties associated with the analysis of environmental media for several of the more recently introduced, low-dosage pesticides, such as the sulfonylurea and imidazolinone herbicides, also appear to have limited the attention devoted to the occurrence of these compounds in ground water (Kolpin, 1995a).

6.6.2 INFLUENCE OF ANALYTE DETECTION LIMITS

The frequency with which a specific contaminant is detected in a particular medium increases as the analytical detection limit decreases. Kolpin and others (1994) observed a nonlinear, inverse correlation of this nature for atrazine in ground water, both among data from their own investigation (the MCPS), and among those from previous monitoring studies in the midcontinental United States. Additional work indicates that this relation holds for both unconsolidated and bedrock aquifers in the midcontinent (Figure 6.12); the smoother trend noted for the unconsolidated aquifers was probably a result of the larger sample size (Kolpin and others, 1995). Inverse relations between detection frequencies and detection limits have also been

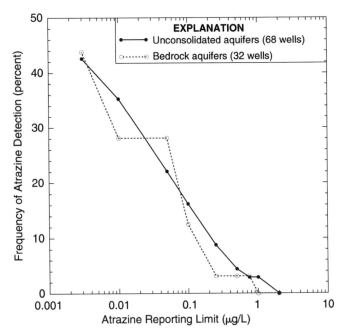

Figure 6.12. Frequency of detection of atrazine as a function of reporting limit in near-surface aquifers (unconsolidated and bedrock) of the midcontinental United States, based on the 1992 sampling of the Midcontinent Pesticide Study. Redrawn with permission from Kolpin and others (1995).

reported by studies of atrazine occurrence in ground waters of Ohio (Baker and others, 1989), Nebraska (Barrett and Williams, 1989), and Wisconsin (LeMasters and Doyle, 1989).

Data from other monitoring studies that sampled randomly selected, small-capacity wells provide additional evidence for this relation. The midcontinental United States is the most appropriate region for conducting this analysis, because of the extensive geographic overlap in this area between several statewide surveys of randomly selected wells and the three random multistate studies that have been carried out to date (MCPS, NPS, and NAWWS). Table 6.2 summarizes the frequencies of pesticide detection and analytical detection limits reported by the three random multistate studies (also given in Table 3.3), and the corresponding data from random, statewide surveys conducted in Iowa (SWRL), Illinois, Wisconsin, and Kansas. The statewide monitoring studies whose results are displayed in Table 6.2 are those that sampled small-capacity wells in the midcontinent and analyzed for one or more of the compounds detected during the multistate studies. The table therefore excludes statewide surveys of pesticide occurrence in large capacity, community-supply wells (e.g., Frink and Hankin, 1986; Ellingson and Redding, 1988; and Thamke and Clark, 1988), including those sampled for the NPS (U.S. Environmental Protection Agency, 1990a); nonrandom studies that targeted areas where pesticide use or presumed ground-water vulnerability were substantial (e.g., Klaseus and others, 1988; Neil and others, 1989; and Walker and Porter, 1990); or investigations that focused on other compounds, such as aldicarb (e.g., Scarano, 1986; and Miller and others, 1989).

Figures 6.13 and 6.14 display relations between detection frequencies in ground waters of the midcontinent and analytical detection limits (for the triazine and acetanilide herbicides, respectively), based on the data from the multistate studies compiled in Table 6.2. The figures indicate that much of the variability in detection frequencies for a given pesticide or

Table 6.2. Pesticide detection frequencies reported by random sampling of small-capacity wells in the midcontinental United States in relation to analytical detection or reporting limits

[Pesticide or Transformation Product: Transformation products are indented. National Pesticide Survey: Data for rural domestic wells only. ND, not detected. SWRL, Iowa State-Wide Rural Well-Water Survey; USEPA, U.S. Environmental Protection Agency. Blank cells indicate no information applicable or available. <, less than; μg/L, micrograms per liter]

Frequencies of Detection

Wells with detections as a percentage of all wells sampled (upper number)
Analyte detection or reporting limits (μg/L) given in italics (lower number)

Pesticide or Transformation Product	Midcontinent Pesticide Study Pre-Planting	Post-Planting 1991	Post-Planting 1992	National Pesticide Survey	National Alachlor Well Water Survey	Iowa (SWRL)	Illinois	Wisconsin (Dairy Farms)	Kansas
Atrazine	14.7 / *0.05*	20.4 / *0.05*	43.0 / *0.005*	0.7 / *0.12*	11.68 / *0.03*	4.4 / *0.13*	2.1 / *0.43*	12.5 / *0.15*	3.9 / *1.2*
Deethyl atrazine	15.4 / *0.05*	21.1 / *0.05*	31.0 / *0.015*	ND / *2.2*		3.5 / *0.10*			
Deisopropyl atrazine	4.0 / *0.05*	7.5 / *0.05*	18.2 / *0.050*			3.4 / *0.10*			
Metolachlor	3.0 / *0.05*	2.5 / *0.05*	11.0 / *0.002*	ND / *0.75*	1.02 / *0.03*	1.5 / *0.04*	0.3 / *1.4*	0.2 / *0.15*	ND / *0.25*
DCPA "acid metabolites"			15.6 / *0.010*	2.5 / *0.10*					
Prometon	4.0 / *0.05*	6.1 / *0.05*	9.0 / *0.010*	0.2 / *0.15*			1.2 / *0.37*		
Simazine	0.7 / *0.05*	1.4 / *0.05*	13.0 / *0.005*	0.2 / *0.38*	1.60 / *0.03*		0.2 / *0.12*	ND / *0.15*	
Metribuzin	0.7 / *0.05*	1.4 / *0.05*	1.0 / *0.005*	ND / *0.60*		1.9 / *0.01*	0.1 / *0.43*	0.4 / *0.15*	
Alachlor			5.0 / *0.002*	<0.1 / *0.50*	0.78 / *0.03*	1.2 / *0.02*	0.7 / *1.3*	0.8 / *0.15*	1.0 / *0.25*
Cyanazine	0.3 / *0.05*	1.1 / *0.05*	3.0 / *0.008*	ND / *2.4*	0.28 / *0.1*	1.2 / *0.12*		ND / *0.15*	

Table 6.2. Pesticide detection frequencies reported by random sampling of small-capacity wells in the midcontinental United States in relation to analytical detection or reporting limits—*Continued*

[Frequencies of Detection / Wells with detections as a percentage of all wells sampled (upper number) / Analyte detection or reporting limits (µg/L) given in italics (lower number)]

Pesticide or Transformation Product	Midcontinent Pesticide Study — Pre-Planting	Post-Planting 1991	Post-Planting 1992	National Pesticide Survey	National Alachlor Well Water Survey	Iowa (SWRL)	Illinois	Wisconsin (Dairy Farms)	Kansas
Hexachlorobenzene				ND / *0.060*				ND / *0.15*	
1,2-Dibromo-3-chloropropane (DBCP)				0.4 / *0.010*					
1,2-Dibromoethane (EDB)				0.2 / *0.010*					
Lindane (γ-HCH)				0.1 / *0.043*				ND / *0.15*	ND / *0.025*
Ethylene thiourea (degradate of EBDC fungicides)				0.1 / *4.5*					
Bentazon			ND / *0.10*	0.1 / *0.25*			1.4 / *0.21*		
Dinoseb				ND / *1.3*			3.7 / *0.16*		
References	Burkart and Kolpin, 1993	Kolpin and others, 1993		USEPA, 1990a, 1992a	Monsanto Agricultural Company, 1990; Holden and others, 1992	Kross and others, 1990	Goetsch and others, 1992	LeMasters and Doyle, 1989	Koelliker and others, 1987

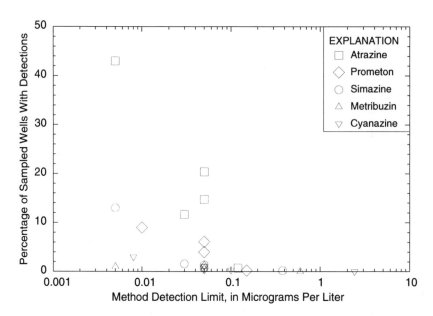

Figure 6.13. Frequencies of detection versus method detection limit for individual triazine herbicides from multistate monitoring studies involving the sampling of randomly selected, small-capacity wells in the midcontinental United States (data from Table 6.2).

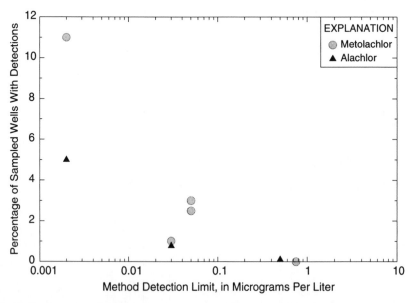

Figure 6.14. Frequencies of detection versus method detection limit for individual acetanilide herbicides from multistate monitoring studies involving the sampling of randomly selected, small-capacity wells in the midcontinental United States (data from Table 6.2).

transformation product among monitoring studies in this region may have resulted from variations in detection limits. For most of the triazine and acetanilide herbicides, detection frequencies generally decreased with increasing analytical detection limits. (This relation was less clearly evident for both herbicide groups among the statewide studies than among the multistate studies.) These results demonstrate that analytical detection limits must always be taken into account when comparing detection frequencies among different studies for a given compound.

6.6.3 INFLUENCE OF IMPROVEMENTS OR VARIATIONS IN ANALYTICAL METHODS

The lowering of analyte detection limits and the development of methods for the analysis of hydrophilic transformation products (e.g., alcohols and carboxylic acids) are among the most significant impacts of technological advances on the detection of pesticides and other contaminants in environmental media. During the MCPS, for example (Kolpin and others, 1995), when the same routine analytical methods were used for a set of 13 pesticides and transformation products (Figure 6.15), the frequency of detection in ground water in 1992 (29.0 percent) was virtually identical to that observed for the same compounds in 1991 (28.4 percent). However, the lowering of analytical detection limits (by as much as an order of magnitude for some compounds), the analysis of many additional parent compounds and transformation products, and the introduction of tritium analyses (see Section 9.2.2) provided further insight on pesticide occurrence—or nonoccurrence—in an additional 49.0 percent of the samples obtained in 1992.

Parsons and Witt (1989) have noted that the use of stricter quality-control procedures, as well as advances in sampling and analytical techniques, may in some cases lead to reductions in pesticide detection frequencies. Improvements in quality control during sample acquisition and shipping may decrease the number of analyte detections in a particular study area by reducing cross-contamination among sites or samples. In the laboratory, the use of higher-resolution chromatographic columns, more structure-specific detectors (e.g., Koskinen, 1989), and more stringent requirements for verifying detections can reduce the incidence of "false positives."

Difficulties associated with the detection of ethylene thiourea (ETU, a carcinogenic degradate of the ethylene bis-dithiocarbamate, or EBDC, fungicides) in Maine ground waters illustrate this point. Although nearly half of the ground-water samples obtained from agricultural areas in Maine by Neil and others (1989) tested positive for ETU, an extensive examination of the quality-control data by the authors cast sufficient doubt on the reliability of the original analytical results to cause exclusion of the ETU data from the final report. By contrast, the ETU detections reported in ground waters by the NPS (Table 3.3) were obtained using an analytical method that was considerably more reliable than that employed for the Maine study (Neil and others, 1989).

In some cases, reported detection frequencies may not accurately represent the occurrence of a specific compound because of ambiguities regarding the precise nature of the analytes in question. For example, whereas extensive data have been generated on the occurrence of "aldicarb residues" in ground waters, the parent compound, as noted earlier in Section 3.1.1, is detected much less frequently than its sulfoxide and sulfone transformation products (Jones, 1986; U.S. Environmental Protection Agency, 1992b). Similar uncertainties accompany reports of the frequencies with which "residues" of other highly reactive pesticides are detected, including DCPA (U.S. Environmental Protection Agency, 1990a; Istok and others, 1993; Kolpin and others, 1993), fenamiphos (Weaver and others, 1988a; Loague and others, 1994), and phorate

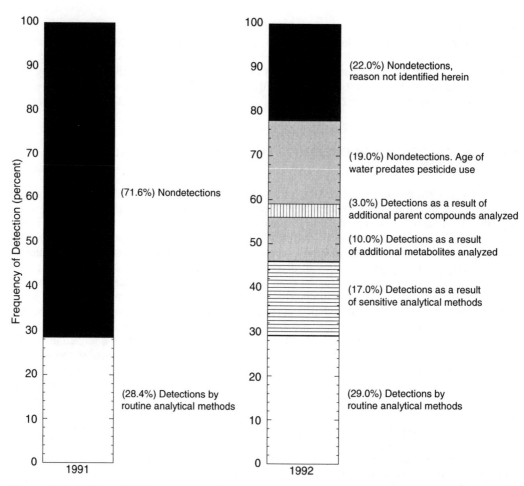

Figure 6.15. Additional insight into pesticide occurrence in near-surface aquifers of the midcontinental United States from the 1992 sampling of the Midcontinent Pesticide Study, relative to the 1991 sampling. Redrawn with permission from Kolpin and others (1995).

(Weaver and others, 1988b). The reporting of pesticide occurrence in terms of the sum of parent and product concentrations increases the difficulty of assessing the environmental and public-health significance of the detections since, as noted earlier (Section 4.4), parent compounds and their products typically differ in their physical, chemical, and toxicological properties.

The manner in which results from immunoassay analyses are sometimes reported offers a similar example. Substantially higher rates of "alachlor" detection were reported by Wallrabenstein and Baker (1992) during the CPWTP than were observed for alachlor by either the NPS (U.S. Environmental Protection Agency, 1990a) or the NAWWS (Holden and others, 1992). Baker and others (1994) attributed these differences to the fact that the NPS and NAWWS both involved the use of gas chromatography (with or without mass spectrometric detection), while the CPWTP employed the less compound-specific immunoassay procedure. Indeed, subsequent confirmatory analyses using gas chromatography resulted in the actual detection of alachlor in only 18 percent of the 126 samples for which the "alachlor" immunoassay test registered a detection (Baker and others, 1993). Based on these and other observations, most of the "alachlor" detections reported during the CPWTP are presumed to have resulted from the presence of alachlor ethanesulfonic acid (ESA)—one of the principal transformation products of alachlor—rather than the parent compound (Baker and others, 1994). The most recent update of the results from the CPWTP (Baker and others, 1994) has clarified this issue by reporting the results as "acetanilide" and "triazine" concentrations, rather than as "alachlor" and "atrazine." Other uncertainties associated with the use of immunoassay analyses were discussed in Section 2.8.2.

6.6.4 TARGETED VERSUS RANDOM SAMPLING

The frequencies with which pesticides are detected in ground water are likely to be lower during studies that sample wells selected at random than during studies that specifically target areas of high pesticide use or locations where ground water is deemed to be more susceptible to pesticide contamination. This pattern has been noted in Iowa (Hallberg, 1989; Glanville and others, 1995) and Ontario, Canada (Rudolph and others, 1992), as well as during investigations of atrazine occurrence in ground waters of the midcontinent (Barrett and Williams, 1989). It is also evident from the pronounced contrast in aldicarb detection rates between targeted (e.g., Baier and Robbins, 1982b; Zaki and others, 1982; Zaki, 1986) and nontargeted sampling in Long Island, New York (Eckhardt and others, 1989a), discussed earlier in Section 5.4.2.

A comparison between the analytical results for metolachlor from the Metolachlor Monitoring Study (Roux and others, 1991a) and those from the three nontargeted multistate investigations illustrates this point over a relatively large scale. Although the detection limit used for metolachlor by Roux and others fell within the range of those employed by the other studies (Figure 6.16), the frequency with which the herbicide was detected by Roux and others was substantially higher than those reported for the other investigations—by at least an order of magnitude in all but one case. The fact that Roux and others focused their sampling on areas with high metolachlor use and high (presumed) susceptibility to ground-water contamination was likely to have been the main reason why their results deviated so extensively from the relation between detection frequency and detection limit established by the surveys of randomly selected wells.

The fact that the PGWDB (U.S. Environmental Protection Agency, 1992b) currently excludes data from the NPS (U.S. Environmental Protection Agency, 1990a) provides an unprecedented opportunity to examine the effects of targeted sampling, relative to the sampling of randomly selected wells, on pesticide detection frequencies on a nationwide scale. Table 6.3, derived from the data summarized in Table 3.2, compares data from the NPS—the most spatially extensive sampling of randomly selected wells conducted to date—with those from the PGWDB, which includes results from both targeted and nontargeted studies across the United States. Each of the 117 compounds for which both the NPS and the PGWDB reported analyses is categorized in the table according to whether its detection was reported by: (1) both the NPS and the PGWDB; (2) only one of the two sources; or (3) neither.

The observations tallied in Table 6.3 are consistent with the expected influence of targeted sampling on compound detections. All of the 15 compounds detected during the sampling of randomly selected wells (NPS) were also detected during more targeted sampling (PGWDB), while none of the 40 that were not detected during more targeted sampling were detected during the nontargeted sampling. Also in accord with expectation is the fact that most of the compounds that were detected during the more targeted sampling (62 of 77, or 81 percent)

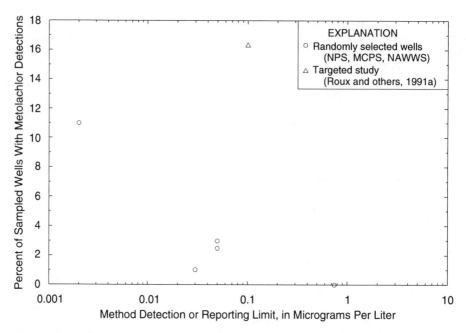

Figure 6.16. Effect of targeted sampling on pesticide detection frequencies from multistate studies. Percentages of wells with metolachlor detections from sampling of randomly selected wells (National Pesticide Survey [NPS], Midcontinent Pesticide Study [MCPS], and National Alachlor Well Water Survey [NAWWS], and sampling of wells in areas with high metolachlor use (Roux and others, 1991a), in relation to method detection or reporting limits (data from Table 3.3).

Table 6.3. Influence of targeted sampling on pesticide detections

[Comparison of the compounds detected or not detected during the National Pesticide Survey (U.S. Environmental Protection Agency, 1990a) with those detected or not detected during the studies compiled in the Pesticides In Ground Water Database (U.S. Environmental Protection Agency, 1992b) among the 117 compounds examined by both sources]

		NUMBER OF COMPOUNDS IN EACH CATEGORY AMONG THE 117 EXAMINED BY BOTH STUDIES *(Parents/Products)*	
PESTICIDES IN GROUND WATER DATABASE (Targeted and Nontargeted Sampling)	*DETECTED*	15 (13/2)	62 (55/7)
	NOT DETECTED	0	40 (30/10)
		DETECTED	*NOT DETECTED*

NATIONAL PESTICIDE SURVEY
(Nontargeted Sampling)

were not detected during the nontargeted sampling (NPS). These observations, however, may have been partially related to the fact that the analytical detection limits employed by the NPS were higher than those used by many of the monitoring studies summarized in the PGWDB.

6.6.5 INFLUENCE OF VARIATIONS IN SAMPLING METHODS

Variations in the procedures employed for obtaining a ground-water sample can substantially affect the concentrations of the analytes detected in it (e.g., Gillham and others, 1983). Authors of several of the reviewed studies noted a variety of aspects of ground-water sampling operations that may influence pesticide concentrations. These include whether samples are filtered in the field (Cohen and others, 1990; Blum and others, 1993), and whether samples are obtained upstream or downstream from pressure tanks (Roaza and others, 1989), cisterns, storage tanks, water conditioners (Mitchem and others, 1988), or any other apparatus that may modify the chemical composition of the sampled water. Although quality-assurance procedures are probably used extensively to assess the potential effects of sampling practices on the final analytical results during many monitoring studies, detailed descriptions of such measures are seldom provided. Exceptions include the quality-assurance measures described for studies by Kross and others (1990), the U.S. Environmental Protection Agency (1990a), Kolpin and Burkart (1991), and Koterba and others (1993).

6.6.6 DEGREE OF PUBLIC ATTENTION

The spatial extent of ground-water contamination observed for an individual pesticide is often related to the amount of effort invested in finding it. The contrast between aldicarb and the herbicide DCPA (dacthal) is instructive in this regard. Data compiled in the PGWDB indicate that by 1991, as a result of widespread concern over their potential presence in ground water—particularly in Long Island, New York—aldicarb or its transformation products had been sampled for in 119,031 wells in 28 states, and detected in 13,063 wells in 14 states (U.S. Environmental Protection Agency, 1992b). During the NPS, however, DCPA and its hydrolysis products (or "acid metabolites") were the compounds most frequently detected nationwide (Table 3.3), while aldicarb was not detected in any of the wells sampled (Table 3.2). (The dramatic contrast in aldicarb detection rates between the NPS and the PGWDB provides another illustration of the effects of targeted sampling on pesticide detection rates, discussed previously in Section 6.6.4.) In addition, the 1992 sampling of the MCPS detected DCPA (and its hydrolysis products) more frequently in ground water than any other pesticide except for atrazine and its transformation products, and the two principal alachlor degradates (Table 3.3).

Despite the pronounced difference in detection frequencies for DCPA (and its hydrolysis products) and aldicarb during the NPS, ground-water monitoring for the herbicide has been carried out in half as many states as for aldicarb. Ground-water sampling for DCPA or its hydrolysis products had been carried out in only 709 wells in 8 states by 1988 (Parsons and Witt, 1989) and 2,231 wells in 13 states by 1991 (U.S. Environmental Protection Agency, 1992b). In both compilations, however, the herbicide or its hydrolysis products were detected in at least half of the states where sampling was carried out. The difference in sampling effort between aldicarb and DCPA is driven largely by the contrast in the degree of regulatory oversight devoted to the two compounds (U.S. Environmental Protection Agency, 1992b) which, in turn, is a consequence of the marked difference in their respective toxicities (U.S. Environmental Protection Agency, 1990c).

The amount of attention devoted to sampling for pesticides in ground water is also related to the amount of agricultural activity in a given state. Data from the PGWDB (U.S. Environmental Protection Agency, 1992b) suggest that, except for extensive monitoring in Rhode Island and Connecticut, the number of wells sampled in a given state tends to be higher with more extensive pesticide use (Figure 6.17). Different states have devoted widely varying degrees of effort in monitoring for pesticides in ground water; Figure 6.17 indicates that the number of wells sampled from 1971 to 1991 varied among different states by four orders of magnitude.

While variations in sampling effort may contribute significantly to differences in the number of wells with pesticide detections in different states, such variations do not have a clear influence over the frequencies of pesticide detection (as noted earlier in Section 3.5.1) or exceedance of water-quality criteria. According to data from the PGWDB, neither pesticide detection frequencies nor the frequencies with which pesticide concentrations have been found to exceed Health Action Levels (HAL) in ground water show any obvious correlations with sampling effort (Figure 6.18), although variability in both parameters appears to diminish as more wells are sampled. Basic statistical principles support these observations.

6.6.7 USE OF OTHER INDICATORS OF PESTICIDE CONCENTRATIONS IN SHALLOW GROUND WATERS

It is occasionally suggested that the concentrations of pesticides in surface waters during base flow (Section 6.4.2), tile line effluents, lysimeter leachates, or springs may be used as indicators of pesticide levels in shallow ground waters, thereby reducing the need to install or sample wells during field investigations. These approaches assume that the concentrations of pesticides in these other media are identical to those that would be measured in ground waters withdrawn from shallow wells. The relatively few studies that have examined the issue, however, provide only marginal support for this assumption.

The concentrations of pesticides and other surface-derived solutes have been measured in tile drainage from agricultural fields for a variety of purposes, such as: (1) to determine their potential effects on downstream receiving waters (e.g., Johnston and others, 1967); (2) to estimate the proportions of applied compounds that percolate to shallow ground waters (Table 4.1); and (3) to understand the processes controlling solute movement through the unsaturated zone (e.g., Muir and Baker, 1976; Mansell and others, 1980; Bottcher and others, 1981; Hallberg and others, 1986; Klaseus and others, 1988; Richard and Steenhuis, 1988; Keim and others, 1989; and Seigley and Hallberg, 1991). The various ways that tile drainage may be used to evaluate the effects of agricultural practices on ground-water quality have been examined in detail by Hallberg and others (1984, 1986).

Among the studies reviewed for this report, however, only one conducted systematic comparisons between pesticide concentrations in tile effluents and those in shallow wells. At two sites in northwestern Ohio, Keim and others (1989) monitored the concentrations of alachlor, atrazine, and cyanazine over a full growing season in tile effluents and in observation wells installed with their screens immediately adjacent to the tile lines. Similarities were observed between the tile effluents and shallow ground waters at one of the sites—with respect to both concentrations and concentration histories—but substantial differences in concentrations (by an order of magnitude) and concentration histories were evident between the two media at the other. Thus, tile effluents may not always be reliable indicators of water quality in shallow wells. This is not entirely unexpected, given that tile drainage systems integrate water quality over a much larger area than low-capacity wells.

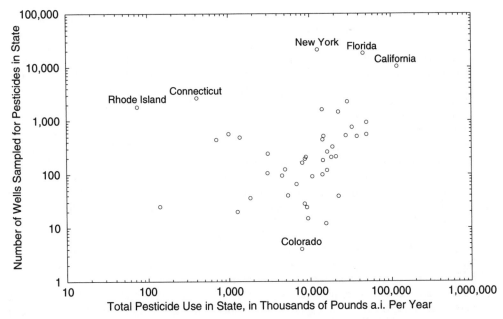

Figure 6.17. Number of wells sampled for pesticides in individual states, based on the Pesticides In Ground Water Database (U.S. Environmental Protection Agency, 1992b), versus statewide pesticide use (herbicides, insecticides, and fungicides combined), based on Gianessi and Puffer (1990,1992a,b). a.i., active ingredient.

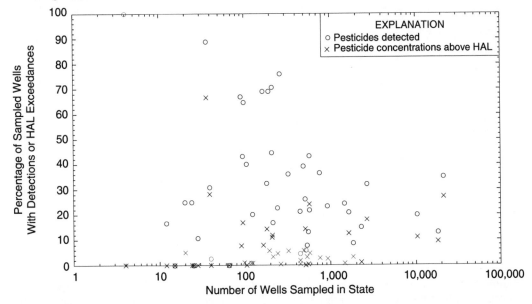

Figure 6.18. Percentages of sampled wells in individual states with pesticide detections or pesticide concentrations above Health Action Levels (HAL) from 1971 to 1991, versus the number of wells sampled, based on the Pesticides In Ground Water Database (data from U.S. Environmental Protection Agency, 1992b).

Shallow lysimeters have also been considered as potential indicators of shallow ground-water quality (e.g., Bergström, 1990). Three of the studies examined for this book compared pesticide concentrations in buried lysimeters with those in other subsurface media (Harkin and others, 1986; Kubiak and others, 1988; and Steenhuis and others, 1990), but only the most recent investigation directly compared pesticide concentrations in shallow lysimeters with those in shallow wells. For alachlor, atrazine, and carbofuran, Steenhuis and others (1990) reported substantial differences between the pesticide concentration histories observed in the lysimeters and those observed in the wells.

Springs have also been considered as a potential indicator of shallow ground-water quality. The most recent data from the CPWTP (Baker and others, 1994), however, showed marked differences in herbicide detection frequencies between springs and shallow wells (<50 ft depth), and opposing trends for the two herbicide groups examined (Figure 6.19). These contrasts in detection frequencies may reflect the different hydrologic settings in which springs and shallow wells are located. Whereas water sampled from a small-screen well represents essentially a point measurement within a given flow system (flow lines are generally parallel), that obtained from a spring—like tile-line effluent—represents an integration of water quality over a larger area (flow lines converge, as they do for any discharge area). The larger contributing areas for springs may account for the observation that detection frequencies in springs were so much higher for the triazines than for the less persistent acetanilides.

The available evidence thus shows that neither surface waters at base flow, tile lines, lysimeters, nor springs provide consistent or comparable indications of pesticide occurrence in water withdrawn from shallow wells. Some knowledge of the hydrogeologic characteristics of

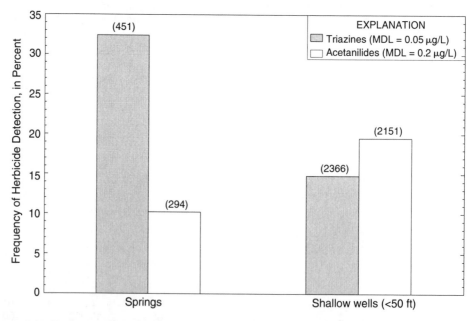

Figure 6.19. Frequencies of herbicide detection (triazines and acetanilides) in springs versus shallow wells (<50 ft) from the Cooperative Private Well Testing Program (data from Baker and others, 1994). Number of springs and wells sampled for each herbicide class given in parentheses. MDL, Method Detection Limit.

the system of interest is required to utilize such data. Indeed, given the various ways in which the physical and chemical characteristics of a well can influence the chemical and hydrologic regime surrounding the well screen, it is not necessarily clear that shallow wells provide a more accurate picture of shallow ground-water quality than these other modes of access to the subsurface. Further study is required to ascertain the degree to which each of these media is representative of shallow ground-water chemistry.

6.7 SUMMARY

The results discussed in this chapter demonstrate the variety of natural and anthropogenic factors that influence the frequencies of pesticide detection in ground water. Data from large-scale monitoring investigations indicate that contrasts among different pesticides with respect to their physical and chemical properties (particularly their soil dissipation half-lives, K_{oc} values, and water solubilities) cannot, by themselves, accurately distinguish compounds that are detected in ground water from those that are not, even among the pesticides examined during a single study.

The effects of edaphic and hydrogeologic setting on pesticide detection frequencies are generally in accord with the basic principles of solute transport in the subsurface (described in Chapter 4), but the evidence for many of these effects does not appear to be extensive. Although pesticide occurrence is generally more frequent beneath high-permeability soils with low organic matter content, the separate influences of these two factors have not been explored, and are difficult to examine in the field. Pesticide detections are also more frequent in more permeable hydrogeologic settings, such as unconsolidated formations or karst areas, and generally lower in other bedrock aquifers. Detection frequencies tend to be higher in areas with shallower water tables, shallower aquifers, or extensive interaction between surface waters containing appreciable pesticide concentrations and their associated aquifers.

A variety of anthropogenic factors, in addition to agricultural practices, can also affect pesticide detection frequencies. Detections become more common with decreasing well depth, or decreasing depth of the well screen below the water table. The likelihood of detecting pesticides is higher for dug or bored wells than for drilled wells, and for older, uncased, or inadequately sealed wells. Pesticide detections do not show consistent differences among different well types, however, perhaps as a result of the varying (and sometimes competing) influences of well construction and well use. Commonly the variables of well age, depth, construction, and use are all interrelated, and often related, in turn, to hydrogeologic setting, further confounding the overall variability found within and among studies that sample existing wells. Several aspects of study design also exert a pronounced influence over pesticide detection frequencies, including analytical detection limits, the procedure used to select wells for sampling (targeted versus nontargeted), and the potential use of other media, such as surface waters, tile drainage, lysimeter leachate, or spring effluents, as indicators of shallow ground-water quality.

CHAPTER 7

Occurrence of Pesticides in Ground Waters in
Non-Agricultural Settings

As with studies of agricultural impacts on ground-water quality, relations between non-agricultural activities and the occurrence of pesticides in ground water have been examined over a considerable range of spatial scales. These include process and matrix-distribution studies linked to specific non-agricultural applications, as well as monitoring studies investigating potential correlations between land use and ground-water quality over regional scales. This chapter focuses primarily on widely distributed (nonpoint) sources of pesticides used for non-agricultural purposes, such as turfgrass, rights-of-way, and timber production. Non-agricultural point sources, such as waste disposal sites and commercial agrichemical handling facilities, are examined in Chapter 8.

7.1 TURFGRASS PESTICIDES DETECTED IN SUBSURFACE WATERS

Table 7.1 summarizes the pesticides and transformation products that have been detected or not detected in subsurface waters following application to turfgrass. Only the study by Cohen and others (1990) examined ground-water quality beneath extensive areas of turfgrass—in this case, golf courses in Cape Cod, Massachusetts. As noted earlier (Section 3.4.4), the results from this investigation demonstrated a positive relation between the rates of pesticide application and the frequencies of detection in ground water beneath different areas of the golf courses examined, including tees, greens, fairways, and untreated locations.

During the Midcontinent Pesticide Study (MCPS), prometon was detected with substantially greater frequency (80 percent) than other herbicides (36 percent) in wells located within 400 m of private residences or within 3.2 km of golf courses (Burkart and Kolpin, 1993; Kolpin and others, 1994). Although its subsurface behavior following application was not examined by any of the process and matrix-distribution studies reviewed, this nonselective, non-agricultural herbicide has been encountered in ground waters in many different regions of the United States. In addition to its detection in ground waters during the MCPS, prometon was encountered in wells during the National Pesticide Survey, or NPS (Table 3.3), and in ground water in Florida (Rutledge, 1987), California (Sitts, 1989), New Jersey (Louis and Vowinkel, 1989), Illinois (Goetsch and others, 1992), Nebraska (Exner and Spalding, 1990), the Delmarva Peninsula (Koterba and others, 1993), Oklahoma City, Oklahoma, and Carson City, Nevada (Christenson and Rea, 1993).

As noted earlier (Table 3.3), the pesticide-related compounds detected most frequently during the NPS were derived from the hydrolysis of DCPA (dacthal), an herbicide commonly used on golf courses (Figure 3.9), as well as in urban and agriculture areas (Table 3.6). For both

Table 7.1 Pesticides and their transformation products detected or not detected in leachates from turfgrasses

[Compounds Detected: See Table 3.2 for explanations of abbreviations. Blank cells indicate no information applicable or available]

Reference	Study Location	Compounds Detected	Compounds Analyzed For But Not Detected
Hotzman and Mitchell, 1977	Delaware— Newark	Dicamba, Vel-4207	
Watschke and Mumma, 1989; Harrison and others, 1993	Pennsylvania— University Park	2,4-D, 2,4-DP, and dicamba	Pendimethalin, chlorpyrifos, 2,4-D BEE (butoxyethyl ester), and 2,4-DP BEE
Cohen and others, 1990	Massachusetts— Cape Cod golf courses	Chlordane, heptachlor epoxide, chlorothalonil, chlorpyrifos, 3,5,6-trichloro-2-pyridinol (chlorpyrifos degradate), DCPA/TPA, dicamba, 2,4-D, 2,4-dichlorobenzoic acid (suspected impurity), and isofenphos	Mecoprop (MCPP), siduron, pentachlorophenol, anilazine, iprodione, and diazinon
Smith and others, 1991b	Georgia—Tifton	Atrazine	Alachlor
Stahnke and others, 1991	Nebraska—Mead	Pendimethalin	
Branham and others, 1993	Laboratory microcosms	DCPA, MTP, TPA, isazofos, and CGA 17193	
Gold and Groffman, 1993	Rhode Island— Kingston	2,4-D and dicamba	
Ojima and others, 1993; Ojima, 1995	Japan—Golf course near Fukuyama City	Fenitrothion and flutlanil [sic]	
Smith and others, 1993	Various locations throughout the United States	Chlorpyrifos, diazinon, and dichlorvos	
Smith and Tillotson, 1993	Georgia— Griffin	2,4-D	
Odanaka and others, 1994	Japan—Ibaraki prefecture	Flutolanil, iprodione, isoprothiolane, metalaxyl, and trichlorfon	Bensulide, fenarimol, fenitrothion, isofenphos, isoxathion, oxine-copper, pendimethalin, prothiofos, pyridaphenthion, tetrachlorvinphos, toclofos-methyl, and triflumizole
Petrovic and others, 1994a	New York— Ithaca	Isazofos	
Petrovic and others, 1994b	New York— Ithaca	Triadimefon and triadimenol	
Smith and Bridges, 1995	Georgia— Griffin	2,4-D and mecoprop	

urban pesticide use and applications to golf courses, the NPS results provide "strong evidence of a positive relationship between the rate of DCPA use...and the probability of detecting DCPA acid metabolites in both [community water system] and rural domestic wells" (U.S. Environmental Protection Agency, 1992a). Similarly, TPA, produced from the hydrolysis of DCPA (Table 4.2) was one of the compounds most frequently detected during the 1992 sampling of the MCPS (Kolpin and others, 1995). These observations are consistent with the detection of DCPA or its transformation products in subsurface waters by both of the reviewed process and matrix-distribution studies that examined its behavior following application to turfgrass (Table 7.1).

7.2 OCCURRENCE OF PESTICIDES IN GROUND WATER IN RESIDENTIAL AND COMMERCIAL SETTINGS

Table 7.2 summarizes results from studies that specifically examined the occurrence of pesticides (and some of their transformation products) in ground waters beneath residential or commercial settings. Somasundaram and others (1993) compiled a more extensive list of similar investigations, incorporating studies of ground-water contamination by a variety of inorganic and organic constituents beneath urban areas, but did not specifically mention pesticides in their summary. Table 7.2 indicates that the pesticides detected in urban areas fall into two categories: (1) chlorinated insecticides and their transformation products; and (2) herbicides. The earlier studies focused primarily on the chlorinated insecticides, while more recent investigations examined a broader range of compound classes.

7.2.1 ORGANOCHLORINE INSECTICIDES

Table 7.2 demonstrates that the organochlorine insecticides used widely in homes and commercial buildings have also been detected in ground water. In Nassau County (Katz and Mallard, 1981), Ontario, Canada (Frank and others, 1987b) and Oklahoma City (Christenson and Rea, 1993), the insecticides detected in ground water (or, for the transformation products, their parent compounds) were known to have been used in buildings in the study areas of interest. For the Oklahoma City investigation, the insecticide-derived compounds found in ground water were a reflection of applications carried out several years earlier, since use of the parent compounds had been discontinued in the United States by the time of the study. Several of the pesticides detected in ground waters during the Oklahoma City study, however, were found to have been stored within the immediate vicinity of the wellheads at some of the sampling sites. The authors note the possibility that one or more of the compounds encountered in the subsurface during this investigation may have entered the well directly, or through an incomplete seal at the wellhead (Christenson and Rea, 1993).

The potential for residential insecticides to contaminate the subsurface is most clearly shown by the fact that one or more components or degradates of technical chlordane (either chlordane itself, heptachlor, or heptachlor epoxide) were detected in ground waters by five of the six studies in Table 7.2 that looked for them; the study by Rutledge (1987) was the sole exception. Such findings are consistent with the results from a study conducted at the University

Table 7.2. Pesticides and their transformation products detected or not detected in ground waters beneath residential and commercial areas

[Specific isomer designations for individual compounds are excluded for brevity. Blank cell indicates no information applicable or available]

Reference	Study Site Location	Land-Use Settings	Compounds Detected	Compounds Analyzed For But Not Detected
Katz and Mallard, 1981	New York—Nassau County	Residential; golf course	Dieldrin and heptachlor epoxide	Aldrin, chlordane, DDT, DDE, DDD, endosulfan, endrin, heptachlor, lindane, methoxychlor, mirex, perthane, toxaphene, diazinon, ethion, malathion, methyl parathion, methyl trithion, parathion, trithion, 2,4-D, 2,4-DP, silvex, and 2,4,5-T
Greenberg and others, 1982	New Jersey—Statewide	Residential	Aldrin, HCH, chlordane, DDT, DDE, DDD, dieldrin, endrin, heptachlor, heptachlor epoxide, lindane, and mirex	Not given
Frank and others, 1987b	Canada—Ontario (Province-wide)	Residential	Chlordane (from termite treatment), 2,4-D, mecoprop, and dicamba (from lawns)	Not given
Rutledge, 1987	Florida—Orlando area	Urban	Silvex	Numbers of parent compounds and products: 15 organochlorines 8 triazines 7 organophosphorus insecticides 3 phenoxy-acid herbicides 1 fumigant (EDB)
Eckhardt and others, 1989a	New York—Nassau and Suffolk Counties	High-density residential Medium and low-density residential Commercial Institutional	Chlordane and heptachlor epoxide Chlordane and heptachlor epoxide Chlordane and DDT	Aldicarb, carbofuran, and DDT Aldicarb, carbofuran, chlor-dane, and heptachlor epoxide Aldicarb, carbofuran, and DDT Aldicarb, carbofuran, and heptachlor epoxide
Christenson and Rea, 1993	Oklahoma—Oklahoma City	Urban	Aldrin, chlordane, dieldrin, DDE, 2,4-D, 2,4-DP, 2,4,5-T, picloram, dicamba, atrazine, and prometon	Numbers of parent compounds and products: 20 triazines and other nitrogen-containing herbicides 12 carbamates 11 organochlorines 10 organophosphorus insecticides 1 phenoxy-acid herbicide
Christenson and Rea, 1993	Nevada—Carson City	Urban	Prometon	

of Missouri-Columbia campus, during which over 70 percent of the total mass of the chlordane applied to soils was estimated to be present 7 years after application (Puri and others, 1990). The persistence of these compounds in exposed soils provides ample opportunity for their downward movement to the water table, and is therefore consistent with their frequent detection in subsurface waters.

Eckhardt and others (1989a) provided evidence of a direct link between chlordane use in residential and commercial buildings, and the presence of chlordane and heptachlor epoxide in underlying ground waters. In accord with typical patterns of chlordane use, these compounds were detected in ground water beneath areas classified by the authors as predominantly recreational, institutional, high-density residential, commercial, or transportation-related, but not beneath undeveloped or agricultural areas. Furthermore, among the three categories of population density examined by Eckhardt and others (1989a)—i.e., low-, medium-, and high-density residential areas—chlordane-derived contamination was observed only beneath high-density residential areas, suggesting a relation between population density and the likelihood of detecting chlordane in the subsurface. Similarly, Greenberg and others (1982) observed that residential areas were disproportionately common near wells containing 1-5 parts per billion (ppb) of a variety of organochlorine pesticides in New Jersey.

7.2.2 HERBICIDES

Nine of the pesticides detected in ground waters beneath urban, residential, and commercial areas are herbicides (Table 7.2). These compounds include several chlorophenoxy acids (2,4-D, 2,4-DP, 2,4,5-T, silvex, and mecoprop), triazines (atrazine and prometon) and chlorinated carboxylic acids (dicamba and picloram). The majority of these herbicides can be transported to subsurface waters following application to turfgrasses (Table 7.1). Thus, the presence of these compounds in ground waters beneath urban areas is probably a consequence of their application to turfgrasses, gardens, or rights-of-way. In contrast, neither atrazine, 2,4-D, silvex, 2,4,5-T, dicamba, nor picloram were detected in any of the Iowa State-Wide Rural Well-Water Survey (SWRL) wells located within 0.5 mi of commercial buildings in Iowa (Hallberg and others, 1992b).

7.2.3 AGRICULTURAL VERSUS NON-AGRICULTURAL USE OF DETECTED COMPOUNDS

Some of the pesticides detected in shallow ground waters during the studies listed in Tables 7.1 and 7.2 may have been present because of nearby agricultural use, with two exceptions. Among the herbicides detected, only mecoprop and prometon are or have been used exclusively for non-agricultural purposes. Atrazine, dicamba, picloram, 2,4-D, 2,4-DP, and 2,4,5-T are all used extensively for weed control in a variety of non-agricultural settings, including lawns, gardens, and rights-of-way, but they are also applied to agricultural lands (Beste, 1983; Domagalski and Dubrovsky, 1991; Gianessi and Puffer, 1990; Wauchope and others, 1992; and Christenson and Rea, 1993). Furthermore, in addition to their use in residential settings, aldrin, dieldrin, and DDT were all employed to control insects in agricultural areas prior to the ban on their use (e.g., McKenna and others, 1988). Similarly, while chlordane was used widely to kill termites and ants in homes and buildings prior to its discontinuation in 1987, it was also used to control a variety of insects in agricultural settings (Puri and others, 1990) and on golf courses (Cohen and others, 1990).

Of the two noncrop herbicides detected by the studies listed in Table 7.2, prometon (discussed earlier) has been detected with substantially greater frequency in shallow ground waters. Although van de Weerd and van der Linden (1991) demonstrated that mecoprop migrates readily through the subsurface with little detectable retardation or transformation, the herbicide was not detected by any of the studies included in the Pesticides In Ground Water Database, or PGWDB (460 wells in 5 states), or by any of the monitoring studies reviewed for this report (Table 3.2). However, the compound has been detected beneath sanitary landfills in six states (Gintautas and others, 1992), and in rural wells contaminated by spills, surface runoff, and spray drift in Ontario (Frank and others, 1987b).

7.3 RIGHTS-OF-WAY

As with other non-agricultural settings, the effects of pesticide applications to railroad, powerline, pipeline, and roadway rights-of-way on ground-water quality have received only minor attention in the published literature. About half of the pesticides detected in subsurface media as a presumed result of their application to rights-of-way in the United States and other countries (Table 7.3) are known to have been applied in these settings (Table 3.8). Bromacil and picloram are among the compounds most commonly detected in such areas. In addition, many of the detections of prometon in ground waters beneath different areas of the United States (Table 3.3) are also likely to have arisen from the application of this non-agricultural herbicide to rights-of-way.

It is often difficult to ascribe pesticide detections in the subsurface solely to their use along rights-of-way if the compounds are also used extensively in agriculture (Table 3.8), a point noted in Section 7.2.3 with regard to pesticide detections in ground waters beneath urban areas. In such cases, rights-of-way or other non-agricultural areas can only be identified as the source of contamination on a case-by-case basis, rather than simply from the nature of the compound alone. For example, as of 1990, 28 percent of all atrazine detections in the ground waters of East Anglia, England, were thought to have been the result of applications of the herbicide for weed control on roads, railways, and factory yards rather than on agricultural lands (Smith, 1990). Consistent with this is the fact that British Rail was the single largest user of atrazine in England at the time (Cartwright and others, 1991). In many rural areas, however, rights-of-way are surrounded by agricultural lands, making it difficult to distinguish contamination arising from pesticide use in the two different settings.

7.4 RANGELANDS

Among the pesticides employed to control the growth of unwanted vegetation on rangelands, tebuthiuron is one of the most commonly used. Results from several studies, most of them conducted in the semiarid areas of the western United States and Mexico, have shown that this compound may persist in rangeland soils for several years following application, and can migrate to depths of several feet below the ground surface (Johnsen and Morton, 1989; Summit and others, 1989). Among the studies reviewed, only one investigated ground-water contamination from a pesticide applied widely to rangelands. Lym and Messersmith (1988) observed more frequent detections of picloram in ground waters beneath areas of North Dakota where the herbicide had been applied most extensively. Several of these detections occurred in rangeland areas that were known to have been treated with picloram.

Table 7.3. Pesticides detected or not detected in the subsurface following their (presumed) application to rights-of-way

[Compounds listed do not include those that may have been applied to crops, rather than to rights-of-way. Blank cells indicate no information applicable or available]

Reference	Study Location	Application Setting	Medium Sampled	Compounds Detected	Compounds Not Detected
Zandvoort and others, 1980	The Netherlands—Maarn	Railroad beds	Soil and leachate	Bromacil	
Segawa and others, 1986	California—Glenn County	Miscellaneous rights-of-way (unspecified)	Ground water	Prometon	
Troiano and Segawa, 1987	California—Tulare County		Ground water	Atrazine	Prometon
Frank and others, 1987b	Canada—Ontario	Roadways and power lines	Ground water	Bromacil, 2,4-D, dicamba, dichlorprop, pentachloro-phenol, and picloram	
McKinley and Arron, 1987	Canada—Ontario	Roadways	Soil and surface water	2,4-D and picloram	
Lagas and others, 1989	The Netherlands—Utrecht	Railroad beds	Ground water	Amitrol and diuron	
Watson and others, 1989	Montana—Northern Rocky Mountains	Roadways	Soil and vegetation	Picloram	
			Ground water and surface water		Picloram
Smith, 1990	England—East Anglia	Roads, railways, and factory yards	Ground water	Atrazine	
Allender, 1991	Australia—New South Wales	Gas pipeline valve site	Soil	Bromacil and hexazinone	

7.5 PUBLIC GARDENS AND COMMERCIAL FLOWER PRODUCTION FACILITIES

Pesticides used for maintaining public gardens and for the production of ornamental plants have been detected in the subsurface beneath areas where they have been applied, although the number of studies that have examined this issue is quite limited. Table 7.4 lists the compounds that have been detected in these settings, as well as the media and locations in which they were detected.

7.6 TIMBER PRODUCTION AND PROCESSING OPERATIONS

Table 7.5 provides an overview of the results from several studies that have examined the occurrence of pesticides in surficial and subsurface materials (leaf litter, soils, soil water, or ground water) in timber production areas and at wood treatment facilities. Hexazinone, picloram, glyphosate, 2,4-D, and sulfometuron methyl have received the most attention among the

Table 7.4. Pesticides and their transformation products detected or not detected in the subsurface beneath public gardens and commercial flower production facilities

[Transformation products are indented beneath their parent compounds, unless otherwise indicated. Blank cells indicate no information applicable or available]

Reference	Study Site Location	Land-Use Setting	Medium Sampled	Compounds Detected	Compounds Not Detected

Weaver and others, 1988a	California— Del Norte County,	Lily bulb cultivation	Soil	Fenamiphos
				  Fenamiphos sulfoxide
				  Fenamiphos sulfone	

Weaver and others, 1988b	California— Del Norte and Humboldt Counties	Daffodil bulb cultivation	Soil	Phorate
				  Phorate sulfoxide
				  Phorate sulfone
				Ethoprop	

Lagas and others, 1989	The Netherlands	Flower bulb cultivation	Shallow ground water	1,3-Dichloropropene 1,2-Dichloropropane Methyl isothiocyanate[1] Ethylene thiourea (ETU)[1]	1,2,3-Trichloro-propane Fluazifop Fluazifop-butyl Linuron Metamitron

		Public garden	Shallow ground water	Dichlobenil
				  2,6-Dichlorobenzamide
				    (BAM)
				Simazine
				  Deisopropyl atrazine
				    (DIA)	

Lorber and others, 1990	California— Del Norte County	Lily bulb cultivation	Ground water	Aldicarb	
	Florida— Volusia County	Fernery	Ground water	Aldicarb	

[1]See Table 3.4 for inferred parent compounds.

pesticides used to control vegetation on timberlands. As noted in Section 3.3.4, 2,4-D, glyphosate, hexazinone, and picloram were also among the compounds used most extensively in forestry during the 1970's and 1980's. In contrast, the principal compounds examined at wood treatment facilities and sawmills have been the polychlorophenol fungicides.

Among the studies listed in Table 7.5, the environmental medium most commonly sampled for pesticides has been soil. Pesticide residues were detected in soils, leaf litter and soil waters during all investigations that examined these media. This is not unexpected, since these are also the media to which the compounds are applied, and in which their pesticidal properties are designed to be expressed. By contrast, approximately half of the investigations that looked for pesticides in ground waters beneath timber production and treatment areas did not detect them. This may have been related to the comparatively high limits of detection used for subsurface waters (>1 µg/L) during most of these studies. In addition, the relative scarcity of analyses for transformation products during these studies makes it difficult to determine the degree to which the general absence of detectable pesticide residues in ground waters was due to transformation of the parent compounds, as opposed to their transport away from the application sites.

Table 7.5. Pesticides and their transformation products examined in the subsurface within the vicinity of timber production and wood processing activities

[Blank cells indicate no information applicable or available]

Reference	Location	Application/Setting	Media Sampled	Compounds Detected	Compounds Not Detected
Neary, 1983	Central Tennessee	Forested watershed	Ground water (springs)		Hexazinone and two unidentified degradates
Neary and others, 1983	Georgia— Clarksville	Forested watersheds	Soil and leaf litter	Hexazinone and two unidentified degradates	
Valo and others, 1984	Central Finland	Sawmills	Soil water and soil	Di-, tri- tetra- and penta-chlorophenols	
Bouchard and others, 1985	Northwestern Arkansas	Forested watershed	Soil and leaf litter	Hexazinone	
Goerlitz and others, 1985; Godsy and others, 1992	Florida— Pensacola,	Wood treatment facility	Ground water	Pentachloro-phenol and creosote	
Neary and others, 1985	Western North Carolina	Mixed hardwood forest	Soil water, spring water, and soil	Picloram	
Cavalier and Lavy, 1987; Cavalier and others, 1989	Arkansas— Ouachita National Forest	Forested watershed	Ground water (springs and wells)		2,4-D 2,4-D ester (Weedone) Hexazinone Picloram
Segal and others, 1987	Florida— Gainesville,	Coastal Plain flatwoods	Ground water		Sulfometuron methyl
Feng and others, 1989	Canada— Grande Prairie, Alberta	Clearcut forest	Soil water	Hexazinone	
Lavy and others, 1989	North-central West Virginia	Forested watershed	Soil and leaf litter	Hexazinone	
Michael and others, 1989	Alabama— Tuskegee National Forest	Coastal Plain forest	Soil water and soil	Picloram	
Neary and Michael, 1989	Florida— Gainesville	Coastal Plain flatwoods	Ground water		Sulfometuron methyl

Table 7.5. Pesticides and their transformation products examined in the subsurface within the vicinity of timber production and wood processing activities—*Continued*

Reference	Location	Application/Setting	Media Sampled	Compounds Detected	Compounds Not Detected
Bush and others, 1990	South Carolina— Barnwell Florida— Hughes Island	Coastal Plain forest	Ground water	Hexazinone	
Feng and Thompson, 1990	Canada— Carnation Creek, Vancouver Island, British Columbia	Forested watershed	Soil and leaf litter	Glyphosate AMPA (glyphosate degradate)	
McNeill, 1990	England— Tyock Burn	Wood treatment facility	Soil	Dieldrin and pentachloro-phenol	
Goerlitz, 1992	California— Visalia	Wood treatment facility	Ground water	Pentachloro-phenol and creosote	

To date, the most severe contamination of ground waters beneath timber production or processing areas appears to have been associated with wood treatment facilities. Goerlitz and others (1985) documented the presence of a broad array of polynuclear aromatic hydrocarbons and alkylated phenols in ground waters contaminated by creosote and pentachlorophenol (PCP) beneath a wood-preserving facility in Pensacola, Florida. Although high PCP concentrations were measured in ground water immediately beneath the contaminant source, it was not detected in wells farther downgradient, despite the fact that the pH of the ground water was above the acid-base dissociation constant (pK_a) of the compound (which would cause the PCP to be present mostly in its anionic, and hence more mobile form), and that column studies by the authors had demonstrated a relatively high mobility of PCP through the aquifer materials (Goerlitz and others, 1985).

7.7 SUMMARY

The results from the studies discussed in this chapter have demonstrated that pesticides used in non-agricultural settings may be detected in the underlying ground water. For most of these settings, however, the data on pesticide occurrence in the subsurface—particularly in ground water—are relatively scarce. Given the substantial quantities of pesticides used in some of these settings (Section 3.3), more extensive data are needed to assess the impact of pesticide use on ground-water quality in non-agricultural areas. During monitoring studies, however, the attribution of pesticide detections in ground water to the application of these compounds in non-agricultural settings may be confounded by the fact that many of them are also used in agriculture, and by the fact that non-agricultural application areas are often surrounded by agricultural activities in rural areas.

CHAPTER 8

Pesticide Contamination from Point Sources

Several studies have examined the impact of known point sources of pesticide contamination on the quality of the underlying ground water. An early summary of this subject by Young (1981) provided a useful overview of the principal chemical characteristics of ground waters contaminated by various types of point sources, but did not examine pesticides or other specific organic compounds. Different types of point sources give rise to different assemblages of pesticides and levels of contamination in the subsurface. Most of the following discussion focuses on data from investigations that examined general patterns of ground-water contamination associated with a particular type of point source at more than one location, rather than on those describing the nature and removal of contamination at a specific site, such as a contaminated well (e.g., Lewallen, 1971; Frank and others, 1979, 1987b), a chemical plant (e.g., Jürgens and Roth, 1989), or a waste disposal site (e.g., Kerdijk, 1981; Dahl, 1986; Stark and others, 1987; and Goerlitz, 1992).

8.1 CHEMICAL MANUFACTURING AND COMMERCIAL SUPPLY FACILITIES

Few studies have assessed pesticide concentrations and distributions in the subsurface beneath pesticide manufacturing and distribution facilities. Furthermore, except for a study currently in progress in 11 counties in Arkansas (Senseman and others, 1990) and one completed in Hawaii (Miles and others, 1990), the investigations that have been completed to date have been confined primarily to states in the midcontinent (Iowa, Illinois, Minnesota, Ohio, and Wisconsin). The analytical scope of some of these studies, however, has been extensive, in some cases exceeding 50 pesticide compounds and transformation products (Long, 1989; Krapac and others, 1993).

8.1.1 SEVERITY AND FREQUENCY OF CONTAMINATION

Table 8.1 (at end of chapter) summarizes the maximum concentrations of individual pesticides and transformation products reported in ground water and surface soils at agrichemical facilities in Illinois, Iowa, Wisconsin, Ohio, Minnesota, and Hawaii. (Maximum concentrations are presented in Table 8.1, rather than mean or median values, because of the three parameters, only maximum concentrations were reported by all of the studies of interest.) To date, the most extensive work has been carried out in Illinois. Table 8.1 includes results from four different

323

summaries of data from this state, but specific sources for some of the reported data could not be determined. To provide a more representative assessment of typical levels of pesticides at these facilities, data for locations where acute spills were known to have occurred were excluded from Table 8.1. Despite the exclusion of data from acute spill areas, however, the table indicates that pesticide concentrations above 100 µg/L are not uncommon in ground waters beneath agrichemical facilities. In many cases, such high concentrations exceed water-quality standards for the protection of human health or aquatic ecosystems. In surficial soils, maximum concentrations typically exceed 1,000 µg/kg for most of the compounds detected.

Pesticide concentrations in the subsurface beneath agrichemical facilities thus tend to be substantially higher than those encountered beneath agricultural fields. Krapac and others (1993) compared the concentrations of pesticides measured in surface soils at 49 agrichemical facilities in Illinois with the corresponding "application rate equivalent" (ARE) for each compound. The defining equation for this parameter (Roy and others, 1993) indicates that the ARE for a given compound is the soil concentration expected following application (at its average rate for Illinois in 1990) and incorporation to a depth of 3 in. (8 cm). Although the median concentrations for individual pesticides in the soils of the 49 Illinois facilities were between 10 percent and 50 percent of their respective ARE values, the maximum soil concentrations were found to exceed the ARE values by up to three orders of magnitude (Krapac and others, 1993). Similarly, the metolachlor concentration measured by Anderson and others (1994) in the soil at an agrichemical dealership in Iowa was nearly eight times higher than that anticipated from normal field applications. Indeed, Hallberg (1985) noted that pesticide concentrations in soils and ponded surface waters near handling, loading, and equipment rinsing areas at agrichemical handling facilities in Iowa are often as high as those in the original formulations.

In addition to those measured in on-site wells (Table 8.1), relatively high concentrations of pesticide residues have been measured in water-supply wells in the vicinity of agrichemical facilities. In Wisconsin, 49 percent of the water-supply wells sampled near 20 mixing-and-loading facilities contained detectable levels of pesticides (Habecker, 1989). Furthermore, among the water-supply wells found by Habecker (1989) to contain detectable levels of pesticides, 94 percent exceeded the "Preventative Action Limits" (PALs) specified by the state of Wisconsin (although it should be noted that the Wisconsin PALs are much stricter than the corresponding U.S. Environmental Protection Agency water-quality standards). In Illinois (Long, 1989), pesticide concentrations in wells located near agrichemical facilities have, in some cases, exceeded levels measured in wells located on-site. Hallberg (1985) noted that in the vicinity of some farm-chemical supply dealerships in Iowa, "the chemical concentrations in the local groundwater/drinking water generally have increased 10-fold for nitrate, and *100 times* [emphasis provided by original author] for pesticides, over the background concentrations in the area."

Table 8.2 summarizes the frequencies with which different pesticides have been detected in ground waters beneath agrichemical handling facilities in Wisconsin and Illinois, the states where, as noted earlier, the most extensive investigations have been carried out. The five pesticides detected most frequently—atrazine, alachlor, metolachlor, cyanazine, and metribuzin—also have been among those used most extensively. Consistent with these observations, metolachlor was found to be the most frequently detected pesticide in ground waters beneath 16 mixing-and-loading facilities in 11 counties in Arkansas (Senseman and others, 1990; Senseman, 1993). Several other studies, discussed in Section 8.10, have noted direct relations between the frequencies of pesticide detection in wells and the proximity of the wells to agrichemical facilities (e.g., Kross and others, 1990; Holden and others, 1992; and Baker and others, 1994).

Table 8.2. Frequencies of occurrence (among sites) of pesticides and their transformation products in ground waters beneath pesticide manufacturing and commercial supply facilities in Wisconsin and Illinois

[Pesticide or Transformation Product: Compounds listed in order of decreasing maximum frequency of detection in ground water. Transformation products are indented. Sources for Use in State: Gianessi and Puffer (1990 [herbicides], 1992b [insecticides], 1992a [fungicides]); U.S. Environmental Protection Agency (1990b [Discontinued compounds]). NG, Detection limit not given. ND, not detected. lb a.i./yr, pounds of active ingredient per year; μg/L, micrograms per liter]

Pesticide or Transformation Product	State	Frequency of Detection in Ground Water (among sites, in percent)	Number of Sites Examined	Detection Limit (μg/L)	Use In State (lb a.i./yr)	Reference
			Herbicides			
Atrazine	Wisconsin	90	[1]20	NG	2,684,832	Habecker, 1989
	Illinois	38	56	NG	8,503,397	Long, 1989
	Illinois	9.6	52	0.43		Goetsch and others, 1993
Alachlor	Wisconsin	80	[1]20	NG	1,284,545	Habecker, 1989
	Illinois	61	56	NG	7,960,274	Long, 1989
	Illinois	ND	52	1.3		Goetsch and others, 1993
Metolachlor	Wisconsin	65	[1]20	NG	1,379,421	Habecker, 1989
	Illinois	57	56	NG	8,082,584	Long, 1989
	Illinois	5.8	52	1.4		Goetsch and others, 1993
Cyanazine	Wisconsin	60	[1]20	NG	1,658,279	Habecker, 1989
	Illinois	41	56	NG	3,071,283	Long, 1989
Metribuzin	Wisconsin	30	[1]20	NG	78,712	Habecker, 1989
	Illinois	55	56	NG	562,153	Long, 1989
	Illinois	ND	52	0.43		Goetsch and others, 1993
Trifluralin	Wisconsin	ND	[1]20	NG	81,716	Habecker, 1989
	Illinois	27	56	NG	2,888,020	Long, 1989
	Illinois	9.6	52	0.003		Goetsch and others, 1993
Butylate	Wisconsin	10	[1]20	NG	322,351	Habecker, 1989
	Illinois	21	56	NG	3,067,417	Long, 1989
	Illinois	ND	52	0.65		Goetsch and others, 1993
Dicamba	Wisconsin	20	[1]20	NG	350,364	Habecker, 1989
	Illinois	1.9	52	0.16	884,140	Goetsch and others, 1993
Pendimethalin	Wisconsin	ND	[1]20	NG	209,808	Habecker, 1989
	Illinois	18	56	NG	1,302,973	Long, 1989

Table 8.2. Frequencies of occurrence (among sites) of pesticides and their transformation products in ground waters beneath pesticide manufacturing and commercial supply facilities in Wisconsin and Illinois—*Continued*

Pesticide or Transformation Product	State	Frequency of Detection in Ground Water (among sites, in percent)	Number of Sites Examined	Detection Limit (µg/L)	Use In State (lb a.i./yr)	Reference
Bentazon	Illinois	ND	56	NG	1,713,775	Long, 1989
	Illinois	11.5	52	0.21		Goetsch and others, 1993
2,4-D	Wisconsin	10	[1]20	NG	241,941	Habecker, 1989
	Illinois	ND	52	0.57	1,214,689	Goetsch and others, 1993
EPTC	Wisconsin	10	[1]20	NG	1,926,542	Habecker, 1989
	Illinois	ND	56	NG	3,131,595	Long, 1989
	Illinois	ND	52	0.68		Goetsch and others, 1993
Linuron	Wisconsin	10	[1]20	NG	73,873	Habecker, 1989
	Illinois	ND	56	NG	321,202	Long, 1989
Simazine	Wisconsin	10	[1]20	NG	1,268	Habecker, 1989
	Illinois	ND	56	NG	309,041	Long, 1989
	Illinois	ND	52	0.12		Goetsch and others, 1993
Chloramben	Wisconsin	5	[1]20	NG	37,450	Habecker, 1989
	Illinois	ND	56	NG	179,949	Long, 1989
	Illinois	1.9	52	0.43		Goetsch and others, 1993
Dinoseb	Wisconsin	5	[1]20	NG	([2])	Habecker, 1989
	Illinois	ND	52	0.16	([2])	Goetsch and others, 1993
Prometon	Wisconsin	5	[1]20	NG	([3])	Habecker, 1989
	Illinois	3.8	52	0.37	([3])	Goetsch and others, 1993
Acifluorfen	Illinois	3.8	52	0.68	140,301	Goetsch and others, 1993
Acephate	Illinois	ND	56	NG	([3])	Long, 1989
Bromacil	Illinois	ND	56	NG	([2])	Long, 1989
	Illinois	ND[4]	52	17.0		Goetsch and others, 1993
2,4-DB	Illinois	ND	52	4.0	9,824	Goetsch and others, 1993
Fluchloralin	Illinois	ND	56	NG	([2])	Long, 1989
Glyphosate	Wisconsin	ND	[1]20	NG	149,406	Habecker, 1989
Picloram	Illinois	ND	56	NG	([2])	Long, 1989

Table 8.2. Frequencies of occurrence (among sites) of pesticides and their transformation products in ground waters beneath pesticide manufacturing and commercial supply facilities in Wisconsin and Illinois—*Continued*

Pesticide or Transformation Product	State	Frequency of Detection in Ground Water (among sites, in percent)	Number of Sites Examined	Detection Limit (µg/L)	Use In State (lb a.i./yr)	Reference
Propachlor	Illinois	ND	56	NG	204,266	Long, 1989
	Illinois	ND	52	0.11		Goetsch and others, 1993
Tebuthiuron	Illinois	ND	56	NG	(²)	Long, 1989
Triclopyr	Illinois	ND	56	NG	(²)	Long, 1989
Vernolate	Illinois	ND	52	0.65	(²)	Goetsch and others, 1993

			Insecticides			
Chlordane (α,γ)	Illinois	45	56	NG	Discontinued compound	Long, 1989
Chlorpyrifos	Illinois	30	56	NG	1,244,347	Long, 1989
Heptachlor	Illinois	ND	56	NG	Discontinued compound	Long, 1989
	Illinois	ND	52	0.005		Goetsch and others, 1993
Heptachlor epoxide	Illinois	23	56	NG		Long, 1989
	Illinois	ND	52	0.004		Goetsch and others, 1993
Carbaryl	Wisconsin	15	[1]20	NG	102,227	Habecker, 1989
Carbofuran	Wisconsin	10	[1]20	NG	72,809	Habecker, 1989
Dimethoate	Wisconsin	10	[1]20	NG	36,342	Habecker, 1989
	Illinois	ND	56	NG	20,969	Long, 1989
Diazinon	Illinois	9	56	NG	3,265	Long, 1989
Dieldrin	Illinois	4	56	NG	Discontinued compound	Long, 1989
	Illinois	5.8	52	0.004		Goetsch and others, 1993
Terbufos	Wisconsin	5	[1]20	NG	448,214	Habecker, 1989
	Illinois	ND	56	NG	747,795	Long, 1989
Lindane	Illinois	2	56	NG	(²)	Long, 1989
α-BHC	Illinois	ND	56	NG	Discontinued compound	Long, 1989
Aldrin	Illinois	ND	56	NG	Discontinued compound	Long, 1989
	Illinois	1.9	52	0.004		Goetsch and others, 1993
Endrin	Illinois	ND	56	NG	Discontinued compound	Long, 1989
	Illinois	1.9	52	0.006		Goetsch and others, 1993

Table 8.2. Frequencies of occurrence (among sites) of pesticides and their transformation products in ground waters beneath pesticide manufacturing and commercial supply facilities in Wisconsin and Illinois—*Continued*

Pesticide or Transformation Product	State	Frequency of Detection in Ground Water (among sites, in percent)	Number of Sites Examined	Detection Limit (µg/L)	Use In State (lb a.i./yr)	Reference
Endrin aldehyde	Illinois	ND	52	0.009		Goetsch and others, 1993
Cypermethrin	Illinois	ND	56	NG	[2]	Long, 1989
DDT (*o,p; p,p'*)	Illinois	ND	56	NG	Discontinued	Long, 1989
DDD (*p,p'*)	Illinois	ND	56	NG	compound	Long, 1989
DDE (*p,p'*)	Illinois	ND	56	NG		Long, 1989
DDVP (dichlorvos)	Illinois	ND	56	NG	[3]	Long, 1989
Dicofol	Illinois	ND	56	NG	[2]	Long, 1989
Disulfoton	Illinois	ND	56	NG	[2]	Long, 1989
Ethoprop	Illinois	ND	56	NG	[2]	Long, 1989
	Illinois	ND	52	0.15		Goetsch and others, 1993
Fenthion	Illinois	ND	56	NG	[3]	Long, 1989
Fenvalerate	Illinois	ND	56	NG	1395	Long, 1989
Fonofos	Wisconsin	ND	[1]20	NG	112,763	Habecker, 1989
	Illinois	ND	56	NG		Long, 1989
Isofenphos	Illinois	ND	56	NG	[2]	Long, 1989
Malathion	Illinois	ND	56	NG	7,708	Long, 1989
Methoxychlor	Illinois	ND	56	NG	[2]	Long, 1989
Naled	Illinois	ND	56	NG	[2]	Long, 1989
Parathion	Wisconsin	ND	[1]20	NG	[3]	Habecker, 1989
Phorate	Illinois	ND	56	NG	363,173	Long, 1989
Propetamphos	Illinois	ND	56	NG	[3]	Long, 1989
Sumithrin	Illinois	ND	56	NG	[3]	Long, 1989
Fungicides						
Captan	Illinois	ND	56	NG	69,262	Long, 1989
Captafol	Illinois	ND	56	NG	[3]	Long, 1989

[1]Unclear from original reference (Habecker, 1989) whether ground waters from all 20 sites were analyzed for all compounds, but this was assumed to have been the case.

[2]No use of this compound reported by Gianessi and Puffer (1990, 1992a,b) for this state.

[3]Data for use of this compound not available from Gianessi and Puffer (1990, 1992a,b).

[4]Bromacil detection frequency may represent an underestimate, since recoveries of the herbicide from spiked water samples were found by Goetsch and others (1993) to have been negligible.

8.1.2 SPATIAL AND TEMPORAL PATTERNS OF CONTAMINATION

Pesticide contamination is usually widely dispersed throughout the areas occupied by individual agrichemical handling facilities (Habecker, 1989; Krapac and others, 1993). The severity of soil and ground-water contamination, however, is highly variable among different locations at a given facility, depending on the nature of the activities involved. Pesticide concentrations beneath 20 mixing-and-loading facilities in Wisconsin (Habecker, 1989) varied by up to three orders of magnitude in soils among different locations at individual sites. Krapac and others (1993) observed pesticide concentrations to vary among the soils at 49 agrichemical facilities in Illinois by six orders of magnitude.

At Wisconsin facilities, Habecker (1989) found the highest residue levels in soils where acute spills had occurred and within mixing-and-loading areas, while the lowest concentrations tended to be in equipment washing areas, intermittently ponded depressions, and drainageways. Elevated pesticide concentrations also occurred beneath burn piles and storage areas for empty containers. In agreement with the latter observation, Hallberg and others (1992b) found the frequencies of pesticide detection in rural domestic wells in Iowa to be higher on farms where empty pesticide containers were either burned or left for pick-up by municipal refuse disposal services, than on farms where neither of these practices were employed. Some of the Iowa observations, however, may have resulted from autocorrelation with other factors related to pesticide use and handling on-site.

Spatial patterns of soil contamination documented by Krapac and others (1993) for the 49 Illinois facilities—based on frequencies of pesticide detection, rather than concentrations— differed markedly from those reported by Habecker (1989) for the Wisconsin facilities. Krapac and others (1993) encountered the highest frequencies of pesticide detection in surface drainage areas, somewhat lower frequencies beneath mixing-and-loading and equipment rinsing areas, and the lowest frequencies beneath burn piles and pesticide storage areas. Although the reasons for the differences between the spatial patterns observed by the Illinois and Wisconsin studies remain unclear, the data suggest that, as noted by Krapac and others (1993), pesticide contamination of the subsurface at agrichemical facilities is sufficiently acute and widespread to merit investigation at all on-site locations where pesticides are handled or surface waters are encountered.

Pesticide concentrations in soils beneath agrichemical facilities generally decrease with increasing depth. Studies in Wisconsin (Habecker, 1989), Illinois (Krapac and others, 1993), and Hawaii (Miles and others, 1990) have all observed this trend. Results reported for Wisconsin suggest a similar pattern for ground waters beneath such facilities. Habecker (1989) assessed pesticide concentrations in both monitoring and water-supply wells at or near each of the Wisconsin facilities. Although the depths of the screened intervals of the sampled wells were not provided, monitoring wells are typically screened at shallower depths than are water-supply wells in a given area (e.g., Koterba and others, 1993). At the Wisconsin sites, pesticide concentrations were typically higher in monitoring wells than in nearby water-supply wells (Habecker, 1989), suggesting that pesticide contamination of the ground water was less pronounced at greater depths than near the surface.

In general, contamination beneath mixing-and-loading areas at many of these sites appears to be chronic, rather than episodic. After site remediation was completed and normal operations resumed following incidents of particularly severe contamination, residue levels in soils beneath mixing-and-loading areas in Wisconsin were found by Habecker (1989) to be higher than or comparable to those measured prior to the commencement of cleanup operations. The study concluded that "spillage is an ongoing problem especially in the mixing/loading and pesticide equipment parking areas." The chronic nature of pesticide contamination at these sites may help to explain why, as noted earlier (Section 6.2), pesticide detections in surficial soils at the Wisconsin facilities were found to be unreliable predictors of pesticide contamination in nearby wells.

8.1.3 RELATIONS BETWEEN PESTICIDE CONTAMINATION AND USE

Patterns of soil contamination at agrichemical facilities indicate that the frequencies with which pesticides have been detected at these sites are related to both the variety and the total amounts of the pesticides handled. Krapac and others (1993) observed the number of "pesticide detections" in soils (presumed to mean the total number of times any pesticide was detected in any soil) at the 49 facilities in Illinois to increase with the number of compounds detected (Figure 8.1). Figure 8.2 indicates that compounds handled at more sites in the state were also detected more often. Four of the compounds shown in Figure 8.2, however (alachlor, bromacil, EPTC, and trifluralin), were detected at several sites where they were not known to have been handled.

The frequency of pesticide contamination in ground waters beneath agrichemical facilities is also related to the amounts of product handled at such sites. In addition to the pesticide detection frequencies reported for mixing-and-loading facilities in different states, Table 8.2 also lists the statewide rates of use for each compound. Figure 8.3, based on data from Table 8.2, shows the significant relations ($\alpha = 0.05$; simple linear regression) observed between the maximum percentages of sites at which individual pesticides have been detected in ground waters and their respective statewide use rates in Illinois and Wisconsin.

8.1.4 RELATIONS BETWEEN PESTICIDE CONTAMINATION AND FACILITY CHARACTERISTICS

Neither the age nor the size of a given agrichemical handling facility appear to be related to the severity of subsurface contamination at the site. Although the twenty facilities examined in Wisconsin (Habecker, 1989) had been in operation for periods of 2 to 34 years, no correlation was evident between the age of the facility and the concentrations of atrazine (the compound detected in ground waters at the highest proportion of sites) in nearby water-supply wells. The work of Krapac and others (1993) yielded the same conclusion with respect to pesticide detections in soils at the Illinois facilities, where the likelihood of detecting pesticides in soils was found to be independent not only of the age, but also of the size of the site (Figure 8.4). These observations are consistent with the chronic nature of pesticide contamination noted by Habecker (1989) at such facilities (Section 8.1.2).

8.1.5 VOLATILE ORGANIC CONTAMINANTS

Pesticides are not the only organic contaminants that may be encountered in ground waters at agrichemical mixing-and-loading facilities. Chemical analysis of water taken from a well at one such facility in Illinois (Long, 1989) led to the detection of 12 volatile organic compounds, none of which were typically used as pesticide active ingredients (Table 8.3). As noted by the author, these compounds may have been present as a "result of carrier solvents, fuels, oils, or other materials used and spilled on-site." The use of organic solvents as "inert ingredients" in pesticide formulations (Section 3.1.3) thus poses a potential threat to ground-water quality at agrichemical handling sites, as well as beneath the agricultural lands to which the products are routinely applied.

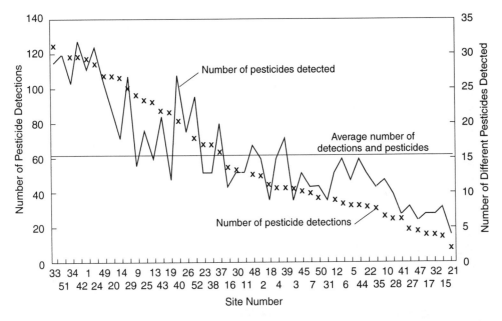

Figure 8.1. Relation between the number of pesticide detections (presumed to mean the total number of times any pesticide was detected in any sample) and the number of different pesticides detected in soils at each of 49 agrichemical facilities in Illinois. Redrawn with permission from Krapac and others (1993).

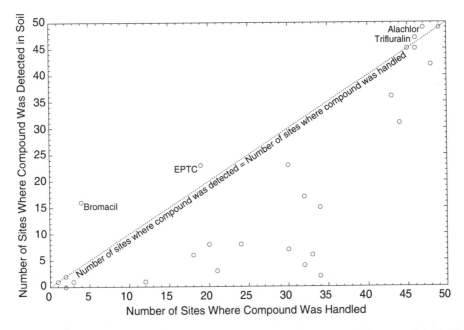

Figure 8.2. Number of agrichemical facilities in Illinois where individual pesticides were detected in soils versus the number of facilities where they were handled. Data from Krapac and others (1993).

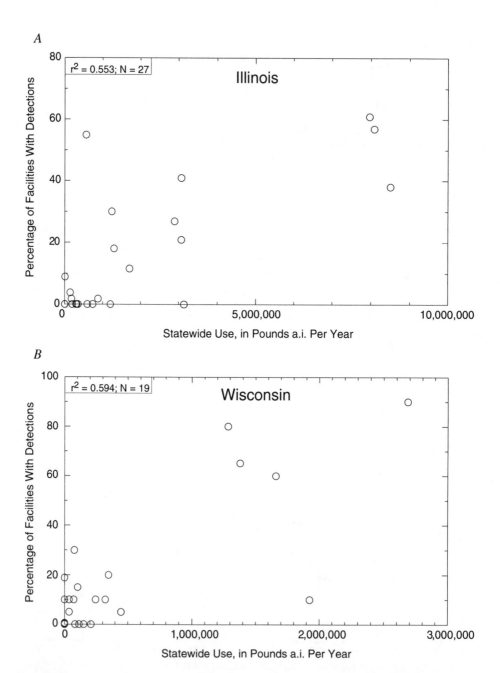

Figure 8.3. Maximum percentages of agrichemical mixing-and-loading facilities with detections of individual pesticides in ground water, versus statewide use of each compound in (*A*) Illinois; and (*B*) Wisconsin (data from Table 8.2). Both relations found to be statistically significant ($\alpha = 0.05$). a.i., active ingredient.

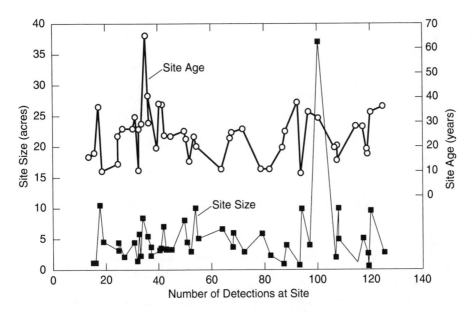

Figure 8.4. Pesticide detections in soils at 49 commercial agrichemical facilities in Illinois, in relation to the age and size of the facilities. Redrawn with permission from Krapac and others (1993).

Table 8.3. Volatile organic compounds identified in a well at an agrichemical facility in Illinois

[Data from Long, 1989. µg/L, micrograms per liter]

Compound	Concentration in Ground Water (µg/L)
1-Heptene	1.38
Ethyl benzene	1.39
Xylene	15.50
Trimethyl heptane	5.48
Dichloromethane	9.90
3-Propoxy-1-propene	0.50
2-Methyl-1-nitropropane	0.22
2-Methyl-3-propyl-*cis*-oxirane	2.00
2-Methylpropyl oxirane	2.50
Trichloroethylene	2.30
Tetradecane	0.03
1,1,1-Trichloroethane	28.40

The results from the studies cited in this section show that a variety of factors govern the likelihood, severity, and spatial extent of subsurface contamination by pesticides beneath agrichemical facilities. Such contamination appears to be controlled more by operational factors than by physical or chemical circumstances. Indeed, the available evidence—both direct and indirect—indicates that the quantities and number of different pesticides handled at an individual site provide more reliable indications of the potential for subsurface contamination than the size, age, or hydrogeologic setting of the facility. At a given facility, this potential is also dependent on the nature of the activities at different on-site locations. Once a pesticide spill occurs, the affinity of the compound for organic matter in the soil influences the likelihood of it being detected in ground water (see Figure 6.3), although the resulting concentration in the surface soil is an unreliable predictor of ground-water contamination downgradient.

8.2 DOMESTIC AND INDUSTRIAL WASTE DISPOSAL SITES

Plumb (1991) has summarized the occurrence of 208 different organic contaminants, including pesticides, in ground water at 479 hazardous-waste disposal sites across the United States. This compilation was based on data from previous investigations conducted either to monitor compliance with U.S. Environmental Protection Agency (USEPA) regulations (principally the Comprehensive Environmental Response, Compensation and Liability Act [CERCLA, or "Superfund"] or the Resource Conservation and Recovery Act [RCRA]), or as part of sanitary- or municipal-landfill monitoring operations. A similar summary was provided earlier by Plumb (1985), based on monitoring data from 358 hazardous-waste disposal sites. Figure 8.5 shows the locations of the sites from which data for the 1985 compilation were obtained (no such map was provided for the 1991 summary). As might be expected, most of the disposal sites are located in or near heavily urbanized and industrialized areas of the nation.

Table 8.4, compiled from the data reported by Plumb (1985, 1991), includes most of the compounds that have been classified as "pesticides" among those for which routine monitoring in ground water is required under CERCLA and RCRA. Excluded from the tabulation, however, were the polychlorinated biphenyls (PCBs) and 2,3,7,8-tetrachlorodibenzo-p-dioxin (TCDD). Although classified as "pesticides" (presumably because, as polychlorinated organic molecules, they were analyzed using techniques similar to those employed for the chlorinated insecticides), these compounds were not known to have been routinely employed for pest control in the United States. Table 8.4 also includes several constituents that have been used as pesticides, but are operationally categorized as "non-priority pollutant," "volatile," or "acid-extractable" compounds (once again, because of the analytical techniques used to detect them), rather than "pesticides," under CERCLA and RCRA.

The pesticides detected most often in ground waters beneath domestic and industrial waste disposal sites are those used most widely for pest control in residential settings. Figure 8.6, derived from the data in Table 8.4, illustrates the significant relation ($\alpha = 0.05$; simple linear regression) observed between the frequency with which individual pesticides were applied in residential settings in the United States in 1990 (either indoors or outdoors), and the percentage of disposal sites where they were detected in ground water as of 1991. Thus, while pesticides used extensively in residential settings were detected at frequencies as high as 12.7 percent (naphthalene) or 8.6 percent (lindane) in ground waters at waste disposal sites in 1991, those applied primarily in agricultural areas—such as disulfoton, chlorobenzilate, phorate, methyl parathion, parathion, dinoseb, *cis*-1,3-dichloropropene, bromomethane, DBCP, 1,2,3-trichloropropane (TCP), and ethylene dibromide (EDB)—were found at only 1 percent or fewer of the

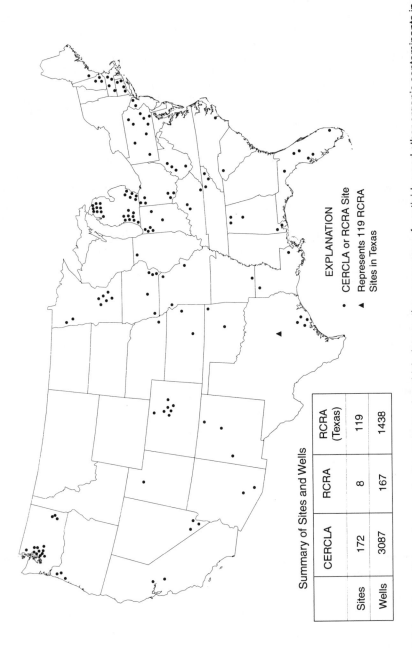

Summary of Sites and Wells

	CERCLA	RCRA	RCRA (Texas)
Sites	172	8	119
Wells	3087	167	1438

EXPLANATION

• CERCLA or RCRA Site

▲ Represents 119 RCRA Sites in Texas

Figure 8.5. Locations of hazardous-waste disposal sites for which data on the occurrence of pesticides and other organic contaminants in ground water have been provided by Plumb (1985). Redrawn from Plumb (1985) by permission of the National Ground Water Association. Copyright 1985. CERCLA, Comprehensive Environmental Response, Compensation and Liability Act (Superfund); RCRA, Resource Conservation and Recovery Act.

Table 8.4. Pesticides detected in ground water beneath waste disposal sites

[Based on monitoring data compiled by Plumb (1985, 1991). This table lists the pesticides and pesticide trans-formation products included among the 208 compounds required for monitoring under "Appendix IX" regulations (Plumb, 1991), given in order of decreasing frequency of occurrence among 479 sites as of 1991. Compound names are reproduced as given by Plumb (1991). List includes compounds used as pesticides, but classified under Appendix IX as "nonpriority pollutants," "volatiles," and "acid-extractable" compounds, as well as "pesticides." Polychlorinated biphenyls (PCBs) and 2,3,7,8-tetrachlorodibenzodioxin (TCDD), though classified as "pesticides," were not included. Source of data on residential applications: National Home and Garden Pesticide Use Survey, Final Report, Volume I (Whitmore and others, 1992). Detection frequencies listed in the last column are for the eleven pesticides detected at 1 percent or more of 183 Superfund (Comprehensive Environmental Response, Compensation, and Liability Act) sites as of 1985, based on data compiled by Plumb (1985). NR, Compound investigated, but no use reported by Whitmore and others (1992). USEPA, U.S. Environmental Protection Agency. Blank cells indicate no information applicable or available]

Pesticide or Transformation Product	Detections at 479 Sites as of 1991				Estimated Annual Applications of Compound During Residential Use (x 1000)		Frequency of Detection at 183 Sites as of 1985 (percent)
	Samples	Sites		USEPA Regions	Inside	Outside	
		No.	Percent				
Naphthalene	369	61	12.7	9	29,357	3,568	4.1
Lindane[1]	149	41	8.6	9	542	1,355	4.8
2,4-Dimethylphenol (xylenol)	159	38	7.9	9	44,217	85	
2,4-D	112	36	7.5	10	NR	[2]10,018	7.7
1,4-Dichlorobenzene	191	34	7.1	9			
1,2-Dichloropropane	158	33	6.9	9			1.7
Pentachlorophenol	77	28	5.8	9	[3]181	[3]89	
Endrin	65	27	5.6	8			
Silvex	62	25	5.2	8	NR	[4]88	2.4
Dimethyl phthalate	31	17	3.5	6			
γ-BHC[1]	73	14	2.9	9			
β-BHC	57	14	2.9	8			3.0
Methoxychlor	33	14	2.9	4			
Δ-BHC	66	13	2.7	8			3.5
Toxaphene	32	13	2.7	3			
Heptachlor	20	13	2.7	7	1,075	177	
Dieldrin	40	12	2.5	7			2.0
Chlordane	23	10	2.1	8	NR	478	1.2
o-Xylene	83	10	2.1	6	[5]148	[5]968	
4,4'-DDE	23	8	1.7	5			1.0
4,4'-DDT	17	7	1.5	4	NR	NR	
4,4'-DDD	18	6	1.3	5			
Heptachlor epoxide	8	5	1.0	5			
Endosulfan I	7	5	1.0	4	NR	561	
2,4,5-T	11	4	0.8	4	[6]181	NR	
cis-1,3-Dichloropropene	9	4	0.8	2			
trans-1,3-Dichloropropene	5	4	0.8	3			
Endosulfan sulfate	32	4	0.8	3			

Table 8.4. Pesticides detected in ground water beneath waste disposal sites—*Continued*

Pesticide or Transformation Product	Detections at 479 Sites as of 1991				Estimated Annual Applications of Compound During Residential Use (x 1000)		Frequency of Detection at 183 Sites as of 1985 (percent)
	Samples	Sites		USEPA Regions	Inside	Outside	
		No.	Percent				
Endrin aldehyde	5	3	0.6	2			
Bromomethane	5	2	0.4	2			
Aldrin	2	2	0.4	2			
Disulfoton	3	1	0.2	1	NR	6,464	
Isodrin	3	1	0.2	1			
Dibromochloropropane (DBCP)	30	1	0.2	1			
Chlorobenzilate	1	1	0.2	1			
Phorate	1	1	0.2	1			
Methyl parathion	1	1	0.2	1			
Parathion	1	1	0.2	1			
Endosulfan II	0	0	0	0			
1,2,3-Trichloropropane	0	0	0	0			
1,2-Dibromoethane (EDB)	0	0	0	0			
Dinoseb	0	0	0	0			
Diallate	0	0	0	0			
Kepone	0	0	0	0			
Phenacetin	0	0	0	0			
Pronamide	0	0	0	0			
α-BHC							4.2

[1]Unclear why different results for lindane and γ-BHC were reported by Plumb (1991), since they are the same compound.

[2]Sum among products containing either the free acid; diethanolamine, octylamine, diethylamine, triethanolamine, or alkanol amine salts; or 2-ethylhexyl, butoxyethyl or isooctyl(2-octyl) esters.

[3]Sum of free phenol and sodium salt.

[4]Sum of silvex, iso-octyl (2-ethylhexyl) ester and polypropoxypropyl ester.

[5]Specific isomer not given.

[6]Sum of free acid and sodium salt.

disposal sites examined. Although such facilities tend to be located in more industrialized and urbanized areas, rather than agricultural areas, it is not known whether the trend shown in Figure 8.6 was related to the land use within the vicinity of the disposal sites examined—rather than the nature of the wastes buried at the sites—because no information was provided by Plumb (1991) on the land use surrounding the sites. In Iowa the presence of a sewage lagoon, landfill, quarry, or strip mine within 0.5 mi was found to be associated with frequencies of detection of atrazine and other pesticides during the Iowa State-Wide Rural Well-Water Survey (SWRL) that were significantly higher ($\alpha = 0.05$) than the statewide averages (Hallberg and others, 1992b).

A relation between the occurrence of pesticides in ground waters beneath waste disposal sites in the United States and their use in household and landscape maintenance products, rather than their application in agricultural settings, is consistent with the urban sources of the wastes

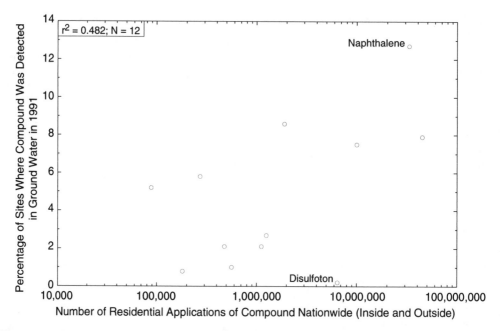

Figure 8.6. Percentages of waste-disposal sites where individual pesticides have been detected in ground water (data from Plumb, 1991), versus their respective frequencies of use in residential settings (inside and outside combined) in the United States (Whitmore and others, 1992). Relation found to be statistically significant ($\alpha = 0.05$).

and the less restrictive economic constraints on pesticide use for the residential user, compared to the farmer. Thus, a considerable quantity of leftover household pesticide products may end up in landfills. Gintautas and others (1992) analyzed leachates from six municipal landfills in six states for a variety of phenoxy acid herbicides. Detectable levels of mecoprop (MCPP), a turf herbicide widely used in residential settings (Wauchope and others, 1992; Whitmore and others, 1992), were found in leachate at all six sites, while 2,4-DP and silvex (2,4,5-TP) were detected at four of the sites and 2,4-DB was detected at three sites. The authors concluded that the comparatively low concentrations measured for the detected herbicides were "consistent with disposal of 'empty' cans of commercial herbicides or with residual herbicides in disposed plant matter."

Despite the fact that it is the phenoxy acid herbicide used most widely for agriculture in the United States (Gianessi and Puffer, 1990), the herbicide 2,4-D was not detected in leachate from any of the six landfills investigated by Gintautas and others (1992). In contrast, Plumb (1991) reported its detection at landfills in all 10 USEPA regions (Table 8.4). Results from an investigation of ground-water contamination at a municipal landfill in Grindsted, Denmark (Rügge and others, 1995) suggest that the apparent absence of 2,4-D in the ground waters at the sites investigated by Gintautas and others (1992) may have been caused by the reductive dechlorination of the herbicide in situ (Section 4.4.1). At the Grindsted site, Rügge and others (1995) detected what appeared to be the products of the complete dechlorination of MCPA and mecoprop in ground water, but not the parent compounds themselves. The fact that Gintautas and others (1992) detected 2,4-DP, silvex, and 2,4-DB at their sites—but not 2,4-D—is consistent with evidence (cited by the authors) that 2,4-D is more readily dechlorinated by microorganisms than the other compounds. While the ground water was hypoxic (i.e., lacking detectable dissolved oxygen) beneath the sites investigated by Gintautas and others (1992), geochemical conditions at the sites where Plumb (1991) reported 2,4-D detections may not have been sufficiently reducing to support the dechlorination of the herbicide in situ.

8.3 SURFACE IMPOUNDMENTS FOR ARTIFICIAL RECHARGE

In many areas of the United States where ground-water withdrawals are substantial, artificial recharge facilities have been established to replenish subsurface water supplies (Pettyjohn, 1981), or to create freshwater barriers to saltwater intrusion (Todd, 1974). Some of these facilities consist of injection wells, while others involve the use of surface-water impoundments (Pinholster, 1995). Since many such impoundments are filled using agricultural return flows (Peterson and Hargis, 1971; Exner, 1990; and Ma and Spalding, 1995), urban runoff (Peterson and Hargis, 1971; Seaburn and Aronson, 1974; and Salo and others, 1986), or reclaimed wastewater (Roberts and others, 1980; Katz and Mallard, 1981), the presence of pesticides and other contaminants in these waters is a source of potential concern with regard to ground-water quality.

Exner (1990) sampled ground water from beneath a 259-acre (105-ha) ground-water recharge structure capturing runoff from row-cropped farmland in southeastern Nebraska. All of the wells known to be receiving recharge from the impoundment were found to contain detectable levels of one or more of the pesticides found in the water retained in the recharge structure itself (atrazine, cyanazine, alachlor, and metolachlor), suggesting that the impoundment was the source of the observed contamination. Similarly, data reported by Ma and Spalding (1995) indicated that atrazine carried in agricultural runoff into a recharge structure near York, Nebraska, could be detected in the shallow ground water near the impoundment. For both of these studies, however, the detected compounds were also present in nearby shallow ground waters, albeit at lower concentrations.

Salo and others (1986) investigated the extent to which metals and synthetic organic compounds in urban runoff were removed by an artificial recharge basin in Fresno, California. A variety of organochlorine and organophosphorus pesticides were present at detectable levels in the runoff influent to the basin, but none of the compounds were detected below the water table (Salo and others, 1986), although detection limits for the target compounds were not reported. Seaburn and Aronson (1974) examined the occurrence of a variety of organochlorine pesticides and their transformation products in the soils beneath three intermittently filled recharge basins receiving urban runoff on Long Island, New York. While the inflows to the basins were found to contain only trace amounts of DDD, DDT and silvex (<0.1 µg/L), soils beneath all three of the basins were found to contain DDD, DDE, DDT, dieldrin, endrin, heptachlor, and lindane at concentrations that, when expressed on a volumetric basis, were several orders of magnitude higher (8-24,000 µg/L). No analyses of ground water from below the basins were reported, but these results indicate that hydrophobic contaminants, even when present in trace quantities in the inflow to such structures, can concentrate several-fold in the soils below, providing a potential source of contamination of the underlying ground water. The detection by Katz and Mallard (1981) of dieldrin and heptachlor epoxide at concentrations ranging from 0.01-1.4 µg/L in ground waters beneath recharge basins receiving treated sewage on Long Island suggests that such contamination is likely.

8.4 DISPOSAL WELLS

In some areas, the principal function of injection wells is not to replenish ground-water supplies per se, but to dispose of agricultural drainage, urban runoff, domestic sewage, industrial waste, oil field brines, or other wastes. Such wells may be sufficiently numerous in some regions to constitute what Hallberg (1986) referred to as "quasi-point sources" of ground-water contamination (Section 4.2). Table 8.5 lists the analytes examined by studies that investigated the impact of waste disposal wells on the chemistry of nearby ground waters.

Investigations by Schneider and others (1970, 1977) provided the clearest demonstration to date that pesticides injected into disposal wells can contaminate nearby ground waters. Following their injection into a sand aquifer through a disposal well over a 10-day period, atrazine, picloram, and trifluralin were all detected in observation wells located 9 m and 20 m from the injection well, but not in those located 45 m away. Most studies have concluded that disposal wells can compromise the quality of nearby ground waters over scales ranging from tens of meters (Schneider and others, 1970, 1977; Graham and others, 1977; Seitz and others, 1977; Graham, 1979; Baker and others, 1985; and Wilson and others, 1990) to several kilometers (Schiner and German, 1983; Baker and others, 1985; and Libra and Hallberg, 1993).

Seven of the 11 investigations listed in Table 8.5 examined the contamination of nearby ground waters by injected pesticides. All of the pesticides detected in monitoring wells were also detected in the injected wastewater by those studies that analyzed the latter for the compounds of interest (Schneider and others, 1970, 1977; Seitz and others, 1977; Schiner and German, 1983; Rutledge, 1987; and Libra and Hallberg, 1993). The consistency of these results provides strong evidence that pesticides injected into disposal wells will contaminate nearby ground waters. Indeed, Libra and Hallberg (1993) concluded that pesticides detected in some parts of the deep, confined, bedrock aquifers beneath Floyd and Mitchell Counties, Iowa, would probably not have been present if they had not been introduced through agricultural drainage wells.

8.5 GRAIN STORAGE FACILITIES

Fumigants used to treat grain in storage facilities have been detected in the underlying ground waters in Iowa (Hallberg, 1989) and Nebraska (Spalding and others, 1989; Exner and Spalding, 1990). These compounds include EDB (which has been used in such facilities for fire suppression, as well as for pest control), carbon tetrachloride, chloroform, and trichloroethylene. EDB, carbon tetrachloride, and chloroform were detected at concentrations above 1 µg/L in wells and seeps downgradient from grain storage facilities at commercial agrichemical dealerships in Iowa, but not in nearby, unaffected ground waters (Hallberg, 1989). Spalding and others (1989) and Exner and Spalding (1990) reported that EDB, carbon tetrachloride, and trichloroethylene used to treat grain in Nebraska storage facilities have been detected in nearby ground waters, but specific data on the concentrations of the individual compounds in the affected areas were not provided.

8.6 LIVESTOCK AND FEEDLOTS

Results from some of the reviewed monitoring studies suggest relations between the proximity of confined feeding operations or barnyards and herbicide detections in ground water

Table 8.5. Pesticides and other chemical constituents monitored in injectate, injection (or disposal) wells, or nearby ground waters during investigations of the influence of agricultural, urban, or domestic waste disposal wells on surrounding ground waters

[Types of Waste Injected: Ag, agricultural drainage; Urb, urban runoff; Dom, domestic sewage; Ind, industrial waste. MBAS, Methylene-blue-active substances (surfactants)]

Reference	Study Location	Types of Waste Injected	Constituents Examined	
			In Injectate or Injection Wells	In Nearby Ground Waters
Schneider and others, 1970, 1977	Texas— Bushland (Ogallala aquifer)	Ag	Pesticides: Picloram, atrazine, and trifluralin Others: Nitrate	Same as for injection wells
Whitehead, 1974	Idaho—Eastern Snake River Plain	Ag, Urb, Dom, Ind	Pesticides: DDT, DDE, dieldrin, chlordane, diazinon, and silvex Others: Major ions, metals, oil & grease, nutrients, bacteria, and sediment	None
Seitz and others, 1977	Idaho— Western Snake River Plain	Ag, Urb, Dom, Ind	Pesticides: Aldrin, chlordane, DDT, DDE, DDD, diazinon, dieldrin, dyfonate, endrin, heptachlor, heptachlor epoxide, lindane, malathion, methyl parathion, parathion, 2,4-D, 2,4,5-T, and silvex Others: Major ions, nutrients, metals, bacteria, and sediment	Same as for injection wells
Graham and others, 1977	Idaho— South-central (near Twin Falls)	Ag	Pesticides: Aldrin, chlordane, DDT, DDE, DDD, dieldrin, diazinon, endrin, heptachlor, heptachlor epoxide, lindane, malathion, methoxychlor, methyl parathion, parathion, toxaphene, 2,4-D, and 2,4,5-T Others: Major ions, metals, nutrients, bacteria, turbidity, and Rhodamine dye	Pesticides: None Others: Major ions, metals, nutrients, bacteria, turbidity, and Rhodamine dye
Graham, 1979	Idaho— Southeast Minidoka County	Ag	Pesticides: None Others: Chloride, nitrate, bacteria, and turbidity	Same as for injection wells
Schiner and German, 1982	Florida— Orlando	Urb	Pesticides: 25 parent compounds and transformation products Others: Major ions, metals, nutrients, bacteria, MBAS, and oil and grease	Same as for injection wells
Baker and others, 1985	Iowa— North-central	Ag	Pesticides: Alachlor, atrazine, cyanazine, dicamba, dieldrin, and metribuzin Others: Major ions, nutrients, and sediment	Pesticides: None Others: Nitrate
Rutledge, 1987	Florida— Orlando	Urb	Pesticides: 35 parent compounds and transformation products Others: 2 trace elements, 26 volatile, 43 base-neutral extractable, and 11 acid-extractable organic compounds	Same as for injection wells

Table 8.5. Pesticides and other chemical constituents monitored in injectate, injection (or disposal) wells, or nearby ground waters during investigations of the influence of agricultural, urban, or domestic waste disposal wells on surrounding ground waters—*Continued*

Reference	Study Location	Types of Waste Injected	Constituents Examined	
			In Injectate or Injection Wells	In Nearby Ground Waters
Kross and others, 1990	Iowa—— Statewide	Ag	None	Pesticides: 27 parent compounds and 5 transformation products Others: Nitrate
Wilson and others, 1990	Arizona—— Tucson	Urb	Not applicable (dry wells only)	Pesticides: Chlordane, chlorpyrifos, 2,4-DB, DDT, and dioxathion Others: Volatile, base/neutral and acid-extractable organic compounds and metals
Libra and Hallberg, 1993	Iowa——Floyd, Humboldt, Pocahontas, and Wright Counties	Ag	Pesticides: Cyanazine, metolachlor, and others (unspecified) Others: Nitrate	Same as for injection wells

(Baker and others, 1989, 1994; Kross and others, 1990; Hallberg and others, 1992b). Figure 8.7 shows these relations for the triazine and acetanilide herbicides, based on the Cooperative Private Well Testing Program, or CPWTP (Baker and others, 1994). In Iowa, wells located in a feedlot or cattle yard exhibited pesticide detection frequencies that were significantly higher ($\alpha = 0.05$) than the statewide average during the SWRL study (Kross and others, 1990).

The reasons for these correlations are unclear, since herbicides are not ordinarily used to treat livestock. One possible explanation is that herbicides may be present in feed or other plant material ingested by livestock, and pass into the manure, thence to leach into the subsurface. The comparative stability of several of the triazine and acetanilide herbicides under reducing conditions, discussed earlier (Section 4.4.4), indicates that the rates of transformation of some of these compounds may be negligible in the ruminant gut. No published data appear to be available on pesticide occurrence in animal manures, but the fact that atrazine concentrations in milk have sometimes been found to exceed acceptable limits in Wisconsin (Hallberg, 1995) suggests that detectable concentrations of pesticides may be present in animal wastes, as well. Alternatively, these correlations may have simply been fortuitous, since confined-feeding operations are frequently located close to pesticide mixing areas (LeMasters, 1994).

Confined animal feeding operations may also represent a potential source of ground-water contamination by pesticides used to treat livestock. Laboratory investigations indicate that the transformation products of avermectin, a potent livestock parasiticide, show appreciable mobility through soil columns (Halley and others, 1989). In addition,

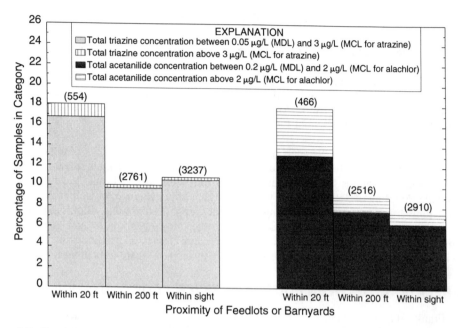

Figure 8.7. Relations between herbicide concentrations in private wells (based on immunoassay analyses) and proximity to feedlots or barnyards, based on the Cooperative Private Well Testing Program. Number of wells sampled in each category given in parentheses (data from Baker and others, 1994). MCL, Maximum Contaminant Level; MDL, Method Detection Limit.

pharmaceutical compounds produced for human use have been detected in ground waters contaminated by municipal landfill leachate (Holm and others, 1995). These observations suggest that contamination of ground water by livestock pharmaceuticals or their transformation products may occur in areas receiving extensive deposition of animal wastes. However, none of the studies examined for this book analyzed for the presence of these compounds in ground water downgradient from feedlots.

8.7 SEPTIC SYSTEMS

Whereas pesticide detection frequencies in ground water may be related to the proximity of feedlots, an analogous relation has not been observed for septic systems. No relation was observed between the proximity of septic systems and detections of herbicides in rural wells in Iowa during the SWRL study (Kross and others, 1990; Hallberg and others, 1992b). Similarly, while data from ground-water sampling in Ohio by Baker and others (1989) showed that 63 percent (10/16) of the wells containing herbicide concentrations higher than 1 µg/L were located near septic systems, this proportion is identical to the 63 percent of all of the sampled wells (8497/13841) that were within 200 ft of a septic system. These observations are consistent with the fact that agricultural herbicides are not generally used in homes. Septic systems are potential sources of ground-water contamination by pesticides used in homes, but this issue was not examined by any of the reviewed studies.

8.8 ACCIDENTAL RELEASES

Pesticide contamination of ground water in agricultural areas has often been attributed to various types of accidental releases, such as spills, back-siphonage accidents, and entry (or "run-in") of contaminated surface waters into open or improperly sealed wells. The basis on which these associations have been established, however, has been neither clear nor well substantiated in most cases (see Section 8.9). An exception is a series of extensive studies of well contamination incidents throughout Ontario, Canada, carried out by Frank and others (1979; 1987b).

Frank and others (1979) analyzed for herbicides in water samples obtained from 237 wells whose owners had notified the Ontario Provincial government about incidents of accidental releases of the compounds to the subsurface. The follow-up study (Frank and others, 1987b) involved the sampling of 359 additional wells throughout the Province, as well as the analysis of a larger number of pesticides, including fungicides, insecticides, and additional herbicides. Taken together, the two studies have provided the most systematic and comprehensive assessment available regarding the behavior of different classes of pesticides in ground water following their inadvertent entry into the subsurface through or near wells in agricultural areas.

8.8.1 RELATIONS BETWEEN ROUTE AND SEVERITY OF CONTAMINATION

Frank and others (1979; 1987b) examined four general types of accidental release: (1) spills of pesticide concentrates directly into or within the immediate vicinity of individual wells; (2) back-siphoning of dilute pesticide solutions from spray equipment directly into wells; (3) releases arising from overfilling, emptying, or rinsing of spray equipment near wells; and (4) pesticides entering open wells from spray drift or from the run-in of surface water draining near-well areas or nearby fields on which pesticides had been recently applied or spilled. Following each contamination incident, the well in question was sampled for the pesticides involved. In some cases, pesticides were detected that had not been among those associated with the incidents of interest.

Tables 8.6 and 8.7 summarize the compounds involved in the contamination incidents investigated by Frank and others (1979, 1987b). For each compound, the distributions of initial concentrations have been tallied, along with the types of release that were reported to have occurred.

The concentration distributions for 1969-1978 (Table 8.6) show distinctions among the different mechanisms of accidental release that are consistent with the nature of the incidents. The lowest pesticide concentrations (0.01-1.0 µg/L) were associated primarily with more dilute sources of contamination, such as spray drift and runoff. In contrast, the most severe contamination of ground water was caused by the spillage of concentrated or dilute pesticide solutions directly into, or in the immediate vicinity of, the sampled wells. All occurrences of pesticide concentrations exceeding 1,000 µg/L from 1969 to 1978 were associated with such releases. Intermediate levels of contamination (1-1,000 µg/L) were associated with all five routes of entry.

Fewer of the patterns noted for 1969-78 (Table 8.6) are evident for 1979-84 (Table 8.7). While the higher concentrations (>100 µg/L) were, once again, almost exclusively the result of contamination by pure or diluted formulation, these sources were also implicated for many of the lowest concentrations.   As with the earlier data, intermediate concentration levels (1.0-100 µg/L) were associated with all of the routes of interest. The data reported by Frank and others (1979, 1987b) thus show that high pesticide concentrations in wells are more likely to be

Table 8.6. Number of wells (out of a total of 237 sampled and 159 containing at least one pesticide residue) contaminated by individual pesticides as a result of accidental releases in Ontario, Canada, from 1969 to 1978

[From Frank and others (1979). In cases where multiple sampling was conducted, the concentration used was the first value measured. Codes used to denote routes of herbicide entry into wells (adapted from Frank and others, 1979): I, Spill of herbicide concentrate entering well "directly or indirectly"; II, Entry of diluted herbicide solution into well caused by either (1) back-siphoning from spray equipment, or (2) direct or indirect entry following overfilling, emptying or rinsing of spray equipment; III, Entry of spray drift into well; IV, Entry, during runoff events, of surface water carrying herbicides spilled on or applied to nearby fields or near-well areas. V, "Subterranean movement into a well from normally used or spilled herbicides." Blank cells indicate no information applicable or available. >, greater than; µg/L, micrograms per liter]

Pesticide Compound	Number of Wells Contaminated Within Given Concentration Range (µg/L) (Route of entry to wells given in parentheses)							Total No. of Wells
	0.01-0.1	[1]0.1-1.0	1.1-10	11-100	101-1,000	1,001-10,000	>10,000	
2,4-D	3 (III,IV)	28 (II-IV)	12 (II-IV)	10 (I-III)	6 (I-III)	1 (I)	1 (II)	61
Atrazine		21 (II-IV)	7 (II-IV)	6 (I-IV)	11 (I-IV)	4 (I,II)	1 (I)	50
Cl-phenols[2]	12	16	4	1 (I)				33
2,4,5-T		13 (III,IV)	7 (II-IV)	3 (I-III)	1 (I)	1 (II)		25
Dicamba	1 (IV)	5 (III,IV)	1 (IV)	1 (II)	1 (II)	1 (I)		10
Fenoprop		6 (III)		2 (II)	1 (I)			9
Mecoprop	2 (III)	2 (III)		4 (II,III)			1 (I)	9
Dinoseb	2 (III)		1 (IV)	1 (IV)	1 (II)	1 (I)		6
Picloram		3 (III,V)	2 (IV,V)	1 (III)				6
Alachlor		1 (IV)	2 (IV)		2 (II)	1 (II)		6
Simazine		3 (IV)	1 (II)				1 (II)	5
Amitrole		1 (III)				2 (II)		3
Pebulate			3 (IV)					3
Cyanazine		1 (IV)	1 (II)					2
Dichlorprop			1 (III)		1 (II)			2
MCPA			1 (III)			1 (I)		2
Paraquat				1 (II)		1 (I)		2
Butylate		1 (II)						1
Dalapon		1 (II)						1
Linuron			1 (IV)					1
Prometone				1 (II)				1
TCA[3]		1 (IV)						1

[1]Heading given as 0.1-1.0 in Frank and others (1979), but inferred to be 0.11-1.0.
[2]Routes of entry not ascertained for most of the contamination incidents involving chlorinated phenols.
[3]Identity of compound not given, but presumed to be trichloroacetic acid.

Table 8.7. Number of wells (out of a total of 359 sampled, and 104 containing at least one pesticide residue) contaminated by individual pesticides as a result of accidental releases in Ontario, Canada, from 1979 to 1984

[Data compiled from Frank and others (1987a,b). In cases where multiple sampling was conducted, the concentration used was the first value measured. Codes used to denote routes of pesticide entry into wells (adapted from Frank and others, 1979): I, Spill of pesticide concentrate entering well "directly or indirectly;" II, Entry of diluted pesticide solution into well caused by back-siphoning from spray equipment (IIA) or direct or indirect entry following overfilling, emptying or rinsing of spray equipment (IIB); III, Entry of spray drift into well; IV, Entry, during runoff events, of surface water carrying pesticides spilled on or applied to nearby fields or near-well areas. VI, Entry of pesticides into well from either spray drift or runoff (not specified). Use: Pesticide use in Ontario during the study period, expressed as the percentage of 91 farms sampled during a rural well survey by Frank and others (1987a) that applied the compound in question in 1984. MDL, Method detection limit for compound of interest. Blank cells indicate no information applicable or available. <, less than; >, greater than; µg/L, micrograms per liter]

Pesticide Compound	Use (percent of farms)	MDL (µgL)	Number of Wells Contaminated Within Given Concentration Range (µg/L) (Route of entry to wells given in parentheses)						
			<MDL	0.1-1.0	1.1-10	11-100	101-1,000	1,001-10,000	>10,000
Atrazine	75	0.1	2 (IIB)	1 (I) 6 (IIB) 10 (III) 2 (IV) 9 (VI)	2 (IIA) 8 (IIB) 4 (III) 1 (IV)	3 (I) 3 (IIA) 4 (IIB)	2 (IIA)	1 (IIA)	
2,4-D	51	0.1	8 (I) 3 (IIB)	1 (I) 1 (IIA) 1 (IIB) 7 (III) 6 (IV) 2 (VI)	5 (I) 1 (IIB) 2 (III) 1 (IV)	1 (IIA) 4 (IIB) 1 (III) 1 (VI)	1 (IIB) 1 (III) 1 (VI)		
Alachlor	43	0.1	2 (IIB)	1 (IIA) 2 (IIB)	1 (IIB) 1 (III)	1 (IIA) 1 (IIB)	1 (IIB)		
Cyanazine	29	0.1		1 (IIA) 1 (III)		2 (IIA) 1 (IIB)			1 (I) 1 (IIA)
Metolachlor	27	0.1		1 (IIB)	1 (IIA) 1 (IIB)	1 (IIA) 2 (IIB)	2 (IIB)		
Dicamba	23	0.1		1 (IIB) 1 (III) 3 (IV)	2 (IIB)	1 (I) 3 (IIB)			
Linuron	23	0.1		1 (III)	1 (IIA)				
Metribuzin	18	0.1				1 (IIB)	1 (I)		
Diazinon	18	(^{1})		1 (I)	1 (I)				
Mecoprop	18	0.1	7 (I)	1 (IIB) 1 (IV)	3 (I) 1 (IV)	1 (IIA)			
MCPA	11	0.1	2 (IIB)	1 (IIA)			1 (IIA)		
2,4-DB	9	0.1			1 (I)		1 (IIB)		

Table 8.7. Number of wells (out of a total of 359 sampled, and 104 containing at least one pesticide residue) contaminated by individual pesticides as a result of accidental releases in Ontario, Canada, from 1979 to 1984—*Continued*

Pesticide Compound	Use (percent of farms)	MDL (µg/L)	Number of Wells Contaminated Within Given Concentration Range (µg/L) (Route of entry to wells given in parentheses)						
			<MDL	0.1-1.0	1.1-10	11-100	101-1,000	1,001-10,000	>10,000
Bromoxynil	4	0.1			1 (IIA)				
Butylate	4	1.0							1 (IIA)
Metalaxyl	4	0.1	1 (IIB)						
EPTC	3	1.0	1 (IIB)			1 (IIB)			
Oxydemeton methyl	2	0.5	1 (I)						
Simazine	1	0.1	2 (IIB)	1 (IIB) 1 (III) 1 (VI)	1 (I) 1 (IIA) 1 (IIB) 2 (VI)			1 (IIA)	
Diuron	1	0.1					1 (IIB)		
Dichlorprop		0.1		1 (I) 1 (III) 1 (VI)	3 (III) 2 (IV)	1 (IIB) 2 (III) 2 (VI)			
Tebuthiuron		0.1	1 (IIB)		1 (III)	1 (IIB) 1 (VI)	2 (IIB)		
Dinoseb		0.05	4 (I)						1 (I)
Diphenamid		0.2	1 (IIB)		1 (IIB)	1 (IIA)			
Chlordane		0.001		1 (VI)[2]		1 (I)			
Diclofop-methyl		0.1			1 (IIA)			1 (IIA)	
Bromacil		0.1			1 (VI)				
Crufomate		5.0	1 (I)						
p,p'-DDT		0.005			1 (I)				
Endosulfan		0.005	1 (IIB)						
Methamidophos		(1)		1 (IIB)					
Monuron		0.1				1 (IIB)			
Pentachlorophenol		0.001			1 (IV)				
Phosmet		0.5			1 (III)				
Picloram		0.1				1 (III)			
Quintozene		0.001					1 (IIA)		
Terbacil		0.1			1 (IIB)				

[1]Method detection limit not provided by authors.
[2]Chlordane concentration: 0.01 µg/L.

associated with spills of concentrated or dilute pesticide solutions than with contamination from spray drift, runoff, or adjacent nonpoint sources. In addition, Frank and others (1987b) found that pesticides applied to adjacent rights-of-way, parking lots, buildings, landfills, recreation areas, and nurseries—as well as orchards and cropped fields—may enter nearby wells as a result of the "run-in" of surface waters from these areas or the entry of spray drift during application.

Butler and others (1981) investigated pesticide levels in soils following simulated spills at a field site in Washington state. They found the concentrations of methyl parathion in the uppermost 2.5 cm of the soil immediately following such spills to be about three times higher when the spill involved a 51 percent emulsifiable concentrate than when a dilute drum rinse of the concentrate was released. The highest levels of the insecticide in the soil—an order of magnitude higher than those arising from the emulsifiable concentrate—were observed when a 22 percent microencapsulated formulation was "spilled" on the field plots.

8.8.2 RELATION OF USE TO FREQUENCIES OF ACCIDENTAL RELEASE

As essentially random events, accidental releases of pesticides are more likely to occur for compounds that are used more frequently. Figure 8.8 displays the relation between the total number of accidental releases of a given compound in Ontario from 1979 to 1984 and its respective use in Ontario in 1984 (from Table 8.7) for those pesticides for which use data were available. Simple linear regression of the data in Figure 8.8 demonstrated a significant correlation between the two parameters ($\alpha = 0.05$). As noted in Table 8.7, estimates of pesticide use in

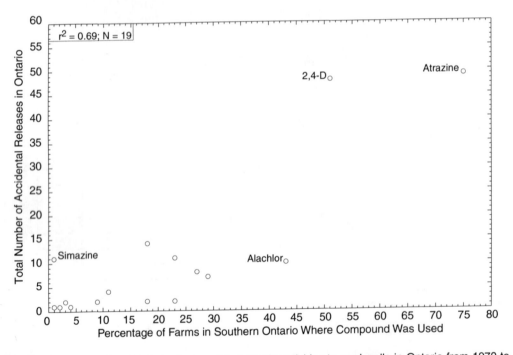

Figure 8.8. Number of accidental releases of individual pesticides to rural wells in Ontario from 1979 to 1984 (Frank and others, 1987b) versus their respective use, expressed as the percentage of 91 farms on which each compound was applied in southern Ontario in 1984 (Frank and others, 1987a).

Ontario between 1979 and 1984 were expressed as the percentage of farms in southern Ontario (out of a total of 91 farms examined) where each compound was applied in 1984, based on data provided by Frank and others (1987a).

8.8.3 CHANGES IN PESTICIDE CONCENTRATIONS IN WELLS DURING POST-SPILL REMEDIATION

During the two studies of pesticide contamination of ground waters by accidental releases in Ontario (Frank and others, 1979; 1987b), pesticide concentrations were monitored over time in several contaminated wells during clean-up operations. These operations involved pumping or scraping the well, or removing the soil surrounding the well. Pesticide concentrations were not always observed to decrease monotonically during clean-up of these wells. Transient concentration maxima were often noted during the course of remediation. In some instances, abrupt increases in concentration were observed following an extended interruption in pumping—a phenomenon frequently reported during field investigations of solute transport through the subsurface (see Section 5.1.3).

Frank and others (1979) found that the time required to reduce pesticide concentrations to acceptable levels during pumping was controlled more by the route of contamination than by the initial concentrations involved. Contamination arising from the direct entry of pesticides into wells (either in concentrate, application solutions, or spray drift), even if severe, was usually removed faster than that which entered by infiltration from the ground surface. Contaminants entering a well from the surrounding aquifer are likely to be more widely dispersed than those entering from the well bore itself, and hence, require the flushing of a substantially larger volume of the subsurface in order to be removed.

These observations are corroborated by data from two studies conducted in the southern United States. Pesticides injected into a sand aquifer in Texas (atrazine, picloram, and trifluralin) over a 10-day period and detected in nearby monitoring wells by Schneider and others (1970, 1977) were subsequently removed to concentrations close to their limits of detection in the monitoring wells after the injection well was then pumped for a similar period of time. A valuable counter-example was provided by Lewallen (1971). Following installation, part of the annular space surrounding a farmstead well in the southeastern Coastal Plain had been back-filled using soil from a nearby area where spraying equipment had been filled and rinsed. Contamination of the well by DDT, DDE, and toxaphene was found to persist over four years of sampling, albeit at comparatively low levels (< 0.3 μg/L for DDT and DDE, < 4 μg/L for toxaphene). By analogy with the findings of Frank and others (1979), the continued detection of these persistent compounds over such a long period was probably because the contaminants were introduced through the soil surrounding the well, rather than directly into the well.

8.8.4 PROXIMITY TO ON-SITE MIXING AREAS

Data from the CPWTP (Baker and others, 1994) show consistent, direct relations between the proximity of individual private wells to on-site pesticide mixing areas and frequencies of detection or exceedance of Health Action Levels (HALs) for the triazine and the acetanilide herbicides (Figure 8.9). Similar observations were reported by Kross and others (1990) with respect to herbicide detections during the SWRL study in Iowa (see Section 8.10).

In contrast with these results, however, Glanville and others (1995) observed significantly lower triazine herbicide concentrations ($\alpha = 0.05$) in rural Iowa wells located within 100 ft of pesticide storage, mixing, application, or disposal areas than in wells that were more

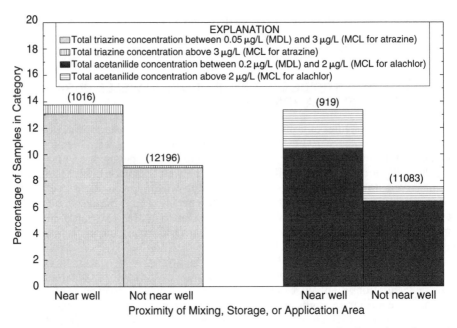

Figure 8.9. Relations between herbicide concentrations in private wells (based on immunoassay analyses) and proximity to on-site mixing, storage, or application areas, based on the Cooperative Private Well Testing Program. Number of wells sampled in each category given in parentheses (data from Baker and others, 1994). MCL, Maximum Contaminant Level; MDL, Method Detection Limit.

than 100 ft from such areas, but no significant relation for the acetanilide herbicides. These observations led the authors to conclude that, "although proximity to point sources has undisputed common sense implications for water quality, [pesticide handling] practices are only a part of the total picture in predicting the likelihood of well contamination. Clearly, many well- and site-related factors affect the outcome."

8.9 DISTINGUISHING BETWEEN POINT-SOURCE AND NONPOINT-SOURCE CONTAMINATION

There is no clear consensus on how ground-water contamination from point sources can be distinguished from that derived from nonpoint sources without direct observational evidence such as that provided by Frank and others (1979; 1987b). Many authors have ascribed pesticide contamination to point or nonpoint sources without providing specific justification for these assignments (e.g., Brown and others, 1986; Ames and others, 1987; Cardozo and others, 1988, 1989; Walker and Porter, 1990; Exner and Spalding, 1990; Miller and others, 1990; Maes and others, 1991). The three most commonly used approaches for distinguishing contamination derived from point and nonpoint sources in ground water have been based on either: (1) the spatial distributions of contaminants; (2) the transiency of contaminant concentrations; or (3) the severity of contamination. Because of the importance of this issue to ground-water management, these and other methods employed for drawing these distinctions are examined below.

8.9.1 SPATIAL DISTRIBUTIONS OF CONTAMINANTS

The most common approach taken to distinguish between point and nonpoint sources is based on the spatial configuration of pesticide detections. The California Department of Pesticide Regulation drew the following distinction between the two types of sources (Maes and others, 1991):

> [Ground water] contamination from a *point source* [authors' emphasis], such as a spill or at a waste site, is initially deposited and concentrated in a small, well-defined area... The contamination can be traced to its point of origin by locating a specifically-shaped pattern of residues in the ground water called a plume. In contrast, contamination from a *non-point source*, such as applications of agricultural chemicals to crops, cannot be traced to a single, definable location. Instead, the contaminants are dispersed over a large, poorly-defined area. When a non-point source results in contamination, locating a distinct residue plume is not possible and contaminant movement is very difficult to predict or trace to its source.

Similarly, Young (1981) has stated that "point sources of pollution are differentiated primarily from diffuse sources by the precision with which the source of contamination can be identified."

The California Department of Food and Agriculture (CDFA) has used the approach described by Maes and others (1991) to determine whether or not well-water contamination by pesticides has occurred as a result of legal agricultural use—i.e., from nonpoint sources—at a variety of locations in California (Segawa and others, 1986; Cardozo and others, 1989; Sitts, 1989). The spatial resolution of most ground-water sampling, however, is rarely sufficient to delineate individual contaminant plumes in three dimensions in the subsurface. Typically, data of this spatial density are only produced during intensive investigations of ground-water contamination, such as those conducted at some waste storage sites (e.g., Cardozo and others, 1988), in areas where subsurface contamination is particularly acute (e.g., McConnell, 1988), or during large-scale solute-transport experiments (e.g., Cherry, 1983; Roberts and others, 1990; Garabedian and LeBlanc, 1991). This approach is too costly for distinguishing between point- and nonpoint-source contamination of ground waters on a routine basis.

As part of a discussion of nitrate contamination in Nebraska ground waters, Exner and Spalding (1985) suggested that point and nonpoint sources of contamination may be distinguished based on the degree of spatial heterogeneity exhibited by contaminant distributions in the subsurface, but did not specify the precise degrees of spatial heterogeneity associated with the two types of sources. During a study of aldicarb contamination in ground waters beneath potato-growing areas in the Central Sand Plain of Wisconsin, however, Harkin and others (1986) observed "astoundingly variable" concentrations of the insecticide among wells located immediately downgradient from fields that had received spatially uniform applications. Wells screened at the same depth often exhibited concentration differences of two orders of magnitude or more. Additional work demonstrated that these spatial distributions changed markedly from one month to the next from 1982-1984 (Fathulla and others, 1988). These results demonstrate that nonpoint sources of pesticides can give rise to subsurface distributions of applied compounds that are highly variable in space and time. Nonpoint-source contamination thus may not be easily distinguished from that generated by localized point sources on the basis of spatial heterogeneity alone.

8.9.2 TRANSIENCY OF CONTAMINATION

If contaminants are detected in a particular well on one occasion, but not after subsequent sampling, the initial detections are often ascribed to a transient point source of contamination, regardless of whether the actual source has been identified. This reasoning has been used to attribute pesticide contamination to spills, improperly constructed wells, or other unknown localized sources, rather than routine use (nonpoint sources) in Arkansas (Cavalier and others, 1989); Indiana (Turco and Konopka, 1988); Nebraska (Exner and Spalding, 1985; Spalding and others, 1989; and Exner and Spalding, 1990); Oklahoma (Christenson and Rea, 1993), Georgia, Iowa, Illinois, and Wisconsin (Roux and others, 1991a). In some cases, if a pesticide or other synthetic organic compound was detected on one occasion, but not after subsequent sampling, the initial detection was considered suspect or disregarded entirely (e.g., Libra and others, 1984; Brown and others, 1986; Weaver and others, 1987; Parsons and Witt, 1989; Steichen and others, 1988; and Suffolk County Department of Health Services, 1989).

The use of the transiency of detection as a criterion to distinguish between pesticide contamination from point and nonpoint sources appears to be based on the assumption that the contamination arising from point sources exhibits greater temporal variability in affected wells than the more chronic contamination derived from nonpoint sources. Pesticide concentrations in subsurface waters influenced by agricultural activities, however, have often been found to be highly variable over time, whether the compounds are present as a result of routine agricultural applications (e.g., Wehtje and others, 1984; Schmidt, 1986a; Oki and Giambelluca, 1987; Capodaglio and others, 1988; Fathulla and others, 1988; Klaseus and others, 1988; McConnell, 1988; McKenna and others, 1988; O'Neill and others, 1989; Isensee and others, 1990; Kalkhoff and others, 1992; Priddle and others, 1992; and Richards, 1992a,b) or point-source inputs such as accidental releases (Frank and others, 1979, 1987b). Data presented in earlier chapters demonstrate the considerable degree of temporal variability that pesticide concentrations may exhibit in shallow ground water after a single application (Figures 5.3 and 5.4) or after the compounds of interest have been discontinued (Figures 3.48, 3.49, and 5.2). In addition, as noted earlier, pesticide concentrations in ground water show marked fluctuations during the year because of seasonal patterns of application and recharge (Sections 3.5.2 and 5.1.3).

Temporal variations in pesticide levels, if they occur at concentrations sufficiently close to the detection limit, can cause intermittent detections, even in the apparent absence of nearby point sources of contamination. For example, if the dashed line in Figure 3.48 were the detection limit for DBCP, rather than a state action limit, the concentration fluctuations shown in the figure would have resulted in intermittent detections of the fumigant. Indeed, in reference to the data in Figure 3.48, similar reasoning led Cohen (1986) to suggest that "although the reasons for this variability are not yet clearly understood, these examples demonstrate the need to continue monitoring even after concentrations fall 'below detection limits'."

In areas of routine agricultural use, frequencies of pesticide detection in ground water generally decrease as pesticide concentrations approach their respective limits of detection. This relation has been documented for atrazine, alachlor, and carbofuran in central Maine (Bushway and others, 1992), atrazine in southeastern Minnesota (Klaseus and others, 1988), endrin in Mason County, Illinois (McKenna and others, 1988), the fumigants DBCP and 1,2-dichloropropane in Maryland (Pinto, 1980), EDB in Connecticut (Droste, 1987), aldicarb in the Central Sand Plain of Wisconsin (Harkin and others, 1986), triazine herbicides and metolachlor in rural wells in Iowa (Glanville and others, 1995), and a variety of pesticides detected in ground waters in Kansas (Steichen and others, 1988), Illinois (Voelker, 1989), South

Dakota (Goodman, 1991), and Indiana (Risch, 1994), as well as during investigations of well cleanup operations following accidental releases (Frank and others, 1979, 1987b). During all of these investigations, when pesticides were detected intermittently in individual wells, the measured concentrations were usually within an order of magnitude of their detection limits.

These observations challenge the assumption that intermittent pesticide detections in individual wells are caused by contamination from a point source, while repeated detections arise from a nonpoint source. They also raise some concern regarding the practices, often employed during ground-water monitoring studies, of resampling wells only if they are found to contain detectable levels of pesticides (e.g., Kelley, 1985; Klaseus and others, 1988; Pettit, 1988; Steichen and others, 1988; and Risch, 1994), or disregarding contamination that is initially detected in a given well, but is not found after subsequent sampling (e.g., Brown and others, 1986; Steichen and others, 1988; and Miller and others, 1990).

More conservative approaches to the problem of intermittent detections have been employed in several states. The California Department of Food and Agriculture has classified wells with single, "unconfirmed" detections in a separate category from those in which initial detections were "confirmed" by subsequent detections, rather than to disregard the "unconfirmed" detections altogether (e.g., Ames and others, 1987; Cardozo and others, 1988; Miller and others, 1990; and Maes and others, 1991). Surveys of pesticides in the ground waters of Louisiana (Stuart and Demas, 1990), Oregon (Pettit, 1988), and Indiana (Risch, 1994) have adopted similar approaches, noting those instances where specific pesticides detected during initial sampling were not detected after subsequent resampling. During the Massachusetts Aldicarb Well Water Survey (Scarano, 1986), wells were resampled (1) if they contained detectable levels of aldicarb, or (2) if the insecticide was detected in nearby wells.

8.9.3 SEVERITY OF CONTAMINATION

High pesticide concentrations in ground water have also been cited as evidence for point-, rather than nonpoint-source contamination (e.g., Kelley, 1985; Turco and Konopka, 1988; Baker and others, 1989; Frank and others, 1990; Spalding, 1992; and Glanville and others, 1995). Baker and others (1989) stated that "in general, the pattern of pesticide contamination in wells that is emerging around the Midwest is that, with the exception of highly vulnerable settings, herbicide levels in excess of 1 µg/L rarely occur in association with field application of pesticides." Other authors have also stated that concentrations of pesticides in ground water arising from nonpoint sources are typically less than 1 µg/L, both in the United States (Hallberg, 1987; Barrett and Williams, 1989), and in Europe (Leistra and Boesten, 1989).

The range of pesticide concentrations in ground waters affected by nonpoint-source contamination, however, exhibits a considerable amount of overlap with those contaminated by point sources. Pesticide concentrations above 1 µg/L have been encountered relatively frequently in ground waters beneath areas of routine agricultural activities throughout North America (e.g., Edwards and Glass, 1971; Glass and Edwards, 1974; Muir and Baker, 1976; Hebb and Wheeler, 1978; Rothschild and others, 1982; Cohen and others, 1986; Hance, 1987; Krawchuk and Webster, 1987; Pacenka and others, 1987; Wilson and others, 1987; Isensee and others, 1988; Hall and others, 1989; Roaza and others, 1989; Isensee and others, 1990; Adams and Thurman, 1991; Gish and others, 1991b; Hall and others, 1991; Kladivko and others, 1991; Priddle and others, 1992; and Katz, 1993). Detailed monitoring of ground water immediately downgradient from aldicarb-treated potato fields in Wisconsin, for example, resulted in the frequent detection of the insecticide at concentrations exceeding 100 µg/L (Harkin and others, 1986), and widespread monitoring of ground waters has led to detections of this compound and its degradates at concentrations in excess of 10 μg/L in 31 counties in 11 states (Lorber and others, 1990). In contrast, a considerable number of wells affected by accidental releases of pesticides have been found to contain residue concentrations of 1 μg/L or less (Tables 8.6 and 8.7).

To determine whether the measured concentrations of a pesticide in ground water at a particular location are too high to have arisen as a consequence of "routine use," the initial concentration of the compound in soil water after application (as opposed to its concentration in whole soils) must be known. None of the studies reviewed for this discussion, however, specifically provided this information. Indeed, for compounds applied in a nonaqueous form, the initial concentrations in the soil water can only be estimated using a series of simplifying assumptions regarding the rates and extent of solute dissolution and transport in percolating waters. The complexities associated with this task have been illustrated by attempts to estimate post-application concentrations in recharge waters for aldicarb in Long Island, New York (Enfield and others, 1982), DBCP in Hawaii (Green and others, 1986), and EDB in Florida (Katz, 1993).

In addition to concentrations, the total number of different pesticides detected in a given well has also been used as an indicator of the severity of contamination (e.g., Frank and others, 1979; Neil and others, 1989; Kross and others, 1990; Goodman, 1991; and Kolpin, 1995a). However, the detection of more than one pesticide in ground water may occur as the result of nonpoint-source (Goodman, 1991) or point-source contamination (Frank and others, 1979; Goetsch and others, 1993).

8.9.4 ASSOCIATIONS WITH SPECIFIC SOURCES OR ROUTES OF CONTAMINATION

Several authors have attempted to determine the specific causes for the contamination observed in individual wells (e.g., Ames and others, 1987; Baker and others, 1989; Exner and Spalding, 1990; Kross and others, 1990; Perry, 1990; Domagalski and Dubrovsky, 1991; and Christenson and Rea, 1993). Studies employing this approach have identified possible reasons for the observed contamination, such as improper well construction or a nearby point source, but it is often unclear whether or not, for a given study, other wells with similar characteristics were free of such contamination. The inherent lack of control over confounding variables such as well construction, hydrogeologic regime, and nearby pesticide releases (both intentional and inadvertent) during most monitoring investigations makes it difficult to identify contaminant sources without ambiguity.

Some investigations have inferred associations between the contamination in a given well and potential nearby point sources on the basis of similarities in chemical composition between the ground water and the contamination derived from the sources in question (e.g., Droste, 1987; Pignatello and others, 1990). However, much of the value of the well-contamination studies by Frank and others (1979; 1987b) lies in the fact that the authors approached this problem from the opposite perspective, by sampling wells whose owners had specifically reported accidental releases on their property (Section 8.8). This strategy increased the likelihood of establishing a direct causal link between the pesticides detected in each contaminated well and the incident of interest, although, as noted in Section 8.8.1, some of the compounds detected were not among those known to have been accidentally released.

To distinguish point- from nonpoint-source contamination, several investigators have looked for associations between pesticide contamination in individual wells and the proximity of potential point sources nearby. The results from these studies are discussed in Section 8.10.

8.9.5 USE OF PRODUCT-TO-PARENT CONCENTRATION RATIOS

Several studies have used the deethyl atrazine-to-atrazine concentration ratio, or DAR (Sections 4.4.1 and 4.4.2), to differentiate between point- and nonpoint-source contamination of either ground water or surface water by atrazine (e.g., Adams and Thurman, 1991; Levy and others, 1993; Pereira and Hostettler, 1993; and Schottler and others, 1994). Because soil microflora are largely responsible for the production of deethyl atrazine (DEA) in situ, this approach assumes that higher DAR values in natural waters are associated with substantially longer residence times in soils, and thus with nonpoint-, rather than point-source contamination. By contrast, the entry of atrazine into the subsurface from point sources is presumed to occur sufficiently rapidly that the contact time with soil microorganisms is comparatively brief, giving rise to low DAR values.

The use of the DAR shows promise for differentiating point- from nonpoint-source contamination, but the reliability of this approach remains to be established. In particular, it has not been demonstrated that the formation of DEA from atrazine does not take place to an appreciable extent below the soil zone, although indirect evidence of this has been reported by Levy and others (1993). Furthermore, the utility of the DAR is limited by the observation that DEA exhibits nearly three times the mobility through the subsurface as atrazine; water-solids partition coefficients (K_p) for DEA and atrazine in a silt-loam soil ($f_{oc} = 0.01$) were determined by Mills and Thurman (1994b) to be 0.4 and 1.1 mL/g, respectively. Thus, the DAR is likely to increase not only with time, but with increasing travel distance, as well.

8.9.6 SUMMARY

An ability to distinguish reliably between point- and nonpoint-sources of ground-water contamination would be a valuable tool for understanding the controls on pesticide concentrations in the subsurface. However, subsurface contamination arising from the two types of sources appears to be too similar—either with respect to its spatial inhomogeneity, temporal variability, absolute concentrations, or multiplicity of detected pesticides—to be distinguished using any of these four criteria alone. Differences between point and nonpoint sources with respect to these characteristics may be common, but they are not diagnostic. Furthermore, although the DAR may prove to be of value for distinguishing between point- and nonpoint-source contamination of ground water by atrazine, its use for this purpose is dependent on some simplifying assumptions whose validity remains to be demonstrated.

The preceding discussion indicates that multiple lines of evidence are needed—including the investigation of all likely contaminant sources nearby, point and nonpoint alike—to determine the causes of pesticide contamination in a given well. An investigation of acute ground-water contamination by atrazine at a Wisconsin dairy farm by Levy and others (1993) is an example of such a multi-faceted investigation. This approach is more complex, and hence more expensive than the others described above, but it also runs less risk of misidentifying contaminant sources.

8.10 RELATIONS BETWEEN PESTICIDE CONTAMINATION AND PROXIMITY TO KNOWN OR SUSPECTED POINT SOURCES

Several studies have examined correlations between pesticide detections in ground water and the proximity of wells to potential point sources. A number of different types of point sources have been studied in this manner, but most of the attention has focused on pesticide handling areas, either on individual farms or at commercial agrichemical facilities. Baker and others (1994) examined relations between herbicide detections in private wells and the distances to nearby point sources in at least eight states. As noted earlier, they observed the frequencies of herbicide detections in private wells to increase with greater proximity to agrichemical handling areas (Figure 8.9), animal feedlots, or barnyards (Figure 8.7).

In agreement with the observations reported by Baker and others (1994), results from the SWRL study in Iowa (Kross and others, 1990) suggest an association between the detection of pesticides in rural wells and the handling of pesticides in the vicinity of the wells (Table 8.8). Relative to the statewide average for all of the wells sampled, the frequencies of pesticide detection were found to be significantly higher ($\alpha = 0.05$) for wells located: (1) within 0.5 mi of chemical storage or handling areas; (2) within 15 ft of areas where herbicides were mixed or pesticide application equipment or containers were rinsed; or (3) on farms where herbicides were mixed in the fields where they were applied. Pesticides were also detected much more frequently in wells where spills or back-siphoning accidents had occurred. Consistent with these results, pesticides detected in individual wells during the Oklahoma City study were, as noted earlier, frequently found to have been stored near the wells (Christenson and Rea, 1993). The National Alachlor Well Water Survey (Holden and others, 1992) also indicated higher frequencies of herbicide detection in wells located within 0.5 mi of "pesticide dealers, formulators or applicators" compared to all wells sampled throughout the "alachlor use area" (Table 8.9), but such wells represented only about 2 percent of all those sampled.

Although several studies indicate that the likelihood of detecting pesticides in private rural wells increases with increasing proximity to chemical handling areas on farms, Goetsch and others (1993) did not observe this trend for wells located on or immediately adjacent to commercial agrichemical sites in Illinois. They examined relations between pesticide detection frequencies in ground water and both: (1) the distances of the sampled wells from on-site areas where chemicals were handled, and (2) the depth to ground water in each of the sampled wells (Table 8.10). Depth to ground water was used as an indicator of ground-water susceptibility to surface-derived contamination, based on earlier work by McKenna and Keefer (1991). The data in Table 8.10 suggest that at the commercial facilities examined, the likelihood of detecting pesticides in ground water was more strongly controlled by the depth of the water table than by the distance of the sampled well from chemical handling areas. The strength of the evidence, however, was limited by the comparatively small number of wells sampled for some of the categories in the table, as well as by the broad range of well types examined (drilled, sand-point, dug, and large-diameter bored wells). As noted in Section 6.5.4, variations in well construction can exert a considerable influence over pesticide detection frequencies.

Results from the study by Krapac and others (1993) of pesticide residues in soils at 49 commercial agrichemical facilities in Illinois suggest why the relation observed between pesticide detections and the proximity of nearby agrichemical handling areas on farms (Kross and others, 1990; Baker and others, 1994) is not readily discernible at commercial facilities. Among 189 soil samples analyzed from the upper 60 cm of the soil column at the Illinois facilities, 89 percent contained pesticide concentrations that exceeded soil cleanup objectives set by the Illinois Environmental Protection Agency for protecting ground water (Krapac and others, 1993). As noted earlier (Section 8.1.2), widespread contamination of soils at commercial mixing-and-loading sites has also been documented in Wisconsin (Habecker, 1989).

The broad distribution of contaminants in surficial soils at commercial agrichemical handling facilities, coupled with the considerable likelihood that some of these soil-associated contaminants may be carried to adjacent off-site areas by winds and runoff, may explain the

Table 8.8. Influence of on-site chemical handling activities or other possible point sources of contamination on percentages of wells in which either any pesticide or atrazine was detected during the Iowa State-Wide Rural Well-Water Survey

[Data from Kross and others (1990) and Hallberg and others (1990). SWRL, Iowa State-Wide Rural Well-Water Survey. Percentage of Wells With Detections: Percentage of sampled wells that contained detectable levels of either any pesticide or atrazine, and that exhibited the characteristic indicated in the first column; "+" and "-" indicate relative proportions that are between 1 and 5 percent above or below the statewide 95 percent confidence intervals, respectively; "++" and "--" denote proportions that are more than 5 percent above or below these confidence intervals, respectively. ft, foot; mi, mile; <, less than]

Site Characteristic	Percentage of Wells With Detections	
	Any Pesticide	Atrazine
• All wells in SWRL survey (statewide)	13.6	8.0
Well location relative to potential point sources		
• Well "isolated from routine traffic/activity"[1]	[1]15	[1]8
• Well located <15 ft from a chemical storage or handling area	[2]0	[2]0
• Well located <0.5 mi from a chemical storage or handling area	24 ++	14 +
• Well located in feedlot or cattle yard	17 +	17 ++
• Well less than 50 ft from manure storage area	17	8
• Well more than 50 ft from manure storage area	14	8
• Well less than 50 ft from septic system	10	6
• Well more than 50 ft from septic system	14	9
Herbicide handling practices		
• All herbicides mixed on the property	15	9
• Some, but not all herbicides mixed on the property	18 +	13 +
• All herbicides mixed within 15 ft of well	23 ++	13 +
• All herbicides mixed at hydrant, more than 15 ft from well	16	8
• Herbicides mixed in field where applied	18 +	12 +
• Herbicides mixed in some other location	14	9
Herbicide application equipment cleaning practices		
• Application equipment or containers rinsed within 15 ft of well	18 +	12 +
• Rinsed at hydrant, more than 15 ft from well	18 +	12 +
• Rinsed in field where applied	15	11
• Rinsed in some other location	13	9
• Nonfarm areas: Application equipment or containers rinsed at faucet near house, or using hose attached to faucet	6 --	0 --
Accidental releases		
• Sites with reported pesticide spills or back-siphoning accidents	29 ++	14 +
Most recent herbicide application		
• Herbicides applied during most recent growing season	15	10
• Herbicides not applied during most recent growing season	11 -	4 -

[1]Specific meaning of "routine traffic/activity" (Hallberg and others, 1990) unclear. While 64 percent of all wells sampled during the SWRL were classified in this manner, the percentages of wells described as being in the immediate vicinity of areas where pesticides were likely to have been used were all substantially higher than the remaining 36 percent, i.e., pasture (47 percent), rowcrop (72 percent), and farmland (79 percent).

[2]No pesticides detected in these wells, but sample sizes were too small to determine significance of detection frequency.

Table 8.9. Frequencies with which herbicides were detected in rural domestic wells located within 0.5 miles of pesticide handling facilities during the National Alachlor Well Water Survey, in comparison with the overall detection frequencies throughout the Alachlor Use Area

[Data from Holden and others (1992). Wells within 0.5 mi of pesticide dealers, formulators, or applicators represented approximately 2 percent of the entire population of sampled wells. No error estimates were provided for these detection frequencies, however. Blank cell indicates no information applicable or available. mi, mile]

Herbicide	Frequency of Detection Within 0.5 mi of Pesticide Dealers, Formulators or Applicators (percent)	Frequency of Detection in Alachlor Use Area (± Standard Error; percent)
Alachlor	12	0.78 (± 0.29)
Atrazine	19	11.68 (± 5.13)
Cyanazine		0.28 (± 0.20)
Metolachlor	11	1.02 (± 0.38)
Simazine	11	1.60 (± 0.87)

Table 8.10. Relations among the frequencies of pesticide detection in ground water, the distances of the sampled wells from chemical handling activities, and the depth of the water table at agrichemical facilities in Illinois

[Data compiled from Goetsch and others (1993). Blank cell indicates no information applicable or available. <, less than; >, greater than. ft, foot]

Depth to Ground Water (ft)	Frequency of Detection of One Or More Pesticides as a Function of Distance of Sampled Wells from Chemical Handling Areas (percentage of wells in category)				Total Number of Wells in Each Depth Range
	<10 ft	10-99 ft	100-1,000 ft	Overall	
0-5	([1])	0	100	66	3
5-20	([1])	0	50	33	3
20-50	100	([1])	40	50	6
>50	20	12	24	18	39
Overall Frequencies of Detection	33	11	35	25	
Total Number of Wells in Each Distance Range	6	19	26		51

[1]None of the sampled wells fell into this category.

apparent absence of any clear relations between pesticide detections in wells and the distance of the wells from currently active chemical handling areas over relatively short distances (<1,000 ft), despite the larger-scale correlations observed over distances of 0.5 mi (Tables 8.8 and 8.9). Furthermore, the fact that such relations have been observed on private farms may reflect the substantially lower levels of soil and ground-water contamination generally encountered on farms, relative to commercial facilities. As with many other issues discussed in this book, however, this conclusion is based on correlations, rather than a direct demonstration of causal relations between pesticide distributions in soils and ground waters.

8.11 SUMMARY

Pesticides derived from a variety of point sources may contaminate ground water. Most of the published research on point sources of pesticide contamination has focused on commercial agrichemical handling facilities, domestic and industrial waste disposal sites, and agrichemical handling areas on farms. Pesticide contamination of soils and ground water at commercial agrichemical facilities is common, widely distributed, and chronic, and may affect water-supply wells nearby. Although contamination at these sites becomes more likely as larger quantities and a wider variety of pesticides are handled, the severity of the contamination is not related to the age or size of the facilities. In addition to pesticides, volatile organic compounds, used for a variety of purposes—including fuels, lubricants, and "inert ingredients" in pesticide formulations—have also been detected in ground water at agrichemical facilities. An extensive data base on contamination at industrial and municipal waste disposal facilities indicates that a considerable number of pesticides have been detected in ground water beneath these sites, but that they are predominantly those used in private residences, rather than in agriculture. Investigations of accidental releases on farms have shown that pesticides may enter farmstead wells by a variety of different pathways, including spray drift from nearby applications, as well as back-siphonage, spills during the mixing of the chemicals, and surface-water "run-in." The ease with which the contamination is removed from wells affected by these releases has been found to depend on the route by which the chemicals entered the well.

More limited work has examined pesticide contamination derived from artificial recharge impoundments, disposal wells for agricultural and urban runoff, feedlots, and septic systems. With the exception of septic systems, all of these other sources have been found to be associated with pesticide contamination of ground water.

Although the detection of pesticides in ground water is frequently ascribed to either point or nonpoint sources, the criteria upon which this assignment is based are either not given or, in most cases, not diagnostic. Neither the transiency nor the severity of contamination—except in the extreme—can reliably distinguish point- from nonpoint-source contamination in most instances. Detailed examinations of spatial distributions of contamination are more valuable in this regard, but they are often prohibitively costly. The use of product-to-parent ratios may be a useful tool for this purpose, but the number of compounds for which it can be used is limited; to date, the technique has been applied only for atrazine.

Pesticide contamination of ground waters by most private or commercial point sources becomes more likely with decreasing distance of a given well from the source in question. General relations have been documented for distances of up to 0.5 mi. Data from commercial agrichemical handling facilities, however, indicate that within the confines or immediate vicinity of the sites themselves, pesticide contamination of ground water or soil is sufficiently widespread that such patterns are not observed.

Table 8.1. Maximum concentrations of pesticides and their transformation products detected in the subsurface beneath pesticide manufacturing and commercial mixing-and-loading facilities

[Data from onsite areas with known, acute spills excluded, where possible. Pesticide or Transformation Product: Compounds in each use class are listed in order of decreasing maximum concentrations in ground water. Transformation products are indented. ND, not detected. Limits of detection for the study by Goetsch and others (1993) are provided in Table 8.2. Specific detection limits not provided by any of the other studies listed, except for Miles and others (1990). Blank cells indicate no information applicable or available. <, less than; >, greater than; μg/L, micrograms per liter; μg/kg, micrograms per kilogram]

Pesticide or Transformation Product	State	No. Sites	Maximum Concentration Ground Water (μg/L)	Maximum Concentration Surface Soil (μg/kg)	Reference
Herbicides					
Alachlor	Iowa	8	145	[1]270,000	Hallberg, 1985
	Wisconsin[2]	20	23,900	5,900,000	Habecker, 1989
	Ohio	2	10		Habecker, 1989
	Minnesota	2	0.1		Habecker, 1989
	Illinois[3]	6	2,370	24,000,000	Habecker, 1989
	Illinois	56	1,300		Long, 1989
	Illinois	49		16,290,700	Krapac and others, 1993
	Illinois	52	ND		Goetsch and others, 1993
	Illinois	4	19		Goetsch and others, 1993
Atrazine	Iowa	8	65	[1]70,000	Hallberg, 1985
	Wisconsin[2]	20	3,000	3,562,000	Habecker, 1989
	Minnesota	2	0.23		Habecker, 1989
	Illinois[3]	6	1,972	1,800,000	Habecker, 1989
	Illinois	56	220		Long, 1989
	Hawaii	2	3.4	3,472,000	Miles and others, 1990
	Illinois	49		405,940	Krapac and others, 1993
	Illinois	52	0.90		Goetsch and others, 1993
	Illinois	4	61		Goetsch and others, 1993
Deethyl atrazine	Hawaii	2	1.9		Miles and others, 1990
	Illinois	49		517	Krapac and others, 1993
	Illinois	4	1.2		Goetsch and others, 1993
Deisopropyl atrazine	Illinois	49		691	Krapac and others, 1993
Metolachlor	Iowa	8	50	[1]270,000	Hallberg, 1985
	Wisconsin[2]	20	2,640	7,720,000	Habecker, 1989
	Ohio	2	1.4		Habecker, 1989
	Illinois[3]	6	2,238	4,000,000	Habecker, 1989
	Illinois	56	2,100		Long, 1989
	Illinois	52	22.0		Goetsch and others, 1993
	Illinois	4	60		Goetsch and others, 1993
Dinoseb	Wisconsin[2]	20	2,100	240,000	Habecker, 1989
	Illinois	52	ND		Goetsch and others, 1993

Table 8.1. Maximum concentrations of pesticides and their transformation products detected in the subsurface beneath pesticide manufacturing and commercial mixing-and-loading facilities--*Continued*

Pesticide or Transformation Product	State	No. Sites	Maximum Concentration Ground Water (µg/L)	Maximum Concentration Surface Soil (µg/kg)	Reference
Pendimethalin	Wisconsin[2]	20		590,000	Habecker, 1989
	Minnesota	2		1,700	Habecker, 1989
	Illinois[3]	6	8.8		Habecker, 1989
	Illinois	56	1,300		Long, 1989
	Illinois	49		2,591,700	Krapac and others, 1993
	Illinois	4	2.2		Goetsch and others, 1993
Metribuzin	Iowa	8	8.0	[1]52,000	Hallberg, 1985
	Wisconsin[2]	20	940	43,000	Habecker, 1989
	Illinois	56	240		Long, 1989
	Illinois	49		8,400	Krapac and others, 1993
	Illinois	52	ND		Goetsch and others, 1993
	Illinois	4	28		Goetsch and others, 1993
Metribuzin DA	Illinois	49		4,560	Krapac and others, 1993
	Illinois	4	5.4		Goetsch and others, 1993
Cyanazine	Iowa	8	36	[1]225,000	Hallberg, 1985
	Wisconsin[2]	20	880	871,000	Habecker, 1989
	Minnesota	2		1,090,000	Habecker, 1989
	Illinois	56	69		Long, 1989
	Illinois	49		83,000	Krapac and others, 1993
	Illinois	4	20		Goetsch and others, 1993
Dicamba	Wisconsin[2]	20	360	5,600	Habecker, 1989
	Minnesota	2	0.1		Habecker, 1989
	Illinois	52	12.0		Goetsch and others, 1993
Bentazon	Illinois	52	197		Goetsch and others, 1993
2,4-D	Wisconsin[2]	20	100	77,000	Habecker, 1989
	Minnesota	2	0.19	9,800	Habecker, 1989
	Illinois	52	ND		Goetsch and others, 1993
Propazine	Illinois[3]	6	35		Habecker, 1989
	Illinois	49		5,520	Krapac and others, 1993
	Illinois	4	1.2		Goetsch and others, 1993
Linuron	Wisconsin[2]	20	34		Habecker, 1989
	Illinois	49		2,930	Krapac and others, 1993
	Illinois	4	ND		Goetsch and others, 1993
Butylate	Wisconsin[2]	20	7.5	24,000	Habecker, 1989
	Illinois	56	28		Long, 1989
	Illinois[3]	6	28		Habecker, 1989
	Illinois	49		460,000	Krapac and others, 1993
	Illinois	52	ND		Goetsch and others, 1993
	Illinois	4	0.66		Goetsch and others, 1993

Table 8.1. Maximum concentrations of pesticides and their transformation products detected in the subsurface beneath pesticide manufacturing and commercial mixing-and-loading facilities--*Continued*

Pesticide or Transformation Product	State	No. Sites	Maximum Concentration Ground Water (µg/L)	Maximum Concentration Surface Soil (µg/kg)	Reference
Simazine	Wisconsin[2]	20	27	117,000	Habecker, 1989
	Illinois	49		72,100	Krapac and others, 1993
	Illinois	4	ND		Goetsch and others, 1993
Trifluralin	Iowa	8	0.2	[1]>1,000	Hallberg, 1985
	Wisconsin[2]	20		1,340,000	Habecker, 1989
	Minnesota	2		76,000	Habecker, 1989
	Illinois[3]	6	1.2		Habecker, 1989
	Illinois	56	10		Long, 1989
	Illinois	49		190,000	Krapac and others, 1993
	Illinois	52	0.012		Goetsch and others, 1993
	Illinois	4	0.84		Goetsch and others, 1993
Chloramben	Wisconsin[2]	20	6.6		Habecker, 1989
	Illinois	52	1.3		Goetsch and others, 1993
Bromacil	Hawaii	2	< 0.7	1,660	Miles and others, 1990
	Illinois	49		15,500	Krapac and others, 1993
	Illinois	52	ND[4]		Goetsch and others, 1993
	Illinois	4	[4]5.9		Goetsch and others, 1993
Prometon	Wisconsin[2]	20	4.8		Habecker, 1989
	Illinois[3]	6	1.0		Habecker, 1989
	Illinois	49		1,605	Krapac and others, 1993
	Illinois	52	3.2		Goetsch and others, 1993
	Illinois	4	ND		Goetsch and others, 1993
Acifluorfen	Illinois	52	3.4		Goetsch and others, 1993
EPTC	Wisconsin[2]	20	2.7		Habecker, 1989
	Illinois	49		9,681	Krapac and others, 1993
	Illinois	52	ND		Goetsch and others, 1993
	Illinois	4	1.5		Goetsch and others, 1993
Ametryn	Hawaii	2	1.4	17,664,000	Miles and others, 1990
Deethyl ametryn	Hawaii	2	0.3		Miles and others, 1990
Picloram	Minnesota	2	0.63		Habecker, 1989
MCPA	Minnesota	6	0.37		Habecker, 1989
Terbacil	Hawaii	2	< 1.1	3,380	Miles and others, 1990
Glyphosate	Wisconsin[2]	20		3,200	Habecker, 1989
Hexazinone	Hawaii	2	< 0.3	2,190	Miles and others, 1990
Diuron	Hawaii	2	< 3.0	1,260	Miles and others, 1990
Propachlor	Illinois	49		420	Krapac and others, 1993
	Illinois	52	ND		Goetsch and others, 1993
	Illinois	4	ND		Goetsch and others, 1993

Table 8.1. Maximum concentrations of pesticides and their transformation products detected in the subsurface beneath pesticide manufacturing and commercial mixing-and-loading facilities--*Continued*

Pesticide or Transformation Product	State	No. Sites	Maximum Concentration Ground Water (µg/L)	Maximum Concentration Surface Soil (µg/kg)	Reference
Demeton	Illinois	49		370	Krapac and others, 1993
	Illinois	4	ND		Goetsch and others, 1993
Insecticides					
Terbufos	Wisconsin[2]	20	52		Habecker, 1989
	Illinois	49		70	Krapac and others, 1993
	Illinois	4	ND		Goetsch and others, 1993
Carbaryl	Wisconsin[2]	20	50		Habecker, 1989
Diazinon	Illinois	56	1.1		Long, 1989
	Illinois	49		270	Krapac and others, 1993
	Illinois	4	1.8		Goetsch and others, 1993
Chlordane	Illinois	56	1.7		Long, 1989
	Illinois	49		79,000	Krapac and others, 1993
	Illinois	4	ND		Goetsch and others, 1993
Dimethoate	Wisconsin[2]	20	1.5		Habecker, 1989
	Illinois	49		870	Krapac and others, 1993
Fonofos	Iowa	8	1.3	[1]>1,000	Hallberg, 1985
	Wisconsin[2]	20		800	Habecker, 1989
	Illinois	49		4,300	Krapac and others, 1993
	Illinois	4	ND		Goetsch and others, 1993
Toxaphene	Illinois	49		1,743	Krapac and others, 1993
	Illinois	4	1		Goetsch and others, 1993
Chlorpyrifos	Wisconsin[2]	20		41,000	Habecker, 1989
	Illinois	56	0.5		Long, 1989
	Illinois	49		26,000	Krapac and others, 1993
	Illinois	4	ND		Goetsch and others, 1993
Lindane	Illinois	56	0.4		Long, 1989
	Illinois	49		ND	Krapac and others, 1993
	Illinois	4	ND		Goetsch and others, 1993
Heptachlor	Illinois	49		2,900	Krapac and others, 1993
	Illinois	4	ND		Goetsch and others, 1993
Heptachlor epoxide	Illinois	56	0.38		Long, 1989
	Illinois	49		230	Krapac and others, 1993
Dieldrin	Illinois	56	0.05		Long, 1989
	Illinois	49		17,000	Krapac and others, 1993
	Illinois	52	0.016		Goetsch and others, 1993
	Illinois	4	0.18		Goetsch and others, 1993
Permethrin	Illinois	49		422,198	Krapac and others, 1993
	Illinois	4	0.05		Goetsch and others, 1993

Table 8.1. Maximum concentrations of pesticides and their transformation products detected in the subsurface beneath pesticide manufacturing and commercial mixing-and-loading facilities--*Continued*

Pesticide or Transformation Product	State	No. Sites	Maximum Concentration		Reference
			Ground Water (µg/L)	Surface Soil (µg/kg)	
Endrin	Illinois	49		72	Krapac and others, 1993
	Illinois	52	0.024		Goetsch and others, 1993
	Illinois	4	ND		Goetsch and others, 1993
Endrin aldehyde	Illinois	49		> 1.6	Krapac and others, 1993
	Illinois	52	ND		Goetsch and others, 1993
	Illinois	4	ND		Goetsch and others, 1993
Endrin ketone	Illinois	49		372	Krapac and others, 1993
	Illinois	4	ND		Goetsch and others, 1993
Aldrin	Illinois	49		10,849	Krapac and others, 1993
	Illinois	52	0.016		Goetsch and others, 1993
	Illinois	4	ND		Goetsch and others, 1993
Carbofuran	Iowa	8	ND	[1]>1,000	Hallberg, 1985
	Illinois	49		2,500	Krapac and others, 1993
	Illinois	4	ND		Goetsch and others, 1993
DDT	Hawaii	2		6,339,000	Miles and others, 1990
	Illinois	49		684	Krapac and others, 1993
	Illinois	4	ND		Goetsch and others, 1993
DDD	Hawaii	2		1,867	Miles and others, 1990
	Illinois	49		308	Krapac and others, 1993
	Illinois	4	ND		Goetsch and others, 1993
DDE	Hawaii	2		570	Miles and others, 1990
	Illinois	49		224	Krapac and others, 1993
	Illinois	4	ND		Goetsch and others, 1993
Parathion	Wisconsin[2]	20		1,900,000	Habecker, 1989
Ethion	Illinois	49		6,300	Krapac and others, 1993
Ethyl parathion	Illinois	49		5,540	Krapac and others, 1993
	Illinois	4	ND		Goetsch and others, 1993
Disulfoton	Illinois	49		4,800	Krapac and others, 1993
	Illinois	4	ND		Goetsch and others, 1993
Fenthion	Illinois	49		1,200	Krapac and others, 1993
	Illinois	4	ND		Goetsch and others, 1993
Azinphos methyl	Illinois	49		878	Krapac and others, 1993
	Illinois	4	ND		Goetsch and others, 1993
Malathion	Illinois	49		690	Krapac and others, 1993
	Illinois	4	ND		Goetsch and others, 1993
Methoxychlor	Illinois	49		630	Krapac and others, 1993
	Illinois	4	ND		Goetsch and others, 1993
Phorate	Illinois	49		560	Krapac and others, 1993
	Illinois	4	ND		Goetsch and others, 1993

Table 8.1. Maximum concentrations of pesticides and their transformation products detected in the subsurface beneath pesticide manufacturing and commercial mixing-and-loading facilities--*Continued*

Pesticide or Transformation Product	State	No. Sites	Maximum Concentration		Reference
			Ground Water (µg/L)	Surface Soil (µg/kg)	
α-BHC	Illinois	49		140	Krapac and others, 1993
	Illinois	4	ND		Goetsch and others, 1993
β-BHC	Illinois	49		220	Krapac and others, 1993
	Illinois	4	ND		Goetsch and others, 1993
Δ-BHC	Illinois	49		119	Krapac and others, 1993
	Illinois	4	ND		Goetsch and others, 1993
Methyl parathion	Illinois	49		113	Krapac and others, 1993
	Illinois	4	ND		Goetsch and others, 1993
Endosulfan I	Illinois	49		70	Krapac and others, 1993
	Illinois	4	ND		Goetsch and others, 1993
Endosulfan II	Illinois	49		17	Krapac and others, 1993
	Illinois	4	ND		Goetsch and others, 1993
Endosulfan sulfate	Illinois	4	ND		Goetsch and others, 1993
Fenamiphos	Hawaii	2	< 0.5	< 10	Miles and others, 1990
Fumigants					
Carbon tetrachloride	Iowa	8	66	[1]10 - 100	Hallberg, 1985
Chloroform	Iowa	8	4.0	[1]10 - 100	Hallberg, 1985
1,2-dichloroethene (1,2-DCE)	Iowa	8	2.0	[1]10 - 100	Hallberg, 1985
	Hawaii	2		< 20	Miles and others, 1990
EDB	Iowa	8	1.0	[1]10 - 100	Hallberg, 1985
Fungicides					
Captan	Illinois	49		903	Krapac and others, 1993
	Illinois	4	ND		Goetsch and others, 1993

[1]Soil concentrations given by Hallberg (1985) in units of µg/L.

[2]Six of the twenty sites studied in Wisconsin were investigated in response to known spills. Data from locations of known spills at individual sites, however, were not included in table.

[3]Since the specific sources of data from states other than Wisconsin (i.e. Ohio, Minnesota, and Illinois) were not provided by Habecker (1989), the degree of overlap with data reported for Illinois by Long (1989) was unknown. Data found to be identical to those provided by Long (1989) for individual compounds in Illinois were therefore not included.

[4]Bromacil concentrations may represent underestimates, since recoveries of the herbicide from spiked water samples were found by Goetsch and others (1993) to have been negligible.

CHAPTER 9

Observation Versus Prediction of Pesticide Occurrence and Behavior in Ground Water

Any phenomenon is best understood when its occurrence and future consequences can be predicted with an acceptable degree of accuracy. The desire to predict contaminant behavior in ground water has motivated attempts to generate mathematical simulations of solute transport and fate through porous media for at least three decades (van Genuchten and others, 1974; Anderson, 1979). In addition to computer simulations, two other approaches have also been employed to predict the likelihood of pesticide contamination in ground water: (1) the use of other solutes as pesticide indicators, and (2) large-scale assessments of the vulnerability of ground waters to pesticide contamination. This chapter examines the extent to which each of these three approaches has improved our ability to predict the occurrence and behavior of pesticides in the subsurface.

9.1 MATHEMATICAL SIMULATIONS OF PESTICIDE MOVEMENT AND FATE IN GROUND WATER

Although considerable effort has been invested in the development of computer models to simulate pesticide movement and fate in the subsurface, several authors have noted that the testing of the predictions of these models against actual field behavior of pesticides has been largely inadequate (e.g., Anderson, 1986; Leistra, 1986; Wagenet, 1986b; Jury and Ghodrati, 1989; Wagenet and Hutson, 1990; Loague and Green, 1991; and Shaffer and Penner, 1991). Indeed, of the 79 computer simulation studies examined for this book, only 40 compared their modeling results with pesticide distributions measured in the field; the remaining 39 studies presented modeling results without the benefit of such comparisons. As noted by Leake and others (1987), however, "predictions of herbicide behavior in soil based solely on laboratory studies and theoretical calculations are of limited value."

Among studies that have compared modeling results to pesticide distributions in the subsurface, computer simulations have been largely unable to successfully predict pesticide concentrations at specific locations in situ. As noted by Anderson (1986), "strictly speaking, field validation [of a computer model] refers to a model prediction made several years into the future, which is later verified in the field. Under this strict definition, no ground water contaminant transport model has been field validated to date." Bredehoeft and Konikow (1993) noted that "the results of the current set of postaudits [of ground water transport model predictions] suggest that extrapolations into the future were rarely very accurate."

Several investigations have demonstrated the limited success with which even widely used models have been able to predict the transport and fate of contaminants in the subsurface. These have included comparisons between the predicted and observed behavior of DDT at a site in Madison, Wisconsin, and aldicarb at a site in Long Island, N.Y. (Enfield and others, 1982), 1,2-dibromo-3-chloropropane (DBCP) beneath pineapple fields on the Hawaiian islands of Maui and Oahu (Green and others, 1986), and atrazine in the Southern Piedmont near Watkinsville, Georgia (Loague and Green, 1991).

Table 9.1 summarizes some of the major characteristics of previous attempts to simulate the movement and fate of pesticides in the subsurface. The table only includes studies that compared observed field behavior—either in undisturbed subsurface systems, buried microcosms, or greenhouses—with the predictions of one or more computer models. Excluded from this summary were studies that either did not include any experimental confirmation, or that involved such confirmation based solely on laboratory-based, rather than field-based observations.

The various models that have been employed by the studies listed in Table 9.1 include the majority of codes in widespread use, as well as programs constructed by some of the authors themselves. The most commonly used simulation code has been the Pesticide Root Zone Model (PRZM), developed by Carsel and others (1985). The principal features of PRZM and several of the other more frequently employed models have been summarized by a number of authors, including Matthies (1987), Crowe and Mutch (1990), Pennell and others (1990), and Shoemaker and others (1990).

9.1.1 RANGE OF COMPOUNDS EXAMINED

Despite a distinct over-representation of aldicarb and its degradates, the range of pesticides examined by the studies listed in Table 9.1 encompasses most of the major classes of active ingredients applied to the land in the United States. In general, however, the principal emphasis has been on the most widely applied herbicides. To an even greater extent than for the process and matrix-distribution studies (Table 2.1), pesticide transformation products are rare among the solutes investigated by modeling studies. This is not unexpected, given that partition coefficients and transformation rate constants, parameters that are indispensable for simulating the subsurface behavior of even moderately hydrophobic solutes, have been measured for relatively few of these products. Such data have been reported, for example, for the transformation products of atrazine (e.g., Schiavon, 1988; Clay and Koskinen, 1990; Adams and Thurman, 1991; Loch, 1991; and Mills and Thurman, 1994b), aldicarb (Moye and Miles, 1988), benazolin-ethyl (Leake and others, 1987), and fenamiphos (Schneider and others, 1990).

9.1.2 LIMITATIONS OF COMMONLY USED MODELS

Solute Behavior Above Versus Below The Water Table

Computer simulations of solute behavior in underground waters have, for the most part, focused on the subsurface regions above or below the water table, but rarely on both. Historically, the primary focus of the agricultural research community has been the modeling of solute movement through the unsaturated zone, motivated primarily by a desire to understand the processes governing the accumulation and transport of the natural and applied solutes to which growing plants are exposed in the root zone. By contrast, the simulation of solute behavior below the water table has been a major focus for investigators in several disciplines, including contaminant hydrogeologists, in their efforts to predict the movement and fate of anthropogenic pollutants in aquifers; petroleum engineers, developing techniques for enhancing the recovery of oil from underground reservoirs; and nuclear waste management engineers, investigating the potential environmental consequences of long-term underground storage of nuclear wastes.

Table 9.1. Comparisons conducted between observed distributions of pesticides in the subsurface and predictions based on model simulations

[All of the cited studies assumed local equilibrium with respect to sorption (Section 9.1.2). Study site locations placed within brackets are inferred, since actual study locations were not given. Systems Examined: Soil-type abbreviations based on those from the Soil Conservation Service, or SCS (U.S. Department of Agriculture, 1991)]. References for Models Employed: Authors', Model constructed or modified by study authors; BAM, Jury and others (1983); 2CM, Hill and Schaalje (1985); CMLS, Nofziger and Hornsby (1986); GLEAMS, Leonard and others (1987); LEACHMP, Wagenet and Hutson (1986); Mackay fugacity model, Mackay and Paterson (1982); MOUSE, Pacenka and Steenhuis (1984); PRZM, Carsel and others (1985); SESOIL, Bonazountas and Wagner (1984). Parameter abbreviations: β, rate coefficient for the approach to sorption equilibrium; D_L, D_T, coefficients of longitudinal and transverse dispersion, respectively; f_{oc}, weight fraction of organic carbon in soil; K_d, water-solids distribution coefficient; K_{oc}, soil organic carbon–water partition coefficient; K-θ-h, soil-water characteristic relationship between hydraulic conductivity (K), soil water content (θ) and soil-water matric potential (h); k_s, k_d, first-order rate constants for sorption and desorption, respectively; k_{soil}, first-order transformation rate constant in surficial soil; Δk_{soil}, time at which transformation rate constant in soil appears to change; k_{subs}, first-order transformation rate constant for subsoil; Q_{ET}, evapotranspiration rate; Q_{irr}, irrigation rate; Q_{pest}, pesticide application rate; Q_{rain}, rainfall; Q_{rech}, recharge; Q_{run}, runoff; R, retardation factor; percent runoff, proportion of incident precipitation entering surface waters (SCS runoff curve number); $t_{1/2}$, pesticide dissipation half-life; V_{gw}, groundwater velocity; z_{pest}, maximum depth of pesticide leaching. NG, not given]

Reference	Study Site Locations	Pesticides Examined	Systems Examined	Models Employed	Parameters Adjusted to Fit Observations	Parameters Examined in Sensitivity Analysis
Osgerby, 1972	France—Versailles Avignon Bordeaux Germany—Geisenheim	2,6-dichloro-benzamide (chlorthiamid degradate)	Vineyards	Authors'	V_{gw}	None
Enfield and others, 1982	New York—Long Island Wisconsin—Madison	Aldicarb DDT	s sil	Authors'	k_{soil}	None None
Baier and Robbins, 1982a,b	New York—Long Island	Aldicarb	s	Authors'	$t_{1/2}$	None
Leistra and others, 1984	The Netherlands	Methomyl	Greenhouse soils (3 sites)	Authors'	None	Q_{irr}; contributions of effluent pipes to drainage
Steenhuis and others, 1984	New York—Long Island	Aldicarb	Potato fields	MOUSE	$t_{1/2}$; D_L	D_L
Bilkert and Rao, 1985	Florida—Gainesville	Aldicarb	s, sl, sicl (buried soil columns)	Rao and others, 1976	K_d	None
Carsel and others, 1985	New York—Long Island	Aldicarb	s	PRZM	Q_{rain}; Q_{run}; Q_{rech}; k_{soil}; k_{subs}; K_d	None
Anderson, 1986	Wisconsin—Central Sand Plain	Aldicarb	NG	Authors'	z_{pest}; Q_{pest}; R; $t_{1/2}$; D_L; D_T; Q_{rech}	D_L; D_T; Q_{pest}; $t_{1/2}$; transient vs. steady-state flow

Table 9.1. Comparisons conducted between observed distributions of pesticides in the subsurface and predictions based on model simulations—Continued

Reference	Study Site Locations	Pesticides Examined	Systems Examined	Models Employed	Parameters Adjusted to Fit Observations	Parameters Examined in Sensitivity Analysis
Bush and others, 1986	North Carolina—Otto	Picloram	stl	PRZM	Percent runoff; K_d; $t_{1/2}$	$t_{1/2}$; K_d; Q_{pest}
Carsel and others, 1986	Florida Maryland	Metalaxyl	fs fsl	PRZM	k_{soil}; Δk_{soil}	None
Green and others, 1986	Hawaii—Oahu, Maui	DBCP	c	Several	K_d	K_d
Jones and others, 1986a	20 sites in 13 States	Aldicarb, aldoxycarb	Various	PRZM	None	V_{gw}
Jones, 1986	Sites in 10 states	Aldicarb, aldicarb sulfone	Various	PRZM	None	$t_{1/2}$; f_{oc}; θ; Q_{rain}; application timing
Jury and others, 1986a,b	California—Etiwanda	Napropamide, bromacil, prometryn	ls	Authors'	R; K_d	None
Lorber & Offut, 1986	Wisconsin— Cameron Hancock North Carolina— Hertford County	Aldicarb	sl sl ls	PRZM	K_d; $t_{1/2}$	Q_{rain}; Q_{irr}; Q_{run}; Q_{ET}; Q_{rech}
Wagenet & Hutson, 1986	New York—Phelps	Aldicarb	s	LEACHMP	k_{soil}; K-θ-h	K-θ-h
Jones and others, 1988	Wisconsin—Hancock	Aldicarb	s, sl	PRZM	$t_{1/2}$	D_L; D_T; $t_{1/2}$
Leonard and others, 1988	Georgia—Tifton	Aldicarb, aldicarb sulfone, atrazine, butylate, fenamiphos	s	GLEAMS	None	None
Garrison and others, 1989	North Carolina—Wayne County	Fomesafen	scl	PRZM	None	None
Shirmohammadi and others, 1989	Maryland—Queenstown	Atrazine, carbofuran, cyanazine, dicamba, metolachlor, simazine	sil	GLEAMS	Q_{rech}	None
Summit and others, 1989	Utah— Hill Air Force Base Shivwitz Idaho—Malad	Tebuthiuron	s fls sil	VIP, CMLS	None	K_{oc}

Table 9.1. Comparisons conducted between observed distributions of pesticides in the subsurface and predictions based on model simulations— *Continued*

Reference	Study Site Locations	Pesticides Examined	Systems Examined	Models Employed	Parameters Adjusted to Fit Observations	Parameters Examined in Sensitivity Analysis
Blair and others, 1990	England— Higham, Bury St. Edmunds Lidgate, Suffolk	Isoproturon	scl c	Authors'	k_s, k_d	None
Clendening and others, 1990	Southern California	Atrazine, bromacil, EPTC, prometon, triallate	sl	Authors'	$t_{1/2}$	None
Leonard and others, 1990	Georgia—Tifton	Fenamiphos, fenamiphos sulfoxide, fenamiphos sulfone	ls	GLEAMS	$t_{1/2}$	None
Pennell and others, 1990	Florida—Davenport	Aldicarb	s	CMLS MOUSE PRZM GLEAMS LEACHMP	None	None
Sauer and others, 1990	Wisconsin—Hancock	Atrazine, metolachlor	s	PRZM	None	None
Brooke and Matthiessen, 1991	England—Herefordshire	Mecoprop, simazine	sil[1]	Mackay fugacity model	$t_{1/2}$	None
Klein, 1991	Germany— Schmallenberg	Cloethocarb, bentazone	NG	SESOIL	Q_{ET}	None
Loague and Green, 1991	Georgia—Watkinsville	Atrazine	l, sl, scl	PRZM	None	None
Obreza and Ontermaa, 1991	Florida—Collier County	1,3-dichloropropene	s	CMLS	None	None
Sabbagh and others, 1991	Louisiana—Baton Rouge Georgia—Tifton	Atrazine and metolachlor Atrazine	cl s	Authors' (EPIC-PST)	None	None
Shaffer and Penner, 1991	Michigan— East Lansing Hickory Corners	Alachlor and metolachlor	scl l	CMLS, PRZM, GLEAMS	$t_{1/2}$	$t_{1/2}$; K_{oc}; θ; wilting point; porosity

Table 9.1. Comparisons conducted between observed distributions of pesticides in the subsurface and predictions based on model simulations—*Continued*

Reference	Study Site Locations	Pesticides Examined	Systems Examined	Models Employed	Parameters Adjusted to Fit Observations	Parameters Examined in Sensitivity Analysis
Sichani and others, 1991	Indiana—North Vernon	Atrazine, carbofuran, cyanazine	sil	GLEAMS	None	Mean daily vs. monthly values for temperature and solar radiation
Smith and others, 1991b	Georgia—Tifton	Atrazine, alachlor	s	PRZM, GLEAMS	None	$t_{1/2}$; K_{oc}
Mueller and others, 1992	Georgia—Midville Athens	Norflurazon Alachlor, metribuzin	ls sl	PRZM, GLEAMS	None	None
Trevisan and others, 1993	Italy—Milano	Atrazine, metolachlor	s	LEACHM, PRZM, BAM	NG	$t_{1/2}$
Mueller, 1994	NG	Dichlorprop, bentazon	c, s	PRZM	θ; K-θ-h; $t_{1/2}$; K_d	None
Loague and others, 1995	Hawaii—Pearl Harbor Basin, Oahu	Chlorpyrifos, diazinon, metribuzin	Oxisol	PRZM	K_{oc}; $t_{1/2}$	None
Poletika and others, 1995	[California—Etiwanda]	Simazine	ls	Authors'	R, β	β
Weed and others, 1995	Iowa—Nashua	Alachlor, atrazine, metribuzin	l	Authors', 2CM	$t_{1/2}$	None

[1]Soil type inferred from percent clay, silt and sand given by authors (surface soil only), using classification diagram provided by Birkeland (1974).

Attempts to simulate solute transport from the land surface to below the water table thus are relatively uncommon. Efforts to do so have typically involved coupling unsaturated- and saturated-zone models together into a single code (e.g., Jones and others, 1988; Enfield and Yates, 1990), such as SUTRA (Voss, 1984) and MOUSE (Moye and Miles, 1988). Commenting on the relatively extensive body of data that have been collected on ground-water contamination in the United States, Jury and Ghodrati (1989) noted that "unfortunately, this ground water monitoring activity has not been complemented to any degree by studies of vadose zone transport or even of surface management practices until recently, so that the origin of many of the observed contaminants is unknown."

Neglect of Preferential Transport

Preferential transport of pesticides and other solutes occurs in the subsurface when water movement is too rapid for local equilibrium to be established with respect to solute partitioning among the aqueous, solid, and gas phases, or between regions of relatively mobile and immobile water (Section 4.3.3). Despite the apparently widespread occurrence of preferential transport, however, all of the simulations listed in Table 9.1 were based on an assumption of local equilibrium, using equilibrium partition coefficients (K_p or K_{oc}) to describe the influence of sorption on the rate of solute migration through the subsurface; none of the models account for the potential influence of preferential transport. Several authors have noted the inability of most commonly used models to account for preferential transport of solutes in the subsurface (Green and others, 1986; Hallberg and others, 1986; Sophocleous and others, 1990; Steenhuis and others, 1990; and Taboada and others, 1994), but this perspective has not been widely expressed in the modeling literature.

The general neglect of preferential transport in the subsurface is of particular concern because the likelihood of ground-water contamination by pesticides or other surface-derived solutes is increased when this phenomenon occurs. Furthermore, recent trends in the regulation of pesticide use involve an increasing reliance on results from solute-transport modeling studies—rather than field studies, which are considerably more expensive to conduct—for reaching regulatory decisions regarding the registration and use of individual compounds (Barbash, 1994; Behl, 1994; U.S. Environmental Protection Agency, 1994a).

The importance of incorporating preferential transport into computer simulations is underscored by the fact that the predictions provided by solute-transport models are generally the least reliable in those parts of the flow system where higher levels of accuracy are most needed. As noted in Section 4.3.3, Steenhuis and others (1990) estimated that the contamination of ground water by as little as 0.1 percent of a 2-kg/ha application of a pesticide could theoretically give rise to concentrations exceeding 1 µg/L below the water table. Boesten (1987), in turn, has noted that for a simulation model to accurately predict the arrival of 0.1 percent of the mass of an applied compound, it must be capable of accounting for the movement and fate of the remaining 99.9 percent. No simulation model to date, however, has demonstrated such accuracy. Indeed, Nicholls (1988) has stated that "concentrations of pesticide in water entering subsoils cannot be reliably simulated to an accuracy of better than an order of magnitude because the complex patterns of water flow and the slow diffusion processes of the pesticide are insufficiently understood." In addition, Dougherty and Bagtzoglou (1993) note that whereas solute transport models are frequently used to make regulatory decisions on the basis of concentrations near contaminant "action levels"—concentrations that are typically two or more orders of magnitude lower than those near the contaminant source—modeling errors are largest where predicted concentrations are lowest.

A variety of statistical techniques have been developed to examine the behavior of the relatively small fraction of solute that first reaches a specific point in the subsurface downgradient from a contaminant source. Efforts by Utermann and others (1990) to characterize solute transit times through the subsurface as probability density functions, rather than single values, represents a promising method for describing this behavior because it acknowledges the inherently stochastic nature of subsurface transport. In situ experiments by White and others (1986) and van de Weerd and van der Linden (1991) have demonstrated the utility of this approach. The use of probability density functions for describing other subsurface characteristics, such as hydraulic conductivity and transformation rates (Carsel and others, 1988a; Moye and Miles, 1988), as well as the use of random-walk models to simulate dispersion (Jones and others, 1988), have also been explored.

There are several reasons why preferential transport is neglected by the transport models used most often to predict solute movement in the subsurface. Perhaps foremost among these is a general perception that it is a relatively rare phenomenon. As noted in Section 4.3.3, however, the effects of preferential transport have been either inferred or directly observed over a wide range of edaphic and hydrogeologic settings, even in coarse-grained unconsolidated materials exhibiting no visible fracturing.

The general reluctance to use nonequilibrium models to describe contaminant behavior in the subsurface also derives from the considerable difficulties associated with estimating many of the input parameters required by such models (Carsel and Jones, 1990; Loague and Green, 1991). These include variables such as the size and porosity of soil particles and aggregates, film transfer coefficients, and diffusion coefficients in different regions of the medium (Crittenden and others, 1986). The use of even marginally satisfactory estimates for some of the more elusive parameters, however, represents an improvement over ignoring their influence altogether. In any event, sensitivity analyses (see Section 9.1.3) can be used to demonstrate the degree to which uncertainties in these parameters influence model predictions.

The avoidance of nonequilibrium models is also a consequence of the added complexity associated with the calculations that the codes require. Given the high speeds with which the more powerful computers are now capable of operating, however, the extensive data requirements of such models can be accommodated. High computing capacities have already made it possible to generate stochastic representations of subsurface properties, contaminant distributions, or climatic conditions using Monte Carlo simulations (Black and Freyberg, 1987; Carsel and others, 1988b; Moye and Miles, 1988; Banton and Villeneuve, 1989; Villeneuve and others, 1990; Isabel and Villeneuve, 1991; McLaughlin and others, 1993; Shrestha and Loganathan, 1994) and other stochastic modeling techniques (van der Zee and Boesten, 1991).

Although relatively few existing models have incorporated physical nonequilibrium to describe the partitioning and transport of solutes in soils, such models have been in existence for at least three decades. As noted in Section 4.3.3, the use of these models has considerably improved the accuracy with which solute transport has been predicted in laboratory microcosms (e.g., Coats and Smith, 1964; van Genuchten and Wierenga, 1976; Crittenden and others, 1986; Goltz and Roberts, 1986; Ball and Roberts, 1991b; Brusseau and Rao, 1991; Dowling and others, 1994). Descriptions of nonequilibrium transport models and their principal features have been provided in reviews by Jury and Ghodrati (1989), Enfield and Yates (1990), Sabatini and Austin(1990), and Wu and others (1994).

Neglect of Physical, Chemical, and Biological Variables

Few of the models utilized by the studies listed in Table 9.1 account for the influence of several fundamental parameters known to affect transformation rates and pathways. These variables include temperature, pH, and the size and species composition of the resident microbial populations (Chapter 4). The use of soil dissipation half-lives to quantify the rates of pesticide transformation (e.g., Jury and others, 1984c) compounds this difficulty because, as noted earlier (Section 2.10.2), this parameter does not distinguish between the disappearance of the parent compound by transport and that from actual transformation in situ.

Parameters Adjusted to Fit Observations

The use of solute-transport models to predict contaminant movement and fate in the subsurface often involves the adjustment of one or more model parameters until a satisfactory match is obtained between model predictions and field observations over a specified period of record. This process has been referred to as "history matching" by Bredehoeft and Konikow (1993), a term suggested as a preferred alternative to the commonly used words "validation" or "verification." Indeed, Oreskes and others (1994) have stated that "verification and validation of numerical models of natural systems is impossible." Nevertheless, a variety of approaches for assessing the extent to which model predictions agree with field data have been proposed (e.g., Parrish and Smith, 1990; Wallach and Goffinet, 1991).

Table 9.1 lists some of the parameters that have been adjusted to achieve closer matches between field results and model predictions. Some of these parameters are indeed difficult to estimate without the use of a transport model. These include the longitudinal and transverse dispersion coefficients (D_L and D_T, respectively), and some of the terms used to quantify water fluxes (e.g., Q_{rech} or Q_{ET}).

A fundamental difficulty with many of the studies summarized in Table 9.1, however, is a tendency to adjust parameters that reflect the physical and chemical properties of the pesticides of interest—parameters that can be measured independently, either in the laboratory or in situ. Such variables include soil-water partition coefficients and rates of transformation. To be sure, several of the simulation studies were careful to ensure that parameter values chosen to match field data fell within the ranges established by previous laboratory or field research (e.g., Enfield and others, 1982; Bilkert and Rao, 1985; Carsel and others, 1985; Lorber and Offutt, 1986). However, most studies that adjusted partition coefficients or dissipation half-lives to match field results did not appear to have taken such precautions.

The use of solute property values that are inconsistent with reliable, independent measurements of those properties, in order to match predictions with field observations, limits the utility of any simulation model. In such cases, substantial modifications of the model itself, rather than of the measured parameter values, may be more appropriate. Alternatively, disparities between prediction and observation may be used to highlight instances where current approaches for measuring parameter values in the laboratory may be inappropriate for quantifying contaminant behavior in situ. Analyses of this type have been carried out for soil sorption coefficients (Green and others, 1986) and transformation rate constants (Harkin and others, 1986).

9.1.3 USE OF SENSITIVITY ANALYSIS

Despite the aforementioned difficulties, mathematical models are powerful tools for assessing the effects of known variations in physical, chemical, and biological parameters on the subsurface behavior of pesticides. Many of the studies listed in Table 9.1 investigated the sensitivity of model predictions to specified variations in one or more of the input parameters.

The range of variables tested in this manner (Table 9.1) encompasses most of those known to exert the greatest influence on the migration and fate of solutes in the subsurface, but a complete summary of these analyses is beyond the scope of this book.

Sensitivity analyses using numerical models have been employed to examine a variety of issues, including (1) the sensitivity of model predictions to variations in the magnitude of input parameters (Leistra and Dekkers, 1976; Jury and others, 1983; Jury and others, 1984a; Leistra and others, 1984a; Helling and Gish, 1986; Donigian and Carsel, 1987; Carsel and others, 1988b; Boesten, 1991; Boesten and van der Linden, 1991; and Isabel and Villeneuve, 1991); (2) the influence of parameter uncertainties on the predicted behavior of pesticides in the subsurface (e.g., Dean and others, 1987; Carsel and others, 1988b; and Loague and others, 1990); (3) the relative propensities of different pesticides to reach the water table after surface application (Rao and others, 1985; McLean and others, 1988); and (4) the effects of different agricultural management practices on the likelihood of a given pesticide reaching the water table (Carsel and others, 1987; Dean and others, 1987; Donigian and Carsel, 1987; Knisel and Leonard, 1989; Leonard and Knisel, 1989; Bretas and Haith, 1990; Shoemaker and others, 1990; and Boesten and van der Linden, 1991). None of these investigations were included in Table 9.1, however, because their predictions were not compared with field observations.

The use of a simulation model to carry out sensitivity analyses requires that the model employed accurately describes all physical, chemical, and biological processes of interest. As discussed earlier, however, there is ample evidence to suggest that most computer models in common use are, in many cases, unable to accurately mimic such processes. Conclusions derived from sensitivity analyses—like other model predictions—must therefore be checked against field observations of pesticide behavior in situ to confirm or refute their accuracy.

9.1.4 PREDICTION OF RELATIVE RATES OF PESTICIDE MOVEMENT

In light of the general inability of commonly used computer models to predict the concentrations or times of arrival of individual solutes at specific points beneath the land surface, it is often stated that these models may be used, nevertheless, to predict relative orders of mobility or persistence among different solutes (e.g., Jury and others, 1984b; McLean and others, 1988; Barrett and Williams, 1989; Gustafson, 1989; and Britt and others, 1992), or for the same compound among different environmental settings or agricultural management approaches (Smith and others, 1991b). Such efforts have, indeed, met with some success (e.g., Jury and others, 1986a; Britt and others, 1992). However, even predictions of relative orders of mobility or persistence among pesticides have been demonstrated to vary among different models (Rao and others, 1985; Funari and others, 1991), or have failed to reproduce the orders of relative mobility observed among different pesticides in the field (Shirmohammadi and others, 1989; Clendening and others, 1990).

Predictions of relative rates of movement that are based on an assumption of uniform transport are of limited utility if preferential transport is significant. Under such circumstances, solutes of widely varying degrees of hydrophobicity may migrate at the same rate, a phenomenon observed during several field experiments (e.g., Bottcher and others, 1981; Everts and others, 1989; Kladivko and others, 1991; and Traub-Eberhard and others, 1994). The neglect of preferential transport, or other phenomena known to significantly influence pesticide transport and fate, can therefore compromise the value of a simulation model for predicting even relative rates of pesticide movement in the subsurface.

9.2 USE OF OTHER SOLUTES AS INDICATORS OF PESTICIDE OCCURRENCE IN GROUND WATER

Laboratory analyses for individual pesticides and transformation products are more expensive than those for many other compounds, particularly in relation to most inorganic analytes, and are likely to remain so for the foreseeable future. As a result, other chemical species have been investigated as possible indicators of the presence of pesticides in subsurface waters potentially affected by agrichemical use. Among the solutes considered for this purpose, the most prominent examples have included nitrate and tritium.

9.2.1 NITRATE

Because fertilizers are frequently applied in conjunction with pesticides during the cultivation of most crops, nitrate is often suggested as an inexpensive indicator of the presence of pesticides in ground waters beneath agricultural areas. (The strong affinity of phosphate for mineral surfaces effectively eliminates its utility for this purpose.) Although nitrate is a natural constituent of soils, nitrate-nitrogen (NO_3^--N) concentrations are generally less than 3 mg/L in subsurface waters subject to minimal anthropogenic impact (Madison and Brunett, 1985). Nitrate-nitrogen levels significantly higher than this, presumed to arise from human influences, have been referred to as "excess nitrate" (Burkart and Kolpin, 1993). The U.S. Environmental Protection Agency (1990a) has established a Maximum Contaminant Limit for nitrate of 10 mg/L, as NO_3^--N. This value was selected primarily to address the risk of methemoglobinemia in infants ("blue-baby syndrome"), but nitrate might also pose other, less well-defined risks to human health (Kross and others, 1990).

Nitrate concentrations are generally unreliable predictors of pesticide concentrations in subsurface waters (e.g., Hallberg and others, 1984; Libra and others, 1984; Klaseus and others, 1988; Detroy and others, 1988; Pionke and others, 1988; Baker and others, 1989; Neil and others, 1989; Kross and others, 1990; Thompson, 1990; Denver and Sandstrom, 1991; Domagalski and Dubrovsky, 1992; U.S. Environmental Protection Agency, 1992a; Koterba and others, 1993; and Glanville and others, 1995). One of the more compelling demonstrations of this poor correlation was provided by a set of comparisons in Minnesota (Klaseus and others, 1988) between atrazine and nitrate concentrations in observation, private (Figure 9.1), and public-supply wells (Figure 9.2). Similarly, Libra and others (1984) observed no relation between the concentrations of pesticides and nitrate in the ground waters of Floyd and Mitchell Counties, Iowa (Figure 9.3). The limited reliability of nitrate concentrations as indicators of pesticide concentrations was also demonstrated by a study of rural well contamination in Iowa. Although Glanville and others (1995) observed significant, positive correlations of mean triazine or acetanilide herbicide concentrations with mean nitrate concentrations among the wells sampled ($\alpha = 0.05$) a significant inverse correlation was also observed between daily nitrate and triazine concentrations in one of the wells sampled.

A few monitoring studies have observed significant correlations between pesticide and nitrate concentrations in ground water. These include investigations conducted in Nebraska (Spalding and others, 1980; Chen and Druliner, 1987; and Druliner, 1989) and Oregon (Pettit, 1988; Istok and others, 1993). Figure 9.4 shows the positive relation observed by Istok and others (1993) between concentrations of DCPA (plus its hydrolysis products) and nitrate in Oregon ground waters. Despite the direct correlations observed in these studies, however, the fact that so many investigations have not observed such relations compromises the value of nitrate concentrations as indicators of pesticide concentrations in ground water.

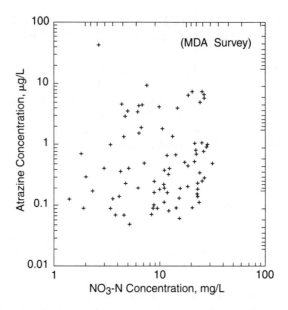

Figure 9.1. Atrazine concentration versus nitrate concentration in both observation and private wells in Minnesota, from a survey conducted by the Minnesota Department of Agriculture (MDA). Redrawn with permission from Klaseus and others (1988).

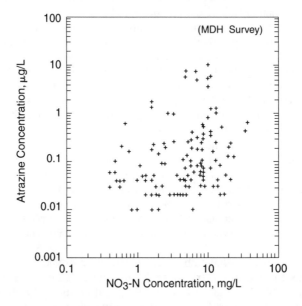

Figure 9.2. Atrazine concentration versus nitrate concentration in public wells in Minnesota, from a survey conducted by the Minnesota Department of Health (MDH). Redrawn with permission from Klaseus and others (1988).

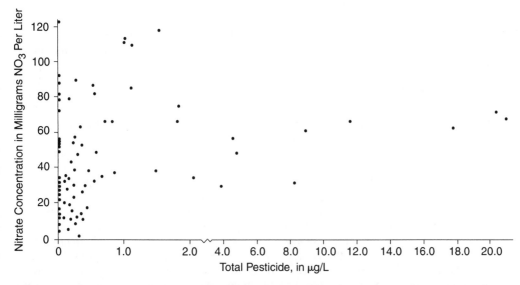

Figure 9.3. Nitrate concentration versus pesticide concentration in ground water in Floyd and Mitchell Counties, Iowa. Note: Concentrations of nitrate expressed as milligrams of nitrate, rather than nitrate-nitrogen, per liter. Modified from Libra and others (1984) by permission of the National Ground Water Association. Copyright 1984.

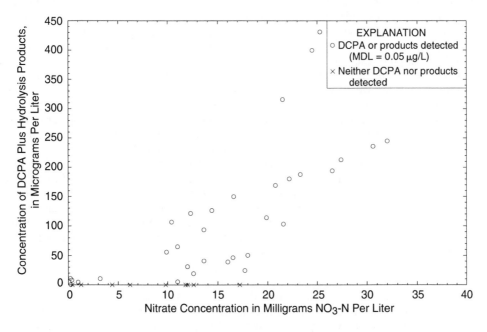

Figure 9.4. Concentration of DCPA (dacthal) plus its hydrolysis products versus nitrate concentration in wells sampled in Oregon (data from Istok and others, 1993). MDL, Method Detection Limit.

By contrast, many studies, including several of those cited above, have found that the likelihood of detecting pesticides in ground waters increases with increasing nitrate concentrations. Thus, although investigations in Minnesota (Figures 9.1 and 9.2), Wisconsin (LeMasters and Doyle, 1989), and New Jersey (Szabo and others, 1994) all observed negligible degrees of correlation between nitrate and pesticide concentrations in ground waters, all three studies found pesticide detection frequencies to exhibit significant increases with increasing nitrate concentrations. Significant relations between the likelihood of encountering high nitrate concentrations and that of detecting pesticides have also been reported in ground waters beneath the Delmarva Peninsula east of Washington D.C. (Pinto, 1980; Koterba and others, 1993), Iowa (Kross and others, 1990), New Jersey (Louis and Vowinkel, 1989), Pennsylvania (Pionke and others, 1988), and Ontario, Canada (Rudolph and others, 1992; 1993), as well as within the multistate areas investigated by the Midcontinent Pesticide Study, or MCPS (Kolpin and others, 1994), and the National Alachlor Well Water Survey, or NAWWS (Holden and others, 1992).

As with the relations based on concentrations alone, however, not all studies observed direct relations between nitrate concentrations and pesticide detection frequencies. Neither a statewide survey in Ohio (Baker and others, 1989), nor a summary of statewide data in Iowa (Detroy and others, 1988), nor the National Pesticide Survey or NPS (U.S. Environmental Protection Agency, 1992a), observed consistently higher frequencies of pesticide detection in the presence of higher nitrate concentrations. From their summary of sampling data from 355 wells throughout Iowa, Detroy and others (1988) found 85 percent of those containing nitrate concentrations above 10 mg/L NO_3^--N to have no detectable pesticides.

Positive correlations between pesticide and nitrate concentrations, or frequencies of nitrate occurrence above some concentration threshold, are most likely to arise from the fact that the concentrations of pesticides and nitrate exhibit inverse relations with the depth of sampling (Detroy and others, 1988; Klaseus and others, 1988; Baker and others, 1989; Druliner, 1989; U.S. Environmental Protection Agency, 1992a; and Burkart and Kolpin, 1993). As noted by Kross and others (1990) from the results of the Iowa State-Wide Rural Well-Water Survey (SWRL), "by far the most significant factor explaining water-quality variations is well depth. An apparent relationship among total coliforms, NO_3^--N, and pesticides is primarily a function of their co-occurrence related to well depth." Szabo and others (1994) noted that the poor correlation they observed between nitrate and pesticide concentrations in ground waters of the southern Coastal Plain of New Jersey was a consequence of a substantial difference in the depths to which the two types of contaminants had penetrated in the aquifer system of interest.

The low frequencies with which pesticide residues are often encountered in ground waters containing excess nitrate (e.g., Detroy and others, 1988) may be accounted for in some settings by contrasts in the migration rates of the different solutes. While most pesticides are slowed during transport in the subsurface due to sorptive interactions with soil organic matter (Sections 4.3.1 and 4.3.2), nitrate behaves conservatively under aerobic conditions. The effects of sorption will thus cause pesticide concentrations to diminish more rapidly than nitrate levels as recharge waters migrate through the subsurface.

In addition to simple differences in the distances of solute migration, the poor correlations that have been observed between nitrate and pesticide concentrations in the subsurface may be driven in part by contrasts in the reactivities of these species under different redox conditions. Under hypoxic conditions (i.e., in the absence of detectable dissolved oxygen), nitrate undergoes microbial reduction to N_2 via N_2O, a process known as denitrification. Evidence for denitrification in ground water has been provided by studies conducted in a variety of settings in Nebraska (Druliner, 1989), Iowa (Thompson and others, 1986; Kross and others, 1990; Thompson, 1990), and the 12 states included in the MCPS (Kolpin and others, 1994). All reported significant correlations between the concentrations of nitrate and dissolved oxygen in ground waters.

Whereas nitrate is readily reduced in the absence of detectable oxygen, some widely used pesticides undergo reductive transformations so slowly that they are effectively recalcitrant under hypoxic conditions, and are therefore likely to persist in denitrifying environments. Of particular interest in this regard is atrazine, which is relatively resistant to reductive transformation (Section 4.4.4). The detection of atrazine residues in the absence of significant nitrate concentrations (e.g., Hallberg and others, 1984; Klaseus and others, 1988; Baker and others, 1989; and Kross and others, 1990), particularly when dissolved oxygen is not detectable (Koterba and others, 1993), may reflect the persistence of atrazine under hypoxic conditions. As noted in Section 4.4.4, the detection of hydrogen sulfide—an indicator of sulfate-reducing conditions—and low nitrate concentrations in the two wells with the highest atrazine concentrations in an agricultural area of Pennsylvania (Pionke and others, 1988) supports this hypothesis. Conversely, since atrazine and other triazines readily undergo dealkylation in the presence of dissolved oxygen (Table 4.2; Section 4.4.4), the dealkylation of atrazine might explain an apparent absence of the parent compound in the presence of substantial nitrate concentrations in some aerobic ground waters. Deethyl atrazine-to-atrazine concentration ratios (DAR) could be used to examine this hypothesis.

The reducing power required for denitrification in the subsurface is provided by organic matter, which is not always in sufficient supply, even in agricultural soils. During their studies of ground-water quality in the Big Spring Basin in Iowa, Hallberg and others (1984) found that hydrogeologic settings where nitrate concentrations were either very low or below detection limits—despite the presence of atrazine—were characterized by relatively abundant organic carbon, and hence, appeared to be areas of active denitrification. Conversely, ground waters with high nitrate concentrations and detectable levels of atrazine were drawn from subsurface environments that were more highly oxidized and contained relatively low levels of sedimentary organic carbon—conditions under which denitrification is less favored.

The associations that have been frequently observed between pesticides and nitrates in ground waters thus do not appear to constitute a sufficient basis for using nitrate as a general indicator for the presence of pesticide residues in the subsurface. While pesticides may, indeed, be detected more frequently in ground waters containing high nitrate concentrations in some areas, the fact that these associations have not been found in many other areas indicates that nitrate is not a reliable predictor of pesticide contamination. Uncertainties associated with using nitrate for this purpose are exacerbated by the substantial contrasts in the environmental behavior of nitrate and many pesticides, both with respect to their susceptibility to retardation and their persistence under different redox conditions.

9.2.2 TRITIUM

Whereas the value of nitrate as an indicator of pesticide contamination in ground water is compromised by the fact that it is depleted under hypoxic conditions, the concentrations of tritium (^3H) are not significantly influenced by chemical reactions in the subsurface. Atmospheric levels of tritium were increased by up to three orders of magnitude as a result of large-scale atmospheric testing of nuclear weapons in the 1950's and early 1960's (Freeze and Cherry, 1979). For reasons described in detail by Plummer and others (1993), ground waters containing tritium concentrations in excess of 0.2 Tritium Units are likely to have entered the subsurface after 1952 (a Tritium Unit, or TU, is defined as 1 ^3H atom in 10^{18} atoms of hydrogen [Freeze and Cherry, 1979], or 3.24 picocuries ^3H/L [Plummer and others, 1993]). Because widespread use of pesticides did not begin until after the Second World War, the presence of tritium at concentrations above 0.2 TU indicates ground water that is "young" enough for the potential presence of pesticides, as well. More important, pesticides would not be expected to be present in ground waters containing tritium concentrations of less than 0.2 TU.

The results from four studies of pesticide contamination in ground waters are consistent with these predictions. In the Central Valley of California (Domagalski and Dubrovsky, 1992), Oklahoma City, Oklahoma (Christenson and Rea, 1993), and the New Jersey Coastal Plain (Szabo and others, 1994), no pesticides were detected in any ground water samples that lacked detectable tritium (^{3}H detection limits: 0.8 TU in California, 0.3 TU in Oklahoma and New Jersey). Many of the samples analyzed during these three studies contained detectable tritium in the absence of measurable levels of pesticides, but this is also consistent with expectation; as a marker of recent recharge, tritium provides a valuable indicator for the potential, but not the guaranteed presence of pesticides in ground water. Similarly, Kolpin and others (1995) reported comparatively low frequencies of herbicide detection in ground waters containing <2 TU during the 1991 sampling of the MCPS (Figure 9.5).

Studies conducted in Iowa reveal the limited utility of tritium as an indicator of pesticide contamination in waters recharged more recently than the 1950's. Two of the repeat samplings of 10 percent of the wells included in the SWRL study in Iowa involved analyses for tritium, in addition to pesticides (Rex and others, 1993; Libra and others, 1993). During both resamplings, however, no clear differences in the detection frequencies for atrazine or any other pesticide were observed between wells found to contain waters considered to have recharged within 20 years ($> 6 \pm 4$ TU) and those with waters that entered the subsurface longer than 20 years ago ($< 6 \pm 4$ TU). The inability of these investigations to distinguish between ground waters with high and low frequencies of pesticide detection on the basis of tritium may have been a consequence of their use of a comparatively recent cutoff date (high tritium detection limit) for

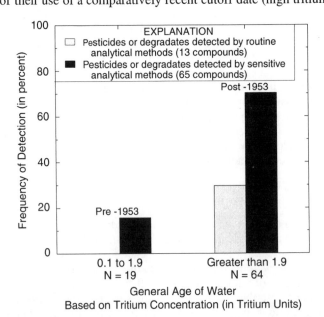

Figure 9.5. Relation between ground-water age, based on tritium concentrations, and the frequency of pesticide detection in near-surface aquifers of the midcontinental United States, based on the Midcontinent Pesticide Study. Modified from Kolpin and others (1995) and published with permission.

this purpose. Atrazine, for example, first came into widespread use in the mid-1970's (Koterba and others, 1993). Use of the herbicide may thus have already been significant in Iowa by the time the "older" ground waters entered the subsurface in the areas investigated.

Tritium concentrations in ground waters from the alluvial aquifer of the West Fork Des Moines River in Iowa also could not be used to distinguish between waters that did or did not contain detectable pesticide residues (Thompson, 1990). This result, however, occurred because the sampled waters were too recently recharged for such a distinction to be possible (18-71 TU), as might have been expected for ground water in a shallow alluvial aquifer.

9.2.3 OTHER PESTICIDES

The presence of one or more pesticides at detectable levels in ground water has also been found to be associated with an increased likelihood that other pesticides will be present. During the NAWWS (Holden and others, 1992), the MCPS (Kolpin and others, 1995), and the study of rural well contamination in Iowa by Glanville and others (1995), the occurrence of atrazine in ground water was found to be a significant indicator of the occurrence of other herbicides as well. Koterba and others (1993) observed that when more than one pesticide was detected in a given well on the Delmarva Peninsula, at least one was commonly a triazine herbicide; the other pesticides typically included dicamba or one of the acetanilide herbicides. Koterba and others (1993) noted that the co-occurrence of several pesticides in individual wells may have been a consequence of the fact that crop rotations among corn, soybeans, and, in some instances, small grains, were common within the vicinity of many of the areas sampled. In contrast, Walker and Porter (1990) suggested that the use of crop rotations, and the resulting variety of pesticides used among different years, may diminish the likelihood of detecting pesticides in ground waters by reducing the average annual amount of a given pesticide entering the subsurface over the entire rotation cycle. Regardless of the reasons for the co-occurrence of different pesticides in individual wells, the comparatively low incidence of pesticide detection in most ground waters, coupled with the high cost of analysis for most of these compounds and their transformation products, makes them even less attractive than nitrate or tritium as indicators for the presence of other pesticides in the subsurface.

9.2.4 ADJUVANTS

As noted in Section 3.1.3, volatile organic compounds (VOCs) are often included among the "inert ingredients" or "adjuvants" used as carriers in pesticide formulations. Surfactants of various types are also employed for this purpose. The relatively slow rates of transformation of most of these compounds, coupled with their comparatively low affinities for soil organic carbon, indicate that adjuvants from these two chemical classes could serve as indicators for pesticide contamination. While the use of nitrate as a potential indicator for pesticide contamination assumes that, in the ideal case, both species are applied in approximately the same location (i.e., the area of recharge for the sampled ground water), the use of adjuvants for this purpose avoids such an assumption, because the latter are part of the pesticide formulation itself. Sampling for either VOCs or surfactants beneath agricultural areas has been limited to date, but several VOCs have been detected in ground waters beneath agricultural areas in various parts of the United States (Section 3.1.3). In addition, chemical analyses for surfactants can be carried out inexpensively in the field. Anionic surfactants are likely to be more valuable for this purpose, because they would be expected to exhibit greater mobility through the subsurface than cationic or nonionic surfactants, which would tend to show greater affinities for clays and soil organic matter, respectively.

The preceding discussions demonstrate that while several chemical species have been investigated for use as indicators of pesticide contamination in ground waters, all have considerable drawbacks. To date, the desire to find such surrogates has been driven largely by the need to avoid the high costs and time delays associated with most pesticide analyses. The recent advent of field-based immunoassay (ELISA) techniques, however, may obviate the need for such surrogates in many situations. Provided that their limitations are fully recognized, ELISA methods appear to represent the most convenient and affordable technology for screening ground waters for the presence of pesticide residues, particularly for identifying ground waters where such residues are not detectable, as noted earlier (Section 2.8.2).

9.3 ASSESSMENTS OF GROUND-WATER VULNERABILITY TO PESTICIDE CONTAMINATION

Over the past decade, extensive time and effort have been expended to predict the vulnerability of ground waters in different locations to contamination by pesticides, as well as nitrate and other agriculturally derived solutes. This section provides an overview of these efforts, and assesses the degree to which they have succeeded in predicting ground-water contamination by agrichemicals. Although the principal focus is on pesticides in this regard, some discussion of work related to nitrate contamination is also included for comparison.

9.3.1 OVERVIEW OF PREVIOUS EFFORTS

To an even greater extent than that observed among modeling studies, few assessments of ground-water vulnerability to agrichemical contamination have been tested against actual ground-water quality data. Of the 63 assessments of ground-water vulnerability reviewed, 41 (65 percent) did not compare their vulnerability predictions against any observations of ground-water contamination by the compounds of interest. Three of the remaining 22 studies incorporated ground-water quality data into the vulnerability assessments themselves.

Table 9.2 summarizes the results of most previous efforts to predict the contamination of ground waters by agriculturally derived solutes through the use of vulnerability assessments. Excluded from the table were the large number of studies that described vulnerability assessment schemes, or predicted the vulnerability of ground waters to agriculturally derived contamination in specific locations, without comparing their predictions against actual field data. Also excluded were studies that used observed contaminant distributions to predict ground-water vulnerability to other types of contamination in the same area, and thus could not predict such distributions a priori (e.g., Anderson and others, 1985; Spalding and others, 1989; Domagalski and Dubrovsky, 1991; Martin and others, 1991).

Table 9.2 does, however, include studies that used existing data on ground-water contamination either to set priorities for additional sampling in the area of interest (Scheibe and Lettenmaier, 1989), or to predict subsurface contamination in nearby areas (Teso and others, 1988; Troiano and Sitts, 1990). Furthermore, although Lorber and others (1989) utilized monitoring data for their nationwide assessment of ground-water vulnerability to aldicarb contamination, this study was also included because the monitoring data could be separated from, and compared to, assessments based on the two other characteristics of interest, hydrogeology and agricultural practices.

Table 9.2. Comparisons between assessments of vulnerability of ground waters or soils to contamination and observed contaminant distributions

[Statistics computed by the present authors, based on data extracted from the original studies, are given in *italics*. Abbreviations and references for ground-water vulnerability (GWV) rating schemes: AF, Attenuation Factor (Rao and others, 1985); Authors', ground-water vulnerability assessment procedure designed by study authors; DRASTIC, Aller and others (1987); GW source, source of ground water, inferred from major-ion chemistry; Hydrogeom.reg., hydrogeomorphic region; LEACH, Dean and others (1984); NLEAP(NL), Shaffer and others (1991); VARSCORE, Alexander and Liddle (1986). CWS, Community water system wells; RD, Rural domestic wells. Parameters Used: conc., constituent concentration in ground water, unless otherwise indicated ("soil" in parentheses denotes concentrations in soil); d.f., frequency of constituent detection; pdf, pesticide detection frequency; VOCs, volatile organic compounds; Methods Used: *Italics* denote tests performed by the present authors. Accuracy of Predictions: r², Unadjusted coefficient of determination for correlation of contaminant concentration or detection frequency with GWV parameter; P, probability that the correlation observed between the GWV parameter and the concentration or detection frequency of the contaminant(s) of interest could have been obtained by chance; N, Number of sampling points (wells, soil sampling sites, regions, or counties) used for assessment; κ, number of vulnerability categories employed for assessment; NDP, value not provided in original study. Correlations significant at the α = 0.05 level are indicated by P values given in **boldface**. USEPA, U.S. Environmental Protection Agency. mg/L, milligrams per liter; <, less than; > greater than; <, less than]

Reference	Study Locations	Hydrogeologic Settings	GWV Assessment Method	Parameters Used to Evaluate Contamination		Methods Used for Comparisons	Accuracy of Predictions		
				Constituent	Variable		r²	P	N/κ
Greenberg and others, 1982	New Jersey—Statewide	Various (but unknown)	Factor analyses based on land use	Organo-chlorine pesticides	d.f.	Mann-Whitney test; Discriminant analysis	NDP; NDP	**0.001**; **0.001**	40/17; 40/17
Curry, 1987	Iowa—Big Spring Basin	Limestone (non-karst to mature karst)	DRASTIC	Nitrate	conc.	Spearman rank correlation coefficient	0.0013 [1](r<0)	[1]>0.05	19/18
			DRASTIC 6/82	Atrazine	conc.		0.18	>0.05	19/18
			7/82				0.033	>0.10	19/18
			LEACH 6/82				0.069 [1](r<0)	[1]>0.05	19/18
			7/82				0.035	>0.05	19/18
Bishop and Lawyer, 1988	California—Statewide	Not applicable (based solely on pesticide properties)	Groundwater Screening Index	Various pesticides	d.f.	*Linear regression*	*0.37*	*0.28*	*5/5*

Table 9.2. Comparisons between assessments of vulnerability of ground waters or soils to contamination and observed contaminant distributions—Continued

Reference	Study Locations	Hydrogeologic Settings	GWV Assessment Method	Parameters Used to Evaluate Contamination		Methods Used for Comparisons	Accuracy of Predictions		
				Constituent	Variable		r^2	P	N/κ
Teso and others, 1988	California—Merced County	Not given	Discriminant analysis	DBCP	d.f. (soil)	Chi-squared goodness-of-fit	NDP	0.025-0.05	199/2
Baker and others, 1989	Ohio—Statewide	Various	DRASTIC (County level)	Nitrate	conc.	Linear regression	0.30	NDP	76/4
Barrett and Williams, 1989	Nebraska—Statewide	Various	DRASTIC (County level)	Atrazine	d.f.	Linear regression	0.037	0.1935	48/4
Khan and Liang, 1989	Hawaii—Oahu	Various	AF	EDB	d.f.	None (insufficient number of GWV classes)	NDP	NDP	5/2
Lorber and others, 1989	Nationwide	Various	Hydrogeology (Authors'/DRASTIC)[2]	Aldicarb	d.f.	Linear regression	0.098	0.191	19/5
			Agricultural practices[2]	Aldicarb	d.f.	Linear regression	0.148	0.104	19/5
Scheibe and Lettenmaier, 1989	Washington—Whatcom County	Not given	Risk-based optimization	EDB	d.f.				
			Parameter set 1			Linear regression	0.1003	0.13	24/3
			Parameter set 2			Linear regression	0.0603	0.25	24/4
			Parameter set 3			Linear regression	0.1228	0.093	24/3
			Parameter set 4			Linear regression	0.0453	0.32	24/3
Meeks and Dean, 1990	California—San Joaquin County	Valley alluvium	Leaching Potential Index	DBCP	d.f.	Linear regression	0.52	0.066	272/6
Troiano and Sitts, 1990	California—Merced County	Not given	Discriminant analysis (Teso and others, 1988)	Nitrate	conc.	Multiple linear regression	0.22	**0.0105**	30/12

Table 9.2. Comparisons between assessments of vulnerability of ground waters or soils to contamination and observed contaminant distributions—*Continued*

Reference	Study Locations	Hydrogeologic Settings	GWV Assessment Method	Parameters Used to Evaluate Contamination — Constituent	Parameters Used to Evaluate Contamination — Variable	Methods Used for Comparisons	Accuracy of Predictions — r^2	Accuracy of Predictions — P	Accuracy of Predictions — N/κ
Roux and others, 1991b	Seven states (California, Delaware, Florida, Illinois, Indiana, Michigan, West Virginia)	Various	VARSCORES	Simazine	conc.	Linear regression	0.41 [1](r<0)	[1]0.002	21/10
Holden and others, 1992	26 states (National Alachlor Well-Water Survey)	Various	Authors' (County level)	Alachlor	d.f.	Weighted logistic regression	NDP	<0.05	1430/2
				Atrazine			NDP	>0.05	1430/2
				Metolachlor			NDP	<0.05	1430/2
				Simazine			NDP	<0.05	1430/2
				Nitrate[3]			NDP	[3]<0.05	1430/2
			DRASTIC (County level)	Alachlor			NDP	>0.05	1430/3
				Atrazine			NDP	>0.05	1430/3
				Metolachlor			NDP	>0.05	1430/3
				Simazine			NDP	<0.05	1430/3
				Nitrate[3]			NDP	[3]<0.05	1430/3
USEPA, 1992a	CWS—50 states RDW—38 states (National Pesticide Survey)	Various	DRASTIC (County level) CWS	Various pesticides	d.f.	Univariate logistic regression	NDP	>0.05	NDP
			CWS	Nitrate			NDP	[1]<0.001	NDP
			RDW	Various pesticides			NDP	>0.05	NDP
			RDW	Nitrate			NDP	>0.05	NDP

Table 9.2. Comparisons between assessments of vulnerability of ground waters or soils to contamination and observed contaminant distributions—*Continued*

Reference	Study Locations	Hydrogeologic Settings	GWV Assessment Method	Parameters Used to Evaluate Contamination — Constituent	Parameters Used to Evaluate Contamination — Variable	Methods Used for Comparisons	Accuracy of Predictions — r^2	Accuracy of Predictions — P	Accuracy of Predictions — N/κ
Koterba and others, 1993	Delmarva Peninsula	Sandy Coastal Plain aquifer	Authors' Local land use	Various pesticides	d.f.	Kruskal-Wallis Test	NDP	**<0.01**	100/2
			Soil group			Kruskal-Wallis Test	NDP	**0.02**	100/2
			Hydrogeom.reg.			Kruskal-Wallis Test	NDP	**0.02**	99/2
			GW source			Kruskal-Wallis Test	NDP	**0.01**	100/2
			DRASTIC Agricultural areas	Various pesticides	d.f.	*Linear regression*	*0.353*	*0.0699*	*51/10*
			High-NO_3^- areas			*Linear regression*	*0.163*	*0.247*	*61/10*
			Non-agricultural areas			*Linear regression*	*0.006*	*0.812*	*51/12*
			Low-NO_3^- areas			*Linear regression*	*0.044*	*0.536*	*41/11*
Krapac and others, 1993	Illinois—59 agrichemical facilities	Various	McKenna and Keefer (1991) All samples	Various pesticides	d.f. (in soils)	*Linear regressions of d.f. (in soils) versus depth of uppermost surface of aquifer materials*	*0.91*	***0.0467***	*784/4*
			0-0.6 m				*0.94*	***0.0281***	*196/4*
			0.6-1.1 m				*0.98*	***0.0095***	*196/4*
			1.1-1.6 m				*0.82*	*0.0958*	*196/4*
			4.0-4.5 m				*0.38*	*0.3874*	*196/4*
Wylie and others, 1993	Colorado—Weld County	Alluvial aquifer	NLEAP (NL)	Nitrate	conc.	Pearson correlation coefficient	0.35	**<0.0001**	108/108
Kalinski and others, 1994	Nebraska—Statewide	Various	DRASTIC	VOCs	d.f.	Linear regression	0.93	**<0.01**	681/7
Vowinkel and others, 1994	New Jersey—Coastal Plain Central	Unconsolidated, semi-confined Sedimentary bedrock	Authors'	Nitrate	conc.	Visual inspection of boxplots	NDP	>0.05	554/6
				Nitrate	conc.	Kruskal-Wallis Test	NDP	0.650	186/3

[1] Inverse correlation observed between contaminant levels and GWV parameter, implying that contamination decreases with increasing GWV.
[2] Ground-water vulnerability rankings of "low," "low-medium," "medium," "medium-high," and "high" assigned numerical values of 1,2,3,4, and 5, respectively, by present authors for purposes of statistical evaluation.
[3] Nitrate considered to have been "detected" if its concentration exceeded 0.1 mg/L.

For a large proportion of the studies listed in Table 9.2, the ability of each vulnerability assessment scheme of interest to predict observed subsurface contamination was evaluated by the study authors in only a qualitative, rather than statistical manner, involving the use of terminology such as "marginal" (Barrett and Williams, 1989) or "relatively good" (Scheibe and Lettenmaier, 1989) to describe the match between prediction and observation. Among studies of this type from which the appropriate data could be extracted, simple linear regressions between the various indices of vulnerability and actual contaminant concentrations or detection frequencies were carried out by the present authors. The results from these statistical tests are given in italics in Table 9.2 (19 out of the 57 statistical tests listed).

There were several instances where correlations between ground-water vulnerability indices and observed contamination were significant at the $\alpha = 0.05$ level (Table 9.2). For two of the studies in question, however (Roux and others, 1991b; U.S. Environmental Protection Agency, 1992a), the relations, though statistically significant, were in the direction opposite to that expected; contamination was found to be more pronounced in areas deemed less vulnerable. In addition, the comparatively small coefficients of determination ($r^2 \leq 0.41$) obtained from all but three of the other studies (Meeks and Dean, 1990; Krapac and others, 1993; and Kalinski and others, 1994) indicate that most of these assessments could account for no more than 40 percent of the observed variability in either contaminant concentrations or frequencies of subsurface contamination.

These low coefficients of determination demonstrate the limited ability of most vulnerability assessment schemes to predict subsurface contamination. For example, Table 9.2 shows that the most widely used assessment scheme employed for this purpose has been the DRASTIC system (Aller and others, 1987). (DRASTIC assigns scores to different subsurface settings on the basis of the seven different factors from which its name is derived: Depth to water; net Recharge; Aquifer media; Soil media; Topography; Impact of the unsaturated zone; and hydraulic Conductivity of the aquifer.) Figure 9.6 shows the low correlation between DRASTIC scores and nitrate concentrations in ground water in 76 counties in Ohio (Baker and others, 1989). Such limitations are a source of concern, given the widespread use of vulnerability assessments by state and federal agencies for setting ground-water protection priorities and designing sampling programs. In fact, the Safe Drinking Water Act permits ground-water vulnerability assessments to be used to set such priorities (Williamson, 1994).

Table 9.2 also excludes studies that compared vulnerability predictions produced by different assessment routines, but did not compare them with actual field data (e.g., Rao and others, 1985; Jones and others, 1987; Banton and Villeneuve, 1989; Funari and others, 1991; and Soller, 1992). However, when Banton and Villeneuve (1989) compared the predictions of vulnerability assessments based on DRASTIC with those based on the PRZM model for a wide range of climatic, edaphic, and hydrogeologic conditions, no significant correlation was observed between the predictions generated by the two systems, even for the case of a conservative solute (Figure 9.7).

There is considerable variation in the degree to which existing vulnerability assessment approaches account for the influence of the physical and chemical characteristics of the solutes and the environmental settings of interest. Several exclusively empirical techniques are based on statistical associations between observed patterns of subsurface contamination and features such as land use (Greenberg and others, 1982), soil taxonomy (Teso and others, 1988; Carter, 1989; and Scheibe and Lettenmaier, 1989), or contaminant occurrence (Anderson and others, 1985; Lorber and others, 1989; and Martin and others, 1991). In contrast, process-based approaches focus on the physical and chemical factors controlling the transport and fate of the solutes of interest in the subsurface (Rao and others, 1985; Miller and others, 1989; and Meeks and Dean,

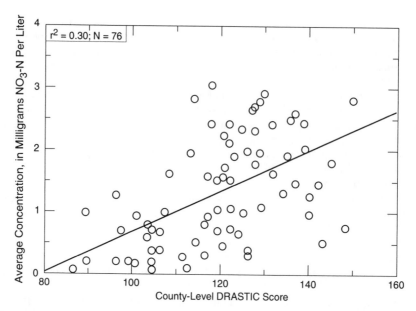

Figure 9.6. Relation between county-averaged nitrate concentrations and county-averaged DRASTIC scores for 76 Ohio counties. Redrawn with permission from Baker and others (1989). Concentration units not given in original figure, but presumed to be milligrams NO$_3$-N per liter.

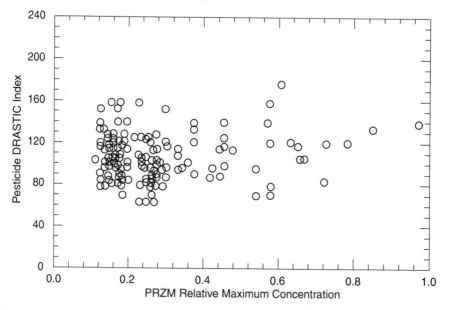

Figure 9.7. Relation between the maximum concentration in ground water predicted by PRZM for a conservative (i.e., nonsorbing, nonreactive) solute and the corresponding values of the pesticide DRASTIC index. Redrawn with permission from Banton and Villeneuve (1989).

1990). Other assessment methods assign semiarbitrary vulnerability scores (e.g., Aller and others, 1987; Roux and others, 1991b) or qualitative vulnerability ratings, such as "low," "moderate," and "high" (e.g., Lorber and others, 1989; Holden and others, 1992), to different areas on the basis of their hydrogeologic, edaphic, or climatic characteristics. The general features of these different approaches are described below.

Empirical Techniques

Empirically based assessment methods place minimal emphasis on the physical and chemical processes that govern the transport and fate of the constituents of interest. Although all three of the studies in Table 9.2 that employed discriminant analysis (Greenberg and others, 1982; Teso and others, 1988; and Troiano and Sitts, 1990) could distinguish pesticide-contaminated areas from uncontaminated areas with greater than 95 percent confidence ($\alpha = 0.05$), approaches of this type cannot predict subsurface contamination a priori. The predictive capabilities of these predominantly statistical methods are strongly dependent on the availability of extensive data on contaminant distributions in the media of interest. In locations where such information is not available, techniques of this type cannot be used.

The parameters whose variations are responsible for the differences in pesticide distributions detected by discriminant analysis (e.g., soil organic carbon, pesticide application rates, and climate) are usually unknown—indeed, this is the principal motivation behind the use of empirical, rather than process-based, methods for vulnerability assessment. However, the variables that govern pesticide distributions in one area may not be the same as those that control pesticide occurrence in other areas. Thus, another reason for the limited utility of discriminant analysis for predicting pesticide occurrence in the subsurface is that differences in the relative importance of specific parameters in governing pesticide occurrence in different locations are not recognized, and hence not accounted for.

Even when data on site properties are included, the chances of obtaining successful predictions may not be improved. The risk-based optimization scheme employed by Scheibe and Lettenmaier (1989) incorporated site-specific data, such as the size of the "capture zones" surrounding individual wells, the depth of the water table, and the distances from nearby wells known to be contaminated. Nevertheless, the accuracy with which it predicted EDB contamination in the ground waters of Whatcom County, Washington., was marginal ($r^2 < 0.13$).

Process-Based Approaches

Process-based vulnerability assessment methods minimize or avoid the use of purely empirical parameters. Systems of this type include the Leaching Potential Index, or LPI (Meeks and Dean, 1990); the pollutant loss-rate model (Bachmat and Collin, 1987); the Attenuation Factor, or AF; LEACH (Rao and others, 1985); and a number of other schemes, several of which have been described in previous reviews (Rao and others, 1985; Matthies, 1987). The LPI, AF, and pollutant loss-rate models account for the hydrogeologic properties of the subsurface, as well as the chemical and physical properties of the solutes of interest. LEACH focuses solely on four properties of the solute; water solubility, dissipation half-life, vapor pressure, and K_{oc}. However, the LPI appears to be the only process-based approach to have been tested against actual contaminant distributions in the field, and was the most successful among all of the vulnerability assessment schemes at predicting pesticide occurrence in ground water.

As with most process-based approaches (e.g., Rao and others, 1985; Bachmat and Collin, 1987; Leonard and Knisel, 1988; and Villeneuve and others, 1990), the LPI is derived from the advection-dispersion equation (see Section 4.3.3). The variables used to compute the LPI include spatially averaged estimates of the soil-organic-carbon content and bulk density of the subsurface solids, soil water content at "field [moisture] capacity," depth to the water table, and annual recharge rates from irrigation and precipitation. The LPI also accounts for the transformation rate and the soil-organic-carbon partition coefficient (K_{oc}) of the solute of interest (Meeks and Dean, 1990).

Predictions of ground-water vulnerability based on the LPI were compared by Meeks and Dean (1990) with the observed distribution of DBCP in ground waters near Stockton, California. Although it employed a comparatively small number of vulnerability categories (6) and was not significant at the $\alpha = 0.05$ level (Table 9.2), the correlation between the LPI and the frequency of occurrence of DBCP in ground water (Figure 9.8) also exhibited less scatter than did any of the other relations listed in Table 9.2 for pesticides ($r^2 = 0.52$), with the exception of the soil contamination study for Illinois agrichemical facilities (Krapac and others, 1993). The correlations shown in the table for the Illinois study, however, were based solely on aquifer depth beneath the sites of interest, and thus did not account for the various climatic, edaphic, and chemical parameters incorporated into the LPI.

Scoring Systems

A variety of methods employ vulnerability scores or qualitative ratings to describe the susceptibility of different areas to ground-water contamination. The most widely used of these systems is the DRASTIC scoring scheme (Aller and others, 1987), described previously. Because of its widespread use, DRASTIC has been subjected to more extensive evaluation than any other vulnerability assessment method (e.g., Alexander and Liddle, 1986; Curry, 1987; McKenna and Keefer, 1991; Soller, 1992; Rosen, 1994). Its application during the NPS and NAWWS studies

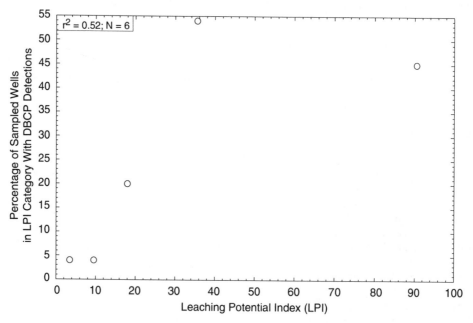

Figure 9.8. Percentage of sampled wells with a given Leaching Potential Index (LPI) in which DBCP was detected near Stockton, California (data from Meeks and Dean, 1990).

constitutes the most spatially extensive testing of any such system to date. As is evident from Table 9.2 and Figure 9.6, however, DRASTIC is not a reliable predictor of ground-water contamination. Various explanations have been proposed to account for the relatively poor fit that has been observed between DRASTIC predictions of ground-water vulnerability and observations of actual contamination. Several of these are examined below.

9.3.2 LIMITATIONS OF EXISTING VULNERABILITY ASSESSMENT METHODS

In order to provide reliable predictions of contaminant occurrence, any system for assessing the susceptibility of ground waters to anthropogenic contamination must: (1) account for the major physical and chemical factors that control the movement and fate of the contaminants of concern; (2) assign appropriate weights to these factors when a weighting scheme is employed; and (3) utilize input data of appropriate scale for its assessments. Further discussion of these requirements is pursued below. These and related issues have also been examined, either directly or indirectly, by other authors, including Alexander and Liddle (1986), Curry (1987), McKenna and Keefer (1991), Soller (1992), and Rosen (1994).

Neglect of Significant Controlling Factors

All of the existing systems for assessing ground-water vulnerability overlook one or more of the physical or chemical processes, or site characteristics, known to influence the occurrence, movement, and fate of pesticides and other contaminants in the subsurface (summarized in Chapter 4). DRASTIC neglects compound-specific properties, such as transformation half-life, K_{oc}, and Henry's Law constant. Meeks and Dean (1990) note that the LPI does not account for hydrodynamic dispersion, vapor-phase transport in the unsaturated zone, or preferential transport. In addition, none of the assessment schemes listed in Table 9.2 specifically incorporate pesticide use as an input variable, either among different areas or among different compounds. In relation to this issue, however, Meeks and Dean (1990) found that "positive detections [of DBCP] in ground water appear to correlate much more readily with the LPI than with pesticide-use records." This is consistent with the generally weak associations observed between pesticide use and the frequencies of pesticide detection in ground water, discussed at length in Section 3.4.

Results from the simazine study by Roux and others (1991b) illustrates the hazards of neglecting potentially important factors in assessing ground-water vulnerability. This investigation employed a scoring system to rank eleven different settings in seven states according to their "aquifer sensitivity," as determined from "soil sensitivity," average depth to water, and average hydraulic conductivity. Differences in recharge, either from precipitation or irrigation, were neglected, despite considerable variations in this parameter among the areas investigated.

Roux and others (1991b) encountered some of the highest simazine concentrations beneath the three sites deemed to be least vulnerable, while two of the three rated to be most vulnerable exhibited low concentrations. Two of the three most contaminated (and least "sensitive") sites, both located in heavily agricultural areas in Tulare and Fresno Counties of California, employed furrow irrigation. All of the other sampled areas employed either sprinkler irrigation, center-pivot irrigation, or no irrigation at all. As noted earlier, a study conducted by Troiano and others (1990)—also in Fresno County, California—showed that the rates of pesticide transport under furrow irrigation are substantially higher than those under sprinkler irrigation. Thus, differences in water flux through the surface, though not accounted for by the vulnerability assessment method of Roux and others (1991b), may have been responsible for their apparently anomalous results.

Many of the disparities observed between vulnerability predictions and observed contamination patterns (Table 9.2) may also result from the fact that none of the assessment methods appear to account for variations in well construction and well depth. As discussed in Section 6.5, the construction and depth of a well are two of the most important factors that can influence the likelihood of detecting pesticides in areas where the compounds are handled. Nevertheless, the use of these assessment methods presumes that a deeper, drilled well is as vulnerable to contamination as a shallow, dug well screened in the same hydrogeologic unit.

A vulnerability assessment may also fail to predict contamination encountered in a ground-water derived water supply if the effort focuses on a hydrogeologic unit other than that from which most of the used water is withdrawn. This point is demonstrated by the manner in which the DRASTIC system was used during the NPS to predict agrichemical contamination of ground water in northwestern Iowa. The DRASTIC assessment carried out for the NPS determined that the vulnerability of ground water to such contamination was "low" in this region of the state (Alexander and Liddle, 1986). The SWRL survey, however (Kross and others, 1990), observed higher frequencies of pesticide detection in the northwest than anywhere else in Iowa. (As noted in Section 3.4.4, descriptions of the spatial distributions of pesticide detections from the NPS do not appear to be available.) This pronounced disparity between predicted and observed patterns of pesticide contamination probably arose because the DRASTIC assessment focused on the principal bedrock aquifer underlying northwestern Iowa (the Dakota Sandstone). Most of the ground water used in the region, however, is drawn from shallow "seepage" wells screened in the surficial glacial drift, rather than in the regional aquifer (Hallberg, 1995). Such wells are much more vulnerable to surface-derived contamination than deep, drilled wells (Section 6.5), thus explaining the high rates of pesticide detection during the SWRL study.

This example illustrates the need for vulnerability assessments, such as DRASTIC, to incorporate more locally-based information, a point also raised by the Scientific Advisory Panel for the NPS (Hallberg and others, 1992a). For example, a map of the "groundwater vulnerability regions" of Iowa (Hoyer and Hallberg, 1991) has not yet been tested against field data on contaminant occurrence (although it is based, in part, on water-quality data), but it focuses more specifically on the hydrogeologic units from which most ground water is drawn in the state, and hence indicates a much higher degree of ground-water vulnerability in northwestern Iowa than did the DRASTIC assessment for the NPS.

Difficulties related to the neglect of significant controlling factors are not confined to arbitrary scoring systems. In their review of process-based vulnerability assessment schemes, Rao and others (1985) found that the relative rankings assigned to different pesticides with respect to their predicted tendencies to reach ground water depended, in part, on the parameters from which the rankings were derived.

Inappropriate Weighting Among Different Factors

As noted earlier, different environmental factors exert varying degrees of influence over the transport and fate of agrichemicals in the subsurface. Ground water vulnerability scoring systems account for these varying influences through the use of one or both of two semiarbitrary techniques: (1) the assignment of numerical values to specific environmental circumstances (e.g., different hydrogeologic settings), and (2) the assignment of different numerical weights to each of the factors of interest to reflect their relative levels of perceived importance to contaminant transport in the subsurface (e.g., hydrogeology versus recharge). Thus, even if an assessment scheme incorporates all of the variables that control the transport and fate of the contaminants of interest in the subsurface, its predictive power can be diminished—or eradicated—if the values selected for the individual parameters, or the relative weights assigned to them, are inappropriate.

Results from the monitoring study conducted in the Delmarva Peninsula by Koterba and others (1993) illustrate the potential errors associated with the assignment of inaccurate numerical scores to individual parameters. The poor correlations observed between pesticide detection frequencies in ground water and overall DRASTIC scores (Table 9.2) were echoed by nonsignificant correlations ($\alpha = 0.05$) between these detection frequencies and each of the seven component DRASTIC scores examined individually (data not shown). Similarly, Barrett and Williams (1989) found DRASTIC to be an inadequate predictor of ground-water contamination by atrazine in Nebraska, regardless of whether overall DRASTIC scores or values for individual DRASTIC components were used to map vulnerability. For both of these studies, the numerical scores used by DRASTIC to quantitatively describe the physical attributes of the systems of interest failed to provide accurate predictions of pesticide occurrence in ground water.

Even if different environmental circumstances are accurately represented by their assigned index values, results from the NPS (U.S. Environmental Protection Agency, 1992a) suggest that inappropriate weighting among different factors may cause the final DRASTIC scores to be unrepresentative of actual vulnerability. During the NPS, a significant, direct association was detected ($\alpha = 0.05$) between pesticide detection frequencies in community water-system wells and the DRASTIC vulnerability ratings based solely on depth to water. The latter parameter has also been found to exert a major influence over pesticide behavior during other studies (Section 6.3.5), including several in which contamination was successfully predicted from subsurface vulnerability (Meeks and Dean, 1990; Wylie and others, 1993; and Krapac and others, 1993). However, when the overall, county-level DRASTIC scores were compared with pesticide detection frequencies in the community water-system wells sampled during the NPS, no significant correlation was observed (Table 9.2). The simplest explanation for the disparity between the results observed for overall and individual DRASTIC ratings during the NPS is that the effect of water-table depth on ground-water vulnerability was obscured by the influence of other, less significant parameters, perhaps because one or more of the latter were weighted too heavily.

While the numerical weights assigned to individual parameters by DRASTIC are informed by "professional judgement," they remain unavoidably empirical in nature. As noted by Rosen (1994), "the qualitative weight functions in DRASTIC result in a simplistic index of unclear meaning that is less useful and less distinctive than desired." Furthermore, because of the relatively arbitrary manner in which individual scores are assigned, there are no fundamental physical, chemical, or biological bases upon which to assume that the resulting numerical ratings will be proportional to the likelihood of encountering contamination (Curry, 1987).

In contrast, the use of process-based vulnerability assessment systems does not involve any such weighting schemes. For example, the LPI (Meeks and Dean, 1990) and AF (Rao and others, 1985) are simply quantitative expressions of the influence of various physical and chemical parameters on pesticide transport and fate in the subsurface—subject to the simplifying assumptions that undergird the equations from which they were derived. Poor matches between the predictions supplied by these indices and observed contamination patterns must be addressed through the refinement of their defining equations and improvements in the reliability and completeness of the input data, rather than through the empirical adjustment of arbitrary ratings or parameter weights.

Contrasting Spatial Scales of Assessment and Observation

Many previous attempts to compare predictions of ground-water vulnerability with geographic distributions of observed contamination have confronted pronounced disparities in the spatial scales at which the two types of information are typically available. As noted in Section 5.4.2, the contamination of shallow ground water by surface-derived solutes is typically presumed to be associated with land use within 1 mi of individual wells. In contrast, many of the studies that have compared observed and predicted patterns of subsurface contamination have used county-based DRASTIC scores to characterize ground-water vulnerability (Table 9.2). Because the areas of individual counties in the United States may range from hundreds to tens of thousands of square miles, the scale on which ground-water vulnerability is assessed in these instances is at least two orders of magnitude larger than the spatial scales over which land use is believed to influence shallow ground-water quality. Such disparities in scale between assessments of vulnerability and contamination may help to explain why county-level DRASTIC scores have generally failed to predict the frequencies with which ground-water contamination is observed in different locations (Table 9.2).

The limited ability of DRASTIC to predict pesticide occurrence in ground water is not necessarily improved, however, when the scales at which vulnerability and contamination are assessed are both sufficiently small. Koterba and others (1993) found DRASTIC scores to be unreliable predictors of pesticide contamination in ground waters of the Delmarva Peninsula, even when they were computed for the 60 ha surrounding each well, an area that was close to the minimum of 40 ha (100 acres) recommended for these computations by the creators of DRASTIC (Aller and others, 1987). Similarly, during the NPS, sub-county level DRASTIC scores—in combination with estimates of cropping intensity on the sub-county level—were not found to be more closely associated with the frequency of nitrate contamination in rural domestic wells than were the county-level values (U.S. Environmental Protection Agency, 1992a). Furthermore, in the Big Spring Basin of Iowa, Curry (1987) failed to observe any significant relations between agrichemical concentrations and either the DRASTIC or the LEACH vulnerability indices (Table 9.2), despite the fact that these indices were computed on the basis of the subsurface characteristics within the 40 ha immediately surrounding each of the 19 wells of interest.

In contrast, process-based vulnerability assessments have provided relatively accurate predictions of subsurface contamination using data collected at relatively small spatial scales. The LPI values used by Meeks and Dean (1990) to predict ground-water contamination by DBCP near Stockton, California (Figure 9.7), were computed for individual 1-mi^2 sections. Similarly, using data collected for individual 0.8-ha areas, Wylie and others (1993) found that the process-based NLEAP(NL) model was able to predict nitrate concentrations in ground water in Weld County, Colorado, that were significantly correlated with observed concentrations (Table 9.2).

9.3.3 COMPARISONS AMONG DIFFERENT ASSESSMENT APPROACHES USED TO PROTECT GROUND WATER

A recent report by the National Research Council (NRC) Committee on Techniques for Assessing Ground Water Vulnerability (National Research Council, 1993) summarized six examples of efforts to evaluate the susceptibility of ground waters to contamination from surficial sources. As with the results compiled in Table 9.2, these case studies illustrate the range of approaches adopted in the United States for assessing ground-water vulnerability and, in some cases, for regulating land use on the basis of these assessments.

Table 9.3. Information employed for assessing ground-water vulnerability to contamination in six areas of the United States, as described by the National Research Council Committee on Techniques for Assessing Ground Water Vulnerability

[Adapted from National Research Council, 1993. USDA, U.S. Department of Agriculture]

Study Area	Information Employed for Ground-Water Vulnerability Assessment				Other References Consulted (where applicable)
	Output from Solute Transport Simulations	Vulnerability Indexing Schemes	Overlays of Ancillary Data (Hydrogeology, depth to water, soil type, etc.)	Measurements of Contaminant Concentrations in Ground Water	
California—San Joaquin Valley				X	
Florida—Statewide		X	X	X	
Iowa—Statewide			X		Hoyer and Hallberg, 1991
Nationwide (USDA)		X	X		Kellogg and others, 1992
Hawaii—Oahu	X	X	X		
Massachusetts—Cape Cod	X	X	X		

Table 9.3 illustrates the types of information used in each case for evaluating the susceptibility of ground waters to surface-derived contamination. Consistent with the trend identified earlier (Section 9.3.1), in only two cases (California and Florida) were measurements of contaminant occurrence in ground waters used for these assessments. In the other four cases, the assessments were based on the spatial variations of one or more subsurface properties, without the added benefit of monitoring data (National Research Council, 1993). With the exception of the work in Iowa, all of the approaches that employed overlays of spatially distributed data did so in conjunction with the use of a vulnerability indexing scheme, such as DRASTIC. In two cases (Oahu and Cape Cod), the use of indexing schemes was supplemented by predictions from solute-transport models.

Only one of the evaluations (Florida) integrated the results from ground-water sampling with predictions of ground-water vulnerability derived from other assessment techniques. This approach is similar to that employed by Lorber and others (1989) in their nationwide assessment for aldicarb. Each of the remaining five evaluations was based on either the direct observation (San Joaquin Valley) or the prediction of contamination, but not both. The paucity of effort devoted to basing these assessments on actual ground-water quality data, however, likely results from the high cost of obtaining such information, rather than from a perception that it is not valuable.

The strategies adopted in California and Iowa suggest a wariness of vulnerability assessments based on indexing schemes or computer simulations. With regard to California, the NRC noted that "the suitability and reliability of databases available for producing vulnerability assessments was a great concern before passage of the California Pesticide Contamination Prevention Act (PCPA) in 1985." In addition, "the use of models [during the California assessment] was not considered appropriate, given the available data and because no single model could cope with the circumstances in which contaminated ground-water sources were being discovered in the state." Furthermore, the NRC committee stated that:

the potential vagaries and uncertainties associated with more scientific approaches to vulnerability assessment, given the tools available when the PCPA was enacted, presented too large a risk for managers to consider endorsing their use. In contrast, the basic definition of the Pesticide Management Zone is difficult to challenge (pesticide contamination has been detected or not detected) in the legal sense. And the logic of investing economic resources in areas immediately surrounding areas of acknowledged contamination are [sic] relatively indisputable (National Research Council, 1993).

Similarly, in Iowa, the NRC report noted that "although more sophisticated approaches were investigated for use in the assessment, ultimately no complex process models of contaminant transport were used."

9.3.4 ITERATIVE APPROACHES TO VULNERABILITY ASSESSMENT

To date, some of the most successful and informative attempts to assess the vulnerability of ground waters to pesticide contamination over large geographic areas have involved the straightforward, iterative examination of the individual factors deemed most important in controlling pesticide behavior in the subsurface (e.g., Kross and others, 1990; Hallberg and others, 1992b; U.S. Environmental Protection Agency, 1992a; Koterba and others, 1993; and Baker and others, 1994). These approaches do not produce a single, numerical index for ranking different areas according to their vulnerability to contamination. Instead, their multi-tiered nature acknowledges that our understanding of the factors governing contamination is too rudimentary for a reliable index of this type to be devised at the present time.

Based on the data from their Cooperative Private Well Testing Program (CPWTP), Baker and others (1994) showed that the disparities in frequencies of herbicide occurrence above Health Advisory Levels between highly vulnerable and relatively protected ground waters increased monotonically as the effects of more vulnerability factors were accounted for. This trend was observed for both the triazine (Figure 9.9) and the acetanilide herbicides (Figure 9.10). Furthermore, all of the factors found by these authors to be significant in this regard have also been shown by other work to exert significant controls over pesticide occurrence and movement in the subsurface (see Sections 5.4.1, 6.2, 6.5.2, 8.8.4, and 8.10). The strength of the conclusions reached by Baker and others (1994) is further reinforced by the large number of wells sampled (Figures 9.9 and 9.10).

Koterba and others (1993) recorded similar observations during their statistical analysis of factors associated with pesticide detections in shallow ground waters of the Delmarva Peninsula. They found the likelihood of detecting pesticides to increase significantly ($\alpha = 0.05$) when local land use, soil type, and the depth of the well screen below the water table were taken into account (Table 9.2). Despite some differences, the parameters found by Koterba and others (1993) to be important to their analysis were consistent with those determined to be significant during the CPWTP (Figures 9.9 and 9.10).

9.3.5 FUTURE PROSPECTS

The advantages of a ground-water vulnerability assessment system that can accurately account for all of the factors deemed to be of importance through a single susceptibility index are

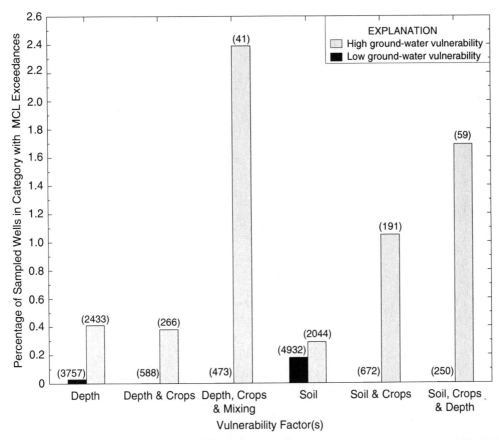

Figure 9.9. Percentage of wells sampled during the Cooperative Private Well Testing Program in which triazine herbicide concentrations exceeded the Maximum Contaminant Level (MCL) for atrazine (3 µg/L), relative to combinations of different vulnerability factors. Key to vulnerability factors (low versus high): "Depth," greater than 100 ft versus less than 50 ft; "Crops," cropland not within sight versus within 20 ft; "Mixing," chemicals not mixed near well versus mixed nearby; "Soil," clayey versus sandy. Number of wells in each category given in parentheses (data from Baker and others, 1994).

indisputable. However, despite its inherent appeal, a system of this type has not yet been devised. Nevertheless, systems designed to predict ground-water vulnerability to pesticide contamination are currently needed to safeguard ground-water quality for current and future use. The results discussed in this section suggest that process-based vulnerability assessment schemes based on rigorously derived, quantitative physical and chemical models of contaminant transport and fate—however imperfect—must be relied on for this purpose, rather than systems that quantify vulnerability on the basis of empirical, numerical indices, regardless of the degree to which the latter may be informed by "professional judgement." Results to date also indicate that the reliability of process-based vulnerability assessments improves when the data used for the assessments are collected over smaller spatial scales.

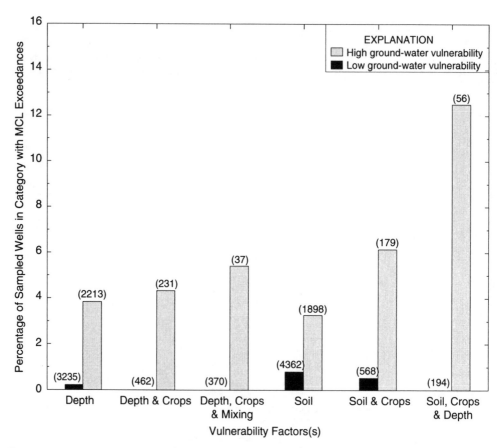

Figure 9.10. Percentage of wells sampled during the Cooperative Private Well Testing Program in which acetanilide herbicide concentrations exceeded the Maximum Contaminant Level (MCL) for alachlor (2 μg/L), relative to combinations of different vulnerability factors. Key to vulnerability factors (low versus high): "Depth," greater than 100 ft versus less than 50 ft; "Crops," cropland not within sight versus within 20 ft; "Mixing," chemicals not mixed near well versus mixed nearby; "Soil," clayey versus sandy. Number of wells in each category given in parentheses (data from Baker and others, 1994).

9.4 SUMMARY

This chapter has examined the three principal approaches that have been employed to predict the occurrence and spatial distributions of pesticides in the subsurface: (1) mathematical simulations of transport and fate; (2) the use of other solutes as indicators for the presence of pesticides in ground water; and (3) ground-water vulnerability assessments. Extensive effort has been invested in the mathematical simulation of pesticide transport and fate in the subsurface, but the computer models in common use for this purpose neglect several important factors known to exert significant control over pesticide behavior, particularly preferential transport and temperature. This may partly explain the relatively limited success with which such models have predicted pesticide transport and fate in the subsurface.

Several studies have examined the potential use of other solutes, such as nitrate and tritium, as indicators of pesticide contamination in ground water. Although some investigators have observed significant relations between nitrate and pesticide occurrence, the fact that many have failed to detect such relations limits the value of nitrate in this regard. Tritium is a reliable indicator of the potential presence of pesticides, but can only be used to distinguish ground waters that recharged before 1953—and are therefore unlikely to contain pesticides—from more recently recharged waters that may contain detectable pesticide concentrations.

Assessments of ground-water vulnerability to contamination have not yet proven to be reliable tools for identifying areas where pesticides are likely to be present in ground water. Although there are a number of possible reasons for this, quantitative assessment indices derived from the specific parameters known to control pesticide movement and fate in the subsurface have exhibited greater success at predicting subsurface contamination than scoring systems based on semiarbitrary evaluations of the importance of different controlling factors. Regardless of the approach used, however, it is clear that data on the physical and chemical characteristics of the subsurface environment (preferably over relatively small spatial scales), as well as on the physical and chemical properties of the solutes themselves, are indispensable for the accurate assessment of ground-water vulnerability to pesticide contamination.

CHAPTER 10

Environmental Significance

The contamination of ground water by pesticides is of concern for two principal reasons. First, elevated concentrations of pesticides in ground water may limit its use for drinking water. Second, pesticide-contaminated ground water may be a source of pesticides to surface waters, which support aquatic ecosystems and are also used as drinking-water supplies in many areas. Although either of these circumstances may be predominant in a given hydrologic system, the potential impacts on ground water tapped by drinking-water wells is of greatest public concern.

10.1 CONTRIBUTION TO SURFACE WATER

Available information on the contribution of ground water derived pesticides to surface water, discussed in Section 6.4.2, is limited only to a few hydrologic systems. Studies conducted to date, primarily in Iowa, indicate that the discharge of ground water to streams in agricultural areas can maintain consistently low concentrations of persistent dissolved pesticides in the streams during seasonal periods dominated by baseflow (Hallberg, 1987; Squillace and others, 1993). This may be an important factor affecting surface-water quality in regions where intensively farmed areas are located near streams that have seasonal periods in which shallow ground water is a dominant source of streamflow. The geographic extent of such settings is not well known, but these general conditions are common for lowland streams in many of the most intensively farmed agricultural areas of the United States. Furthermore, because pesticides are also used extensively and have been detected in ground water in a variety of non-agricultural settings (Section 3.3 and Chapter 7), ground waters may also discharge pesticides to surface waters in non-agricultural areas. The timing and degree of influence of pesticide contributions to streams by ground water can be expected to vary greatly among different systems depending on the hydrogeologic conditions, the amount and timing of recharge due to precipitation or irrigation, the types and degree of pesticide use, and seasonal patterns in the amounts and sources of streamflow.

For streams that experience low, but consistent concentrations of pesticides contributed from ground water during baseflow periods, the environmental significance of these contributions may be difficult to assess. Generally, concentrations in stream water during these periods will seldom exceed water-quality criteria for either aquatic life or drinking water. However, long-term chronic effects of year-round exposure on aquatic life have not been studied extensively. The potential significance to drinking-water quality is similarly difficult to judge. In some cases, annual mean concentrations computed for regulatory purposes may be affected and could trigger management decisions. Some individuals who presently choose to avoid drinking water from certain supplies during high-use seasons, when pesticides are known to be present but below regulatory standards, may choose to avoid the use of the water for a longer period of time during the year.

10.2 IMPORTANCE TO DRINKING-WATER QUALITY

Most concern about pesticides in ground water stems from their potential impact on drinking water supplied from wells. Table 10.1 compares the concentrations of individual pesticides detected in ground water during the multistate monitoring studies and the studies summarized in the Pesticides In Ground Water Database, or PGWDB (U.S. Environmental Protection Agency, 1992b) with water-quality criteria that have been established by the U.S. Environmental Protection Agency for the protection of human health (Nowell and Resek, 1994). The table also summarizes the frequencies with which either Maximum Contaminant Levels (MCLs) or lifetime health advisory levels (HAs) have been exceeded for individual compounds among the studies of interest (frequencies of pesticide detection by these studies were summarized earlier, in Tables 3.2 and 3.3.). Each water-quality criterion given in the table is the most recent one reported by Nowell and Resek (1994) for which a published rationale is available. With only one apparent exception (the PGWDB data for simazine), the MCL or HA values used by the different investigations to compute exceedances were identical to those reported by Nowell and Resek (1994). Because water-quality criteria are typically established for specific pesticides, rather than for pesticide groups, Table 10.1 excludes the results from the Cooperative Private Well Testing Program (Baker and others, 1994), which has used the less compound-specific immunoassay techniques to analyze ground waters for triazine and acetanilide herbicides (Section 2.8.2). As might be anticipated from the scope of the PGWDB in relation to the individual multistate studies (Table 2.4), the majority of the pesticide concentration data in Table 10.1 are derived from the PGWDB.

With few exceptions, pesticide concentrations measured during the studies listed in Table 10.1 exceeded MCL or HA values in fewer than 1 percent of the wells sampled for each compound of interest. Pesticides that exceeded MCL or HA criteria in more than one percent of the sampled wells were those for which extensive sampling has been carried out in areas of known or suspected contamination, i.e., aldicarb and its transformation products, arsenic, 1,2-dibromo-3-chloropropane (DBCP), and 1,2-dibromoethane (EDB). As with pesticide detections in general (Section 6.6.4), exceedances of water-quality criteria are expected to be more common if sampling is targeted toward such areas than if it is carried out in wells selected at random. This also explains why the frequencies of MCL exceedance for aldicarb, DBCP, and EDB were so much higher for the PGWDB data than for the National Pesticide Survey (NPS) results.

The data in Table 10.1 thus indicate that the concentrations of most agricultural pesticides have been found to be below existing criteria established for protecting human health in over 99 percent of the wells sampled in agricultural areas. However, this analysis is limited in several important ways. First, MCLs or other water-quality criteria have not been established for many pesticides and most transformation products, and existing criteria may be revised as more is learned about the toxicity of these compounds. Second, MCLs and other criteria are currently based on individual pesticides and do not account for possible cumulative effects if several different pesticides are present in the same well. Third, chemical analysis for many pesticides and most transformation products have not been carried out widely in ground water. Finally, very little sampling has been done in urban and suburban areas, where pesticide use is often high (see Sections 3.3.1 and 7.2). The widespread detection of pesticides in ground water at levels below current MCLs—particularly high-use compounds in vulnerable areas—therefore suggests that exceedances of water-quality criteria are likely to increase if existing criteria are lowered, as criteria are established for more compounds, as a wider range of pesticides and their transformation products are analyzed for, and as sampling expands to include more non-agricultural areas.

Table 10.1. Pesticide concentrations measured in ground water during multistate studies (Table 3.3) and studies from the Pesticides In Ground Water Database in relation to drinking-water-quality criteria

[Pesticide or Transformation Product: Compounds listed are those for which one or more of the multistate studies or the studies included in the Pesticides In Ground Water Database conducted analyses, and for which one or more of the indicated criteria were reported by Nowell and Resek (1994); Transformation products are indented. Criteria listed do not account for the protection of aquatic organisms. Ranges of Observed Concentrations: Numbers of significant figures given for concentrations in the original references have been retained; Concentration data for the Pesticides In Ground Water Database exclude values given in original reference (U.S. Environmental Protection Agency, 1992b) as "0," "trace" or "< X," where X is some maximum value. CWS, Community water system wells; HA, lifetime health advisory level for a 70-kg adult; MCL, maximum contaminant level (MCL values given are the most recent ones for which, according to Nowell and Resek [1994], a published rationale is available); MCPS, Midcontinent Pesticide Study (Kolpin and others, 1995; Kolpin, 1995b; data shown are based on all three years of sampling reported to date [1991–1993], unless otherwise indicated); MMS, Metolachlor Monitoring Study (Roux and others, 1991a); NA, not applicable; NAS, National Academy of Sciences; NAWWS, National Alachlor Well Water Survey (Holden and others, 1992; Klein, 1993); ND, not detected; NPS, National Pesticide Survey (U.S. Environmental Protection Agency, 1990a); NR, health advisories for exposure over the longer term are not recommended due to the carcinogenic risk associated with this compound; nsr, no standard or guideline reported by Nowell and Resek (1994) for this compound, unless stated otherwise; PGWDB, Pesticides In Ground Water Database (U.S.Environmental Protection Agency, 1992b); Q, Compound may have been present, but "[could not] be quantified or reliably detected" (U.S.Environmental Protection Agency, 1990a); RD, Rural domestic wells; USEPA, U.S. Environmental Protection Agency; Blank cells indicate no information applicable or available; <, less than; kg, kilograms; µg/L, micrograms per liter]

| Pesticide or Transformation Product | USEPA Water-Quality Criteria for Drinking Water (in µg/L) | | | | Ranges of Observed Concentrations (in µg/L) in Ground Water | | | | | | Frequencies of Exceedance of MCL or HA (if MCL not available) Among Sampled Wells, in percent | | | | | |
	MCL	HA	10-kg Child Acute (1 day)	10-kg Child Chronic (7 years)	MCPS	NPS CWS	NPS RD	NAWWS	MMS	PGWDB	MCPS	NPS CWS	NPS RD	NAWWS	MMS	PGWDB
Acifluorfen	nsr	nsr	2,000	100	ND	Q[1]				0.003-0.025	ND	NA	NA			NA
Alachlor	2	nsr	100	NR	0.003-4.270	ND	4.2	0.0385-6.185		0.006-3,000	[2]0.91	ND	0.13	0.02		0.38
Alachlor ESA	nsr	nsr	nsr	nsr	0.100-8.630						NA					
2,6-Diethyl-aniline	nsr	nsr	nsr	nsr	0.002-0.022					ND	NA					ND
Hydroxy-alachlor	nsr	nsr	nsr	nsr						0.910						NA
Aldicarb	3	1	1	1		ND	ND			0.08-1,264.00		ND	ND			4.6
Aldicarb sulfone	2	42	60	60		ND	ND			0.01-153.00		ND	ND			12

Table 10.1. Pesticide concentrations measured in ground water during multistate studies (Table 3.3) and studies from the Pesticides In Ground Water Database in relation to drinking-water-quality criteria—*Continued*

Pesticide or Transformation Product	USEPA Water-Quality Criteria for Drinking Water (in μg/L)				Ranges of Observed Concentrations (in μg/L) in Ground Water						Frequencies of Exceedance of MCL or HA (if MCL not available) Among Sampled Wells, in percent					
	MCL	HA	10-kg Child		MCPS	NPS		NAWWS	MMS	PGWDB	MCPS	NPS		NAWWS	MMS	PGWDB
			Acute (1 day)	Chronic (7 years)		CWS	RD					CWS	RD			
Aldicarb sulfoxide	4	9	10	10		ND	ND			0.01-1030.00		ND	ND			9.2
Aldrin	nsr	nsr	0.3	0.3		ND	ND			0.0052-21		ND	ND			[3]0.033
Ametryn	nsr	60	9,000	900	ND	ND	ND			0.01-0.200	ND	ND	ND			0
Arsenic	[4]50									1.6-680.0						18
Atrazine	3	3	100	50	0.003-2.090	Q-0.92	Q-7.0	0.03-6.719		0.001-1,500	0	0	0.13	0.1		0.64
Deethyl atrazine	nsr	nsr	nsr	nsr	0.002-2.320		ND			0.05-2.860	NA	ND	ND			NA
Deisopropyl atrazine	nsr	nsr	nsr	nsr	0.050-1.170					0.100-3.540	NA					NA
Baygon (propoxur)	nsr	3	40	40		ND	ND			2.0-35.0		ND	ND			0.019
Bentazon	nsr	20	300	300	ND	ND	2.9			0.10-41.89	ND	ND	0			0.28
α-BHC	nsr	nsr	50	50	ND	ND	ND			[5]0.0014-0.16	ND	ND	ND			NA
β-BHC	nsr	nsr	nsr	nsr		ND	0.04					ND	NA			
δ-BHC	nsr	nsr	nsr	nsr		Q[1]						NA	NA			
γ-BHC (lindane)	0.2	0.2	1,200	33		ND	Q-0.42			0.0006-180.000	0	ND	0.13			0.045
Bromacil	nsr	90	5,000	3,000		ND	ND			0.03-951.6		ND	ND			0.12
Butylate	nsr	350	2,000	1,000	ND	ND	ND			0.87-2.23		ND	ND			0

Table 10.1. Pesticide concentrations measured in ground water during multistate studies (Table 3.3) and studies from the Pesticides In Ground Water Database in relation to drinking-water-quality criteria—*Continued*

| Pesticide or Transformation Product | USEPA Water-Quality Criteria for Drinking Water (in µg/L) | | | | Ranges of Observed Concentrations (in µg/L) in Ground Water | | | | | | Frequencies of Exceedance of MCL or HA (if MCL not available) Among Sampled Wells, in percent | | | | | |
| | | | 10-kg Child | | | NPS | | | | | | NPS | | | | |
	MCL	HA	Acute (1 day)	Chronic (7 years)	MCPS	CWS	RD	NAWWS	MMS	PGWDB	MCPS	CWS	RD	NAWWS	MMS	PGWDB
Carbaryl	nsr	700	1,000	1,000	ND	ND	ND			0.03-610.00	ND	ND	ND			0
Carbofuran	40	36	50	50	ND	ND	ND			0.01-176.00	ND	ND	ND			0.26
Carboxin	nsr	700	1,000	1,000		ND	ND			ND		ND	ND			ND
Chloramben	nsr	100	3,000	200		Q[1]				1.00		0	0			0
α-Chlordane	[6]2	nsr	[6]60	[6]0.5		Q-0.01	Q-0.01			[6]0.01-20.000		[6]0	[6]0			[6]0.20
γ-Chlordane		nsr				ND	Q-0.01									
Chlorothalonil	nsr	nsr	200	200						0.140-1.100						[3]0.18
Chlorpyrifos	nsr	20	30	30	0.005-0.024					0.05-0.654	0					0
Cyanazine	nsr	1	100	20	0.010-0.880	ND	ND	0.1205-0.1485		0.002-29.0	0	ND	ND	0		0.29
Cyanazine amide	nsr	nsr	nsr	nsr	0.050-0.550						NA					
2,4-D	70	70	1,000	100	0.100-0.890	ND	ND			0.0079-57.1	0	ND	ND			[3]0.33
Dacthal (DCPA)	nsr	4,000	80,000	5,000	ND	ND	ND			0.010-300.0	ND	ND	ND			0
DCPA hydrolysis products	nsr	nsr	nsr	nsr	0.010-2.220	Q-7.2	Q-2.4			0.21-431.0	NA	NA	NA			NA
Dalapon	200	200	3,000	300		Q[1]				ND		0	0			ND
Diazinon	nsr	0.6	20	5	ND	Q[1]	ND			0.01-3.2	ND	ND	0			0.051

Table 10.1. Pesticide concentrations measured in ground water during multistate studies (Table 3.3) and studies from the Pesticides In Ground Water Database in relation to drinking-water-quality criteria—*Continued*

| Pesticide or Transformation Product | USEPA Water-Quality Criteria for Drinking Water (in µg/L) | | | | Ranges of Observed Concentrations (in µg/L) in Ground Water | | | | | | Frequencies of Exceedance of MCL or HA (if MCL not available) Among Sampled Wells, in percent | | | | | |
| | MCL | HA | 10-kg Child | | MCPS | NPS | | NAWWS | MMS | PGWDB | MCPS | NPS | | NAWWS | MMS | PGWDB |
			Acute (1 day)	Chronic (7 years)		CWS	RD					CWS	RD			
1,2-Dibromo-3-chloropropane (DBCP)	0.2	nsr	200	NR		Q	0.48-0.71			0.001-8000.00		0	0.27			5.5
Dicamba	nsr	200	300	300	0.100	ND	ND			0.006-44.0	0	ND	ND			0
1,3-Dichloropropene	nsr	nsr	[6]30	[6]30		ND	ND			[6]0.279-140		ND	ND			NA
Dieldrin	nsr	nsr	0.5	0.5	ND	ND	ND			0.001-2.600	ND	ND	ND			[3]0.095
Dinoseb	7	7	300	10		3.5	ND			0.008-47.00		0	ND			0.59
Diphenamid	nsr	200	300	300		ND	ND			ND		ND	ND			ND
Diquat	20	20	nsr	nsr						ND						ND
Disulfoton	nsr	0.3	10	3	ND	Q[1]				0.04-100.00	ND	0	0			0.61
Diuron	nsr	10	1,000	300		ND	ND			0.01-5.37		ND	ND			0
Endothall	100	140	800	200						ND						ND
Endrin	2	2	20	3		ND	ND			0.001-3.5		ND	ND			0.024
Ethylene dibromide (EDB)	0.05	nsr	8	NR		ND	0.29			0.001-15,772.4		ND	0.13			10.6
Ethylene thiourea (ETU)[7]	nsr	nsr	300	100		ND	Q-16			0.725		ND	NA			NA
Fenamiphos	nsr	2	9	5		ND	ND			ND		ND	ND			ND
Fluometuron	nsr	90	2,000	2,000		ND	ND			0.8-5.000		ND	ND			0

Table 10.1. Pesticide concentrations measured in ground water during multistate studies (Table 3.3) and studies from the Pesticides In Ground Water Database in relation to drinking-water-quality criteria—*Continued*

| Pesticide or Transformation Product | USEPA Water-Quality Criteria for Drinking Water (in µg/L) | | | | Ranges of Observed Concentrations (in µg/L) in Ground Water | | | | | | Frequencies of Exceedance of MCL or HA (if MCL not available) Among Sampled Wells, in percent | | | | | |
| | MCL | HA | 10-kg Child | | MCPS | NPS | | NAWWS | MMS | PGWDB | MCPS | NPS | | NAWWS | MMS | PGWDB |
			Acute (1 day)	Chronic (7 years)		CWS	RD					CWS	RD			
Fonofos	nsr	10	20	20	ND					0.007-0.90	ND					0
Glyphosate	700	700	20,000	1,000						0.004-150.0						0
Heptachlor	0.4	nsr	10	5.0		ND	ND			0.001-0.8		ND	ND			0.12
Heptachlor epoxide	0.2	nsr	nsr	0.1		ND	ND			0.01-0.22		ND	ND			0.032
Hexachlorobenzene	1	nsr	50	50		Q-0.17	ND			0.0039-0.0056		0	ND			0
Hexazinone	nsr	200	3,000	3,000		ND	ND			0.060-0.720		ND	ND			0
Malathion	nsr	200	200	200	ND					0.007-6.17	ND					0
MCPA	nsr	10	100	100						0.13-5.5						0
Methomyl	nsr	200	300	300		ND	ND			1.0-20.00		ND	ND			0
Methoxychlor	40	40	50	50		ND	ND			0.01-0.312		ND	ND			0
Methyl parathion	nsr	2	300	30	ND					0.01-0.256	ND					0
Metolachlor	nsr	100	2,000	2,000	0.003-1.460	ND	ND	0.0375-3.805	(8)	0.02-157.00	0	ND	ND	0	0	0.013
Metribuzin	nsr	200	5,000	300	0.050-0.220	ND	ND			0.001-25.10		ND	ND			0
Metribuzin DA	nsr	nsr	nsr	nsr		ND	ND			ND		ND	ND			ND

Table 10.1. Pesticide concentrations measured in ground water during multistate studies (Table 3.3) and studies from the Pesticides In Ground Water Database in relation to drinking-water-quality criteria—*Continued*

Pesticide or Transformation Product	USEPA Water-Quality Criteria for Drinking Water (in µg/L)				Ranges of Observed Concentrations (in µg/L) in Ground Water						Frequencies of Exceedance of MCL or HA (if MCL not available) Among Sampled Wells, in percent					
			10-kg Child			NPS						NPS				
	MCL	HA	Acute (1 day)	Chronic (7 years)	MCPS	CWS	RD	NAWWS	MMS	PGWDB	MCPS	CWS	RD	NAWWS	MMS	PGWDB
Metribuzin DADK	nsr	nsr	nsr	nsr		Q¹				ND		NA	NA			ND
Metribuzin DK	nsr	nsr	nsr	nsr		Q¹				ND		NA	NA			ND
Oxamyl	200	200	200	200		ND	ND			0.01-395.00		ND	ND			0.013
Paraquat	nsr	30	100	50						0.01-100.0						0.21
Pentachloro-phenol	1	NR	1,000	300		ND	ND			0.001-0.64	0	ND	ND			0
Picloram	500	500	20,000	700	0.010-0.030	ND	ND			0.01-30.0	0	ND	ND			0
Prometon	nsr	100	200	200	0.050-1.350	Q	Q-0.57			0.05-29.6	0	0	0			0
Pronamide	nsr	50	800	800	ND	Q¹				ND	ND	0	0			ND
Propachlor	nsr	90	500	100	0.002	ND	ND			0.02-3.5	0	ND	ND			0
Propazine	nsr	10	1,000	500	ND	ND	ND			0.01-0.20	ND	ND	ND			0
Propham	nsr	100	5,000	5,000	ND	ND	ND			6.000	0	ND	ND			0
Simazine	4	4	500	50	0.002-0.270	Q-0.76	Q	0.043-8.359		0.001-67.0	0	0	0	< 0.01		[9]0.40
2,4,5-T	nsr	70	800	300	0.020	ND	ND			0.01-2.99	0	ND	ND			0
Tebuthiuron	nsr	500	3,000	700	0.050	ND	ND			20.700-380.0	0	ND	ND			0
Terbacil	nsr	90	300	300	ND	ND	ND			0.3-8.9	ND	ND	ND			0

Table 10.1. Pesticide concentrations measured in ground water during multistate studies (Table 3.3) and studies from the Pesticides In Ground Water Database in relation to drinking-water-quality criteria—*Continued*

Pesticide or Transformation Product	USEPA Water-Quality Criteria for Drinking Water (in µg/L)		10-kg Child		Ranges of Observed Concentrations (in µg/L) in Ground Water	NPS					Frequencies of Exceedance of MCL or HA (if MCL not available) Among Sampled Wells, in percent	NPS				
	MCL	HA	Acute (1 day)	Chronic (7 years)	MCPS	CWS	RD	NAWWS	MMS	PGWDB	MCPS	CWS	RD	NAWWS	MMS	PGWDB
Terbufos	nsr	0.9	5	1	ND	Q[1]				0.02-20.0	ND	0	0			0.19
Toxaphene	3	nsr	500	nsr						1.15-18.000						0.070
2,4,5-TP (silvex)	50	50	200	70	ND	ND	ND			0.002-1.4	ND	ND	ND			0
Trifluralin	nsr	5	80	80	0.008-0.016	ND	ND			0.0018-14.890	0	ND	ND			0.018

[1]Detection reported in the overall summary of analytical results (U.S. Environmental Protection Agency, 1990a), but not in the data reported for the individual wells sampled for the NPS (U.S. Environmental Protection Agency, 1994b). No distinction made between detections in CWS and RD wells.

[2]Value for frequency of exceedance of MCL for alachlor pertains to 1993 sampling only (Kolpin and Thurman, 1995). No samples exceeded the MCL for alachlor in 1991 or 1992 (Kolpin and others, 1995).

[3]Basis for reported MCL or HA exceedance not provided in original reference (U.S. Environmental Protection Agency, 1992b).

[4]Source: U.S. Environmental Protection Agency (1992b).

[5]Original reference did not differentiate between α, β, and δ isomers.

[6]No isomers specified.

[7]Ethylenebis(dithiocarbamate) fungicide transformation product.

[8]Compound detected, but full concentration range not reported.

[9]Exceedance frequency based on an MCL (1 µg/L) that has been superceded by a higher value of 4 µg/L (Nowell and Resek, 1994) since the publication of the Pesticides In Ground Water Database (U.S. Environmental Protection Agency, 1992b).

CHAPTER 11

Summary and Implications

Over the past two decades, the occurrence, spatial distributions, transport, and fate of pesticides in the subsurface have received much attention. This book summarizes current understanding regarding pesticides in the ground waters of the United States, and the various natural and anthropogenic factors that control their movement and fate in the subsurface. The discussion also examines the current status of efforts to predict the occurrence and behavior of pesticides in ground water.

Previous reviews have generally adopted one of two approaches. Many summarized the pesticides detected in ground waters, the general areas where these detections have occurred, and the concentrations measured, but with little emphasis on the specific natural and anthropogenic factors likely to have given rise to these observations. Others summarized the various physical, chemical, and biological processes believed to control the subsurface transport and fate of pesticides, but with only minor attention to the experimental and field data that illustrate, or fail to illustrate, these processes.

The present work combines these two approaches by using the results from large-scale sampling surveys (monitoring studies), field-scale investigations (process and matrix-distribution studies), and in some instances, laboratory studies, to gain a better understanding of the natural and anthropogenic factors governing pesticide occurrence, movement, and fate in the subsurface. Despite the value of laboratory experiments and computer simulations for formulating and testing hypotheses, conclusions regarding the influence of each of these factors on pesticide behavior in the subsurface must ultimately be derived from field observations. However, because of the difficulty of controlling for the confounding effects of other variables in the field, such conclusions are necessarily based on a preponderance of evidence from studies conducted at different spatial scales and in a variety of different settings, rather than on unambiguous demonstrations of cause and effect.

The influence of such confounding factors is often not accounted for in existing investigations because of: (1) study design (i.e., the factor was not considered or could not be examined as a result of time or budget constraints); (2) limitations on available information; or (3) unavoidable, natural covariation with the parameter of interest. Variables whose influence was often unaccounted for because of study design or limitations on available information include pesticide application rates, agricultural management practices, well characteristics (e.g., well depth, well construction, and depth of screen below the water table), hydrogeologic setting, and analytical detection limits. Variables that are so intimately correlated with other factors that their separate influence has not yet been clearly discerned include the organic carbon content, permeability and microbial activity of soils; the timing of recharge and temperature variations during application periods (particularly during the spring); and land-use patterns in areas dominated by a single predominant landscape (e.g., agriculture or urban development). These and other examples are summarized later in this chapter.

11.1 SPATIAL AND TEMPORAL PATTERNS OF PESTICIDE OCCURRENCE IN GROUND WATER

A considerable body of data has been collected over the past decade on the occurrence of pesticides in ground waters of the United States. At least four monitoring studies have documented subsurface contamination by pesticides across large, multistate regions. In addition, many states have completed or are carrying out multicounty or statewide surveys of pesticide residues in ground water. Results from these and other monitoring efforts indicate that pesticides from every major chemical class have been detected in the ground waters of the United States.

However, substantial variations in study design among different monitoring investigations hinder any attempt to obtain a consistent picture of the spatial distributions of pesticides in ground waters across the nation at the present time. Because they have been carried out independently, different monitoring investigations have exhibited considerable variations in several key design features, including: (1) the spatial extent of sampling; (2) the types and number of compounds examined; (3) the analytical detection limits employed for a given compound; and (4) the criteria used for well selection. Such variations in design make it difficult to compare results among different studies because each of these parameters can affect the frequencies with which the target analytes are detected. Both the spatial extent of ground-water contamination by pesticides and the number of different pesticide-derived compounds detected in the subsurface (though not necessarily detection frequencies) tend to be greater in areas where greater effort has been expended to find them. Among studies conducted within the same region, detection frequencies for a given pesticide generally increase with decreasing analytical detection limits. Furthermore, studies that focus their sampling in areas of known or suspected contamination yield higher frequencies of detection than those that sample randomly selected wells.

The designs of previous monitoring studies also make it difficult to determine whether the severity or the extent of ground-water contamination by pesticides have changed significantly over the past two decades. This difficulty arises from a paucity of long-term studies designed specifically to address either question. In addition, long-term trends in pesticide occurrence in ground water may be obscured by the substantial seasonal variations in pesticide detection frequencies and concentrations during the year; minimum values of each are typically encountered before, and maximum values after, spring pesticide applications. The separate influences of temperature, recharge, and pesticide applications in controlling these seasonal variations have not been elucidated, primarily because spring applications often coincide with periods of low temperature and significant recharge (from precipitation or irrigation).

11.1.1 OCCURRENCE AND DISTRIBUTION OF PESTICIDES IN RELATION TO AGRICULTURAL USE

In a very general sense, the likelihood of detecting agricultural pesticides in ground water tends to increase directly with an increase in their use on crops. Frequencies with which a given pesticide is detected in ground water are typically low in areas where its use is limited. This is to be expected, given the fact that pesticides, as synthetic compounds, have no natural sources; natural substances that have been used as pesticides, such as copper, sulfur, and various arsenic compounds, were largely excluded from this book. Conversely, for a given pesticide, areas where detection frequencies are high are frequently those of greater use. Detections of triazine and acetanilide herbicides and their transformation products, for example, are widespread in the ground waters of the corn and soybean regions of the midcontinent, where these compounds are used most extensively.

Low frequencies of pesticide detection are often encountered, however, in areas of high use. Extensive pesticide use is therefore a necessary, but not sufficient, condition for encountering substantial pesticide contamination in the subsurface. Other factors such as study design, physical and chemical properties of the compounds of interest, environmental setting, and well construction may exert significant influences over the likelihood with which pesticides are detected in ground water. Deviations from direct relations between occurrence and use may also be related to the incomplete and often inconsistent nature of pesticide use data for most of the nation.

11.1.2 PESTICIDE TRANSFORMATION PRODUCTS IN GROUND WATER

It is now widely recognized that a decrease in the measured concentration of a pesticide in the subsurface over time—i.e., its "dissipation"—is rarely caused by its complete mineralization to simple substances such as water, carbon dioxide, inorganic ions, and nitrogen gas. Instead, the dissipation of a pesticide in situ arises from the simultaneous action of a variety of processes that usually lead to relatively minor modifications in its structure, or its transport away from the point of measurement. Because of the value of transformation products for understanding the ultimate fate of applied pesticides, as well as their potential impact on ground-water quality, the importance of determining the occurrence and distribution of these compounds has gained wider recognition in recent years. Consequently, the analysis of pesticide transformation products, though relatively rare among previous ground-water monitoring studies, is becoming more common. With this expanded effort has come the discovery that in many areas, it is the transformation products, rather than the parent compounds, that are detected most frequently in ground water.

11.1.3 PESTICIDES IN GROUND WATER BENEATH NON-AGRICULTURAL AREAS

The contamination of subsurface waters by pesticides and their transformation products arises from the use of the parent compounds not only in agricultural areas, but in a wide variety of non-agricultural settings, as well. These include lawns and golf courses; residential, commercial, and industrial areas; rights of way; timber production and processing areas; public gardens; flower production operations; and rangelands. Contrasts in the assemblages of pesticides detected in each of these settings reflect the variations in the types of compounds used.

Non-agricultural pesticide use may be substantial in some areas. The amounts of pesticides applied to lawns and golf courses in the United States, for example, may be comparable to the quantities used in most agricultural settings, suggesting a similar impact on ground-water quality. In addition to use, another indication of the potential significance of non-agricultural applications is the fact that the pesticide detected most frequently during the U.S. Environmental Protection Agency's (USEPA) National Pesticide Survey (NPS) was dacthal (DCPA), an herbicide applied primarily for non-agricultural purposes.

11.1.4 GROUND-WATER CONTAMINATION BY POINT SOURCES

Studies of ground-water quality near point sources have shown subsurface contamination by pesticides to be acute in the vicinity of agrichemical production and handling facilities, as well as near locations on farms where accidental releases have occurred. The frequencies of pesticide detection in ground water at agrichemical facilities and the frequencies of accidental release on farms were both found to be directly related to the rates at which these compounds are used in agriculture. By contrast, the pesticides found to contaminate ground waters beneath waste disposal sites in the United States are predominantly those used in homes and gardens, rather than in agriculture; the more frequently a given pesticide is applied in residential settings, the larger the proportion of waste disposal sites at which it has been detected in ground water.

In addition to agrichemical handling facilities, locations where accidental releases have occurred, and waste disposal sites, several other types of point sources have been found to be associated with pesticide contamination of ground water. The use of artificial-recharge structures to replenish ground-water supplies may lead to subsurface contamination by pesticides if these compounds are present in the recharge water. In addition, wells used for the disposal of agricultural drainage or urban runoff also appear to be responsible for pesticide contamination in nearby ground waters. In predominantly agricultural areas, some point sources—most notably, agrichemical handling facilities and agricultural drainage wells—are sufficiently numerous that their impact on ground-water quality may be widespread. In these situations, the sources in question have been referred to as "quasi-point sources."

The detection of agricultural pesticides in ground water becomes more likely with increasing proximity to a variety of localized sources on a farm—including agricultural fields and agrichemical handling areas—as well as with increases in the intensity of nearby agricultural activity. In predominantly agricultural regions, however, these relations may be difficult to detect because nearly all of the land in such areas is (or was) under cultivation.

For reasons that remain unclear, herbicides are detected more frequently with increasing proximity to feedlots and other areas where animal manures accumulate. This trend might have been expected if herbicide residues were commonly present in livestock manure (although data do not appear to be available to address this), but it might also be a spurious correlation arising from the fact that feedlots are often located close to agrichemical mixing and handling areas on farms. The occurrence of veterinary pesticides in ground water near confined feeding areas, though of potential concern, does not appear to have been investigated. In accordance with expectation, however, pesticide detections in ground water do not appear to be spatially correlated with septic systems.

Several monitoring studies have sought to distinguish pesticide detections in ground water arising from point sources from those caused by nonpoint sources of contamination. At least three different criteria have been used to make this distinction: (1) spatial patterns of contamination; (2) transiency of pesticide detections in individual wells; and (3) severity of contamination. Comparisons of point- and nonpoint-source contamination of ground waters by pesticides, however, indicate that none of these criteria can reliably distinguish point- from nonpoint-source contamination.

11.2 PHYSICAL AND CHEMICAL FACTORS GOVERNING TRANSPORT AND FATE

The principal physical and chemical processes that control the transport and fate of pesticides in subsurface waters have been known for at least two decades, primarily on the basis of laboratory studies. Despite the considerable body of research that has identified these processes, however, only moderate efforts have been invested in using these results to explain the behavior of pesticides in situ. This has been especially the case with respect to the preferential transport and transformation of these compounds in the subsurface.

11.2.1 PARTITIONING AND TRANSPORT

As is the case for other semivolatile solutes, the movement of most pesticides in the subsurface is controlled largely by the movement of water, and occurs through the combined action of advection, hydromechanical dispersion, and molecular diffusion. (In contrast, as well as by design, fumigant migration in the unsaturated zone is governed by both aqueous- and vapor-phase transport.) Soil organic matter slows the migration of pesticides in proportion to their hydrophobicity. The greater affinity of most pesticides for organic matter, relative to water, is demonstrated by the observation that the fractions of applied pesticides observed to be associated with soils and subsurface solids are generally one to two orders of magnitude larger than those remaining in the aqueous phase. Patterns of pesticide detection at agrichemical handling facilities are consistent with this pattern; the pesticides associated only with soils at such facilities exhibit consistently higher soil organic-carbon partition coefficient (K_{oc}) values than compounds detected in the ground water.

Most efforts to predict ground-water contamination or to simulate subsurface transport have been based on the assumption that solutes migrate through the soil and vadose zones in "local equilibrium" with respect to partitioning between the aqueous, solid (i.e., sorbed), and gas phases in the subsurface. Preferential transport occurs when the partitioning of the solute of interest fails to reach equilibrium with respect to all phases present (aqueous, solid, and vapor) as it migrates down-gradient. Much of the aqueous phase in subsurface systems, particularly in fine-grained deposits, may be "immobile," and hence relatively inaccessible to solutes being carried in more "mobile" zones. As a result, the actual volume of water in which solutes are transported through the subsurface may be much smaller than originally thought, leading to what appear to be anomalously rapid migration rates. Failure to account for preferential transport may explain many of the discrepancies that have been encountered between predicted and observed pesticide behavior in the subsurface.

11.2.2 CHEMICAL TRANSFORMATIONS

As they migrate through the subsurface, pesticides may undergo chemical transformation, with or without the aid of microorganisms. Extensive work with laboratory microcosms has established that environmental factors such as temperature, pH, redox regime, nutrient concentrations, the presence of other reactive chemical species in solution, the size and composition of the resident microbial population, and the physical and chemical properties of aquifer solids may all influence the rates and pathways of these reactions in situ.

Despite the recent interest in the occurrence of pesticide transformation products in ground waters, most studies of pesticide behavior and fate in the subsurface, including modeling investigations, have given only minor attention, if any, to the various biotic and abiotic factors that govern the rates and pathways of transformation of these compounds in situ. Monitoring and process and matrix-distribution studies, for example, seldom report the temperatures at which their observations are conducted, and only a few studies to date have attempted to provide a complete mass balance among parent compound and transformation products in situ. The general neglect of the factors that control transformation rates, coupled with the relative scarcity of data on transformation products in the systems of interest, restrict efforts to reach a complete understanding of the subsurface fate of pesticides applied to the land.

11.2.3 COMPARISONS OF LABORATORY-BASED OBSERVATIONS WITH MONITORING RESULTS

It is commonly assumed that pesticides exhibiting greater mobility or persistence in the subsurface are more likely to be detected in ground water, while those that are more hydrophobic or reactive are less apt to be detected. However, neither hydrophobicity, nor soil dissipation half-life—nor the two parameters in combination—can successfully distinguish the compounds detected in ground water from those that have not been detected during monitoring studies that have examined broad ranges of pesticides across extensive areas. Water solubility is also an unreliable predictor of pesticide mobility or occurrence in the subsurface, although it has been invoked repeatedly for these purposes.

The fact that pesticides from every major chemical class have been detected in ground water casts additional doubt on the assumption that only pesticides that are more mobile in the aqueous phase (i.e., more hydrophilic) or more persistent are likely to reach the water table in detectable concentrations. These discrepancies between prediction and observation may be caused by several factors, including: (1) preferential transport; (2) the substantial variability of published values for K_{oc} and dissipation half-lives; and (3) variability in the initial concentrations of the pesticides of interest following application (caused mainly by variations in application rates), in relation to their respective analytical detection limits.

11.3 INFLUENCE OF ENVIRONMENTAL SETTING ON PESTICIDE BEHAVIOR

The effects of climate, soil type, and hydrogeologic setting on pesticide occurrence are generally consistent with the basic principles of pesticide behavior in the subsurface, although such effects are often difficult to discern from available information. In addition, despite the obvious importance of these factors in controlling the transport and fate of pesticides, they have received surprisingly limited attention in the published literature.

11.3.1 CLIMATE

The climatic variables of greatest importance in controlling pesticide movement and fate in the subsurface are temperature and precipitation. Although, as noted earlier, few field investigations have reported temperature data, geographic patterns of dissipation rates across the United States suggest that pesticide persistence in the subsurface increases with decreasing temperature, as might be expected. Also consistent with expectation is the observation that pesticide contamination of ground water becomes more likely with increased precipitation, particularly if recharge occurs soon after application. The importance of precipitation in facilitating the entry of pesticides into the subsurface appears to be more pronounced for more persistent compounds.

11.3.2 SOIL CHARACTERISTICS

The two soil properties that exert the strongest influence on pesticide movement in the subsurface are permeability and organic carbon content (f_{oc}), parameters that typically exhibit an inverse correlation with one another. Definitive conclusions regarding the separate influence of permeability and organic carbon content on pesticide behavior in the subsurface are elusive, however—none of the reviewed studies examined the effect of one parameter while specifically controlling for the other, presumably because of their mutual interdependence in natural systems. Nevertheless, the limited evidence suggests that pesticide contamination of ground water is more likely to occur where surficial soils exhibit high permeability (and, typically, lower f_{oc}) than where they exhibit lower permeability (and higher f_{oc}). Here, the influence of organic carbon may arise not only from its effect on pesticide mobility, but also from the fact that microbial populations, and hence biotransformation rates, in the subsurface typically increase with f_{oc}. More data on transformation product occurrence are needed to determine the relative importance of transport and transformation in this regard.

11.3.3 HYDROGEOLOGIC SETTING

Hydrogeologic features affect pesticide detection frequencies primarily through their influence on the time required for applied compounds to travel from their point of entry at the surface to the well screen; the longer this time interval, the less likely that pesticides will be detected in ground water. Most of the available data suggest that, as might be expected, pesticide contamination of ground water is more common in unconsolidated deposits than in bedrock at similar depths. (The principal exception to this pattern was observed in Maine where the opposite pattern emerged, perhaps because of the influence of fracture flow.) Pesticide contamination of ground water also appears to be relatively common in agricultural areas underlain by karst. By contrast, pesticides are generally detected in ground water less frequently beneath thick sequences of low permeability materials such as glacial tills and lacustrine clays.

Temporal variations in pesticide detection frequencies also appear to be lower in bedrock than in unconsolidated deposits and, as a general rule, decrease with increasing depth in the subsurface. These observations are in accord with expectation; with increasing residence time in the subsurface, fluctuations in pesticide concentrations are likely to be "damped" by the averaging effects of hydrodynamic dispersion.

Pesticide contamination becomes more likely with decreases in any of the following depth-related parameters: aquifer depth, water-table depth, and depth of the well screen below the water table, but such relations are not always obvious. The clarity with which these relations are observed among existing studies appears to increase with improvements in the quality of the hydrogeologic data, or with increases in the quantitative range of the individual depth parameters examined.

11.3.4 INTERACTIONS BETWEEN GROUND WATER AND SURFACE WATER

The movement of water between rivers and their associated alluvial aquifers leads to a dynamic exchange of pesticides between ground and surface waters if these compounds are present. As a result, pesticide contamination of rivers may lead to ground-water contamination by these compounds in agricultural areas during periods of recharge, particularly following spring applications, when pesticide loads and river stages reach annual maxima. Conversely, the lower levels of pesticides in an alluvial aquifer caused by infiltration recharge give rise to more chronic, albeit less severe, contamination of the associated river during periods dominated by ground-water discharge (i.e., baseflow conditions).

Studies conducted in the midcontinent indicate that the exchange of water between rivers and their alluvial aquifers influences ground-water quality within a limited distance from the river banks—on the order of tens of meters in the areas investigated. These studies also suggest that several months are required following significant discharge events before bank storage is returned to the river and the pesticide concentrations in the river again approximate the concentrations in the ground water within the adjacent alluvium. In addition, investigations carried out in several areas of the United States and Europe indicate that bank filtration may be ineffective for removing pesticides from water that is withdrawn from supply wells completed in alluvial aquifers adjacent to pesticide-contaminated rivers.

11.4 EFFECTS OF VARIATIONS IN AGRICULTURAL MANAGEMENT PRACTICES

Extensive efforts to understand the influence of different agricultural management practices on ground-water quality indicate that the most effective ways to reduce the likelihood of ground-water contamination by pesticides applied in agricultural settings involve: (1) minimizing the flux of water through the land surface, especially following pesticide application; (2) applying the compounds in smaller amounts and over more limited areas; and (3) slowing the rate at which the active ingredients are released to the soil. Efforts to achieve the first objective affect not only irrigation techniques and the timing of applications, but also the manner in which the soil is tilled.

Pesticide concentrations in ground water have been found to vary directly with the intensity and the proximity of irrigation. Irrigation methods that increase the rate of water movement through the soil accelerate solute migration through the subsurface, thereby increasing the likelihood of ground-water contamination by applied agrichemicals. The negative impacts of irrigation on ground-water quality may, therefore, be reduced through the use of low-volume irrigation methods. In addition, increasing the time interval between pesticide application and irrigation can reduce the amount of parent compound migrating downward, particularly for pesticides that are less persistent in the soil. Subsurface contamination by transformation products remains a concern in this context, but has received only minor attention to date.

Recent research has provided relatively consistent evidence that reductions in soil tillage, introduced to reduce losses of agrichemicals and soil to surface waters, typically lead to increases in water flux through the soil, although some of these effects may be seasonal. Consequently, most of this research has indicated that reductions in tillage may facilitate the migration of pesticides to ground water. The use of no-till or reduced-till agriculture thus poses an increased threat to ground-water quality if pesticides continue to be applied at conventional rates. However, because many management practices may change with reductions in tillage (particularly the pesticides used and their mode and rates of application), these effects are complex and difficult to generalize.

Practices employed to reduce the concentrations of pesticide active ingredients in subsurface media have included: (1) the use of controlled-release formulations to diminish the rate at which the active ingredient is administered; (2) the introduction of more potent pesticides to reduce the amount needed to control target pests; and (3) the use of application banding, which substantially reduces the area of soil to which pesticides are applied, as well as the total mass applied. Although several studies indicate that controlled-release formulations decrease the rate of pesticide introduction to the subsurface, a clear influence of such formulations on pesticide mobility or persistence has not been demonstrated. With regard to the second strategy, the application of smaller amounts of more potent compounds may reduce the mass of chemicals released to the environment, but if the pesticides are more toxic, this approach may not reduce their harmful effects on nontarget species. Reductions in overall pesticide use may also be achieved through the use of integrated pest management techniques, such as scouting, to ensure that pesticides are delivered only where—and when—they are needed.

11.5 INFLUENCE OF WELL CHARACTERISTICS AND WELL USE ON PESTICIDE DETECTIONS

Of all the factors examined in relation to pesticide contamination of ground water, the influence of well depth was both the most frequently studied and the most consistent. Data from a large number of investigations showed inverse relations between well depth and pesticide concentrations, or the likelihood of detecting pesticides in ground water. This pattern has been observed across a wide range of geographic and hydrogeologic settings in North America and Europe. In most cases, it reflects the fact that pesticide use is a relatively recent phenomenon; because these surface-derived contaminants were only introduced to the environment within the past 50 years, insufficient time has passed for them to have reached the screens of many of the deeper wells sampled during these studies.

The influence of well construction is also one of the most consistent, and frequently examined, relations noted between pesticide occurrence and site characteristics. A relatively large number of investigations have demonstrated that pesticides are more likely to be detected in springs, and dug, bored, or driven wells, than in drilled wells, which are typically deeper, and provide more complete isolation of ground water from surficial contaminant sources. This trend may also be related to the fact that dug, bored, or driven wells, due to their manner of installation, are restricted to unconsolidated formations or highly weathered bedrock, whereas drilled wells may installed in any hydrogeologic setting. A failure to control for variations in well construction may be one of the principal reasons why the effects of other site characteristics on pesticide occurrence have been so difficult to discern from most previous monitoring studies.

Similarly, wells with proper seals, either at the surface or around the well annulus, show consistently lower frequencies of contamination than those with inadequate seals. Bedrock wells equipped with deep casings also exhibit lower incidences of pesticide contamination than boreholes installed without casings. In accord with these observations, pesticide detection frequencies have been shown by several large-scale monitoring studies to vary directly with well age—an indirect indicator of well integrity. As with well construction, however, well age is correlated with depth; older wells are not only less effectively sealed than newer wells, they also tend to be shallower and have shallower casings. In any event, whether it is well construction practices or routine applications that are primarily responsible for pesticide contamination of ground waters in predominantly agricultural regions continues to be a source of some disagreement, a debate similar to that regarding the relative importance of point- versus nonpoint-source contamination in such areas.

By contrast with the influence of well construction, the effects of well type (e.g., domestic, irrigation, municipal, or monitoring) on the likelihood of pesticide contamination are less clear. The inconsistency of relations between well type and pesticide occurrence in ground water may arise from the competing influences of several factors, most notably well depth, well construction, and pumping rate. With regard to the latter parameter, some evidence indicates that pesticide contamination becomes more likely with increasing rates of ground-water withdrawal, but other results suggest that enhanced pumping may reduce detection frequencies by dilution with nearby, uncontaminated water. In accord with expectation, however, high rates of ground-water withdrawal have been observed to accelerate the downward migration of pesticides in the saturated zone. The effects of pumping on pesticide detections, like well construction and age, are thus likely to be related to well depth—particularly the depth of the well screen below the water table.

11.6 PREDICTION VERSUS OBSERVATION OF PESTICIDE CONTAMINATION IN GROUND WATER

Substantial effort has been invested in developing tools for predicting the likelihood of detecting pesticides in the subsurface. Three approaches to this task were examined: (1) mathematical simulations of pesticide transport and fate; (2) the use of other solutes as potential indicators of pesticide contamination; and (3) large-scale assessments of the vulnerability of ground waters to pesticide contamination.

11.6.1 USE OF MATHEMATICAL SIMULATIONS

Efforts to simulate the movement and fate of pesticides in the subsurface using computer models have spanned at least three decades. Despite the extensive literature on the subject, however, much of this effort has been conducted without the benefit of independent confirmation; of the 79 investigations reviewed, only 40 compared their model predictions with field- or laboratory-derived data. Among the results from these studies, the agreement between prediction and observation has generally been marginal, particularly among those that did not adjust model parameters to improve the fit between model results and observational data. Variables that have been adjusted for this purpose have included several that can be (and, indeed, have been) measured independently, such as water-solids partition coefficients and trans-formation rate constants. In addition, studies carried out in this manner often did not ensure that the "best fit" values for these parameters fell within the ranges established by independent (usually laboratory) measurements.

A number of different explanations have been offered to account for the limited abilities of most computer simulations to reproduce observed pesticide behavior. One of the principal reasons for these discrepancies is the neglect of preferential transport by all of the models in widespread use. The general failure of most models to account for the effects of physical, chemical, and biological controls over pesticide transformation in the subsurface, through such basic parameters as temperature or pH, may also have contributed to these deficiencies.

The significance of preferential transport to the subsurface behavior of pesticides has been recognized for at least three decades, and widely acknowledged for the past 10 to 15 years. Nevertheless, computer models that account for the effects of partitioning between "mobile" and "immobile" water and other nonequilibrium phenomena on solute movement remain the exception, rather than the rule. Overlooking these phenomena runs the risk of overestimating the time required for detectable concentrations of pesticides and other surface-derived solutes to appear in ground water. The widespread neglect of preferential transport by most simulations of pesticide movement in the subsurface is a source of concern because of the frequent reliance upon modeling results for regulatory purposes, and because any detection of these compounds in ground water is generally regarded as a potential threat to both human and ecosystem health.

11.6.2 USE OF OTHER INDICATOR SOLUTES

Because the chemical analysis of individual pesticides is expensive and difficult, several other solutes have been examined as indicators for the presence of pesticides in ground water. Among these, nitrate has received the most attention because of its low analytical costs, its widespread application to the land in conjunction with pesticides, and the high frequency with which its concentration is measured during assessments of ground-water quality in agricultural areas. Other solutes that have been examined as potential indicators of pesticide contamination include tritium and some of the more commonly used pesticides, particularly atrazine.

Nitrate concentrations have been used in two different ways as predictors of pesticide contamination of ground water: (1) as indicators of pesticide concentrations, and (2) as indicators of the likelihood of detecting pesticides. A few monitoring studies showed direct relations between nitrate and pesticide concentrations in ground water, but most showed little correlation between the two parameters. By contrast, most studies that examined the concentration of nitrate as an indicator of the likelihood of detecting pesticides in the subsurface observed direct relations between the two parameters. These relations were not observed in large-scale monitoring investigations, however, including the NPS.

Despite the occasional significant correlation found between nitrate concentrations and either pesticide concentrations or pesticide detection frequencies in ground water in selected areas, the overall conclusion is that nitrate cannot be relied on as a predictor of pesticide contamination in the subsurface. Possible reasons for the poor correspondence between nitrate and pesticide occurrence in ground water include: (1) the different rates at which nitrate and pesticides migrate through the subsurface; (2) their differing degrees of chemical or biochemical stability under aerobic and hypoxic conditions (i.e., in the presence and absence of detectable dissolved oxygen, respectively); and (3) the absence of any specific causal relation to account for their co-occurrence, other than the fact that they are both typically (though not always) applied to the land in agricultural settings.

As an indicator of ground water recharged after the early 1950's, tritium has been examined as a predictor of the potential for pesticide contamination in the subsurface. A number of monitoring studies have demonstrated that pesticides are generally not detected in ground waters containing less than 0.2 tritium units (TU), the tritium concentrations predicted for pre-1953 waters. However, several investigations failed to distinguish between ground waters with and without detectable pesticide concentrations on the basis of tritium levels, perhaps because of their use of tritium detection limits that were an order of magnitude higher than the levels expected in pre-1953 water. Nevertheless, tritium is a valuable indicator of recently recharged ground water and, by extension, the potential presence of pesticides, provided its detection limit is sufficiently low.

Several monitoring studies showed that the detection of a pesticide in ground water increases the likelihood that other pesticides will also be detected. In general, however, the relative infrequency with which pesticides are typically detected in ground water, coupled with the high cost of their analysis, makes individual pesticides even less attractive than either nitrate or tritium as indicators of pesticide contamination.

Adjuvants, or "inert ingredients," are another set of possible indicators of pesticide contamination in ground water. Among the wide variety of compounds used as adjuvants in pesticide formulations, those likely to be the most mobile and persistent—and hence the most useful as tracers in the subsurface—are the volatile organic compounds (VOCs) and anionic surfactants. To date, however, no study appears to have evaluated adjuvants in ground water for this purpose. As with pesticides and tritium, the relatively high cost of chemical analyses for VOCs, coupled with the difficulties of sampling for these compounds, also limits their utility as indicators of pesticide contamination.

Advances in the technology of enzyme-linked immunoassay analyses (ELISA) may eliminate the need for surrogate indicators of pesticide contamination in the near future. Although most ELISA methods are not compound-specific at present, their low cost and relatively low detection limits make them valuable tools for screening ground-water samples for the presence of pesticides and transformation products from a given chemical class. In particular, they are especially useful for identifying areas where pesticides from specific chemical classes are not detectable in ground water at the time of sampling.

11.6.3 USE OF GROUND-WATER VULNERABILITY ASSESSMENTS

A third approach to the prediction of pesticide contamination in ground water has involved assessing the likelihood of pesticide occurrence in specific areas. As with the modeling studies, however, few of these assessments have been tested against actual ground-water quality data. Of the 63 vulnerability assessments reviewed, 41 (65 percent) did not compare their predictions with field observations of ground-water contamination.

Statistical comparisons between predicted degrees of vulnerability and actual ground-water contamination indicate that most of these efforts have been unsuccessful in predicting contaminant occurrence in ground water. Nineteen different studies of ground-water susceptibility to surface-derived contamination provided monitoring data against which the predicted vulnerability assessments were (or could be) compared. Significant correlations between predicted vulnerabilities and contaminant occurrence ($\alpha = 0.05$) were detected during only 10 of the 19 studies. Of these 10, two observed relations opposite to that expected, i.e., more severe contamination was encountered in areas deemed to be less vulnerable.

Previous studies have adopted one of three general approaches to the assessment of ground-water vulnerability: (1) discriminant analyses; (2) arbitrary scoring systems; and (3) process-based methods. Discriminant analyses have involved statistical comparisons of pesticide distributions among areas with different land-use, hydrogeologic, or edaphic characteristics. All three investigations using discriminant analyses detected significant differences in the severity of pesticide contamination among the different areas of interest. However, these methods require the prior existence of extensive data on contaminant distributions within the area(s) of interest, and thus cannot be used in the absence of such information.

The most widely used approaches for assessing the susceptibility of ground water to contamination by pesticides or other agrichemicals are the arbitrary scoring methods, the most commonly employed being the DRASTIC system (Aller and others, 1987). Among the 10 reviewed studies that used scoring systems, however, only 3 revealed significant, positive correlations between predicted and observed contamination. A number of reasons may account for the limited success of these scoring systems in predicting ground-water contamination: (1) the neglect of significant physical and chemical processes that influence the transport and fate of pesticides in the subsurface, including preferential transport and transformation; (2) the inappropriate use or weighting of one or more vulnerability factors; and (3) the use of large-scale (e.g., countywide) input data for predicting contamination occurring on a local scale (i.e., within a mile of the affected wells). Inaccurate predictions of contamination may also arise if a vulnerability assessment focuses on a region within the subsurface (typically a major aquifer) that is different from the formation(s) from which most ground water is actually withdrawn in the area of interest, or if the effects of well construction on contaminant occurrence are ignored.

Process-based approaches involve the computation of relative degrees of vulnerability among different areas, or the relative tendencies of different compounds to contaminate ground water, from the numerical values of selected climatic, hydrogeologic, edaphic, or chemical variables. As such, these methods seek to minimize or avoid the use of empirical parameters in assessing vulnerability. Process-based approaches are employed less frequently than scoring systems. One of them, however, the Leaching Potential Index (LPI) of Meeks and Dean (1990), although it was only applied in a limited area, was one of the most successful methods for predicting pesticide contamination of ground water among those examined.

The environmental parameters examined most often by assessments that successfully predicted ground-water vulnerability to pesticide contamination are well depth, aquifer depth, and soil texture. In addition, several monitoring studies have demonstrated significant differences in the severity of subsurface contamination between areas exhibiting contrasts in aquifer confinement, land use, hydrogeomorphology, and proximity to cropland or pesticide mixing areas.

Based on these results, the methods that are the most successful in predicting ground-water vulnerability to pesticide contamination appear to be those that involve the least amount of arbitrary scoring, regardless of the degree to which the assigned scores are informed by "professional judgement." The limited success with which scoring methods have predicted ground-water contamination is a source of considerable concern, given the widespread and increasing dependence on such assessments by state and federal agencies for setting ground-water protection priorities and designing sampling programs. Indeed, uncertainties regarding the reliability of scoring approaches to assessing vulnerability have led at least two states to eschew such methods in favor of actual ground-water monitoring data (California), or overlays of hydrogeologic or edaphic properties in conjunction with monitoring data (Iowa), in their efforts to distinguish between areas of high and low susceptibility to pesticide contamination.

11.7 SIGNIFICANCE TO WATER QUALITY

Most concern about pesticides in ground water stems from their potential impact on drinking water. The USEPA established Maximum Contaminant Levels (MCLs) for pesticides in drinking water. Nationally, fewer than 2 percent of the wells sampled by multistate studies, which mostly focused on agricultural areas, had concentrations that exceeded MCLs. Although this suggests that the problem is small at the national scale, our current ability to assess the significance of pesticides in ground water is limited by several factors. First, MCLs or other water-quality criteria have not been established for many pesticides and most transformation products, and existing criteria may be revised as more is learned about the toxicity of these compounds. Second, MCLs and other criteria are currently based on individual pesticides and do not account for possible cumulative effects if several different pesticides are present in the same well. Finally, many pesticides and most transformation products have not been widely sampled for in ground water and very little sampling has been done in urban and suburban areas, where pesticide use is often high.

The widespread detection of pesticides in ground water at levels below current MCLs—particularly high-use compounds in vulnerable areas—indicates that exceedances of water-quality criteria are likely to increase if existing criteria are lowered, as criteria are established for more compounds, as a wider range of pesticides and their transformation products are analyzed for, and as sampling expands to include more non-agricultural areas. Together, these factors result in uncertainty in our present ability to make strong conclusions about the national significance of pesticide contamination of ground water, and suggest that major data gaps will need to be filled in order to reduce this uncertainty.

11.8 FUTURE WORK

A vast amount of research has been directed toward understanding the occurrence, transport, and fate of pesticides in ground waters of the United States. Monitoring efforts in many states have revealed that pesticides from every major chemical class have been detected in ground water across a diversity of environmental settings, and such sampling activities will undoubtedly continue to proliferate. A coherent picture of pesticide occurrence in ground water across the nation can only be assembled, however, if monitoring investigations conducted in different areas are consistent with respect to a variety of design features, such as the compounds analyzed and their analytical detection limits, the criteria used to select wells for sampling, the timing of sampling with respect to major pesticide application periods, and the methods used for obtaining samples. In addition, variations in well construction and use must be controlled for, so that monitoring studies can provide overviews of ground-water quality—rather than just summaries of well-water quality—across the areas investigated. In order to account for the broad diversity of pesticides in current use, such investigations will also need to analyze for a wide range of compounds in non-agricultural, as well as agricultural settings, including the major transformation products for the parent compounds of interest.

In general, there is an acute need for greater coordination among the various efforts being conducted to understand and predict the subsurface behavior of pesticides at different scales of observation. The design of process and matrix-distribution studies must be more carefully directed toward the testing of conclusions derived from laboratory investigations of the various phenomena that control the fate of pesticides in the subsurface. This is particularly true with regard to the effects of nonequilibrium partitioning and subsurface heterogeneities on transport, and the influence of the physical, chemical, and microbiological environment on the rates and pathways of transformation. Monitoring investigations must incorporate more of the lessons learned from previous studies to maintain better control over the various factors that can interfere with obtaining an accurate picture of the chemical composition of ground water. As with the process and matrix-distribution studies, monitoring investigations should also place greater emphasis on characterizing the physical, chemical, and microbiological properties of the ground waters sampled.

Many of the uncertainties related to the various anthropogenic and natural controls over pesticide occurrence in ground water arise from limitations on available data regarding pesticide use and environmental characteristics. Significant progress in detecting relations between pesticide contamination and use will require that more detailed and consistent data be collected on the rates and locations of pesticide application in both agricultural and non-agricultural settings across the country. Difficulties associated with determining the effects of nearby land and pesticide use, soil and hydrogeologic properties, and contaminant sources on ground-water quality have underscored the importance of gathering data relating to these factors over comparatively small spatial scales, i.e., within approximately 1 mi of individual wells, rather than over large areas, such as entire counties.

Finally, efforts to predict pesticide occurrence and behavior in the subsurface using computer simulations and ground-water vulnerability assessments must always be tested against field data to maximize the reliability of these powerful tools. The need for this testing has become more pronounced in recent years with the growth of interest in using these techniques to support state and federal decisions regarding the registration and use of pesticides. Only by fully incorporating the lessons learned from previous work at all scales of investigation will it be possible to design pesticide management policies that ensure a reliable food supply, while safeguarding the quality of the Earth's land, air, and water resources for the future.

APPENDIX

Glossary Of Common And Chemical Names Of Pesticides And Related Compounds Given In Text

[Chemical Class: miscellaneous N, miscellaneous nitrogen-containing compound; VOC, volatile organic compound. Use class abbreviations: Ac, acaricide; Ad, adjuvant; An, antibiotic; Df, defoliant; Di, disinfectant; Fm, fumigant; H, herbicide; I, insecticide; IGR, insect growth regulator; Ind, industrial; IR, insect repellant; IS, insecticide synergist; Mi, miticide; Mo, molluscicide; N, nematocide; Ni, nitrification inhibitor; PGR, plant growth regulator; R, rodenticide; U, unspecified. CAS, Chemical Abstracts Service. In cases where more than one CAS number was available for a given pesticide, the number pertaining to the more structurally specific chemical nomenclature is given. Principal sources: Windholz (1976), U.S. Environmental Protection Agency (1987; 1992b), Worthing and Walker (1987), Howard and Neal (1992), Meister Publishing Company (1995), and Milne (1995). Blank cell, information not available or given elsewhere in appendix]

Common Name	Use	CAS No.	Chemical Nomenclature	Chemical Class
Abamectin	I,Mi	71751-41-2	mixture of avermectin B,a and avermectin B,b	miscellaneous
Acenaphthene	I,Fn	83-32-9	1,2-dihydroacenaphthylene	polynuclear aromatic hydrocarbon
Acephate	I	30560-19-1	O,S-dimethyl acetylphosphoramidothioate	organophosphorus
Acifluorfen	H	62476-59-9	sodium 5-[2-chloro-4-(trifluoromethyl)phenoxy]-2-nitrobenzoate	benzoic acid derivative
Acrolein	H,R	107-02-8	2-propenal	VOC
Acrylonitrile	Fu	107-13-1	2-propenenitrile	VOC
Alachlor	H	15972-60-8	2-chloro-N-(2,6-diethylphenyl)-N-(methoxymethyl) acetamide	acetanilide
Alachlor ethanesulfonic acid			2-[(2,6-diethylphenyl)(methoxymethyl)amino]-2-oxoethanesulfonic acid	alachlor degradate
Alanap				see Naptalam
Alar				see Daminozide
Aldicarb	I,Ac,N	116-06-3	2-methyl-2-(methylthio)propionaldehyde O-(methyl-carbamoyl)oxime	carbamate

APPENDIX. Glossary Of Common And Chemical Names Of Pesticides And Related Compounds Given In Text—*Continued*

Common Name	Chemical Class	Use	CAS No.	Chemical Nomenclature
Aldicarb sulfone	carbamate, aldicarb degradate	I,N	1646-88-4	2-methyl-2-(methylsulfonyl)propionaldehyde *O*-(methylcarbamoyl)oxime
Aldicarb sulfoxide	aldicarb degradate		1646-87-3	2-methyl-2-(methylsulfinyl)propionaldehyde *O*-(methylcarbamoyl)oxime
Aldoxycarb	see Aldicarb sulfone			
Aldrin	organochlorine	I	309-00-2	(1α,4α,4aβ,5α,8α,8aβ)-1,2,3,4,10,10-hexachloro-1,4,4a,5, 8,8a-hexahydro-1,4:5,8-dimethanonapthylene
Allethrin	pyrethroid	I	584-79-2	(RS)-3-allyl-2-methyl-4-oxocyclopent-2-enyl (1RS)-*cis, trans*-chrysanthemate
Allidochlor	amide	H	93-71-0	*N,N*-diallyl-2-chloroacetamide
Ametryn	triazine	H	834-12-8	2-(ethylamino)-4-isopropylamino-6-methylthio-*s*-triazine
Ametryne	See Ametryn			
Amiben	See Chloramben			
Aminocarb	carbamate	I	2032-59-9	4-(dimethylamino)-3-methylphenyl methylcarbamate
Aminomethylphosphonic acid	glyphosate degradate		1066-51-9	1-aminomethylphosphonic acid
Amino-triazole	See Amitrole			
Amitraz	amide	I,Ac,IS	33089-61-1	*N'*-(2,4-dimethylphenyl)-*N*-[[(2,4-dimethylphenyl)imino]methyl]-*N*-methylmethanimidamide
Amitrole	triazole	H	61-82-5	3-amino-1,2,4-triazole
Ammonium sulfamate	inorganic	H	7773-06-2	ammonium sulfamate

APPENDIX. Glossary Of Common And Chemical Names Of Pesticides And Related Compounds Given In Text—*Continued*

Common Name	Chemical Class	Use	CAS No.	Chemical Nomenclature
AMPA	see Aminomethyl phosphonic acid			
Anilazine	triazine	Fn	101-05-3	4,6-dichloro-*N*-(2-chlorophenyl)-1,3,5-triazin-2-amine
Arsenic acid	inorganic arsenical	H	7778-39-4	orthoarsenic acid
Asulam	carbamate	H	3337-71-1	methyl [(4-aminophenyl)sulfonyl]carbamate
Atraton	triazine	H	1610-17-9	2-(ethylamino)-4-(isopropylamino)-6-methoxy-*s*-triazine
Atratone	see Atraton			
Atrazine	triazine	H	1912-24-9	6-chloro-*N*-ethyl-*N'*-(1-methylethyl)-1,3,5-triazine-2,4-diamine
Azinphos-ethyl	organophosphorus	I	2642-71-9	*O,O*-diethyl *S*-[(4-oxo-1,2,3-benzotriazin-3(4*H*)-yl)methyl] phosphorodithioate
Azinphos-methyl	organophosphorus	I	86-50-0	*O,O*-dimethyl *S*-[(4-oxo-1,2,3-benzotriazin-3(4*H*)-yl) methyl] phosphorodithioate
Azinphos-methyl oxon	azinphos-methyl degradate		961-22-8	*O,O*-dimethyl *S*-[(4-oxo-1,2,3-benzotriazin-3(4*H*)-yl) methyl] phosphorothioate
Azodrin	see Monocrotophos			
Bacillus thuringiensis var. *kurstaki*	bacterial agent	I	68038-71-1	
BAM	see 2,6-Dichloro-benzamide			
Barban	carbamate	H	101-27-9	4-chloro-2-butynyl (3-chlorophenyl)carbamate
BAS 263I	see Cloethocarb			

APPENDIX. Glossary Of Common And Chemical Names Of Pesticides And Related Compounds Given In Text—*Continued*

Common Name	Chemical Class	Use	CAS No.	Chemical Nomenclature
Baygon	carbamate	I	204-043-8	2-(1-methylethoxy)phenyl methylcarbamate
BAY SMY 1500	see Ethiozin			
Benazolin	miscellaneous, benazolin-ethyl degradate	H	3813-05-6	4-chloro-2-oxobenzothiazolin-3-yl acetic acid
Benazolin-ethyl	miscellaneous	H	25059-80-7	ethyl 4-chloro-2-oxobenzothiazolin-3-yl acetate
Bendiocarb	carbamate	I	22781-23-3	2,2-dimethyl-1,3-benzodioxol-4-yl methylcarbamate
Benefin	see Benfluralin			
Benfluralin	dinitroaniline	H	1861-40-1	*N*-butyl-*N*-ethyl-α, α,α-trifluoro-2,6-dinitro-*p*-toluidine
Benomyl	carbamate	Fn	17804-35-2	methyl 1-(butylcarbamoyl)-2-benzimidazol-2-yl carbamate
Bensulfuron-methyl	sulfonylurea	H	83055-99-6	methyl 2-[[[[(4,6-dimethoxypyrimidin-2-yl)amino]-carbonyl]-amino]sulfonyl]methyl]benzoate
Bensulide	organophosphorus	H	741-58-2	*S*-2-benzenesulfonamidoethyl *O*,*O*-di-isopropyl phosphorodithioate
Bentazon	miscellaneous	H	25057-89-0	3-isopropyl-1*H*-2,1,3-benzothiadiazin-4(3*H*)-one 2,2-dioxide
Bentazone	see Bentazon			
Benthiocarb	see Thiobencarb			
Benzothiazolin	benazolin-ethyl degradate			benzothiazolin
α-BHC (α-benzene hexachloride)	see α-HCH			
β-BHC (β-benzene hexachloride)	see β-HCH			
γ-BHC (γ-benzene hexachloride)	see γ-HCH			

APPENDIX. Glossary Of Common And Chemical Names Of Pesticides And Related Compounds Given In Text—*Continued*

Common Name	Chemical Class	Use	CAS No.	Chemical Nomenclature
δ-BHC (δ-benzene hexachloride)	see δ-HCH			
Bifenthrin	pyrethroid	I,Mi	82657-04-3	[1α,3α-(Z)]-(±)-(2 methyl[1,1'-biphenyl]-3-yl) methyl 3-(2-chloro-3,3,3-trifluoro-1-propenyl)-2,2-dimethylcyclopropanecarboxylate
Borax	inorganic	H	1303-96-4	sodium tetraborate decahydrate
Brodifacoum	miscellaneous	R	56073-10-0	3-[3-(4'-bromo[1,1'-biphenyl]-4-yl)-1,2,3,4-tetrahydro-1-naphthalenyl]-4-hydroxy-2H-1-benzopyran-2-one
Bromacil	uracil	H	314-40-9	5-bromo-6-methyl-3-(1-methylpropyl)-2,4-(1H,3H)-pyrimidinedione
Bromadiolone	miscellaneous	R	28772-56-7	3-[3-(4'-bromo[1,1'-biphenyl]-4-yl)-3-hydroxy-1-phenylpropyl]-4-hydroxy-2H-1-benzopyran-2-one
Bromomethane	see Methyl bromide			
Bromoxynil	miscellaneous N	H	1689-84-5	3,5-dibromo-4-hydroxybenzonitrile
Bt	see *Bacillus thuringiensis* var. *kurstaki*			
Bufencarb	carbamate	I	2282-34-0	3-(1-methylbutyl)phenyl methylcarbamate
Butachlor	acetanilide	H	23184-66-9	2-chloro-2',6'-diethyl-N-(butoxymethyl) acetanilide
Butonate	organophosphorus	I	126-22-7	dimethyl 1-butyryloxy-2,2,2-trichloroethylphosphonate
Buturon	urea	H	3766-60-7	3-(4-chlorophenyl)-1-methyl-1-(1-methylprop-2-ynyl)urea
Butylate	thiocarbamate	H	2008-41-5	S-ethyl bis(2-methylpropyl)thiocarbamate
Captafol	imide	Fn	2939-80-2	N-[(1,1,2,2-tetrachloroethyl)thio]-4-cyclohexene-1,2-dicarboximide
Captan	imide	Fn	133-06-2	N-(trichloromethylthio)-4-cyclohexene-1,2-dicarboximide

APPENDIX. Glossary Of Common And Chemical Names Of Pesticides And Related Compounds Given In Text—*Continued*

Common Name	Chemical Class	Use	CAS No.	Chemical Nomenclature
Carbaryl	carbamate	I	63-25-2	1-naphthalenyl-*N*-methylcarbamate
Carbendazim	imidazole	Fn	83601-81-4	2-(methoxycarbonylamino)-benzimidazole
Carbofuran	carbamate	I,N	1563-66-2	2,3-dihydro-2,2-dimethyl-7-benzofuranyl methylcarbamate
Carbofuran phenol	carbofuran degradate		1563-38-8	2,3-dihydro-2,2-dimethyl-7-hydroxy-benzofuran
Carbon disulfide	VOC	Fm,Ad	75-15-0	carbon disulfide
Carbon tetrachloride	VOC	Fm,Ad	56-23-5	tetrachloromethane
Carbophenothion	see Trithion			
Carboxin	amide	Fm	5234-68-4	5,6-dihydro-2-methyl-*N*-phenyl-1,4-oxathiin-3-carboxamide
CDAA	see Allidochlor			
CDEC	thiocarbamate	H	95-06-7	2-chloroallyl diethyldithiocarbamate
CGA 17193	isazofos degradate			5-chloro-3-hydroxy-1-isopropyl-1*H*-1,2,4-triazole
Chloramben	chlorobenzoic acid	H	133-90-4	3-amino-2,5-dichlorobenzoic acid
Chlorbromuron	urea	H	13360-45-7	3-(4-bromo-3-chlorophenyl)-1-methoxy-1-methylurea
Chlordane	organochlorine	I	57-74-9	1,2,4,5,6,7,8,8-octachloro-3a,4,7,7a-tetrahydro-4,7-methanoindan
α-Chlordane	organochlorine	I	5103-71-9	1α,2α,4β,5,6,7β,8,8-octachloro-3aα,4,7,7aα-tetrahydro-4,7-methanoindan
γ-Chlordane	organochlorine	I	5566-34-7	2,2,4,5,6,7,8,8-octachloro-3a,4,7,7a-tetrahydro-4,7-methanoindan
cis-Chlordane	see α-Chlordane			

APPENDIX. Glossary Of Common And Chemical Names Of Pesticides And Related Compounds Given In Text—*Continued*

Common Name	Chemical Class	Use	CAS No.	Chemical Nomenclature
trans-Chlordane	organochlorine	I	5103-74-2	1β,2α,4α,5,6,7α,8,8-octachloro-3aβ,4,7,7aβ-tetrahydro-4,7-methanoindan
Chlordecone	organochlorine	I	143-50-0	1,1a,3,3a,4,5,5,5a,5b,6-decachlorooctahydro-1,3,4-metheno-2*H*-cyclobuta-[c,d]-pentalen-2-one
Chlordimeform	amide	I,Ac	6164-98-3	*N*'-(4-chloro-*o*-tolyl)-*N*,*N*-dimethylformamidine
Chlorfenac	organochlorine	H	85-34-7	(2,3,6-trichlorophenyl)acetic acid
Chlorfenson	organochlorine	Ac	80-33-1	4-chlorophenyl-4-chlorobenzenesulfonate
Chlorfenvinphos	organophosphorus	I	470-90-6	2-chloro-1-(2,4-dichlorophenyl)vinyl diethylphosphate
Chlorflurenol, methyl ester	miscellaneous	H	2464-37-1	methyl 2-chloro-9-hydroxyfluorene-9-carboxylate
Chloridazon	miscellaneous N	H	1698-60-8	5-amino-4-chloro-2-phenyl-3-(2*H*)-pyridazinone
Chlornitrofen	organochlorine	H	1836-77-7	2,4,6-trichlorophenyl-4-nitrophenyl ether
3-Chloroallyl alcohol	1,3-dichloropropene degradate		29560-84-7	3-chloro-2-propen-1-ol
Chlorobenzilate	organochlorine	Ac	510-15-6	ethyl 4,4'-dichlorobenzilate
Chloroform	VOC	Fm,Ad	67-66-3	trichloromethane
p-Chloro-*m*-cresol	organochlorine	Fn,I	59-50-7	4-chloro-3-methylphenol
p-Chloro-*o*-cresol	organochlorine	Fn,I	1570-64-5	4-chloro-2-methylphenol
Chloroneb	organochlorine	Fn	2675-77-6	1,4-dichloro-2,5-dimethoxybenzene
Chloropicrin	VOC	Fm	76-06-2	trichloronitromethane
Chlorothalonil	organochlorine	Fn	1897-45-6	2,4,5,6-tetrachloro-1,3-benzenedicarbonitrile

APPENDIX. Glossary Of Common And Chemical Names Of Pesticides And Related Compounds Given In Text—*Continued*

Common Name	Chemical Class	Use	CAS No.	Chemical Nomenclature
Chloroxuron	urea	H	1982-47-4	3-[p-(p-chlorophenoxy)phenyl]-1,1-dimethylurea
Chlorpyrifos	organophosphorus	I	2921-88-2	O,O-diethyl O-(3,5,6-trichloro-2-pyridinyl) phosphorothioate
Chlorpyrifos-methyl	organophosphorus	I	5598-13-0	O,O-dimethyl O-(3,5,6-trichloro-2-pyridinyl) phosphorothioate
Chlorsulfuron	sulfonylurea	H	64902-72-3	2-chloro-N-[[(4-methoxy-6-methyl-1,3,5-triazin-2-yl)amino]carbonyl]benzenesulfonamide
Chlorthiamid	amide	H	1918-13-4	2,6-dichlorothiobenzamide
Chlorthion	organophosphorus	I	500-28-7	O-(3-chloro-4-nitrophenyl) O,O-dimethyl phosphorothioate
Chlortoluron	urea	H	15545-48-9	N-(3-chloro-4-methylphenyl)-N',N'-dimethylurea
Cloethocarb	carbamate	I,N	51487-69-5	2-(2-chloro-1-methoxyethoxy)phenyl methylcarbamate
Clomazone	miscellaneous N	H	81777-89-1	2-[(2-chlorophenyl)methyl]-4,4-dimethyl-3-isoxazolidinone
Clopyralid	organochlorine	H	1702-17-6	3,6-dichloro-2-pyridinecarboxylic acid
Copper sulfate	inorganic	Fn,H	7758-99-8	copper sulfate
Coumaphos	organophosphorus	I,N	56-72-4	O,O-diethyl O-(3-chloro-4-methyl-2-oxo-2H-1-benzo-pyran-7-yl) phosphorothioate
Crotoxyphos	organophosphorus	I	7700-17-6	α-methylbenzyl 3-hydroxy-cis-crotonate, dimethyl phosphate
Crufomate	organophosphorus	I	299-86-5	4-tert-butyl-2-chlorophenyl N-methyl O-methylphosphoramidate
Cryolite	inorganic	I	15096-52-3	sodium fluoaluminate
Cyanazine	triazine	H	21725-46-2	2-[[4-chloro-6-(ethylamino)-1,3,5-triazin-2-yl]amino]-2-methylpropionitrile

APPENDIX. Glossary Of Common And Chemical Names Of Pesticides And Related Compounds Given In Text—*Continued*

Common Name	Chemical Class	Use	CAS No.	Chemical Nomenclature
Cyanazine amide	cyanazine degradate			2-chloro-4-(1-carbamoyl-1-methylethylamino)-6-ethyl-amino-*s*-triazine
Cyanide	miscellaneous			
Cycloate	thiocarbamate	H	1134-23-2	*S*-ethyl *N*-ethyl *N*-cyclohexylthiocarbamate
Cyfluthrin	pyrethroid	I	68359-37-5	cyano(4-fluoro-3-phenoxyphenyl)methyl 3-(2,2-dichloroethenyl)-2,2-dimethylcyclopropanecarboxylate
Cyhalothrin	pyrethroid	I	91465-08-6	(RS)-α-cyano-3-phenoxybenzyl (Z)-(1R,3R)-3-(2-chloro-3,3,3-trifluoroprop-1-enyl)-2,2-dimethylcyclopropanecarboxylate
λ-Cyhalothrin	see Cyhalothrin			
Cypermethrin	pyrethroid	I	52315-07-8	(±)-α-cyano-3-phenoxybenzyl-(±)-*cis,trans*-3-(2,2-dichlorovinyl)-2,2-dimethylcyclopropanecarboxylate
Cyprazine	triazine	H	22936-86-3	2-chloro-4-(cyclopropylamino)-6-(isopropylamino)-*s*-triazine
Cyromazine	triazine	I	66215-27-8	*N*-cyclopropyl-1,3,5-triazine-2,4,6-triamine
Cythioate	organophosphorus	I	115-93-5	*O*-[4-(aminosulfonyl)phenyl] *O,O*-dimethyl phosphorothioate
2,4-D	chlorophenoxy acid	H	94-75-7	(2,4-dichlorophenoxy)acetic acid
2,4-D, methyl ester	chlorophenoxy acid ester	H	1928-38-7	methyl (2,4-dichlorophenoxy)acetate
2,4-D, BEE	see 2,4-D, butoxyethyl ester			
2,4-D, butoxyethyl ester	chlorophenoxy acid ester	H	1929-73-3	butoxyethyl (2,4-dichlorophenoxy)acetate
2,4-D, DMA salt	see 2,4-D, dimethylamine salt			

APPENDIX. Glossary Of Common And Chemical Names Of Pesticides And Related Compounds Given In Text—*Continued*

Common Name	Chemical Class	Use	CAS No.	Chemical Nomenclature
2,4,-D, dimethylamine salt	chlorophenoxy acid salt	H	2008-39-1	dimethylamine (2,4-dichlorophenoxy)acetate
D2A (diamino atrazine)	see Didealkylatrazine			
DAA (diamino atrazine)	see Didealkylatrazine			
Dacthal	chlorobenzoic acid	H	1861-32-1	dimethyl 2,3,5,6-tetrachloro-1,4-benzenedicarboxylate
Dalapon	organochlorine	H	75-99-0	2,2-dichloropropanoic acid
Daminozide	amide	PGR	1596-84-5	butanedioic acid mono (2,2-dimethyl hydrazide)
Dazomet	miscellaneous N	Fn,H,N	533-74-4	tetrahydro-3,5-dimethyl-2H-1,3,5-thiadiazine-2-thione
2,4-DB	chlorophenoxy acid	H	94-82-6	4-(2,4-dichlorophenoxy)butanoic acid
DBCP	see 1,2-Dibromo-3-chloropropane			
DCP	see 1,2-Dichloropropane			
DCNA	nitroaniline	Fn	99-30-9	2,6-dichloro-4-nitroaniline
DCPA	see Dacthal			
D-D	VOC	Fm,N	8003-19-8	mixture of 1,2-dichloropropane and 1,3-dichloropropene
p,p'-DDD	organochlorine, p,p'-DDT degradate	I	72-54-8	1,1-dichloro-2,2-bis(p-chlorophenyl)ethane
p,p'-DDE	p,p'-DDT degradate	I	72-55-9	1,1-dichloro-2,2-bis(p-chlorophenyl)ethene
o,p'-DDT	organochlorine	I	789-02-6	1,1,1-trichloro-2-(p-chlorophenyl)-2-(o-chlorophenyl) ethane
p,p'-DDT	organochlorine	I	50-29-3	1,1,1-trichloro-2,2-bis(p-chlorophenyl)ethane

APPENDIX. Glossary Of Common And Chemical Names Of Pesticides And Related Compounds Given In Text—*Continued*

Common Name	Chemical Class	Use	CAS No.	Chemical Nomenclature
DDVP	see Dichlorvos			
DEA	see Deethylatrazine			
Decamethrin	see Deltamethrin			
Decyclopropyl cyprazine	cyprazine degradate (identical to DEA)		6190-65-4	2-chloro-4-amino-6-isopropylamino-*s*-triazine
Deet	see Diethyltoluamide			
Deethyl ametryn	ametryn degradate			2-amino-4-isopropylamino-6-methylthio-*s*-triazine
Deethylatrazine	atrazine degradate		6190-65-4	2-chloro-4-amino-6-isopropylamino-*s*-triazine
Deethylcyanazine	cyanazine degradate			2-[(4-chloro-6-amino-1,3,5-triazin-2-yl)amino]-2-methylpropionitrile
Deethylcyanazine amide	cyanazine degradate			2-chloro-4-(1-carbamoyl-1-methylethylamino)-6-amino-*s*-triazine
Deethylsimazine	simazine degradate (identical to DIA)		1007-28-9	2-chloro-4-amino-6-ethylamino-*s*-triazine
DEF	see Tribufos			
Deisopropylatrazine	atrazine degradate		1007-28-9	2-chloro-4-amino-6-ethylamino-*s*-triazine
Deltamethrin	pyrethroid	I	52918-63-5	(S)-α-cyano-(3-phenoxyphenyl) (1R,3R)-3-(2,2-dibromoethenyl)-2,2-dimethylcyclopropanecarboxylate
Demeton	organophosphorus	I,Ac	298-03-3	*O,O*-diethyl *O*-[2-(ethylthio)ethyl] phosphorothioate
Demeton methyl	organophosphorus	I,Ac	8022-00-2	mixture (7:3) of *O,O*-dimethyl *O*-[2-(ethylthio)ethyl] phosphorothioate (Demeton-methyl I) and *O,O*-dimethyl *S*-[2-(ethylthio)ethyl] phosphorothioate (Demeton-methyl II)
Demeton-S	organophosphorus	I,Ac	126-75-0	*O,O*-diethyl *S*-[2-(ethylthio)ethyl] phosphorothioate

APPENDIX. Glossary Of Common And Chemical Names Of Pesticides And Related Compounds Given In Text—*Continued*

Common Name	Chemical Class	Use	CAS No.	Chemical Nomenclature
Demeton-S sulfone	demeton-S degradate			*O,O*-diethyl *S*-[2-(ethylsulfonyl)ethyl] phosphorothioate
Desamino metamitron	metamitron degradate			3-methyl-6-phenyl-1,2,4-triazin-5(4*H*)-one
Dextrin	miscellaneous	Ad	9004-53-9	hydrolyzed starch
DIA	see Deisopropylatrazine			
Diallate	thiocarbamate	H	2303-16-4	*S*-(2,3-dichloroallyl) *N,N*-(di-isopropyl)thiocarbamate
Diamino atrazine	see Didealkylatrazine			
Diazinon	organophosphorus	I,N	333-41-5	*O,O*-diethyl *O*-[6-methyl-2-(1-methylethyl)-4-pyrimidinyl] phosphorothioate
Dibrom	see Naled			
Dibromochloropropane	see 1,2-Dibromo-3-chloropropane			
1,2-Dibromo-3-chloropropane	VOC	Fm	96-12-8	1,2-dibromo-3-chloropropane
Dibutyl phthalate	miscellaneous	IR	84-74-2	dibutyl 1,2-benzenedicarboxylate
Dicamba	chlorobenzoic acid, Vel-4207 degradate	H	1918-00-9	3,6-dichloro-2-methoxybenzoic acid
Dicapthon	organophosphorus	I	2463-84-5	*O*-(2-chloro-4-nitrophenyl) *O,O*-dimethyl phosphorothioate
Dichlobenil	organochlorine	H	1194-65-6	2,6-dichlorobenzonitrile
Dichlone	organochlorine	Fn	117-80-6	2,3-dichloro-1,4-naphthoquinone
2,6-Dichlorobenzamide	dichlobenil degradate, chlorthiamid degradate		2008-58-4	2,6-dichlorobenzamide
1,4-Dichlorobenzene	see *p*-Dichlorobenzene			

APPENDIX. Glossary Of Common And Chemical Names Of Pesticides And Related Compounds Given In Text—*Continued*

Common Name	Chemical Class	Use	CAS No.	Chemical Nomenclature
o-Dichlorobenzene	organochlorine	Fm,H, I,Ad	95-50-1	1,2-dichlorobenzene
m-Dichlorobenzene	organochlorine	U	541-73-1	1,3-dichlorobenzene
p-Dichlorobenzene	organochlorine	Fm,Fn, I,R	106-46-7	1,4-dichlorobenzene
2,4-Dichlorobenzoic acid	chlorobenzoic acid	U	50-84-0	2,4-dichlorobenzoic acid
p,p'-Dichlorobenzophenone	*p,p'*-DDT degradate		90-98-2	4,4'-dichlorobenzophenone
1,2-Dichloroethane	VOC	Fm,Ad	107-06-2	1,2-dichloroethane
1,2-Dichloroethene	VOC	Fm	540-59-0	1,2-dichloroethene
Dichloromethane	see Methylene chloride			
Dichlorophen	organochlorine	Fn,Ad	97-23-4	bis(5-chloro-2-hydroxyphenyl)methane
Dichlorophene	see Dichlorophen			
1,2-Dichloropropane	VOC	Fm,Ad	78-87-5	1,2-dichloropropane
cis-1,3-Dichloropropene	VOC	Fm,N	10061-01-5	*cis*-1,3-dichloropropene
trans-1,3-Dichloropropene	VOC	Fm,N	10061-02-6	*trans*-1,3-dichloropropene
Dichlorprop	see 2,4-DP			
Dichlorvos	organophosphorus	Fm,I, Ad	62-73-7	O,O-dimethyl O-(2,2-dichlorovinyl) phosphate
Diclofop	chlorophenoxy acid	H	40843-25-2	(RS)-2-[4-(2,4-dichlorophenoxy)phenoxy]propionic acid
Diclofop-methyl	chlorophenoxy acid ester	H	51338-27-3	methyl (RS)-2-[4-(2,4-dichlorophenoxy)phenoxy] propionate

APPENDIX. Glossary Of Common And Chemical Names Of Pesticides And Related Compounds Given In Text—*Continued*

Common Name	Chemical Class	Use	CAS No.	Chemical Nomenclature
Dicofol	organochlorine	I	115-32-2	4-chloro-α-(4-chlorophenyl)-α-(trichloromethyl) benzenemethanol
Dicrotophos	organophosphorus	I	141-66-2	(E)-2-dimethylcarbamoyl-1-methylvinyl dimethyl phosphate
Didealkylatrazine	atrazine degradate			2-chloro-4,6-diamino-s-triazine
Dieldrin	organochlorine	I	60-57-1	1,2,3,4,10,10-hexachloro-6,7-epoxy-1,4,4a,5,6,7,8,8a-octahydro (endo,exo) 1,4:5,8-dimethanonaphthalene
Dienochlor	organochlorine	Mi	2227-17-0	decachlorobis(2,4-cyclopentadien-1-yl)
Diethatyl ethyl	amino acid derivative	H	58727-55-8	N-(chloroacetyl)-N-(2,6-diethylphenyl)glycine ethyl ester
2,6-Diethylaniline	alachlor degradate		579-66-8	2,6-diethylaniline
Diethyltoluamide	amide	IR	134-62-3	N,N-diethyl-m-toluamide
Difenoxuron	urea	H	14214-32-5	3-[4-(4-methoxyphenoxy)phenyl]-1,1-dimethylurea
Difolitan	see Captafol			
Diflubenzuron	urea	I	35367-38-5	1-(4-chlorophenyl)-3-(2,6-difluorobenzoyl)urea
Dimethoate	organophosphorus	I	60-51-5	O,O-dimethyl S-methylcarbamoylmethyl phosphorodithioate
N-(1,1-Dimethylacetonyl)-3,5-dichlorobenzamide	pronamide degradate			N-(1,1-dimethylacetonyl)-3,5-dichlorobenzamide
2,4-Dimethylphenol	miscellaneous	Di	105-67-9	2,4-dimethylphenol
Dimethyl phthalate	miscellaneous	IR	131-11-3	dimethyl 1,2-benzenedicarboxylate
Dimilin	see Diflubenzuron			
2,4-Dinitrophenol	nitrophenol	I,Fn, Ac,Ad	51-28-5	2,4-dinitrophenol

APPENDIX. Glossary Of Common And Chemical Names Of Pesticides And Related Compounds Given In Text—*Continued*

Common Name	Chemical Class	Use	CAS No.	Chemical Nomenclature
Dinoben	chlorobenzoic acid salt	H	88-86-8	sodium 2,5-dichloro-3-nitrobenzoate
Dinocap	miscellaneous N	Ac,Fn	39300-45-3	2,4-dinitro-6-octylphenylcrotonate
Dinoseb	nitrophenol	H	88-85-7	2-sec-butyl-4,6-dinitrophenol
Dioctyl phthalate	miscellaneous	Ac	117-84-0	dioctyl-1,2-benzenedicarboxylate
Dioxacarb	carbamate	I	6988-21-2	2-(1,3-dioxolan-2-yl)phenyl methylcarbamate
Dioxathion	organophosphorus	I,Ac	78-34-2	S,S'-(1,4-dioxane-2,3-diyl) O,O,O',O'-tetraethyl bis(phosphorodithioate)
Diphacinone	miscellaneous	I,R	82-66-6	2-(diphenylacetyl)-1,3-indandione
Diphenamid	amide	H	957-51-7	N,N-dimethyl-2,2-diphenylacetamide
Diquat dibromide	miscellaneous N	H	85-00-7	1,1'-ethylene-2,2'-bipyridylium dibromide, monohydrate
Disulfoton	organophosphorus	I	298-04-4	O,O-diethyl S-[2-(ethylthio)ethyl]phosphorodithioate
Disulfoton sulfone	disulfoton degradate		2497-06-5	O,O-diethyl S-[2-(ethylsulfonyl)ethyl]phosphorodithioate
Disulfoton sulfoxide	organophosphorus, disulfoton degradate	I, Ac	2497-07-6	O,O-diethyl S-[2-(ethylsulfinyl)ethyl]phosphorodithioate
Disyston	see Disulfoton			
Diuron	urea	H	330-54-1	3-(3,4-dichlorophenyl)-1,1-dimethylurea
DMPA	see Zytron			
DNBP	see Dinoseb			
DNOC	miscellaneous N	H,I,Ad	534-52-1	4,6-dinitro-o-cresol

APPENDIX. Glossary Of Common And Chemical Names Of Pesticides And Related Compounds Given In Text—*Continued*

Common Name	Chemical Class	Use	CAS No.	Chemical Nomenclature
Dodine	amine	Fn	2439-10-3	1-dodecylguanidine acetate
2,4-DP	chlorophenoxy acid derivative	H	120-36-5	(±)-2-(2,4-dichlorophenoxy)propanoic acid
DSMA	organic arsenical	H	144-21-8	disodium methanearsonate
Dursban	see Chlorpyrifos			
Dyfonate	see Fonofos			
EBDC fungicides	ethylene bis-dithio-carbamates	Fn	111-54-6	
EDB	see Ethylene dibromide			
Endosulfan	organochlorine	I	115-29-7	6,7,8,9,10-hexachloro-1,5,5a,6,9,9a-hexahydro-6,9-methano-2,4,3-benzodioxathiepin-3-oxide
Endosulfan I	organochlorine	I	959-98-8	3α,5aβ,6α,9α,9aβ-6,7,8,9,10-hexachloro-1,5,5a,6,9,9a-hexahydro-6,9-methano-2,4,3-benzodioxathiepin-3-oxide
Endosulfan II	organochlorine	I	33213-65-9	3α,5aα,6β,9β,9aβ-6,7,8,9,10-hexachloro-1,5,5a,6,9,9a-hexahydro-6,9-methano-2,4,3-benzodioxathiepin-3-oxide
Endosulfan sulfate	endosulfan degradate		1031-07-8	3α,5aα,6β,9β,9aβ-6,7,8,9,10-hexachloro-1,5,5a,6,9,9a-hexahydro-6,9-methano-2,4,3-benzodioxathiepin-3,3-dioxide
Endothall	miscellaneous	H	129-67-9	7-oxabicyclo[2,2,1]heptane-2,3-dicarboxylic acid
Endrin	organochlorine	I	72-20-8	1,2,3,4,10,10-hexachloro-6,7-epoxy-1,4,4a,5,6,7,8,8a-octahydro-(endo,endo)-1,4:5,8-dimethanonaphthalene
Endrin aldehyde	endrin degradate		7421-93-4	2.2a,3,3,4,7-hexachlorodecahydro-(2α,3aβ,3bβ,4β,5β,6aβ, pentalene-r-carboxaldehyde
Endrin ketone	endrin degradate		53494-70-5	3b,4,5,6,6,6a-hexachlorodecahydro-(2α,3aβ,3bβ,4β,5β,6aβ, 7α,7aβ,8R*)-2,5,7-metheno-3*H*-cyclopenta(a)pentalen-3-one

APPENDIX. Glossary Of Common And Chemical Names Of Pesticides And Related Compounds Given In Text—*Continued*

Common Name	Chemical Class	Use	CAS No.	Chemical Nomenclature
EPN	organophosphorus	I, Ac	2104-64-5	*O*-ethyl *O*-4-nitrophenyl phenylphosphonothioate
Eptam	thiocarbamate	H	759-94-4	*S*-ethyl dipropylthiocarbamate
EPTC	see Eptam			
ESA	see Alachlor ethanesulfonic acid			
Esfenvalerate	pyrethroid	I	66230-04-4	(*S*)-α-cyano-3-phenoxybenzyl (*S*)-2-(4-chlorophenyl)-3-methylbutyrate
Ethalfluralin	dinitroaniline	H	55283-68-6	*N*-ethyl-*N*-(2-methyl-2-propenyl)-2,6-dinitro-4-(trifluoromethyl)benzenamine
Ethiofencarb	carbamate	I	29973-13-5	2-[(ethylthio)methyl]phenyl methylcarbamate
Ethion	organophosphorus	I,Ac	563-12-2	*S,S*′-methylene bis(*O,O*-diethyl phosphorodithioate)
Ethiozin	triazine	H	64529-56-2	6-*tert*-butyl-4-amino-3-ethylthio-1,2,4-triazin-5(4*H*)-one
Ethofumesate	miscellaneous	H	26225-79-6	(±)-2-ethoxy-2,3-dihydro-3,3-dimethyl-5-benzofuranyl methanesulfonate
Ethoprop	organophosphorus	I,N	13194-48-4	*O*-ethyl *S,S*-dipropyl phosphorodithioate
Ethyl alcohol	miscellaneous	Ad	64-17-5	ethanol
Ethylan	see Perthane			
Ethylbenzene	VOC	Ad	100-41-4	ethylbenzene
Ethylene dibromide	VOC	Fm,I,N	106-93-4	1,2-dibromoethane

APPENDIX. Glossary Of Common And Chemical Names Of Pesticides And Related Compounds Given In Text—*Continued*

Common Name	Chemical Class	Use	CAS No.	Chemical Nomenclature
Ethylene thiourea	urea, ethylene bis-dithiocarbamate (EBDC) degradate	Ad	96-45-7	1,3-ethylene-2-thiourea
Ethyl parathion	organophosphorus	U	56-38-2	*O,O*-diethyl *O*-(*p*-nitrophenyl) phosphorothioate
Etridiazole	miscellaneous N	Fn	2593-15-9	5-ethoxy-3-trichloromethyl-1,2,4-thiadiazole
ETU	see Ethylene thiourea			
Fenac	see Chlorfenac			
Fenamiphos	organophosphorus	N	22224-92-6	*O*-ethyl [3-methyl-4-(methylthio)phenyl] (1-methylethyl) phosphoramidate
Fenamiphos sulfone	fenamiphos degradate		31972-44-8	*O*-ethyl [3-methyl-4-(methylsulfonyl)phenyl] (1-methylethyl) phosphoramidate
Fenamiphos sulfoxide	fenamiphos degradate		31972-43-7	*O*-ethyl [3-methyl-4-(methylsulfinyl)phenyl] (1-methylethyl) phosphoramidate
Fenarimol	miscellaneous N	Fn	60168-88-9	α-(2-chlorophenyl)-α-(4-chlorophenyl)-5-pyrimidine methanol
Fenbutatin oxide	organotin	I,Ac	13356-08-6	bis[tris (2-methyl-2-phenylpropyl)tin] oxide
Fenitrothion	organophosphorus	I,Ac	122-14-5	*O,O*-dimethyl *O*-(3-methyl-4-nitrophenyl)phosphorothioate
Fenoprop	see 2,4,5-TP			
Fenoxycarb	carbamate	IGR	79127-80-3	ethyl [2-(4-phenoxyphenoxy)ethyl]carbamate
Fensulfothion	organophosphorus	I	115-90-2	*O,O*-diethyl *O*-[4-(methylsulfinyl)phenyl] phosphorothioate
Fenthion	organophosphorus	I	55-38-9	*O,O*-dimethyl *O*-[3-methyl-4-(methylthio)phenyl] phosphorothioate

APPENDIX. Glossary Of Common And Chemical Names Of Pesticides And Related Compounds Given In Text—*Continued*

Common Name	Chemical Class	Use	CAS No.	Chemical Nomenclature
Fenuron	urea	H	101-42-8	3-phenyl-1,1-dimethylurea
Fenvalerate	pyrethroid	I	51630-58-1	cyano-(3-phenoxyphenyl)methyl-4-chloro-(1-methylethyl)-benzeneacetate
Ferbam	thiocarbamate	Fn	14484-64-1	ferric dimethyldithiocarbamate
Fluazifop-butyl	miscellaneous N	H	69806-50-4	butyl (RS)-2-[4-[[5-(trifluoromethyl)-2-pyridinyl]oxy]phenoxy]propanoate
Fluchloralin	dinitroaniline	H	33245-39-5	N-propyl-N-(2-chloroethyl)-2,6-dinitro-4-trifluoromethyl aniline
Flumetralin	dinitroaniline	PGR	62924-70-3	2-chloro-N-[2,6-dinitro-4-(trifluoromethyl)-phenyl]-N-ethyl-6-fluorobenzenemethanamine
Fluometuron	urea	H	2164-17-2	1,1-dimethyl-3-(α,α,α-trifluoro-m-tolyl)urea
Fluorodifen	miscellaneous N	U	15457-05-3	4-nitrophenyl α,α,α-trifluoro-2-nitro-p-tolyl ether
Fluridone	miscellaneous N	H	59756-60-4	1-methyl-3-phenyl-5-[3-(trifluoromethyl)phenyl]-4(1H)-pyridinone
Flutolanil	amide	Fn	66332-96-5	3'-isopropoxy-2-(trifluoromethyl)benzanilide
Folpet	imide	Fn	133-07-3	N-[(trichloromethyl)thio]phthalimide
Fomesafen	amide	H	72178-02-0	5-[2-chloro-4-(trifluoromethyl)phenoxy]-N-(methyl sulfonyl)-2-nitrobenzamide
Fomesafen amine	fomesafen degradate			5-[2-chloro-4-(trifluoromethyl)phenoxy]-N-(methyl sulfonyl)-2-aminobenzamide
Fonofos	organophosphorus	I	944-22-9	O-ethyl S-phenyl ethylphosphonodithioate
Formaldehyde	VOC	Fm,Ad	50-00-0	methanal
Formetanate HCl	carbamate	I, Ac	23422-53-9	3-dimethylaminomethyleneaminophenyl methylcarbamate hydrochloride

APPENDIX. Glossary Of Common And Chemical Names Of Pesticides And Related Compounds Given In Text—*Continued*

Common Name	Chemical Class	Use	CAS No.	Chemical Nomenclature
Fosamine	see Fosamine ammonium			
Fosamine ammonium	organophosphorus	PGR	25954-13-6	ammonium ethyl carbamoylphosphonate
Fosetyl-Al	see Fosetyl aluminum			
Fosetyl aluminum	organophosphorus	Fn	39148-24-8	aluminum tris(*O*-ethyl phosphonate)
Furadan	see Carbofuran			
Glyphosate	amino acid derivative	H	1071-83-6	*N*-(phosphonomethyl)glycine, isopropylamine salt
HCB	see Hexachlorobenzene			
α-HCH	organochlorine	I	319-84-6	1α,2α,3β,4α,5β,6β-hexachlorocyclohexane
β-HCH	organochlorine	I	319-85-7	1α,2β,3α,4β,5α,6β-hexachlorocyclohexane
γ-HCH	organochlorine	I	58-89-9	1α,2α,3β,4α,5α,6β-hexachlorocyclohexane
δ-HCH	organochlorine	I	319-86-8	1α,2α,3α,4β,5α,6β-hexachlorocyclohexane
HCH, technical	organochlorine	I	608-73-1	1,2,3,4,5,6-hexachlorocyclohexane (mixture of isomers)
Heptachlor	organochlorine	I	76-44-8	1,4,5,6,7,8,8-heptachloro-3a,4,7,7a-tetrahydro-4,7-methano-1*H*-indene
Heptachlor epoxide	heptachlor degradate		1024-57-3	2,3,4,5,6,7,8-heptachloro-1a,1b,5,5a,6,6a-hexahydro-2,5-methano-2*H*-indeno(1,2b)oxirene
1-Heptene	VOC	U	592-76-7	1-heptene
Hexachlorobenzene	organochlorine	Fn	118-74-1	hexachlorobenzene
Hexazinone	triazine	H	51235-04-2	3-cyclohexyl-6-(dimethylamino)-1-methyl-1,3,5-triazine-2,4(1*H*,3*H*)-dione

APPENDIX. Glossary Of Common And Chemical Names Of Pesticides And Related Compounds Given In Text—*Continued*

Common Name	Chemical Class	Use	CAS No.	Chemical Nomenclature
HOA	see Hydroxyatrazine			
Hydramethylnon	miscellaneous N	I	67485-29-4	tetrahydro-5,5-dimethyl-2(1*H*)-pyrimidinone [3-[4-(trifluoromethyl)phenyl]-1-[2-[4-(trifluoromethyl) phenyl]ethenyl]-2-propenylidene]hydrazone
Hydramethylon	see Hydramethylnon			
Hydroxyalachlor	alachlor degradate			
3-Hydroxyalachlor	alachlor degradate			
Hydroxyatrazine	atrazine degradate		2163-68-0	6-hydroxy-*N*-ethyl-*N'*-(1-methylethyl)-1,3,5-triazine-2,4-diamine
3-Hydroxycarbofuran	carbofuran degradate		16655-82-6	2,3-dihydro-2,2-dimethyl-7-methylcarbamate, 3,7-benzofurandiol
1-Hydroxychlordene	chlordane degradate		24009-05-0	4,5,6,7,8,8-hexachloro-3a,4,7,7a-tetrahydro-(endo,exo)-4,7-methanoinden-1-ol
5-Hydroxydicamba	dicamba degradate			
4-Hydroxymethyl pendimethalin	pendimethalin degradate			*N*-(1-ethylpropyl)-3-methyl-4-hydroxymethyl-2,6-dinitro-benzeneamine
N-(4-hydroxyphenyl)-*N'*-methoxy-*N'*-methylurea	monolinuron degradate			*N*-(4-hydroxyphenyl)-*N'*-methoxy-*N'*-methylurea
Imazapyr	imidazolinone	H	81334-34-1	2-(4-isopropyl-4-methyl-5-oxo-2-imidazolin-2-yl)nicotinic acid
Imazaquin	imidazolinone	H	81335-37-7	(±)-2-[4,5-dihydro-4-methyl-4-(1-methylethyl)-5-oxo-1*H*-imidazol-2-yl]-3-quinolinecarboxylic acid
Imazethapyr	imidazolinone	H	81335-77-5	(±)-2-[4,5-dihydro-4-methyl-4-(1-methylethyl)-5-oxo-1*H*-imidazol-2-yl]-5-ethyl-3-pyridinecarboxylic acid
Imidacloprid	chloronicotinyl	I	138261-41-3	1-[(6-chloro-3-pyridinyl)methyl]-*N*-nitro-2-imidazolidinimine

APPENDIX. Glossary Of Common And Chemical Names Of Pesticides And Related Compounds Given In Text—*Continued*

Common Name	Chemical Class	Use	CAS No.	Chemical Nomenclature
Imidan	see Phosmet			
Iprodione	amide	Fn	36734-19-7	3-(3,5-dichlorophenyl)-*N*-(1-methylethyl)-2,4-dioxo-1-imidazolidinecarboxamide
Isazofos	organophosphate	I,N	42509-80-8	*O*-[5-chloro-1-(methylethyl)-*1H*-1,2,4-triazol-3-yl] *O,O*-diethyl phosphorothioate
Isobornyl thiocyanoacetate	miscellaneous N	I	115-31-1	thiocyanato-1,7,7-trimethylbicyclo-(2,2,1)hept-2-yl-exo-acetate
Isodrin	organochlorine, aldrin isomer	I	465-73-6	1,2,3,4,10,10-hexachloro-1,4,4a,5,8,8a-hexahydro-1,4:5,8-(endo,endo)-dimethanonaphthalene
Isofenphos	organophosphorus	I	25311-71-1	1-methylethyl 2-[[ethoxy[(1-methylethyl)amino] phosphinothioyl]oxy]benzoate
Isophenfos	see Isofenphos			
Isophenphos	see Isofenphos			
Isopropalin	dinitroaniline	H	33820-53-0	2,6-dinitro-*N,N*-dipropylcumidine
Isoprothiolane	miscellaneous	F,I	50512-35-1	di-isopropyl 1,3-dithiolan-2-ylidenemalonate
Isoproturon	urea	H	34123-59-6	3-(4-isopropylphenyl)-*N-N'*-dimethylurea
Isoxathion	organophosphorus	I	18854-01-8	*O,O*-diethyl *O*-(5-phenyl-3-isoxazolyl)phosphorothioate
Kelthane	see Dicofol			
Kepone	see Chlordecone			
3-Keto carbofuran	carbofuran degradate		16709-30-1	2,2-dimethyl-7-[[(methylamino)carbonyl]oxy]-3(*2H*)-benzofuranone
3-Keto carbofuran phenol	carbofuran degradate			

APPENDIX. Glossary Of Common And Chemical Names Of Pesticides And Related Compounds Given In Text—*Continued*

Common Name	Chemical Class	Use	CAS No.	Chemical Nomenclature
2-Keto molinate	molinate degradate			*S*-ethyl hexahydro-1*H*-azepine-2-one-1-carbothioate
4-Keto molinate	molinate degradate			*S*-ethyl hexahydro-1*H*-azepine-4-one-1-carbothioate
Lead arsenate	inorganic arsenical	I	7784-40-9	arsenic acid, lead(II) salt (1:1)
Leptophos	organophosphorus	I	21609-90-5	*O*-(4-bromo-2,5-dichlorophenyl) *O*-methylphenyl phosphonothioate
Limonene	hydrocarbon	Ad	138-86-3	1-methyl-4-(1-methylethenyl)cyclohexene
Lindane	see γ-HCH			
Linuron	urea	H	330-55-2	*N′*-(3,4-dichlorophenyl)-*N*-methoxy-*N*-methylurea
Malaoxon	see Malathion oxon			
Malathion	organophosphorus	I	121-75-5	*O,O*-dimethyl *S*-[1,2-bis(ethoxycarbonyl)ethyl] dithiophosphate
Malathion oxon	malathion degradate		1634-78-2	*O,O*-dimethyl *S*-[1,2-bis(ethoxycarbonyl)ethyl] phosphorothioate
Mancozeb	ethylene bisdithiocarbamate	Fn	8018-01-7	coordination product of zinc ion and manganese ethylene bisdithiocarbamate
Maneb	ethylene bisdithiocarbamate	Fn	12427-38-2	manganese ethylenebisdithiocarbamate
Manzate	see Maneb			
MBC	see Methyl-2-benzimidazole carbamate			
MCPA	chlorophenoxy acid	H	94-74-6	(4-chloro-2-methyl)phenoxyacetic acid
MCPA, dimethylamine salt	chlorophenoxy acid salt	H		dimethylamine (4-chloro-2-methyl)phenoxyacetate

APPENDIX. Glossary Of Common And Chemical Names Of Pesticides And Related Compounds Given In Text—*Continued*

Common Name	Chemical Class	Use	CAS No.	Chemical Nomenclature
MCPA, DMA salt	see MCPA, dimethylamine salt			
MCPB	chlorophenoxy acid	H	94-81-5	4-(4-chloro-2-methylphenoxy)butanoic acid
MCPP	chlorophenoxy acid salt	H	1929-86-8	potassium (RS)-2-(4-chloro-2-methylphenoxy)propanoate
Mecoprop	chlorophenoxy acid	H	7085-19-0	(RS)-2-(4-chloro-2-methylphenoxy)propanoic acid
Mercury	inorganic	Fm,Fn		
Merphos	organophosphorus	Df	150-50-5	*S,S,S*-tributyl phosphorotrithioite
Metabromuron	see Metobromuron			
Metalaxyl	amino acid derivative	Fn	57837-19-1	*N*-(2,6-dimethylphenyl)-*N*-(methoxyacetyl)-DL-alanine methyl ester
Metaldehyde	miscellaneous	Mo	108-62-3	2,4,6,8-tetramethyl-1,3,5,7-tetroxocane
Metamitron	triazine	H	41394-05-2	4-amino-3-methyl-6-phenyl-1,2,4-triazin-5(4*H*)-one
Metam-sodium	thiocarbamate	Fm,Fn, H,I,N	137-42-8	sodium *N*-methyldithiocarbamate
Metasystox	see Demeton methyl			
Methabenzthiazuron	urea	H	18691-97-9	*N*-2-benzothiazolyl-*N,N'*-dimethylurea
Methamidophos	organophosphorus	I	10265-92-6	*O,S*-dimethyl phosphoramidothioate
Methazole	miscellaneous N	H	20354-26-1	2-(3,4-dichlorophenyl)-4-methyl-1,2,4-oxadiazolidine-3,5-dione
Methidathion	organophosphorus	I,Ac	950-37-8	[(5-methoxy-2-oxo-1,3,4-thiadiazol-3(2*H*)-yl)methyl] *O,O*-dimethylphosphorodithioate

APPENDIX. Glossary Of Common And Chemical Names Of Pesticides And Related Compounds Given In Text—*Continued*

Common Name	Chemical Class	Use	CAS No.	Chemical Nomenclature
Methiocarb	carbamate	I,Ac, Mo	2032-65-7	3,5-dimethyl-4-(methylthio)phenyl methylcarbamate
Methiocarb sulfone	methiocarb degradate			3,5-dimethyl-4-(methylsulfonyl)phenyl methylcarbamate
Methomyl	carbamate, thiodicarb degradate	I	16752-77-5	methyl *N*-[[(methylamino)carbonyl]oxy] ethanimidothioate
Methoprene	miscellaneous	IGR	40596-69-8	isopropyl (2E,4E,7S)-11-methoxy-3,7,11-trimethyldodeca-2,4-dienoate
Methoxychlor	organochlorine	I	72-43-5	2,2-bis(4-methoxyphenyl)-1,1,1-trichloroethane
Methylarsonic acid, mono-sodium salt	organic arsenical	H	2163-80-6	monosodium methanearsonate
Methyl 2-benzimidazole carbamate	benomyl degradate		10605-21-7	methyl 2-benzimidazole carbamate
Methyl benzthiazolylurea	methabenzthiazuron degradate			methyl benzthiazolylurea
Methyl bromide	VOC	Fm,Ad	74-83-9	bromomethane
Methylene chloride	VOC	Fm,Ad	79-09-2	dichloromethane
Methyl isothiocyanate	VOC	Fm	556-61-6	methyl isothiocyanate
2-Methyl-1-nitropropane	VOC	U		2-methyl-1-nitropropane
Methyl paraoxon	organophosphorus	I	950-35-6	*O,O*-dimethyl *O*-(4-nitrophenyl) phosphate
Methyl parathion	organophosphorus	I	298-00-0	*O,O*-dimethyl *O*-(4-nitrophenyl) phosphorothioate
2-Methyl-3-propyl-*cis*-oxirane	VOC	U		2-methyl-3-propyl-*cis*-oxirane

APPENDIX. Glossary Of Common And Chemical Names Of Pesticides And Related Compounds Given In Text—*Continued*

Common Name	Chemical Class	Use	CAS No.	Chemical Nomenclature
2-Methylpropyl oxirane	VOC	U		2-methylpropyl oxirane
Methyl trithion	organophosphorus	I,Ac	953-17-3	S-[[(4-chlorophenyl)thio]methyl] O,O-dimethyl phosphorodithioate
Metiram	thiocarbamate	Fn	9006-42-2	tris[amine-[ethylen bis(dithiocarbamato)]zinc(II)] [tetrahydro-1,2,4,7-dithiadiazocine-3,8-dithione] polymer
Metobromuron	urea	H	3060-89-7	3-(4-bromophenyl)-1-methoxy-1-methylurea
Metolachlor	acetanilide	H	51218-45-2	2-chloro-N-(2-ethyl-6-methylphenyl)-N-(2-methoxy-1-methylethyl)acetamide
Metoxuron	urea	H	19937-59-8	N'-(3-chloro-4-methoxyphenyl)-N,N-dimethylurea
Metribuzin	triazine	H	21087-64-9	4-amino-6-(1,1-dimethylethyl)-3-(methylthio)-1,2,4-triazin-5(4H)-one
Metribuzin DA (deaminated metribuzin)	metribuzin degradate			6-(1,1-dimethylethyl)-3-(methylthio)-1,2,4-triazin-5(4H)-one
Metribuzin DADK (deaminated diketo metribuzin)	metribuzin degradate			6-(1,1-dimethylethyl)-1,2,4-triazin-3,5-dione
Metribuzin DK (diketo metribuzin)	metribuzin degradate			4-amino-6-(1,1-dimethylethyl)-1,2,4-triazin-3,5-dione
Metriflufen	fluorophenoxy acid	H		methyl 2-[4-(4-trifluoromethylphenoxy)phenoxy] propanoate
Metsulfuron	sulfonylurea	H	79510-48-8	2-[[[(4-methoxy-6-methyl-1,3,5-triazin-2-yl)-amino] carbonyl]amino]sulfonyl]benzoic acid
Metsulfuron-methyl	sulfonylurea	H	74223-64-6	methyl 2-[[[[(4-methoxy-6-methyl-1,3,5-triazin-2-yl)-amino]carbonyl]amino]sulfonyl]benzoate
Mevinphos	organophosphorus	I,Ac	7786-34-7	methyl 3-[(dimethoxyphosphinyl)oxy]-2-butenoate
Mexacarbate	carbamate	I,Mo, Ac	315-18-4	4-dimethylamino-3,5-xylyl methylcarbamate

APPENDIX. Glossary Of Common And Chemical Names Of Pesticides And Related Compounds Given In Text—*Continued*

Common Name	Chemical Class	Use	CAS No.	Chemical Nomenclature
MGK 264	imide	IS	113-48-4	N-(2-ethylhexyl)bicyclo(2,2,1)-hept-5-ene-2,3-dicarboximide
Mirex	organochlorine	I	2385-85-5	1,1a,2,2,3,3a,4,5,5,5a,5b,6-dodecachlorooctahydro-1,3,4-metheno-1H-cyclobuta[cd]pentalene
Molinate	thiocarbamate	H	2212-67-1	S-ethyl hexahydro-1H-azepine-1-carbothioate
Molinate sulfoxide	molinate degradate			hexahydro-1H-azepine-1-(ethylsulfinyl)-carbamate
Monocrotophos	organophosphorus	I,Ac	6923-22-4	dimethyl (E)-[1-methyl-3-(methylamino)-3-oxo-1-propenyl] phosphate
Monolinuron	urea	H	1746-81-2	N'-(4-chlorophenyl)-N-methoxy-N-methylurea
Monuron	urea	H	150-68-5	N'-(4-chlorophenyl)-N,N-dimethylurea
MSMA	see Methylarsonic acid, monosodium salt			
MTP (monomethyl tetrachloroterephthalate)	dacthal degradate		887-54-7	monomethyl 2,3,5,6-tetrachloro-1,4-benzenedicarboxylate
Myclobutanil	triazole	Fn	88671-89-0	α-butyl-α-(4-chlorophenyl)-1H-1,2,4-triazole-1-propanenitrile
Napropamide	amide	H	15299-99-7	(RS)-N,N-diethyl-2-(1-naphthyloxy)propionamide
Naled	organophosphorus	I	300-76-5	1,2-dibromo-2,2-dichloroethyl dimethyl phosphate
Naphthalene	hydrocarbon	I,Fm	91-20-3	naphthalene
1-Naphthol	carbaryl degradate		90-15-3	1-naphthalenol
Naptalam	amine	H	132-66-1	sodium 2-[(1-naphthalenylamino)carbonyl]benzoate
Neburon	urea	H	555-37-3	1-butyl-3-(3,4-dichlorophenyl)-1-methylurea

APPENDIX. Glossary Of Common And Chemical Names Of Pesticides And Related Compounds Given In Text—*Continued*

Common Name	Chemical Class	Use	CAS No.	Chemical Nomenclature
Nitralin	dinitroaniline	H	4726-14-1	4-methylsulfonyl-2,6-dinitro-*N,N*-dipropylaniline
Nitrapyrin	miscellaneous N	Ni	1929-82-4	2-chloro-6-(trichloromethyl)pyridine
Nitrofen	organochlorine	H	1836-75-5	2,4-dichlorophenyl 4-nitrophenyl ether
4-Nitrophenol	nitrophenol, methyl parathion degradate	Fn,Ad	100-02-7	4-nitrophenol
trans-Nonachlor	organochlorine	I	39765-80-5	1,2,3,4,5,6,7,8,8-nonachloro-2,3,3a,4,7,7a-hexahydro-4,7-methano-1*H*-indene (combined nomenclature for *cis* and *trans* isomers)
Norea	urea	H	18530-56-8	3-(hexahydro-4,7-methanoindan-5-yl)-1,1-dimethylurea
Norflurazon	amine	H	27314-13-2	4-chloro-5-methylamino-2-(α,α,α-trifluoro-m-tolyl) pyridazin-3(2*H*)-one
Octyl bicycloheptene dicarboximide	imide	I,Fn	113-48-4	n-octyl bicyclo-(2.2.1)-5-heptene-2,3-dicarboximide
Oryzalin	dinitroaniline	H	19044-88-3	3,5-dinitro-N_4,N_4-dipropylsulfanilamide
Ovex	see Chlorfenson			
Oxadiazol	oxadiazol	H	19666-30-9	2-*tert*-butyl-4-(2,4-dichloro-5-isopropoxyphenyl)-Δ^2-1,3,4-oxadiazolin-5-one
Oxamyl	carbamate	I,N,Ac	23135-22-0	S-methyl *N',N'*-dimethyl-*N*-(methylcarbamoyloxy)-1-thio-oxaminidate
Oxine-copper	organocopper	Fn	10380-28-6	copper 8-quinolinate
Oxychlordane	chlordane degradate		27304-13-8	2,3,4,5,6,6a,7,7-octachloro-1a,1b,5,5a,6,6-hexahydro-2,5-methano-2*H*-indeno(1,2b)oxirene
Oxydemeton-methyl	organophosphorus	I	301-12-2	S-[2-(ethylsulfinyl)ethyl] *O,O*-dimethyl phosphorothioate

APPENDIX. Glossary Of Common And Chemical Names Of Pesticides And Related Compounds Given In Text—*Continued*

Common Name	Chemical Class	Use	CAS No.	Chemical Nomenclature
Oxydisulfoton	see Disulfoton sulfoxide			
Oxyfluorfen	diphenyl ether	H	42874-03-3	2-chloro-1-(3-ethoxy-4-nitrophenoxy)-4-(trifluoromethyl benzene)
Oxytetracycline	amide	An	79-57-2	4-(dimethylamino)-1,4,4α,5,5α,6,11,12a-octahydro-3,5,6,10,12,12a-hexahydroxy-6-methyl-1,11-dioxo-2-naphthacenecarboxamide
Oxythioquinox	dithiocarbonate	Ac,Fm, Fn	2439-01-2	6-methyl-1,3-dithiolo[4,5-b]quinoxalin-2-one
Paraoxon	organophosphorus	I	311-45-5	*O,O*-diethyl *O*-(4-nitrophenyl) phosphate
Paraquat	miscellaneous N	H	1910-42-5	1,1'-dimethyl-4,4'-bipyridinium ion, dichloride salt
Parathion	organophosphorus	I	56-38-2	*O,O*-diethyl *O*-(4-nitrophenyl) phosphorothioate
Patoran	see Metobromuron			
PCBs	see Polychlorinated biphenyls			
PCNB	organochlorine	Fn	82-68-8	pentachloronitrobenzene
PCP	see Pentachlorophenol			
Pebulate	thiocarbamate	H	1114-71-2	*S*-propyl butyl(ethyl)thiocarbamate
Pendimethalin	dinitroaniline	H	40487-42-1	*N*-(1-ethylpropyl)-3,4-dimethyl-2,6-dinitrobenzeneamine
Pentachlorophenol	organochlorine	Fn,Mo, Ad	87-86-5	pentachlorophenol
Permethrin	pyrethroid	I	52645-53-1	(3-phenoxyphenyl)methyl (±)-*cis,trans*-3-(2,2-dichloroethenyl)-2,2-dimethylcyclopropanecarboxylate

APPENDIX. Glossary Of Common And Chemical Names Of Pesticides And Related Compounds Given In Text—*Continued*

Common Name	Chemical Class	Use	CAS No.	Chemical Nomenclature
cis-Permethrin	pyrethroid	I	61949-76-6	(3-phenoxyphenyl)methyl *cis*-3-(2,2-dichloroethenyl)-2,2-dimethylcyclopropanecarboxylate
trans-Permethrin	pyrethroid	I		(3-phenoxyphenyl)methyl *trans*-3-(2,2-dichloroethenyl)-2,2-dimethylcyclopropanecarboxylate
Perthane	organochlorine	I	72-56-0	1,1-dichloro-2,2-bis(4-ethylphenyl)ethane
Phenacetin	acetanilide	U	62-44-2	*N*-(4-ethoxyphenyl) acetamide
Phorate	organophosphorus	I	298-02-2	*O,O*-diethyl *S*-ethylthiomethyl phosphorodithioate
Phorate sulfone	phorate degradate		2588-04-7	*O,O*-diethyl *S*-ethylsulfonylmethyl phosphorodithioate
Phorate sulfoxide	phorate degradate		2588-05-8	*O,O*-diethyl *S*-ethylsulfinylmethyl phosphorothioate
Phoratoxon	phorate degradate			
Phoratoxon sulfone	phorate degradate			
Phoratoxon sulfoxide	see Phorate sulfoxide			
Phosalone	organophosphorus	Ac,I	2310-17-0	*S*-[(6-chloro-2-oxo-3(2*H*)-benzoxazolyl)methyl] *O,O*-diethyl phosphorodithioate
Phosdrin	see Mevinphos			
Phosmet	organophosphorus	I	732-11-6	*N*-(mercaptomethyl)phthalimide-*S*-(*O,O*-dimethylphosphorodithioate)
Phosphamidon	organophosphorus	I	13171-21-6	*O,O*-dimethyl *O*-(2-chloro-2-diethylcarbamoyl-1-methylvinyl)-phosphate
Picloram	amine	H	1918-02-1	4-amino-3,5,6-trichloropicolinic acid
Pindone	inandione	I,R	83-26-1	2-(2,2-dimethyl-1-oxopropyl)-1*H*-indene-1,3(2*H*)-dione

APPENDIX. Glossary Of Common And Chemical Names Of Pesticides And Related Compounds Given In Text—*Continued*

Common Name	Chemical Class	Use	CAS No.	Chemical Nomenclature
Piperonyl butoxide	miscellaneous	IS	51-03-6	5-[[2-(2-butoxyethoxy)ethoxy]methyl]-6-propyl-1,3-benzodioxole
Pirimicarb	carbamate	I	23103-98-2	2-dimethylamino-5,6-dimethylpyrimidin-4-yl dimethylcarbamate
Pirimicarb sulfone	primicarb degradate			
Polychlorinated biphenyls	organochlorine	Ind	1336-36-3	polychlorobiphenyls
Profenofos	organophosphorus	Ac,I	41198-08-7	O-4-bromo-2-chlorophenyl O-ethyl S-propyl phosphorothioate
Profluralin	dinitroaniline	H	26399-36-0	2,6-dinitro-N-cyclopropylmethyl-N-propyl-4-(trifluoromethyl)benzenamine
Promecarb	carbamate	I	2631-37-0	3-isopropyl-5-methylphenyl methylcarbamate
Prometon	triazine	H	1610-18-0	6-methoxy-N,N'-bis(1-methylethyl)-1,3,5-triazine-2,4-diamine
Prometone	see Prometon			
Prometryn	triazine	H	7287-19-6	N,N'-bis(1-methylethyl)-6-(methylthio)-1,3,5-triazine-2,4-diamine
Prometryne	see Prometryn			
Pronamide	amide	H	23950-58-5	3,5-dichloro-N-(1,1-dimethyl-2-propynyl)benzamide
Propachlor	acetanilide	H	1918-16-7	2-chloro-N-(1-methylethyl)-N-phenylacetanilide
Propanil	amide	H	709-98-8	N-(3,4-dichlorophenyl)propanamide
Propargite	sulfite ester	Ac	2312-35-8	2-[4-(1,1-dimethylethyl)phenoxy]cyclohexyl-2-propynyl sulfite
Propazine	triazine	H	139-40-2	6-chloro-N,N'-bis(1-methylethyl)-1,3,5-triazine-2,4-diamine
Propetamphos	organophosphorus	I	31218-83-4	(E)-O-2-isopropoxycarbonyl-1-methylvinyl O-methyl ethylphosphoramidothioate
Propham	carbamate	H,PGR	122-42-9	1-methylethylphenyl carbamate

APPENDIX. Glossary Of Common And Chemical Names Of Pesticides And Related Compounds Given In Text—*Continued*

Common Name	Chemical Class	Use	CAS No.	Chemical Nomenclature
Propiconazole	triazole	Fn	60207-90-1	1-[[2-(2,4-dichlorophenyl)-4-propyl-1,3-dioxolan-2-yl] methyl]-1*H*-1,2,4-triazole
Propoxur	see Baygon			
3-Propoxy-1-propene	VOC	U	1471-03-0	3-propoxy-1-propene
Prothiophos	organophosphorus	I	34643-46-4	*O*-(2,4-dichlorophenyl) *O*-ethyl *S*-propyl phosphorodithioate
Pyriclor	organochlorine	H	1970-40-7	2,3,5-trichloro-4-pyridinol
Pyridaphenthion	organophosphorus	I	119-12-0	*O*-(1,6-dihydro-6-oxo-1-phenylpyridazin-3-yl) *O,O*-diethyl phophorothioate
Quintozene	see PCNB			
Reldan	see Chlorpyrifos-methyl			
Resmethrin	pyrethroid	I	10453-86-8	(5-benzyl-3-furyl)methyl *cis,trans*-2,2-dimethyl-3-(2-methylpropenyl)-cyclopropanecarboxylate
Ronalin	see Ronilan			
Ronilan	organochlorine	Fn	50471-44-8	3-(3,5-dichlorophenyl)-5-methyl-5-vinyl-1,3-oxazolidine-2,4-dione
Ronnel	organophosphorus	I	299-84-3	*O,O*-dimethyl *O*-(2,4,5-trichlorophenyl)phosphorothioate
Rotenone	miscellaneous	I	83-79-4	1,2,12,12a-tetrahydro-8,9-dimethoxy-2-(1-methylethenyl)-[1]benzopyrano[3,4-b]furo[2,3-h][1]-benzopyran-6(6*H*)-one
Rotenolone	Rotenone degradate			
Roundup	see Glyphosate			
Secbumeton	triazine	H	26259-45-0	2-*sec*-butylamino-4-ethylamino-6-methoxy-1,3,5-triazine

APPENDIX. Glossary Of Common And Chemical Names Of Pesticides And Related Compounds Given In Text—*Continued*

Common Name	Chemical Class	Use	CAS No.	Chemical Nomenclature
Sethoxydim	miscellaneous N	H	74051-80-2	2-[1-(ethoxyimino)butyl]-5-[2-(ethylthio)propyl]-3-hydroxy-2-cyclohexen-1-one
Sevin	see Carbaryl			
Siduron	urea	H	1982-49-6	1-(2-methylcyclohexyl)-3-phenylurea
Silvex	see 2,4,5-TP			
Simazine	triazine	H	122-34-9	2-chloro-4,6-bis(ethylamino)-s-triazine
Simetone	triazine	H	673-04-1	2,4-bis(ethylamino)-6-methoxy-s-triazine
Simetryn	triazine	H	1014-70-6	2,4-bis(ethylamino)-6-methylmercapto-s-triazine
Sodium bromide	inorganic	Fn,H,I	7647-15-6	sodium bromide
Sodium thiocyanate	miscellaneous N	H	540-72-7	sodium thiocyanate
Stirofos	see Tetrachlorvinphos			
Streptomycin	amine	An	57-92-1	O-2-deoxy-2-(methylamino)-α-L-glucopyranosyl-(1→2)-O-5-deoxy-3-C-formyl-α-L-lyxofuranosyl-(1→4)-N,N'-bis(aminoiminomethyl)-D-streptamine
Strobane	organochlorine	I	8001-50-1	mixture of polychlorinated camphene, pinene and related terpenes
Sulfometuron-methyl	sulfonylurea	H	74222-97-2	methyl 2-[[[[(4,6-dimethyl-2-pyrimidinyl)amino]carbonyl]amino]sulfonyl]benzoate
Sulprofos	organophosphorus	I	35400-43-2	O-ethyl O-[(4-methylthio)phenyl] S-propyl phosphorodithioate
Sumithrin	pyrethroid	I	26046-85-5	3-phenoxybenzyl (1R)-cis,trans-chrysanthemate
Swep	carbamate	H	1918-18-9	methyl 3,4-dichlorocarbanilate

APPENDIX. Glossary Of Common And Chemical Names Of Pesticides And Related Compounds Given In Text—*Continued*

Common Name	Chemical Class	Use	CAS No.	Chemical Nomenclature
2,4,5-T	chlorophenoxy acid	H	93-76-5	2,4,5-trichlorophenoxyacetic acid
TBA	see 2,3,6-Trichloro-benzoic acid			
TCA	see Trichloroacetic acid			
TCBC	see Trichlorobenzyl-chloride			
TCDD	impurity in 2,4,5-T and other organochlorines		1746-01-6	2,3,7,8-tetrachlorodibenzo-*p*-dioxin
TCP	see 1,2,3-Trichloro-propane			
Tebuthiuron	urea	H	34014-18-1	*N*-[5-(1,1-dimethylethyl)-1,3,4-thiadiazol-2-yl]-*N,N'*-dimethylurea
Tefluthrin	pyrethroid	I	79538-32-2	1α,3α(Z)-(±)-(2,3,5,6-tetrafluoro-4-methylphenyl)methyl-3-(2-chloro-3,3,3-trifluoro-1-propenyl)-2,2-dimethylcyclopropanecarboxylate
Telone II	VOC	Fm, N	542-75-6	mixture of *cis*- and *trans*-1,3-dichloropropene
Terbacil	uracil	H	5902-51-2	3-*tert*-butyl-5-chloro-6-methyluracil
Terbufos	organophosphorus	I,N	13071-79-9	*S*-[[(1,1-dimethylethyl)thio]methyl] *O,O*-diethyl phosphorodithioate
Terbufos sulfone	terbufos degradate		56070-16-7	*S*-[[(1,1-dimethylethyl)sulfonyl]methyl] *O,O*-diethyl phosphorodithioate
Terbufos sulfoxide	terbufos degradate		10548-10-4	*S*-[[(1,1-dimethylethyl)sulfinyl]methyl] *O,O*-diethyl phosphorodithioate
Terbuthylazine	triazine	H	5915-41-3	2-(*tert*-butylamino)-4-chloro-6-(ethylamino)-*s*-triazine
Terbutryn	triazine	H	886-50-0	2-(*tert*-butylamino)-4-(ethylamino)-6-(methylthio)-*s*-triazine

APPENDIX. Glossary Of Common And Chemical Names Of Pesticides And Related Compounds Given In Text—*Continued*

Common Name	Chemical Class	Use	CAS No.	Chemical Nomenclature
Tetrachloroethylene	VOC	Fm,Ad	127-18-4	tetrachloroethene
Tetrachloroterephthalic acid	see TPA			
Tetrachlorvinphos	organophosphorus	I	22248-79-9	(Z)-2-chloro-1-(2,4,5-trichlorophenyl)vinyl dimethyl phosphate
Tetradecane	VOC	U	629-59-4	tetradecane
Tetradifon	organochlorine	Ac	116-29-0	1,2,4-trichloro-5-[(4-chlorophenyl)sulfonyl]benzene
Tetramethrin	pyrethroid	I	7696-12-0	3,4,5,6-tetrahydrophthalimidomethyl (1RS)-*cis,trans*-chrysanthemate
Thiabendazole	imidazole	Fn	148-79-8	2-(4'-thiazolyl)-benzimidazole
Thiobencarb	thiocarbamate	H	28249-77-6	S-4-chlorobenzyl diethylthiocarbamate
Thiobencarb sulfoxide	thiobencarb degradate			S-4-chlorobenzyl diethylsulfinylcarbamate
Thiodan I	see Endosulfan I			
Thiodan II	see Endosulfan II			
Thiodicarb	carbamate	I	59669-26-0	dimethyl N,N'-[thiobis[(methylimino)carbonyloxy]] bis(ethanimidothioate)
Thiofenate	see Thiophanate			
Thiophanate	carbamate	Fn	23564-06-9	diethyl 4,4'-*o*-phenylenebis(3-thioallophanate)
Thiophanate-methyl	carbamate	Fn	23564-05-8	dimethyl 4,4'-*o*-phenylenebis(3-thioallophanate)
Thiram	thiocarbamate	Fn	137-26-8	bis(dimethylthiocarbamyl) disulfide
Thymol	miscellaneous	Fn	89-83-8	5-methyl-2-(1-methylethyl)phenol
Tolclofos-methyl	organophosphorus	Fn	57018-04-9	O-(2,6-dichloro-4-methylphenyl) O,O-dimethyl phosphorothioate

APPENDIX. Glossary Of Common And Chemical Names Of Pesticides And Related Compounds Given In Text—*Continued*

Common Name	Chemical Class	Use	CAS No.	Chemical Nomenclature
Toxaphene	organochlorine	I	8001-35-2	polychlorinated camphene
2,4,5-TP	chlorophenoxy acid	H	93-72-1	(±)-2-(2,4,5-trichlorophenoxy)propanoic acid
TPA (tetrachloroterephthalic acid)	dacthal degradate	I	2136-79-0	2,3,5,6-tetrachlorobenzene-1,4-dicarboxylic acid
Tralomethrin	pyrethroid	I	66841-25-6	(S)-α-cyano-3-phenoxybenzyl(1R,3S)-3-[(1′RS)(1′,2′,2′,2′-tetrabromoethyl)]-2,2-dimethylcyclopropanecarboxylate
Triadimefon	triazole	Fn	43121-43-3	1-(4-chlorophenoxy)-3,3-dimethyl-1-(1*H*-1,2,4-triazol-1-yl)-2-butanone
Triadimenol	triazole, triadimefon degradate	Fn	55219-65-3	β-(4-chlorophenoxy)-α-(1,1-dimethylethyl)-1*H*-1,2,4-triazole-1-ethanol
Triallate	thiocarbamate	H	2303-17-5	S-(2,3,3-trichloro-2-propenyl) bis(1-methylethyl) thiocarbamate
Triasulfuron	sulfonylurea	H	82097-50-5	2-(2-chloroethoxy)-N-[[4-methoxy-6-methyl-1,3,5-triazin-2-yl)amino]carbonyl]benzenesulfonamide
Triazophos	organophosphorus	I,Mi,N	24017-47-8	O,O-diethyl O-(1-phenyl-1*H*-1,2,4-triazol-3-yl) thiophosphate
Tribufos	organophosphorus, merphos degradate	Df	78-48-8	S,S,S-tributyl phosphorotrithioate
Trichlorfon	organophosphorus	I	52-68-6	dimethyl (2,2,2-trichloro-1-hydroxyethyl) phosphonate
Trichloroacetic acid	organochlorine	H	76-03-9	trichloroacetic acid
Trichlorobenzene	organochlorine	H	12002-48-1	1,2,4-trichlorobenzene
2,3,6-Trichlorobenzoic acid	chlorobenzoic acid	H	50-31-7	2,3,6-trichlorobenzoic acid
Trichlorobenzyl chloride	organochlorine	H	1344-32-7	trichloro(chloromethyl)benzene
1,1,1-Trichloroethane	VOC	Fm,Ad	71-55-6	1,1,1-trichloroethane

APPENDIX. Glossary Of Common And Chemical Names Of Pesticides And Related Compounds Given In Text—*Continued*

Common Name	Chemical Class	Use	CAS No.	Chemical Nomenclature
Trichloroethylene	VOC	Fm,Ad	79-01-6	trichloroethene
Trichloronate	organophosphorus	I	327-98-0	O-ethyl O-2,4,5-trichlorophenyl ethylphosphonothioate
Trichlorophenol	organochlorine	Fn	25167-82-2	trichlorophenol (no isomer specified)
1,2,3-Trichloropropane	VOC, D-D impurity		96-18-4	1,2,3-trichloropropane
3,5,6-Trichloro-2-pyridinol	chlorpyrifos degradate		6515-38-4	3,5,6-trichloro-2-pyridinol
Triclopyr	organochlorine	H	55335-06-3	(3,5,6-trichloro-2-pyridinyloxy)acetic acid
Tricosene	hydrocarbon	I	27519-02-4	cis-tricos-9-ene
Tricyclazole	thiazole	Fn	41814-78-2	5-methyl-1,2,4-triazolo[3,4-b]-benzothiazole
Triflumizole	imidazole	Fn	99387-89-0	(E)-4-chloro-α,α,α-trifluoro-N-(1-imidazol-1-yl-2-propoxyethylidene)-o-toluidine
Trifluralin	dinitroaniline	H	1582-09-8	2,6-dinitro-N,N-dipropyl-4-(trifluoromethyl) benzenamine
Triforine	amide	Fn	26644-46-2	N,N'-[1,4-piperazinediylbis(2,2,2-trichloroethylidene)] bis(formamide)
Trimethacarb	carbamate	I	12407-86-2	3,4,5- (or 2,3,5-)trimethylphenyl methylcarbamate
Trimethyl heptane	VOC	U		trimethyl heptane
Trithion	organophosphorus	I,Ac	786-19-6	S-[[(4-chlorophenyl)thio]methyl] O,O-diethyl phosphorodithioate
UDMH	see Unsymmetrical dimethylhydrazine			
Unsymmetrical dimethyl-hydrazine	daminozide degradate		57-14-7	1,1-dimethylhydrazine

APPENDIX. Glossary Of Common And Chemical Names Of Pesticides And Related Compounds Given In Text—*Continued*

Common Name	Chemical Class	Use	CAS No.	Chemical Nomenclature
Vel-4207	chlorobenzoic acid derivative	H		(phenylimino)di-2,1-ethanediyl bis(3,6-dichloro-2-methoxybenzoate)
Vernolate	thiocarbamate	H	1929-77-7	S-propyl dipropylthiocarbamate
Vinclozolin	see Romilan			
Warfarin	miscellaneous	Ro	81-81-2	3-(α-acetonylbenzyl)-4-hydroxycoumarin
m-Xylene	VOC	Ad	108-38-3	1,3-dimethylbenzene
o-Xylene	VOC	Ad	95-47-6	1,2-dimethylbenzene
p-Xylene	VOC	Ad	106-42-3	1,4-dimethylbenzene
Zectran	see Mexacarbate			
Zineb	thiocarbamate	Fn	12122-67-7	[[1,2-ethanediylbis[carbamodithioato]](-2-)]zinc complex
Ziram	thiocarbamate	Fn	137-30-4	zinc bis(dimethyldithiocarbamate)
Zytron	organophosphorus	I	299-85-4	O-2,4-dichlorophenyl O-methyl isopropylphosphoroamidothioate

REFERENCES

Abernathy, J.R., 1989, Protecting water quality through changes in chemical use and management practices, in *Proceedings of the Great Plains Agricultural Council, Lubbock, Texas*: Great Plains Agricultural Council, Lubbock, Texas, pp. 104-108.

Achari, R.G., Sandhu, S.S., and Warren, W.J., 1975, Chlorinated hydrocarbon residues in ground water: *Bull. Environ. Contam. Toxicol.*, v. 13, no. 1, pp. 94-96.

Adams, C.D., and Thurman, E.M., 1991, Formation and transport of deethylatrazine in the soil and vadose zone: *J. Environ. Qual.*, v. 20, no. 3, pp. 540-547.

Aelion, C.M., Dobbins, D.C., and Pfaender, F.K., 1989, Adaptation of aquifer microbial communities to the biodegradation of xenobiotic compounds: Influence of substrate concentration and preexposure: *Environ. Toxicol. Chem.*, v. 8, pp. 75-86.

Aelion, C.M., Swindoll, C.M., and Pfaender, F.K., 1987, Adaptation to and biodegradation of xenobiotic compounds by microbial communities from a pristine aquifer: *Appl. Environ. Microbiol.*, v. 53, no. 9, pp. 2212-2217.

Aga, D.S., Thurman, E.M., and Pomes, M.L., 1994, Determination of alachlor and its sulfonic acid metabolite in water by solid-phase extraction and enzyme-linked immunosorbent assay: *Anal. Chem.*, v. 66, no. 9, pp. 1495-1499.

Agertved, J., Rügge, K., and Barker, J.F., 1992, Transformation of the herbicides MCPP and atrazine under natural aquifer conditions: *Ground Water*, v. 30, no. 4, pp. 500-506.

Akkari, K.H., Frans, R.E., and Lavy, T.L., 1986, Factors affecting degradation of MSMA in soil: *Weed Sci.*, v. 34, no. 5, pp. 781-787.

Albanis, T.A., Pomonis, P.J., and Sdoukos, A.T., 1988, Movement of methyl parathion, lindane and atrazine through lysimeters in field conditions: *Toxicol. Environ. Chem.*, v. 17, pp. 35-45.

Alexander, M., 1981, Biodegradation of chemicals of environmental concern: *Science*, v. 211, no. 4478, pp. 132-138.

Alexander, W.J., and Liddle, S.K., 1986, Ground water vulnerability assessment in support of the first stage of the National Pesticide Survey, in *Proceedings, the Agricultural Impacts on Ground Water—A conference, Omaha, Nebraska*: National Water Well Association, Omaha, Nebr., pp. 77-87.

Allender, W.J., 1991, Movement of bromacil and hexazinone in a municipal site: *Bull. Environ. Contam. Toxicol.*, v. 46, no. 2, pp. 284-291.

Aller, L., Bennett, T., Lehr, J.H., Petty, R., and Hackett, G., 1987, DRASTIC: A standardized system for evaluating ground water pollution potential using hydrogeologic settings: Robert S. Kerr Environmental Research Laboratory, U.S. Environmental Protection Agency Report EPA/600/2-87/035, 622 p.

Alva, A.K., and Singh, M., 1991a, Sorption-desorption of herbicides in soil as influenced by electrolyte cations and ionic strength: *J. Environ. Sci. Health*, v. 26, no. 2, pp. 147-163.

——1991b, Use of adjuvants to minimize leaching of herbicides in soil: *Environ. Manag.*, v. 15, no. 2, pp. 263-267.

Alvarez-Cohen, L., McCarty, P.L., and Roberts, P.V., 1993, Sorption of trichloroethylene onto a zeolite accompanied by methanotrophic biotransformation: *Environ. Sci. Technol.*, v. 27, no. 10, pp. 2141-2148.

Ames, M., Cardozo, C., Troiano, J., Monk, S., Ali, S., and Brown, S., 1987, Sampling for pesticide residues in California well water: 1987 update, well inventory data base: Second annual report to the Legislature: California Department of Food and Agriculture, Environmental Hazards Assessment Program Report EH 87-05, 150 p.

Anderson, G.D., Opaluch, J.J., and Sullivan, W.M., 1985, Nonpoint agricultural pollution: Pesticide contamination of groundwater supplies: *Am. J. Agric. Econ.*, v. 67, pp. 1238-1243.

Anderson, M.P., 1979, Using models to simulate the movement of contaminants through groundwater flow systems: *CRC Crit. Rev. Environ. Control*, v. 8, pp. 97-156.

——1986, Field validation of ground water models, *in* Garner, W.Y., Honeycutt, R.C., and Nigg, H.N., eds., *Evaluation of pesticides in ground water: Developed from a symposium sponsored by the Division of Pesticide Chemistry at the 189th Meeting of the American Chemical Society, Miami Beach, Florida*: American Chemical Society Symposium Series 315, pp. 396-412.

Anderson, R.L., and Humburg, N.E., 1987, Field duration of chlorsulfuron bioactivity in the central Great Plains: *J. Environ. Qual.*, v. 16, no. 3, pp. 263-266.

Anderson, T.A., Kruger, E.L., and Coats, J.R., 1994, Enhanced degradation of a mixture of three herbicides in the rhizosphere of a herbicide-tolerant plant: *Chemosphere*, v. 28, no. 8, pp. 1551-1557.

Ando, C.M., 1992, Survey for chlorthal-dimethyl residues in well water of seven California counties: State of California Environmental Protection Agency Report EH-92-01, 60 p.

Andrilenas, P., 1974, Farmers' use of pesticides in 1971—Quantities: U.S. Department of Agriculture, Economic Research Service, Agricultural Economic Report 252, 56 p.

Armstrong, D.E., and Chesters, G., 1968, Adsorption catalyzed chemical hydrolysis of atrazine: *Environ. Sci. Technol.*, v. 2, no. 9, pp. 683-689.

Armstrong, D.E., Chesters, G., and Harris, R.F., 1967, Atrazine hydrolysis in soil: *Proc. Soil Sci. Soc. Am.*, v. 31, pp. 61-66.

Artman, J., 1995, personal communication, telephone conference, March, Virginia Department of Forestry, Charlottesville, Va.

Ashworth, R.A., Howe, G.B., Mullins, M.E., and Rogers, T.N., 1988, Air-water partitioning coefficients of organics in dilute aqueous solutions: *J. Hazardous Materials*, v. 18, pp. 25-36.

Aspelin, A.L., 1994, Pesticides industry sales and usage, 1992 and 1993 market estimates: U.S. Environmental Protection Agency, Office of Pesticide Programs, Biological and Economic Analysis Division, Economic Analysis Branch Report 733-K-94-001, 33 p.

Aspelin, A.L., Grube, A.H., and Torla, R., 1992, Pesticides industry sales and usage; 1990 and 1991 market estimates: U.S. Environmental Protection Agency, Office of Pesticide Programs, Biological and Economic Analysis Division, Economic Analysis Branch Report 733-K-92-001, 37 p.

Bachmat, Y., and Collin, M., 1987, Mapping to assess groundwater vulnerability to pollution, *in* van Duivenbooden, W., and van Waegeninger, H.G., eds., *Vulnerability of soil and groundwater to pollutants:* National Institute of Public Health and Environmental Hygiene, Bilthoven, The Netherlands, pp. 297-307.

Baedecker, M.J., and Back, W., 1979, Hydrogeological processes and chemical reactions at a landfill: *Ground Water*, v. 17, no. 5, pp. 429-437.

Baier, J.H., and Robbins, S.F., 1982a, *Report on the occurrence and movement of agricultural chemicals in groundwater: North Fork of Suffolk County*: County of Suffolk, Department of Health Services, Hauppauge, N.Y., 71 p.

——1982b, *Report on the occurrence and movement of agricultural chemicals in groundwater: South Fork of Suffolk County*: County of Suffolk, Department of Health Services, Hauppauge, N.Y., 68 p.

Baker, D.B., 1993, The Lake Erie agroecosystem program: Water quality assessments: *Agri. Ecosys. Envir.*, v. 46, pp. 197-215.

Baker, D.B., Bushway, R.J., Adams, S.A., and Macomber, C., 1993, Immunoassay screens for alachlor in rural wells: False positives and an alachlor soil metabolite: *Environ. Sci. Technol.*, v. 27, no. 3, pp. 562-564.

Baker, D.B., Wallrabenstein, L.K., and Richards, R.P., 1994, Well vulnerability and agrichemical contamination: Assessments from a voluntary well testing program, in *New directions in pesticide research, development, management, and policy: Proceedings of the Fourth National Conference on Pesticides*: Virginia Polytechnic Institute and State University, Virginia Water Resources Center, Blacksburg, Va., pp. 470-494.

Baker, D.B., Wallrabenstein, L.K., Richards, R.P., and Creamer, N.L., 1989, *Nitrate and pesticides in private wells of Ohio: A state atlas*: Heidelberg College, Water Quality Laboratory, Tiffin, Ohio, 304 p.

Baker, J.L., 1987, Hydrologic effects of conservation tillage and their importance relative to water quality, in Logan, T.J., Davidson, J.M., Baker, J.L., and Overcash, M.R., eds., *Effects of conservation tillage on groundwater quality: Nitrates and pesticides*: Lewis Publishers, Chelsea, Mich., pp. 113-124.

Baker, J.L., Kanwar, R.S., and Austin, T.A., 1985, Impact of agricultural drainage wells on groundwater quality. *J. Soil Water Conserv.*, v. 40, no. 6, pp. 516-520.

Ball, W.P., and Roberts, P.V., 1991a, Long-term sorption of halogenated organic chemicals by aquifer material. 1. Equilibrium: *Environ. Sci. Technol.*, v. 25, no. 7, pp. 1223-1236.

——1991b, Long-term sorption of halogenated organic chemicals by aquifer material. 2. Intraparticle diffusion: *Environ. Sci. Technol.*, v. 25, no. 7, pp. 1237-1249.

Bank, S., and Tyrrell, R.J., 1984, Kinetics and mechanism of alkaline and acidic hydrolysis of aldicarb: *J. Agric. Food Chem.*, v. 32, no. 6, pp. 1223-1232.

Banton, O., and Villeneuve, J., 1989, Evaluation of groundwater vulnerability to pesticides: A comparison between the pesticide DRASTIC index and the PRZM leaching quantities: *J. Contam. Hydrol.*, v 4, pp. 285-296.

Barbash, J.E., 1987, The effect of surface-active compounds on chemical reactions of environmental interest in natural waters, in American Chemical Society, *Preprints of papers presented at the 194th National Meeting, New Orleans, LA*: American Chemical Society, Division of Environmental Chemistry, v. 27, no. 2, pp. 58-61.

——1994, Problems associated with the use of solute-transport models and vulnerability assessments for predicting the behavior of pesticides in the subsurface, in *Abstracts of papers, 208th American Chemical Society National Meeting, Washington, D.C.*: American Chemical Society Agrichemicals Division, part 1.

Barbash, J.E., and Reinhard, M., 1989a, Abiotic dehalogenation of 1,2-dichloroethane and 1,2-dibromoethane in aqueous solution containing hydrogen sulfide: *Environ. Sci. Technol.*, v. 23, no. 11, pp. 1349-1358.

——1989b, Reactivity of sulfur nucleophiles toward halogenated organic compounds in natural waters, in Saltzman, E.S., and Cooper, W.J., eds., *Biogenic sulfur in the environment*: American Chemical Society Symposium Series 393, pp. 101-138.

——1992a, Abiotic reactions of halogenated ethanes and ethylenes in buffered aqueous solutions containing hydrogen sulfide, in American Chemical Society, *Preprints of papers presented at the 203rd National Meeting, San Francisco, California*: American Chemical Society, Division of Environmental Chemistry, v. 32, no. 1, pp. 670-673.

——1992b. The influence of pH buffers and nitrate concentration on the rate and pathways of abiotic transformation of 1,2-dibromoethane (EDB) in aqueous solution, *in* American Chemical Society, *Preprints of papers presented at the 203rd National Meeting, San Francisco, California*: American Chemical Society, Division of Environmental Chemistry, v. 32, no. 1, pp. 674-677.

Barceló, D.C., Valverde, A., and Fernandez-Alba, A., 1995, Monitoring ground water pollution by carbamate insecticides and their transformation products, *in* American Chemical Society, *Preprints of papers presented at the 209th National Meeting, Anaheim, California*: American Chemical Society, Division of Environmental Chemistry, v. 35, no. 1, pp. 282-285.

Barcelona, M.J., and Holm, T.R., 1991, Oxidation-reduction capacities of aquifer solids: *Environ. Sci. Technol.*, v. 25, no. 9, pp. 1565-1572.

Barcelona, M.J., Holm, T.R., Schock, M.R., and George, G.K., 1989, Spatial and temporal gradients in aquifer oxidation-reduction conditions: *Water Resour. Res.*, v. 25, no. 5, pp. 991-1003.

Barles, B., and Kotas, J., 1984, Pesticides and the nation's ground water: *EPA Journal*, v. 13, no. 3, pp. 42-43.

Barlow, P.M., 1994, Two- and three-dimensional pathline analysis of contributing areas to public-supply wells of Cape Cod, Massachusetts. *Ground Water*, v. 32, no. 3, p. 399.

Barnes, C.J., Goetz, A.J., and Lavy, T.L., 1989, Effects of imazaquin residues on cotton (*Gossypium hirsutum*): *Weed Sci.*, v. 37, no. 6, pp. 820-824.

Barnes, C.J., Lavy, T.L., and Talbert, R.E., 1992, Leaching, dissipation, and efficacy of metolachlor applied by chemigation or conventional methods: *J. Environ. Qual.*, v. 21, no. 2, pp. 232-236.

Barrett, M.R., and Lavy, T.L., 1983, Effects of soil water content on pendimethalin dissipation: *J. Environ. Qual.*, v. 12, no. 4, pp. 504-508.

——1984, Effects of soil water content on oxadiazon dissipation: *Weed Sci.*, v. 32, no. 5, pp. 697-701.

Barrett, M.R., and Williams, W.M., 1989, The occurrence of atrazine in groundwater as a result of agricultural use, *in* Weigmann, D.L., ed., *Pesticides in terrestrial and aquatic environments: Proceedings of a national research conference:* Virginia Polytechnic Institute and State University, Virginia Water Resources Research Center, Blacksburg, Va., pp. 39-61.

Barringer, T., Dunn, D., Battaglin, W., and Vowinkel, E., 1990, Problems and methods involved in relating land use to ground-water quality: *Water Resour. Bull.*, v. 26, no. 1, pp. 1-9.

Barton, C., Vowinkel, E.F., and Nawyn, J.P., 1987, Preliminary assessment of water quality and its relation to hydrogeology and land use: Potomac-Raritan-Magothy aquifer system, New Jersey: U.S. Geological Survey Water-Resources Investigations Report 87-4023, 79 p.

Basham, G.W., and Lavy, T.L., 1987, Microbial and photolytic dissipation of imazaquin in soil: *Weed Sci.*, v. 35, no. 6, pp. 865-870.

Basham, G.W., Lavy, T.L., Oliver, L.R., and Scott, H.D., 1987, Imazaquin persistence and mobility in three Arkansas soils: *Weed Sci.*, v. 35, no. 4, pp. 576-582.

Basile, M., 1982, Percolazione dell'1,3 dicloropropene in differenti tipi di terreno: *Redia Giornale di Zoologia*, v. 65, pp. 75-80.

Battaglin, W.A., and Goolsby, D.A., 1995, Spatial data in geographic information system format on agricultural chemical use, land use, and cropping practices in the United States: U.S. Geological Survey Water-Resources Investigations Report 94-4176, 87 p.

Bayer, D.E., 1966, Effect of surfactants on leaching of substituted urea herbicides in soil: *Weeds*, v. 3, pp. 249-252.

Becker, R.L., Herzfeld, D., Stamm-Katovich, E.J., and Ostlie, K.R., 1991, A pesticide's risk of leaching or runoff: *American Nurseryman*, v. 173, no. 8, pp. 108-111.

Behl, E., 1994, Groundwater protection in the Office of Pesticide Program's new risk assessment paradigm, in *Abstracts of papers, 208th American Chemical Society National Meeting, Washington, D.C.*: American Chemical Society Agrichemicals Division, part 1.

Behl, E., and Eiden, C.A., 1991, Field-scale monitoring studies to evaluate mobility of pesticides in soils and groundwater, in Nash, R.G., and Leslie, A.R., eds., *Groundwater residue sampling design*: American Chemical Society Symposium Series 465, pp. 27-46.

Bergström, L., 1990, Leaching of chlorsulfuron and metsulfuron methyl in three Swedish soils measured in field lysimeters: *J. Environ. Qual.*, v. 19, no. 4, pp. 701-706.

Berner, R.A., 1981, A new geochemical classification of sedimentary environments: *J. Sedimentary Petrology*, v. 51, no. 2, pp. 359-365.

Berryhill, W.S., Jr., Lanier, A.L., and Smolen, M.D., 1989, The impact of conservation tillage and pesticide use on water quality. Research needs, in Weigmann, D.L., ed., *Pesticides in terrestrial and aquatic environments: Proceedings of a national research conference*: Virginia Polytechnic Institute and State University, Virginia Water Resources Research Center, Blacksburg, Va., pp. 397-404.

Beste, C.E., ed., 1983, *Herbicide handbook of the Weed Science Society of America* (5th ed.): Weed Science Society of America, Champaign, Ill., 515 p.

Beynon, K.I., Stoydin, G., and Wright, A.N., 1972, A comparison of the breakdown of the triazine herbicides cyanazine, atrazine and simazine in soils and in maize: *Pestic. Biochem. Physiol.*, v. 2, pp. 153-161.

Bilkert, J.N., and Rao, P.S.C., 1985, Sorption and leaching of three nonfumigant nematicides in soils: *J. Environ. Sci. Health*, v. B20, no. 1, pp. 1-26.

Birkeland, P.W., 1974, *Pedology, weathering, and geomorphological research*: Oxford University Press, N.Y., 285 p.

Bishop, K.C., III, 1985, Why are chemicals in groundwater? in *Proceedings of the North Central States Weed Control Conference, St. Louis, Missouri*: North Central States Weed Control Conference, St. Paul, Minn., v. 40, pp. 147-151.

Bishop, K.C., III, and Lawyer, A.L., 1988, Proposal for determining the potential of agricultural chemicals to contaminate groundwater, in Summers. J.B., and Anderson, S.S., eds., *Toxic substances in agricultural water supply and drainage: Searching for solutions: Report of the 1987 panel of experts and papers from the 1987 national meeting, Las Vegas, Nevada*: U.S. Committee on Irrigation and Drainage, Denver, Colo., pp. 143-151.

Bjerg, P.L., Rügge, K., Pedersen, J.K., and Christensen, T.H., 1995, Distribution of redox-sensitive groundwater quality parameters downgradient of a landfill (Grindsted, Denmark): *Environ. Sci. Technol.*, v. 29, no. 5, pp. 1387-1394.

Black, T.C., and Freyberg, D.L., 1987, Stochastic modeling of vertically averaged concentration uncertainty in a perfectly stratified aquifer: *Water Resour. Res.*, v. 23, no. 6, pp. 997-1004.

Blair, A.M., Martin, T.D., Walker, A., and Welch, S.J., 1990, Measurement and prediction of isoproturon movement and persistence in three soils: *Crop Protection*, v. 9, pp. 289-294.

Blanchet, P.F., and St.-George, A., 1982, Kinetics of chemical degradation of organophosphorus pesticides: Hydrolysis of chlorpyrifos and chlorpyrifos-methyl in aqueous solution in presence of copper(II): *Pest. Sci.*, v. 13, no. 1, pp. 85-91.

Blodgett, J.E., 1988, Agrichemicals and groundwater: What do we know? in *Agricultural chemicals and groundwater protection: Emerging management and policy: Proceedings of a conference, St. Paul, Minnesota*: Freshwater Foundation, Navarre, Minn., pp. 19-26.

Blum, D.A., Carr, J.D., Davis, R.K., and Pederson, D.T., 1993, Atrazine in a stream-aquifer system: Transport of atrazine and its environmental impact near Ashland, Nebraska: *Ground Water Monitor. Remed.*, v. 13, no. 2, pp. 125-133.

Boesten, J. J. T. I., 1987, Leaching of herbicides to ground water: A review of important factors and of available measurements, in *1987 British Crop Protection Conference—Weeds—1987: Proceedings of a conference held at Brighton Metropole, England*: British Crop Protection Council, v. 2, pp. 559-568.

——1991, Sensitivity analysis of a mathematical model for pesticide leaching to groundwater: *Pest. Sci.*, v. 31, pp. 375-388.

Boesten, J. J. T. I., and van der Linden, A.M.A., 1991, Modeling the influence of sorption and transformation on pesticide leaching and persistence: *J. Environ. Qual.*, v. 20, no. 2, pp. 425-435.

Bonazountas, M., and Wagner, J., 1984, *SESOIL, a seasonal soil compartment model:* U.S. Environmental Protection Agency, Office of Toxic Substances, Washington, D.C., 70 p.

Bond, L.D., 1987, Origins of seawater intrusion in a coastal aquifer—A case study of the Pajaro Valley: *J. Hydrology*, v. 92, nos. 3/4, p. 363.

Bottcher, A.B., Monke, E.J., and Huggins, L.F., 1981, Nutrient and sediment loadings from a subsurface drainage system: *Trans. Am. Soc. Agric. Eng.*, v. 24, no. 5, pp. 1221-1226.

Bouchard, D.C., and Lavy, T.L., 1985, Hexazinone adsorption-desorption studies with soil and organic adsorbents: *J. Environ. Qual.*, v. 14, no. 2, pp. 181-186.

Bouchard, D.C., Lavy, T.L., and Lawson, E.R., 1985, Mobility and persistence of hexazinone in a forest watershed: *J. Environ. Qual.*, v. 14, no. 2, pp. 229-233.

Boul, H.L., Garnham, M.L., Hucker, D., Baird, D., and Aislabie, J., 1994, Influence of agricultural practices on the levels of DDT and its residues in soil: *Environ. Sci. Technol.*, v. 28, no. 8, p. 1397.

Bourg, A.C.M., and Richard-Raymond, F., 1994, Spatial and temporal variability in the water redox chemistry of the M27 experimental site in the Drac River calcareous alluvial aquifer (Grenoble, France): *J. Contam. Hydrol.*, v. 15, nos. 1-2, pp. 93-105.

Bouwer, H., 1990, Agricultural chemicals and groundwater quality: *J. Soil Water Conserv.*, v. 45, no. 2, pp. 184-189.

Bouwer, H., Bowman, R.S., and Rice, R.C., 1985, Effect of irrigated agriculture on underlying groundwater, in Dunin, F.X., Mattess, G., and Gras, R.A., eds., *Relation of groundwater quantity and quality*: International Association of Hydrologic Sciences, Wallingford, Oxfordshire, UK, pp. 13-20.

Bovey, R.W., Burnett, E., Richardson, C., Baur, J.R., Merkle, M.G., and Kissel, D.E., 1975, Occurrence of 2,4,5-T and picloram in subsurface water in the blacklands of Texas: *J. Environ. Sci. Health*, v. 4, no. 1, pp. 103-106.

Bovey, R.W., and Richardson, C.W., 1991, Dissipation of clopyralid and picloram in soil and seep flow in the blacklands of Texas: *J. Environ. Qual.*, v. 20, no. 3, pp. 528-531.

Bowman, B.T., 1988, Mobility and persistence of metolachlor and aldicarb in field lysimeters: *J. Environ. Qual.*, v. 17, no. 4, pp. 689-694.

——1989, Mobility and persistence of the herbicides atrazine, metolachlor and terbuthylazine in Plainfield sand determined using field lysimeters: *Environ. Toxicol. Chem.*, v. 8, pp. 485-491.

——1991, Use of field lysimeters for comparison of mobility and persistence of granular and EC formulations of the soil insecticide isazofos: *Environ. Toxicol. Chem.*, v. 10, no. 7, pp. 873-879.

Boyle, L.C., 1995, Determination of nature and extent of ground water contamination in Boise City and Boise urban planning areas, Ada County, Idaho: Idaho Department of Health and Welfare, Division of Environmental Quality, Water Quality Status Report 114, 24 p.

Brach, J., 1989, *Agriculture and water quality: Best management practices for Minnesota*: Minnesota Pollution Control Agency, St. Paul, Minn., 64 p.

Brandt, E., 1995, personal communication, facsimile, August 15, U.S. Environmental Protection Agency, Office of Pesticide Programs, Washington, D.C.

Branham, B.E., Smitley, D.R., and Miltner, E.D., 1993, Pesticide fate in turf—Studies using model ecosystems, *in* Racke, K.D., and Leslie, A.R., eds., *Pesticides in urban environments: Fate and significance:* American Chemical Society Symposium Series 522, pp. 156-167.

Braverman, M.P., Lavy, T.L., and Barnes, C.J., 1986, The degradation and bioactivity of metolachlor in the soil: *Weed Sci.*, v. 34, no. 3, pp. 479-484.

Bredehoeft, J.D., and Konikow, L.F., 1993, Ground-water models: Validate or invalidate: *Ground Water*, v. 31, no. 2, pp. 178-179.

Brennan, B.M., 1987, Agricultural chemical use in Hawaii, *in* Rao, P.S.C., and Green, R.E., eds., *Toxic organic chemicals in Hawaii's water resources:* Hawaii Institute of Tropical Agriculture and Human Resources, Honolulu, Hawaii, pp. 1-5.

Bretas, F.S., and Haith, D.A., 1990, Linear programming analysis of pesticide pollution of groundwater: *Am. Soc. Agric. Eng.*, v. 33, no. 1, pp. 167-171.

Brewer, F., Lavy, T.L., and Talbert, R.E., 1982, Effects of flooding on dinitroaniline persistence in soybean (*Glycine max*)-rice (*Oryza sativa*) rotations: *Weed Sci.*, v. 30, no. 5, pp. 531-539.

Briggs, G.F., and Fiedler, A.G., eds., 1975, *Ground water and wells: A reference book for the water-well industry*: United Oil Products Company, Johnson Division, St. Paul, Minn., 440 p.

Britt, J.K., Dwinell, S.E., and McDowell, T.C., 1992, Matrix decision procedure to assess new pesticides based on relative groundwater leaching potential and chronic toxicity: *Environ. Toxicol. Chem.*, v. 11, pp. 721-728.

Brooke, D., and Matthiessen, P., 1991, Development and validation of a modified fugacity model of pesticide leaching from farmland: *Pest. Sci.*, v. 31, no. 3, pp. 349-361.

Brown, C.L., 1988, How the environment affects pesticides: *American Nurseryman*, v. 168, no. 9, pp. 77-79.

Brown, M., Cardozo, C., Nicosia, J., Troiano, J., and Ali, S., 1986, Sampling for pesticide residues in California well water: 1986 well inventory data base: First annual report to the Legislature: California Department of Food and Agriculture Report EH 86-04, 207 p.

Brusseau, M.L., and Rao, P.S.C., 1989, The influence of sorbate-organic matter interactions on sorption nonequilibrium: *Chemosphere*, v. 18, nos. 9/10, pp. 1691-1706.

——1991, Influence of sorbate structure on nonequilibrium sorption of organic compounds: *Environ. Sci. Technol.*, v. 25, no. 8, pp. 1501-1506.

Buchanan, G.A., and Hiltbold, A.E., 1973, Performance and persistence of atrazine: *Weed Sci.*, v. 21, no. 5, pp. 413-416.

Buchmiller, R.C., 1992, personal communication, electronic mail, September 25, U.S. Geological Survey, Iowa City, Iowa.

Buhler, D.D., Koskinen, W.C., Schreiber, M.M., and Gan, J., 1994, Dissipation of alachlor, metolachlor, and atrazine from starch-encapsulated formulations in a sandy loam soil: *Weed Sci.*, v. 42, no. 3, pp. 411-417.

Burgard, D.J., Dowdy, R.H., Koskinen, W.C., and Cheng, H.H., 1994, Movement of metribuzin in a loamy sand soil under irrigated potato production: *Weed Sci.*, v. 42, no. 3, pp. 446-452.

Burkart, M.R., and Kolpin, D.W., 1993, Hydrologic and land-use factors associated with herbicides and nitrate in near-surface aquifers: *J. Environ. Qual.*, v. 22, no. 4, pp. 646-656.

Burlinson, N.E., Lee, L.A., and Rosenblatt, D.H., 1982, Kinetics and products of hydrolysis of 1,2-dibromo-3-chloropropane: *Environ. Sci. Technol.*, v. 16, no. 9, pp. 627-632.

Bush, P.B., Michael, J., Neary, D.G., and Miller, K.V., 1990, Effect of hexazinone on groundwater quality in the Coastal Plain, in *Proceedings, Southern Weed Science Society, 43rd annual meeting*: Southern Weed Science Society, Raleigh, N.C., pp. 184-194.

Bush, P.B., Neary, D.G., Dowd, J.F., Allison, D.C., and Nutter, W.L., 1986, Role of models in environmental impact assessment, in *Weed Science and Risk Assessment: Proceedings, Southern Weed Science Society, 39th annual meeting*: Southern Weed Science Society, Nashville, Tenn., pp. 502-512.

Bushway, R.J., Hurst, H.L., Perkins, L.B., Tian, L., Cabanillas, G.C., Young, B.E.S., Ferguson, B.S., and Jennings, H.S., 1992, Atrazine, alachlor, and carbofuran contamination of well water in central Maine: *Bull. Environ. Contam. Toxicol.*, v. 49, no. 1, pp. 1-9.

Bushway, R.J., Litten, W., Porter, K., and Wertam, J., 1982, A survey of azinphos methyl and azinphos methyl oxon in water and blueberry samples from Hancock and Washington Counties of Maine: *Bull. Environ. Contam. Toxicol.*, v. 28, no. 3, pp. 341-347.

Butler, L.C., Staiff, D.C., and Davis, J.E., 1981, Methyl parathion persistence in soil following simulated spillage: *Arch. Environ. Contam. Toxicol.*, v. 10, pp. 451-458.

Buxton, H.T., Reilly, T.E., Pollock, D.W., and Smolensky, D.A., 1991, Particle tracking analysis of recharge areas on Long Island, New York: *Ground Water*, v. 29, no. 1, pp. 63-71.

California Department of Food and Agriculture, 1986, *Pesticide use: Annual report:* California Department of Food and Agriculture, Agricultural Chemicals and Feed, Sacramento, Calif., 110 p.

——1991, *1991 Summary of pesticide use report data*: California Department of Food and Agriculture, Agricultural Chemicals and Feed, Sacramento, Calif.

California Department of Health Services, 1986, *Organic chemical contamination of large public water systems in California*: California Department of Health Services, Sanitary Engineering Branch, 152 p.

Canter, L.W., 1987, Nitrates and pesticides in ground water: An analysis of a computer-based literature search, in Fairchild, D.M., ed., *Ground water quality and agricultural practices*: Lewis Publishers, Chelsea, Mich., pp. 153-174.

Capodaglio, A.G., Baldi, M., Valentini, P., and Bellinzona, G., 1988, Diffused groundwater contamination by herbicides in an agricultural area in northern Italy, in Novotny, V., ed., *Proceedings of the symposium on nonpoint pollution: 1988—Policy, economy, management, and appropriate technology:* American Water Resources Association, Bethesda, Md., p. 313.

Cardozo, C., Moore, C., Pepple, M., Troiano, J., and Weaver, D., 1989, Sampling for pesticide residues in California well water: 1989 update well inventory data base: California Department of Food and Agriculture Report EH 90-1, 134 p.

Cardozo, C.L., Nicosia, S., and Troiano, J., 1985, Agricultural pesticide residues in California well water: Development and summary of a well inventory data base for non-point sources: California Department of Food and Agriculture Report EH 85-01, 65 p.

Cardozo, M., Pepple, J., Troiano, J., Weaver, D., Fabre, B., Ali, S., and Brown, S., 1988, Sampling for pesticide residues in California well water: 1988 update well inventory data base: California Department of Food and Agriculture, EH 88-10, 151 p.

Carsel, R.F., and Jones, R.L., 1990, Using soil and meteorologic data bases in unsaturated zone modeling of pesticides: *Ground Water Monitor. Rev.*, v. 10, no. 4, pp. 96-101.

Carsel, R.F., Jones, R.L., Hansen, J.L., Lamb, R.L., and Anderson, M.P., 1988a, A simulation procedure for groundwater quality assessments of pesticides: *J. Contam. Hydrol.*, v. 2, pp. 125-138.

Carsel, R.F., Mulkey, L.A., Lorber, M.N., and Baskin, L.B., 1985, The pesticide root zone model (PRZM): A procedure for evaluating pesticide leaching threats to groundwater: *Ecolog. Model.*, v. 30, pp. 49-69.

Carsel, R.F., Nixon, W.B., and Ballantine, L.G., 1986, Comparison of pesticide root zone model predictions with observed concentrations for the tobacco pesticide metalaxyl in unsaturated zone soils: *Environ. Toxicol. Chem.*, v. 5, pp. 345-353.

Carsel, R.F., Parrish, R.S., Jones, R.L., Hansen, J.L., and Lamb, R.L., 1988b, Characterizing the uncertainty of pesticide leaching in agricultural soils: *J. Contam. Hydrol.*, v. 2, pp. 111-124.

Carsel, R.F., and Smith, C.N., 1987, Impact of pesticides on ground water contamination, *in* Marco, G.J., Hollingworth, R.M., and Durham, W., eds., *Silent spring revisited:* American Chemical Society, Washington, D.C., pp. 70-83.

Carsel, R.F., Smith, C.N., and Parrish, R.S., 1987, Modeling pesticide leaching differences between disk harrowing and till plant, in *Optimum Erosion Control at Least Cost*: *Proceedings of the National Symposium on Conservation Systems, St. Joseph, Mich.*: American Society of Agricultural Engineers, pp. 262-275.

Carson, R.L., 1962, *Silent spring*: Houghton-Mifflin, Boston, Mass., 368 p.

Carter, A.D., 1989, The use of soil survey information to assess the risk of surface and groundwater pollution from pesticides, *in Brighton Crop Protection Conference—Weeds— 1989*: *Proceedings of an international conference organised by the British Crop Protection Council, held at Brighton Centre and Brighton Metropole, Brighton, English*: British Crop Protection Council, v. 3, pp. 1157-1164.

Cartwright, N., Clark, L., and Bird, P., 1991, The impact of agriculture on water quality: *Outlook on Agriculture*, v. 20, no. 3, pp. 145-152.

Castro, C.E., and Belser, N.O., 1968, Biodehalogenation. Reductive dehalogenation of the biocides ethylene dibromide, 1,2-dibromo-3-chloropropane, and 2,3-dibromobutane in soil: *Environ. Sci. Technol.*, v. 22, no. 3, pp. 298-303.

Cavalier, T.C., and Lavy, T.L., 1987, Eastern Arkansas groundwater tested for pesticides: *Arkansas Farm Research*, v. 36, no. 3, p. 11.

Cavalier, T.C., Lavy, T.L., and Mattice, J.D., 1989, Assessing Arkansas ground water for pesticides: Methodology and findings: *Ground Water Monitor. Rev.*, v. 9, no. 4, p. 159-166.

——1991, Persistence of selected pesticides in ground-water samples: *Ground Water*, v. 29, no. 2, pp. 225-231.

Cessna, A.J., Elliott, J.A., Kerr, L.A., Best, K.B., Nicholaichuk, W., and Grover, R., 1994, Transport of nutrients and post-emergence-applied herbicides during corrugation irrigation of wheat: *J. Environ. Qual.*, v. 23, no. 5, pp. 1038-1045.

Champ, D.R., Gulens, J., and Jackson, R.E., 1979, Oxidation-reduction sequences in ground water flow systems: *Canadian J. Earth Sciences*, v. 16, pp. 12-23.

Chapelle, F.H., McMahon, P.B., Dubrovsky, N.M., Fuji, R.F., Oaksford, E.T., and Vroblesky, D.A., 1995, Deducing the distribution of terminal electron-accepting processes in hydrologically diverse groundwater systems: *Water Resour. Res.*, v. 31, no. 2, pp. 359-371.

Chen, H.-H., and Druliner, A.D., 1987, Nonpoint-source agricultural chemicals in ground water in Nebraska—Preliminary results for six areas of the high plains aquifer: U.S. Geological Survey Water-Resources Investigations Report 86-4338, 68 p.

Cheng, H.H., and Koskinen, W.C., 1986, Processes and factors affecting transport of pesticides to ground water, *in* Garner, W.Y., Honeycutt, R.C., and Nigg, H.N., eds., *Evaluation of pesticides in ground water: Developed from a symposium sponsored by the Division of Pesticide Chemistry at the 189th Meeting of the American Chemical Society, Miami Beach, Florida:* American Chemical Society Symposium Series 315, pp. 2-13.

Cherry, J.A., ed., 1983, Migration of contaminants in groundwater at a landfill: A case study: *J. Hydrology*, v. 63, nos. 1/2, 197 p.

Cherryholmes, K.L., Breuer, G.M., and Hausler, W.J., 1989, *One time testing of Iowa's regulated drinking water supplies:* University of Iowa, Iowa City, Iowa, 14 p.

Chesters, G., 1992, Environmental fate of alachlor, metolachlor, aldicarb and atrazine, *in* Marani, Alessandro, and Rinaldo, eds., *Transport processes and the hydrological cycle:* Venice, Italy: Istituto Veneto di Scienze, Lettere ed Arti, Venice, Italy, pp. 143-208.

——1995, personal communication, peer review, University of Wisconsin Water Resources Center, Madison, Wis.

Chesters, G., and Konrad, J., 1971, Effects of pesticide usage on water quality: *Bioscience,* v. 21, no. 12, pp. 565-569.

Chesters, G., Simsiman, G.V., Levy, J., Alhajjar, B.J., Fathulla, R.N., and Harkin, J.M., 1989, Environmental fate of alachlor and metolachlor: *Rev. Environ. Contam. Toxicol.,* v. 110, pp. 1-74.

Chesters, G., Wood, J.B., and Harkin, J.M., 1991, Research needed to regulate pesticide contamination of ground water, in *Integrated approaches to water pollution problems: Proceedings of the international symposium (SISIPPA) (Lisbon, Portigal)*: Elsevier Applied Science, London, v. 1, pp. 27-46.

Chiou, C.T., Peters, L.J., and Freed, V.H., 1979, A physical concept of soil-water equilibria for nonionic organic compounds: *Science,* v. 206, no. 16, pp. 831-832.

Choquette, A.F., and Katz, B.G., 1989, Grid-based groundwater sampling: Lessons from an extensive regional network for 1, 2-dibromoethane (EDB) in Florida, *in* Ragone, Steven, ed., *Regional Characterization of Water Quality Symposium: Proceedings of a symposium held during the Third Scientific Assembly of the International Association of Hydrological Sciences at Baltimore, Maryland, USA*: International Association of Hydrological Science, Wallingford, Oxfordshire, UK, pp. 79-86.

Christenson, S.C., and Rea, A., 1993, Ground-water quality in the Oklahoma City urban area, *in* Alley, W.M., ed., *Regional ground-water quality*: Van Nostrand Reinhold, N.Y., pp. 589-611.

Cink, J.H., and Coats, J.R., 1993, Effect of concentration, temperature, and soil moisture on the degradation of chlorpyrifos in an urban Iowa soil, *in* Racke, K.D., and Leslie, A.R., eds., *Pesticides in urban environments: Fate and significance*: American Chemical Society Symposium Series 522, pp. 62-69.

Clark, D.W., 1990, Pesticides in soils and ground water in selected irrigated agricultural areas near Havre, Ronan, and Huntley, Montana: U.S. Geological Survey Water-Resources Investigations Report 90-4023, 34 p.

Clay, S.A., Clay, S.E., Koskinen, W.C., and Malzer, G.L., 1992, Agrichemical placement impacts on alachlor and nitrate movement through soil in a ridge tillage system: *J. Environ. Sci. Health*, v. B27, no. 2, pp. 125-138.

Clay, S.A., and Koskinen, W.C., 1990, Adsorption and desorption of atrazine, hydroxyatrazine, and S-glutathione atrazine on two soils: *Weed Sci.,* v. 38, no. 3, pp. 262-266.

Clendening, L.D., Jury, W.A., and Ernst, F.F., 1990, A field mass balance study of pesticide volatilization, leaching, and persistence, *in* Kurtz, D.A., ed., *Long range transport of pesticides*: Lewis Publishers, Chelsea, Mich., pp. 47-60.

Coats, J.R., 1991, Pesticide degradation mechanisms and environmental activation, *in* Somasundaram, L., and Coats, J.R., eds., *Pesticide transformation products: Fate and significance in the environment:* American Chemical Society Symposium Series 459, pp. 10-31.

Coats, K.H., and Smith, B.D., 1964, Dead-end pore volume and dispersion in porous media: *Soc. Pet. Eng. .J.*, v. 4, pp. 73-84.

Cohen, D.B., 1986, Ground water contamination by toxic substances—A California assessment, *in* Garner, W.Y., Honeycutt, R.C., and Nigg, H.N., eds., *Evaluation of pesticides in ground water: Developed from a symposium sponsored by the Division of Pesticide Chemistry at the 189th Meeting of the American Chemical Society, Miami Beach, Florida:* American Chemical Society Symposium Series 315, pp. 499-529.

Cohen, S.Z., 1990a, Pesticides in ground water: An overview, *in* Hutson, D.H., and Roberts, T.R., eds., *Environmental fate of pesticides:* Wiley, Chichester, England, pp. 13-25.

——1990b, What is a leacher? *Ground Water Monitor. Rev.*, v. 10, no. 3, pp. 72-75.

——1993, EPA releases national ground water database: *Ground Water Monitor. Remed.*, v. 13, no. 3, pp. 99-102.

Cohen, S.Z., Creeger, S.M., Carsel, R.F., and Enfield, C.G., 1984, Potential pesticide contamination of groundwater from agricultural uses, *in* Krueger, R.F., and Seiber, J.N., eds., *Treatment and disposal of pesticide wastes: Based on a symposium sponsored by the Division of Pesticide Chemistry at the 186th meeting of the American Chemical Society, Washington, D.C.:* American Chemical Society Symposium Series 259, pp. 297-325.

Cohen, S.Z., Eiden, C., and Lorber, M.N., 1986, Monitoring ground water for pesticides, *in* Garner, W.Y., Honeycutt, R.C., and Nigg, H.N., eds., *Evaluation of pesticides in ground water: Developed from a symposium sponsored by the Division of Pesticide Chemistry at the 189th Meeting of the American Chemical Society, Miami Beach, Florida*: American Chemical Society Symposium Series 315, pp. 170-196.

Cohen, S. Z., Nickerson, S., Maxey, R., Dupuy, A., Jr., and Senita, J. A., 1990, A ground water monitoring study for pesticides and nitrates associated with golf courses on Cape Cod: *Ground Water Monitor. Rev.*, v. 10, no. 1, pp. 160-173.

Colborn, T., vom Saal, F.S., and Soto, A.M., 1993, Developmental effects of endocrine-disrupting chemicals in wildlife and humans: *Environmental Health Perspectives*, v. 101, no. 5, pp. 378-384.

Cooke, A.R., 1966, Controlled studies on the interaction of rainfall and pre-emergence herbicide activity: *Mededelingen van de Rijksfaculteit Landbouwwetenschappen te Gent*, v. 31, no. 4, pp. 1165-1170.

Cooper, J.F., and Zheng, S.Q., 1994, Behaviour of metolachlor in tropical and Mediterranean plain field conditions: *Sci. Total Environ.*, v. 153, pp. 133-139.

Copin, A., Deleu, R., Salembier, J.F., and Belien, J.M., 1985, Etude en conditions naturelles du transfert dans le sol de trois pesticides de solubilité différente, *in* Hascoet, M., Schuepp, H., and Steen, E., eds., *Behavior and side effects of pesticides in soil:* Institut National de la Recherche Agronomique, Versailles, France, pp. 47-56.

Corbin, B.R., Jr., McClelland, M., Frans, R.E., Talbert, R.E., and Horton, D., 1994, Dissipation of fluometuron and trifluralin residues after long-term use: *Weed Sci.*, v. 42, no. 3, pp. 438-445.

Cox, C., 1991, Pesticides on golf courses—Mixing toxins with play? *J. Pesticide Reform*, v. 11, no. 3, pp. 2-4.

Creeger, S.M., 1986, Considering pesticide potential for reaching ground water in the registration of pesticides, *in* Garner, W.Y., Honeycutt, R.C., and Nigg, H.N., eds., *Evaluation of pesticides in ground water: Developed from a symposium sponsored by the Division of Pesticide Chemistry at the 189th Meeting of the American Chemical Society, Miami Beach, Florida:* American Chemical Society Symposium Series 315, pp. 548-557.

Cress, D.C., 1994, *Kansas agricultural chemical usage: 1993 rangeland and pasture summary*: Kansas State University, Cooperative Extension Service, Manhattan, Kans.

Crittenden, J.C., Hutzler, N.J., Geyer, D.G., Oravitz, J.L., and Friedman, G., 1986, Transport of organic compounds with saturated groundwater flow: Model development and parameter sensitivity: *Water Resour. Res.*, v. 22, no. 3, pp. 271-284.

Croll, B.T., 1974, The impact of organic pesticides and herbicides upon groundwater pollution, *in* Cole, J.A., ed., *Groundwater pollution in Europe: Proceedings of a conference organized by the Water Research Association in Reading, England:* Water Information Center, Port Washington, N.Y., pp. 350-364.

Crowe, A.S., and Mutch, J.P., 1990, Assessing the migration and transformation of pesticides in the subsurface: The role of expert systems: *Water Pollution Research Journal, Canada*, v. 25, no. 3, pp. 293-323.

Crutchfield, S., Hansen, L., and Ribaudo, M., 1993, *Agricultural and water-quality conflicts: Economic dimensions of the problem*: U.S. Department of Agriculture, Economic Research Service, Washington, D.C., 18 p.

Curry, D.S., 1987, Assessment of empirical methodologies for predicting ground water pollution from agricultural chemicals, *in* Fairchild, D.M., ed., *Ground water quality and agricultural practices*: Lewis Publishers, Chelsea, Mich., pp. 227-245.

Curtis, G.P., 1991, Reductive dehalogenation of hexachloroethane and carbon tetrachloride by aquifer sand and humic acid: Stanford University, Department of Civil Engineering, Stanford, Calif., Ph.D. dissertation, 228 p.

Curtis, G.P., and Reinhard, M., 1994, Reductive dehalogenation of hexachloroethane, carbon tetrachloride, and bromoform by anthrahydroquinone disulfonate and humic acid: *Environ. Sci. Technol.*, v. 28, no. 13, pp. 2393-2401.

Curtis, G.P., Reinhard, M., and Roberts, P.V., 1986, Sorption of hydrophobic organic compounds by sediments, *in* Davis, J.A., and Hayes, K.F., eds., *Geochemical processes at mineral surfaces*: American Chemical Society Symposium Series 323, pp. 191-216.

Czapar, G.F., Kanwar, R.S., and Fawcett, R.S., 1994, Herbicide and tracer movement to field drainage tiles under simulated rainfall conditions: *Soil Tillage Res.*, v. 30, pp. 19-32.

Dahl, T.O., 1986, Occidental Chemical Company at Lathrop, California, A groundwater/soil contamination problem and a solution, *in* Assink, J.W., and van den Brink, W.J., eds., *Contaminated soil: First International TNO Conference on Contaminated Soil, Utrecht, The Netherlands*: Martinus Nijhoff Publishers, Dordrecht, Lancaster, pp. 793-806.

Dao, T.H., 1987, Behavior and subsurface transport of agrochemicals in conservation systems, *in* Fairchild, D.M., ed., *Ground water quality and agricultural practices*: Lewis Publishers, Chelsea, Mich., pp. 175-184.

Dean, J.D., Jowist, P.P., and Donigian, A.S., Jr., 1984, *Leaching evaluation of agricultural chemicals (LEACH) handbook*: U.S. Environmental Protection Agency, Office of Research and Development, Environmental Research Laboratory Report EPA-600/3-84-068, 407 p.

Dean, J.D., Strecker, E.W., Salhotra, A.M., and Mulkey, L.A., 1987, *Exposure assessment modeling for aldicarb in Florida*: U.S. Environmental Protection Agency, Office of Research and Development, Environmental Research Laboratory Report EPA/600/3-85/051, 374 p.

Deeley, G.M., Reinhard, M., and Stearns, S.M., 1991, Transformation and sorption of 1,2-dibromo-3-chloropropane in subsurface samples collected at Fresno, California: *J. Environ. Qual.*, v. 20, no. 3, pp. 547-556.

Dehart, B.A., Lavy, T.L., and Mattice, J.D., 1991, Monitoring northwest Arkansas springs for herbicides, nitrates and phosphates: *Arkansas Farm Research*, v. 40, no. 1, p. 9.

Dejonckheere, W., Steurbaut, W., Melkebeke, G., and Kips, R.H., 1983, Leaching of aldicarb and thiofanox, and their uptake in soils by sugarbeet plants: *Pest. Sci.*, v. 14, pp. 99-107.

Deleur, R., Copin, A., Delmarcelle, J., Dreze, P., and Renaud, A., 1980, Dispersion verticale dans le sol de trois herbicides de la famille des urées substituées: Le neburon, le chlortoluron et la metoxuron: *Med.Fac.Landbouww.Rikjsuniv.Gent*, v. 45, no. 4, pp. 1037-1046.

DeMartinis, J.M., 1989, Small-scale retrospective ground water monitoring studies for agricultural chemicals: Study design and site selection: *Ground Water Monitor. Rev.*, v. 9, no. 4, pp. 167-176.

Demon, M., Schiavon, M., Portal, J.M., and Munier-Lamy, C., 1994, Seasonal dynamics of atrazine in three soils under outdoor conditions: *Chemosphere*, v. 28, no. 3, pp. 453-466.

Denver, J.M., and Sandstrom, M.W., 1991, Distribution of dissolved atrazine and two metabolites in the unconfined aquifer, southeastern Delaware, *in* Mallard, G.E., and Aronson, D.A., eds., U.S. Geological Survey Toxic Substances Hydrology Program—Proceedings of the technical meeting, Monterey, California, March 11-15, 1991: U.S. Geological Survey Water-Resources Investigations Report 91-4034, pp. 314-318.

Detroy, M.G., 1986, Areal and vertical distribution of nitrate and herbicides in the Iowa River alluvial aquifer, Iowa County, Iowa, *in Proceedings, the Agricultural Impacts on Ground Water—A conference, Omaha, Nebraska*: National Water Well Association, Omaha, Nebr., pp. 381-398.

Detroy, M.G., Hunt, P.K., and Holub, M.A., 1988, Ground-water-quality-monitoring program in Iowa: Nitrate and pesticides in shallow aquifers: U.S. Geological Survey Water-Resources Investigations Report 88-4123, 31 p.

Detroy, M.G., and Kuzniar, R.L., 1988, Occurrence and distribution of nitrate and herbicides in the Iowa River alluvial aquifer, Iowa—May 1984 to November 1985: U.S. Geological Survey Water-Resources Investigations Report 88-4117, 93 p.

Deubert, K.H., 1990, Environmental fate of common turf pesticides—Factors leading to leaching: *U.S.G.A. Green Section Record*, pp. 5-8.

Dick, W.A., and Daniel, T.C., 1987, Soil chemical and biological properties as affected by conservation tillage: Environmental implications, *in* Logan, T.J., Davidson, J.M., Baker, J.L., and Overcash, M.R., eds., *Effects of conservation tillage on groundwater quality: Nitrates and pesticides*: Lewis Publishers, Chelsea, Mich., pp. 125-137.

Dion, N.P., 1971, Some effects of land-use changes on the shallow ground-water system in the Boise-Nampa area, Idaho: Idaho Department of Water Administration Water Information Bulletin 24, 47 p.

Doane Marketing Services, Inc., 1992, *1992 pesticide profile*: Doane Marketing Services, St. Louis, Mo.

Domagalski, J.L., and Dubrovsky, N.M., 1991, Regional assessment of nonpoint-source pesticide residues in ground water, San Joaquin Valley, California: U.S. Geological Survey Water Resources Investigations Report 91-4027, 64 p.

——1992, Pesticide residues in ground water of the San Joaquin Valley, California: *J. Hydrology*, v. 130, nos. 1-4, pp. 299-338.

Donigian, A.S., Jr., and Carsel, R.F., 1987, Modeling the impact of conservation tillage practices on pesticide concentrations in ground and surface waters: *Environ. Toxicol. Chem.*, v. 6, pp. 241-250.

Dougherty, D.E., and Bagtzoglou, A.C., 1993, A caution on the regulatory use of numerical solute transport models: *Ground Water*, v. 31, no. 6, pp. 1007-1020

Dowling, K.C., Costella, R.G., and Lemley, A.T., 1994, Behavior of the insecticides ethoprophos and carbofuran during soil-water transport: *Pest. Sci.*, v. 41, pp. 27-33.

Droste, E.X., 1987, Ground water contamination from the agricultural use of ethylene dibromide (EDB) in Simsbury, Connecticut, in *Proceedings of the Fourth Annual Eastern Regional Ground Water Conference, Burlington, Vermont*: National Water Well Association, Dublin, Ohio, pp. 372-401.

Druliner, A.D., 1989, Overview of the relations of nonpoint-source agricultural chemical contamination to local hydrogeologic, soil, land-use, and hydrochemical characteristics of the High Plains Aquifer of Nebraska, *in* Mallard, G.E., ed., U.S. Geological Survey Toxic Substances Hydrology Program—Proceedings of the technical meeting, Phoenix, Arizona, September 26-30, 1988: U.S. Geological Survey Water-Resources Investigations Report 88-4220, pp. 411-435.

Dubrovsky, N.M., Deverel, S.J., and Gilliom, R.J., 1993, Multiscale approach to regional ground-water-quality assessment: Selenium in the San Joaquin Valley, California, *in* Alley, W.M., ed., *Regional ground-water quality*: Van Nostrand Reinhold, N.Y., pp. 537-562.

Dunkle, S.A., Plummer, L.N., Busenberg, E., Phillips, P.J., Denver, J.M., Hamilton, P.A., Michel, R.L., and Coplen, T.B., 1993, Chlorofluorocarbons (CCl_3F and CCl_2F_2) as dating tools and hydrologic tracers in shallow groundwater of the Delmarva Peninsula, Atlantic Coastal Plain, United States: *Water Resour. Res.*, v. 29, no. 12, pp. 3837-3860.

Eckhardt, D.A., Flipse, W.J., and Oaksford, E.T., 1989a, Relation between land use and ground-water quality in the upper glacial aquifer in Nassau and Suffolk Counties, Long Island, New York: U.S. Geological Survey Water-Resources Investigations Report 86-4142, 35 p.

Eckhardt, D.A., Siwiec, S.F., and Cauller, S.J., 1989b, Regional appraisal of ground-water quality in five different land-use areas, Long Island, New York, *in* Mallard, G.E., ed., U.S. Geological Survey Toxic Substances Hydrology Program—Proceedings of the technical meeting, Phoenix, Arizona, September 26-30, 1988: U.S. Geological Survey Water-Resources Investigations Report 88-4220, pp. 397-403.

Edwards, W.M., and Glass, B.L., 1971, Methoxychlor and 2,4,5-T in lysimeter percolation and runoff water: *Bull. Environ. Contam. Toxicol.*, v. 6, no. 1, pp. 81-84.

Eichers, T., Andrilenas, P., Blake, H., Jenkins, R., and Fox, A., 1970, *Quantities of pesticides used by farmers in 1966*: U.S. Department of Agriculture, Economic Research Service, Washington, D.C., 61 p.

Ellingson, C.T., 1987, Irrigation impact issues at the proposed Grant County waste management site, in *Proceedings of the NWWA FOCUS Conference on Northwestern Ground Water Issues, Portland, Oregon*: National Water Well Association, Dublin, Ohio, p. 590.

Ellingson, S.B., and Redding, M.B., 1988, Random survey of VOC's, pesticides, and inorganics in Arizona's drinking water wells, in *Proceedings of FOCUS Conference on Southwestern Ground Water Issues, Albuquerque, New Mexico*: National Water Well Association, Dublin, Ohio, pp. 223-247.

Ellington, J.J., Stancil, F.E., Jr., and Payne, W.D., 1986, *Measurement of hydrolysis rate constants for evaluation of hazardous waste land disposal*: U.S. Environmental Protection Agency, Environmental Research Laboratory Report EPA/600/3-86/043, v. 1, 122 p.

Ellington, J.J., Stancil, F.E., Jr., Payne, W.D., and Trusty, C., 1987, *Measurement of hydrolysis rate constants for evaluation of hazardous waste land disposal*: U.S. Environmental Protection Agency Report EPA/600/3-87/019, v. 2, 152 p.

Enfield, C.G., Carsel, R.F., Cohen, S.Z., Phan, T., and Walters, D.M., 1982, Approximating pollutant transport to ground water: *Ground Water*, v. 20, no. 6, pp. 711-722.

Enfield, C.G., and Yates, S.R., 1990, Organic chemical transport to groundwater, *in* Cheng, H.H., ed., *Pesticides in the soil environment: Processes, impacts, and modeling*: Soil Science Society of America, Madison, Wis., pp. 271-302.

Erickson, L.E., and Kuhlman, D., 1987, What you need to know about groundwater concerns: *Grounds Maintenance*, v. 22, no. 2, pp. 74-75, 128.

Everts, C.J., and Kanwar, R.S., 1988, Quantifying preferential flow to a tile line with tracers, in *International Winter Meeting of the American Society of Agricultural Engineers*: American Society of Agricultural Engineers, St. Joseph, Mich., 15 p.

Everts, C.J., Kanwar, R.S., Alexander, E.C., Jr., and Alexander, S.C., 1989, Comparison of tracer mobilities under laboratory and field conditions: *J. Environ. Qual.*, v. 18, no. 4, pp. 491-498.

Exner, M.E., 1990, Pesticide contamination of ground water artificially recharged by farmland runoff: *Ground Water Monitor. Rev.*, v. 10, no. 1, pp. 147-159.

Exner, M.E., and Spalding, R.F., 1985, Ground-water contamination and well construction in southeast Nebraska: *Ground Water*, v. 23, no. 1, pp. 26-34.

——1990, *Occurrence of pesticides and nitrate in Nebraska's ground water*: University of Nebraska, Water Center, Institute of Agriculture and Natural Resources, Lincoln, Nebr., 34 p.

Fairchild, D.M., 1987, A national assessment of ground water contamination from pesticides and fertilizers, in Fairchild, D.M., ed., *Ground water quality and agricultural practices:* Lewis Publishers, Chelsea, Mich., pp. 273-294.

Fathulla, R.N., Jones, F.A., Harkin, J.M., and Chesters, G., 1988, Distribution and persistence of aldicarb residues in the sand-and-gravel aquifer of central Wisconsin. 1. Relationship between aldicarb residue concentration and groundwater chemistry, in Marani, A., ed., *Advances in environmental modelling*: Elsevier, Amsterdam, The Netherlands, pp. 59-84.

Fawcett, R., 1990, Contamination of wells by pesticides and nitrate: *Water Well Journal*, v. 44, no. 7, pp. 42-44.

——1995, Where is the atrazine? *Farm Journal*, v. 119, p. 5.

Felsot, A., 1990, What is the real problem with pesticides in groundwater? in *Proceedings of the 16th Annual Illinois Crop Protection Workshop*: University of Illinois at Urbana-Champaign, Urbana. Ill., pp. 32-42.

Felsot, A., Wei, L., and Wilson, J., 1982, Environmental chemodynamic studies with terbufos (Counter) insecticide in soil under laboratory and field conditions: *J. Environ. Sci. Health*, v. B17, no. 6, pp. 649-673.

Feng, J.C., Sidhu, S.S., Feng, C.C., and Servant, V., 1989, Hexazinone residues and dissipation in soil leachates: *J. Environ. Sci. Health*, v. B24, no. 2, pp. 131-143.

Feng, J.C., and Thompson, D. G., 1990, Fate of glyphosate in a Canadian forest watershed. 2. Persistence in foliage and soils: *J. Agric. Food Chem.*, v. 38, no. 4, pp. 1118-1125.

Fishel, D.K., and Lietman, P.L., 1986, Occurrence of nitrate and herbicides in ground water in the upper Conestoga River Basin, Pennsylvania, in *Proceedings, the Agricultural Impacts on Ground Water—A conference, Omaha, Nebraska*: National Water Well Association, Omaha, Nebr., pp. 317-323.

Flury, M., Fluhler, H., Jury, W.A., and Leuenberger, J., 1994, Susceptibility of soils to preferential flow of water: A field study: *Water Resour. Res.*, v. 30, no. 7, pp. 1945-1954.

Flury, M., Leuenberger, J., Studer, B., and Flühler, H., 1995, Transport of anions and herbicides in a loamy and a sandy field soil: *Water Resour. Res.*, v. 31, no. 4, pp. 823-835.

Foy, C.L., and Hiranpradit, H., 1989, Movement of atrazine by water from application sites in conventional and no-tillage corn production, in Weigmann, D.L., ed., *Pesticides in terrestrial and aquatic environments: Proceedings of a national research conference:* Virginia Polytechnic Institute and State University, Virginia Water Resources Research Center, Blacksburg, Va., pp. 355-377.

Frank, R., Braun, H.E., Clegg, B.S., Ripley, B.D., and Johnson, R., 1990, Survey of farm wells for pesticides, Ontario, Canada, 1986 and 1987: *Bull. Environ. Contam. Toxicol,* v. 44, no. 3, pp. 410-419.

Frank, R., Clegg, B.S., and Patni, N.K., 1991a, Dissipation of atrazine on a clay loam soil, Ontario, Canada, 1986-90: *Arch. Environ. Contam. Toxicol.,* v. 21, pp. 41-50.

——1991b, Dissipation of cyanazine and metolachlor on a clay loam soil, Ontario, Canada, 1987-1990: *Arch. Environ. Contam. Toxicol.,* v. 21, pp. 253-262.

Frank, R., Clegg, B.S., Ripley, B.D., and Braun, H.E., 1987b, Investigations of pesticide contaminations in rural wells, 1979-1984, Ontario, Canada: *Arch. Environ. Contam. Toxicol.,* v. 16, pp. 9-22.

Frank, R., Ripley, B.D., Braun, H.E., Clegg, B.S., Johnston, R., and O'Neill, T.J., 1987a, Survey of farm wells for pesticides [sic] residues, southern Ontario, Canada, 1981-1982, 1984: *Arch. Environ. Contam. Toxicol.,* v. 16, p. 8.

Frank, R., Sirons, G.J., and Ripley, B.D., 1979, Herbicide contamination and decontamination of well waters in Ontario, Canada, 1969-1978: *J. Pest. Monitor.,* v. 13, no. 3, pp. 120-127.

Freeze, R.A., and Cherry, J.A., 1979, *Groundwater:* Prentice-Hall, Englewood Cliffs, N.J., 604 p.

Freitag, D., and Scheunert, I., 1990, Fate of [^{14}C]monolinuron in potatoes and soil under outdoor conditions: *Ecotoxicol. Environ. Safety,* v. 20, no. 3, pp. 256-268.

Frink, C.R., and Hankin, L., 1986, Pesticides in ground water in Connecticut: *The Connecticut Agricultural Experiment Station Bulletin 839,* October, pp. 1-10.

Fryer, J.D., Smith, P.D., and Ludwig, J.W., 1979, Long-term persistence of picloram in a sandy loam soil: *J. Environ. Qual.,* v. 8, no. 1, pp. 83-86.

Funari, E., Bottoni, P., and Giuliano, G., 1991, Groundwater contamination by herbicides. Measured and simulated runoff volumes and peak discharges for all storms used in calibration and verification of the 1990-93 rainfall-runoff model at basin 9, Perris Valley. Processes and evaluation criteria, *in* Richardson, M.L., ed., *Chemistry, agriculture and the environment*: The Royal Society of Chemistry, Cambridge, England, pp. 235-254.

Gamerdinger, A.P., Lemley, A.T., and Wagenet, R.J., 1991, Nonequilibrium sorption and degradation of three 2-chloro-*s*-triazine herbicides in soil-water systems: *J. Environ. Qual.,* v. 20, no. 4, pp. 815-822.

Garabedian, S.P., and LeBlanc, D.R., 1991, Overview of research at the Cape Cod site: Field and laboratory studies of hydrologic, chemical, and microbiological processes affecting transport in a sewage-contaminated sand and gravel aquifer, *in* Mallard, G.E., and Aronson, D.A., eds., U.S. Geological Survey Toxic Substances Hydrology Program—Proceedings of the technical meeting, Monterey, California, March 11-15, 1991: U.S. Geological Survey Water-Resources Investigations Report 91-4034, pp. 1-9.

García, S. C., García, S. R., and Pérez, X. M., 1984, Estudio de lavado de herbicidas en un suelo Pardo con Carbonato: *Centro Agricola,* v. 11, no. 1, pp. 27-34.

Garrett, D., Maxey, F. P., and Katz, H., 1976, The impact of intensive application of pesticides and fertilizers on underground water recharge areas which may contribute to drinking water supplies—A preliminary review: U.S. Environmental Protection Agency, Office of Toxic Substances Report EPA 560/3-75-006, 100 p.

Garrison, L.J., Rose, R.P., Rigsby, D.W., Sanders, G., Pay, J., Johnson, T.D., Riggle, B.D., Kirsch, O.H., Howell, J.R., Ekoniak, P., Hendley, P., Hatfield, M.W., Travis, K., Hill, I.R., and Simmons, N.D., 1989, *FOMESAFEN: USA groundwater study*: ICI Agrochemicals Products, Jealotts Hill Research Station Laboratory Project PP021/BD/13 No. RJ0751B, Bracknell, Berkshire, England.

Gascón, J., Durand, G., and Barceló, D., 1995, Pilot survey for atrazine and total chlorotriazines in estuarine waters using magnetic particle-based immunoassay and gas chromatography-nitrogen/phosphorus detection: *Environ. Sci. Technol.*, v. 29, no. 6, pp. 1551-1556.

Gaston, L.A., and Locke, M.A., 1995, Fluometuron sorption and transport in Dundee soil: *J. Environ. Qual.*, v. 24, no. 1, pp. 29-36.

Geddes, J.D., Zive, D.M., and Miller, G.C., 1994, Hydrolysis of methyl isothiocyanate, *in* American Chemical Society, *Preprints of papers presented at the 208th National Meeting, San Diego, California*: American Chemical Society, Division of Environmental Chemistry, v. 34, no. 2, pp. 56-59.

German, E.R., 1989, Quantity and quality of stormwater runoff recharged to the Floridan Aquifer system through two drainage wells in the Orlando, Florida, area: U.S. Geological Survey Water Supply Paper 2344, 51 p.

Gianessi, L.P., and Puffer, C., 1990, *Herbicide use in the United States*: Resources for the Future, Quality of the Environment Division, Washington, D.C., 128 p.

——1992a, *Fungicide use in U.S. crop production*: Resources for the Future, Quality of the Environment Division, Washington, D.C., variously paged.

——1992b, *Insecticide use in U.S. crop production*: Resources for the Future, Quality of the Environment Division, Washington, D.C., variously paged.

Gillham, R.W., Robin, M.J.L., Barker, J.F., and Cherry, J.A., 1983, Groundwater Monitoring and sample bias: University of Waterloo, Department of Earth Sciences, American Petroleum Institute Publication 4367, 206 p.

Gilliom, R.J., ed., 1989, Preliminary assessment of sources, distribution, and mobility of selenium in the San Joaquin Valley, California: U.S. Geological Survey Water-Resources Investigations Report 88-4186, 129 p.

Gilliom, R.J., Alexander, R.B., and Smith, R.A., 1985, Pesticides in the nation's rivers, 1975-1980, and implications for future monitoring: U.S. Geological Survey Water-Supply Paper No. 2271, 26 p.

Gintautas, P.A., Daniel, S.R., and Macalady, D.L., 1992, Phenoxyalkanoic acid herbicides in municipal landfill leachates: *Environ. Sci. Technol.*, v. 26, no. 3, pp. 517-521.

Gish, T.J., Helling, C.S., and Mojasevic, M., 1991a, Preferential movement of atrazine and cyanazine under field conditions: *Trans. Am. Soc. Agric. Eng.*, v. 34, no. 4, pp. 1699-1705.

Gish, T.J., Isensee, A.R., Nash, R.G., and Helling, C.S., 1991b, Impact of pesticides on shallow groundwater quality: *Trans. Am. Soc. Agric. Eng.*, v. 34, no. 4, pp. 1745-1753.

Gish, T.G., Shirmohammadi, A., and Wienhold, B.J., 1994, Field-scale mobility and persistence of commercial and starch-encapsulated atrazine and alachlor: *J. Environ. Qual.*, v. 23, no. 2, pp. 355-359.

Glanville, T., Baker, J., and Newman, J., 1995, *Understanding and reducing pesticide contamination in rural wells—Project summary*: Iowa Department of Agriculture and Land Stewardship, 41 p.

Glass, B.L., and Edwards, W.M., 1974, Picloram in lysimeter runoff and percolation water: *Bull. Environ. Contam. Toxicol.*, v. 11, no. 2, pp. 109-112.

——1979, Dicamba in lysimeter runoff and percolation water: *J. Agric. Food Chem.*, v. 27, no. 4, pp. 908-909.

Godsy, E.M., Goerlitz, D.F., and Grbić-Galić, D., 1992, Methanogenic biodegradation of creosote contaminants in natural and simulated ground-water ecosystems: *Ground Water*, v. 30, no. 2, pp. 232-242.

Goerlitz, D.F., 1992, A review of studies of contaminated groundwater conducted by the U.S. Geological Survey Organics Project, Menlo Park, California, 1961-1990, *in* Lesage, S., and Jackson, R.E., eds., *Groundwater contamination and analysis at hazardous waste sites*: Marcel Dekker, N.Y., pp. 295-355.

Goerlitz, D.F., Troutman, D.E., Godsy, E.M., and Franks, B.J, 1985, Migration of wood-preserving chemicals in contaminated groundwater in a sand aquifer at Pensacola, Florida: *Environ. Sci. Technol.*, v. 19, no. 10, pp. 955-961.

Goetsch, W.D., Kirbach, G.C., and Black, W.F., 1993, Pesticides in well water and groundwater at agrichemical facilities in Illinois: An initial investigation, *in Agrichemical facility site contamination study*: Illinois Department of Agriculture, Springfield, Ill., variously paged.

Goetsch, W.D., McKenna, D.P., and Bicki, T.J., 1992, *Statewide survey for agricultural chemicals in rural, private water-supply wells in Illinois*: Illinois Department of Agriculture, Bureau of Environmental Programs, Springfield, Ill., 4 p.

Goetz, A.J., Lavy, T.L., and Gbur, E.E., Jr., 1990, Degradation and field persistence of imazethapyr: *Weed Sci.*, v. 38, nos. 4-5, pp. 421-428.

Gold, A.J., and Groffman, P.M., 1993, Leaching of agrichemicals from suburban areas, *in* Racke, K.D., and Leslie, A.R., eds., *Pesticides in urban environments: Fate and significance*: American Chemical Society Symposium Series 522, pp. 182-190.

Goltz, M.N., and Roberts, P.V., 1986, Three-dimensional solutions for solute transport in an infinite medium with mobile and immobile zones: *Water Resour. Res.*, v. 22, no. 7, pp. 1139-1148.

Gomme, J., Shurvell, S., Hennings, S.M., and Clark, L., 1992, Hydrology of pesticides in a chalk catchment: Groundwaters: *J.IWEM*, v. 6, pp. 172-178.

Goodman, J., ed., 1991, *Ten-year report: Oakwood Lakes-Poinsett rural clean water program, 1981-1991*: Oakwood Lakes-Poinsett Rural Clean Water Program, Pierre, S.Dak., variously paged.

Goolsby, D.A., Boyer, L.L., and Battaglin, W.A., 1994, Plan of study to determine the effect of changes in herbicide use on herbicide concentrations in midwestern streams, 1989-94: U.S. Geological Survey Open-File Report 94-347, 14 p.

Goolsby, D.A., Thurman, E.M., and Kolpin, D.W., 1991, Geographic and temporal distribution of herbicides in surface waters of the upper midwestern United States, 1989-90, *in* Mallard, G.E., and Aronson, D.A., eds., U.S. Geological Survey Toxic Substances Hydrology Program—Proceedings of the technical meeting, Monterey, California, March 11-15, 1991: U.S. Geological Survey Water-Resources Investigations Report 91-4034, pp. 183-188.

Gossett, J.M., 1987, Measurement of Henry's Law Constants for C-1 and C-2 chlorinated hydrocarbons: *Environ. Sci. Technol.*, v. 21, no. 2, pp. 202-208.

Goswami, K.P., and Green, R.E., 1971, Microbial degradation of the herbicide atrazine and its 2-hydroxy analog in submerged soils: *Environ. Sci. Technol.*, v. 5, no. 5, pp. 426-429.

Grady, S.J., 1989, Statistical comparison of ground-water quality in four land-use areas of stratified-drift aquifers in Connecticut: U.S. Geological Survey Water-Resources Investigations Report 88-4220, pp. 473-481.

Graetz, D.A., Chesters, G., Daniel, T.C., Newland, L.W., and Lee, G.B., 1970, Parathion degradation in lake sediments: *J. Water Pollution Control Federation*, v. 42, no. 2, pp. R76-R94.

Graham, R.C., Ulery, A.L., Neal, R.H., and Teso, R.R., 1992, Herbicide residue distributions in relation to soil morphology in two California vertisols: *Soil Sci.*, v. 153, no. 2, pp. 115-121.

Graham, W.G., 1979, The impact of intensive disposal well use on the quality of domestic groundwater supplies in southeast Minidoka County, Idaho: Idaho Department of Water Resources, Boise, Idaho, 36 p.

Graham, W.G., Clapp, D.W., and Putkey, T.A., 1977, Irrigation wastewater disposal well studies—Snake Plain Aquifer: Environmental Protection Agency Ecological Research Series EPA-600/3-77-071, 52 p.

Green, R.E., Liu, C.C.K., and Tamrakar, N., 1986, Modeling pesticide movement in the unsaturated zone of Hawaiian soils under agricultural use, *in* Garner, W.Y., Honeycutt, R.C., and Nigg, H.N., eds., *Evaluation of pesticides in ground water: Developed from a symposium sponsored by the Division of Pesticide Chemistry at the 189th Meeting of the American Chemical Society, Miami Beach, Florida:* American Chemical Society Symposium Series 315, pp. 366-383.

Greenberg, M., Anderson, R., Keene, J., Kennedy, A., and Page, G.W., 1982, Empirical test of the association between gross contamination of wells with toxic substances and surrounding land use: *Environ. Sci. Technol.*, v. 16, no. 1, pp. 14-19.

Gruessner, B., Shambaugh, N.C., and Watzin, M.C., 1995, Comparison of an enzyme immunoassay and gas chromatography/mass spectrometry for the detection of atrazine in surface waters: *Environ. Sci. Technol.*, v. 29, no. 1, pp. 251-254.

Grundl, T., 1994, A review of the current understanding of redox capacity in natural, disequilibrium systems: *Chemosphere*, v 28, no. 3, p. 613-626.

Gu, B., West, O.R., and Siegrist, R.L., 1995, Using ^{14}C-labeled radiochemicals can cause experimental error in studies of the behaviour of volatile organic compounds: *Environ. Sci. Technol.*, v. 23, no. 5, pp. 1210-1214.

Gusmão, H.C., Lord, K.A., and Rüegg, E.F., 1981, The persistence, leaching and volatilization of [^{14}C]aldrin in two Brazilian soils: *Ciencia e Cultura*, v. 33, no. 1, pp. 101-105.

Gustafson, D.I., 1989, Groundwater Ubiquity Score: A simple method for assessing pesticide leachability: *Environ. Toxicol. Chem.*, v. 8, pp. 339-357.

——1993, *Pesticides in Drinking Water*: Van Nostrand Reinhold, N.Y., 241 p.

Haag, W.R., and Mill, T., 1988a, Some reactions of naturally occurring nucleophiles with haloalkanes in water: *Environ. Toxicol. Chem.*, v. 7, pp. 917-924.

——1988b, Effect of a subsurface sediment on hydrolysis of haloalkanes and epoxides: *Environ. Sci. Technol.*, v. 22, no. 6, pp. 658-663.

Haaser, C.A., 1994, personal communication, electronic data transfer, May 16, U.S. Environmental Protection Agency, Office of Pesticide Programs, Washington, D.C.

Haaser, C.A., and Waldman, E., 1994, Discussion of "EPA releases national ground water database" (Ground Water Monitoring & Remediation, vol. 13, no. 3, p. 99-102, Summer 1993): *Ground Water Monitor. Remed.*, v. 14, no. 1, p. 152.

Habecker, M.A., 1989, *Environmental contamination at Wisconsin pesticide mixing/loading facilities: Case study, investigation and remedial action evaluation*: Wisconsin Department of Agriculture, Trade and Consumer Protection, Agricultural Resource Management Division, Madison, Wis., 80 p.

Hall, D.G.M., and Webster, C.P., 1993, An amended functional leaching model applicable to structured soils. II. Model application: *J. Soil Sci.*, v. 44, no. 4, pp. 589-600.

Hall, J.K., and Hartwig, N.L., 1978, Atrazine mobility in two soils under conventional tillage: *J. Environ. Qual.*, v. 7, no. 1, pp. 63-68.

Hall, J.K., and Mumma, R.O., 1994, Dicamba mobility in conventionally tilled and non-tilled soil: *Soil Tillage Res.*, v. 30, pp. 3-17.

Hall, J.K., Mumma, R.O., and Watts, D.W., 1991, Leaching and runoff losses of herbicides in a tilled and untilled field: *Agric. Ecosys. Environ.*, v. 37, pp. 303-314.

Hall, J.K., Murray, M.R., and Hartwig, N.L., 1989, Herbicide leaching and distribution in tilled and untilled soil: *J. Environ. Qual.*, v. 18, no. 4, pp. 439-445.

Hallberg, G.R., 1985, Groundwater quality and agricultural chemicals: A perspective from Iowa, in *Proceedings of the North Central States Weed Control Conference, St. Louis, Missouri*: North Central States Weed Control Conference, St. Paul, Minn., v. 40, pp. 130-147.

——1986, From hoes to herbicides—Agriculture and groundwater quality: *J. Soil Water Conserv.*, v. 41, no. 6, pp. 357-364.

——1987, Agricultural chemicals in ground water: Extent and implications: *Am. J. Altern. Agric.*, v. 11, no. 1, pp. 3-15.

——1989, Pesticide pollution of groundwater in the humid United States: *Agric. Ecosys. Environ.*, v. 26, pp. 299-367.

——1995, peer review, University of Iowa, University Hygienic Laboratory, Iowa City, Iowa.

Hallberg, G.R., Baker, J.L., and Randall, G.W., 1986, Utility of tile-line effluent studies to evaluate the impact of agricultural practices on ground water, in *Proceedings, the Agricultural Impacts on Ground Water—A conference, Omaha, Nebraska*: National Water Well Association, Omaha, Nebr., pp. 298-326.

Hallberg, G.R., Hoyer, B.E., Bettis, A.E., III, and Libra, R.D., 1983, Hydrogeology, water quality, and land management in the Big Spring Basin, Clayton County, Iowa: Iowa Geological Survey Open File Report 83-3, 191 p.

Hallberg, G.R., Kross, B.C., Libra, R.D., Burmeister, L.F., Weih, L.M.B., Lynch, C.F., Bruner, D.R., Lewis, M.Q., Cherryholmes, K.L., Johnson, J.K., and Culp, M.A., 1990, The Iowa State-wide rural well-water survey design report: A systematic sample of domestic drinking water quality: Iowa Department of Natural Resources Technical Information Series 17, 135 p.

Hallberg, G.R., Libra, R.D., Bettis, E.A.I., and Hoyer, B.E., 1984, Hydrogeologic and water quality investigations in the Big Spring Basin, Clayton County, Iowa; 1983 water-year: Iowa Geological Survey Open File Report 84-4, 231 p.

Hallberg, G.R., Libra, R.D., Quade, D.J., Littke, J., and Nations, B., 1989, Groundwater monitoring in the Big Spring Basin 1984-1987: A summary review: Iowa Department of Natural Resources Technical Information Series 16, 68 p.

Hallberg, G.R., Helling, C., Jabine, T., Jaeger, R.B., Moody, D., and Porter, K., 1992a. Review and comments on "Another look: National survey of pesticides in drinking water wells, phase II report": A report from the FIFRA Scientific Advisory Panel Subpanel on the National Pesticide Survey to the U.S. Environmental Protection Agency Office of Pesticide Programs, 29 p.

Hallberg, G.R., Woida, K., Libra, R.D., Rex, K.D., Sesker, K.D., Kross, B.C., Nicholson, H.F., Johnson, J.K., and Cherryholmes, K.L., 1992b, The Iowa state-wide rural well-water survey: Site and well characteristics and water quality: Iowa Department of Natural Resources Technical Information Series 23, 43 p.

Halley, B.A., Jacob, T.A., Lu, A.Y.H., and Sharp, M., 1989, The environmental impact of the use of ivermectin: Environmental effects and fate: *Chemosphere*, v. 18, nos. 7-8, pp. 1543-1563.

Hance, R.J., 1987, Herbicide behaviour in the soil, with particular reference to the potential for ground water contamination, in Hutson, D.H., and Roberts, T.R., eds., *Herbicides:* Wiley, Chichester, England, pp. 223-247.

Hansch, C., and Leo, A., 1979, *Substituent constants for correlation analysis in chemistry and biology*: Wiley, N.Y., 339 p.

Hansen, J.L., and Spiegel, M.H., 1983, Hydrolysis studies of aldicarb, aldicarb sulfoxide and aldicarb sulfone: *Environ. Toxicol. Chem.*, v. 2 pp. 147-153.

Haque, R., and Freed, V.H., 1973, Behavior of pesticides in the environment: "Environmental chemodynamics": *Residue Rev.*, v. 52, pp. 89-116.

Hardy, M.A., Leahy, P.P., and Alley, W.M., 1989, Well installation and documentation, and ground-water sampling protocols for the pilot National Water-Quality Assessment program: U.S. Geological Survey Open-File Report 89-0396, 36 p.

Harkin, J.M., Jones, F.A., Fathulla, R.N., and Kroll, D.G., 1986, Fate of aldicarb in Wisconsin ground water, *in* Garner, W.Y., Honeycutt, R.C., and Nigg, H.N., eds., *Evaluation of pesticides in ground water: Developed from a symposium sponsored by the Division of Pesticide Chemistry at the 189th Meeting of the American Chemical Society, Miami Beach, Florida:* American Chemical Society Symposium Series 315, pp. 219-255.

Harmon, T.C., Ball, W.P., and Roberts, P.V., 1989, Nonequilibrium transport of organic contaminants in groundwater, *in* Sawhney, B.L., and Brown, K., eds., *Reactions and Movement of Organic Chemicals in Soils:* Soil Science Society of America and American Society of Agronomy, Madison, Wis., pp. 405-437.

Harper, S.S., Moorman, T.B., and Locke, M.A., 1990, Pesticide biodegradation in the subsurface terrestrial environment and impact on groundwater pollution, *in* Wick, C.B., ed., *Forty-Sixth General Meeting of the Society for Industrial Microbiology:* Elsevier, Amsterdam, Holland, pp. 65-73.

Harris, C.I., 1967, Fate of 2-chloro-*s*-triazine herbicides in soil: *J. Agric. Food Chem.*, v. 15, no. 1, pp. 157-162.

Harris, C.K., 1965, Hydroxysimazine in soil: *Weed Resear.*, v. 5, pp. 275-276.

Harris, G.L., Nicholls, P.H., Bailey, S.W., Howse, K.R., and Mason, D.J., 1994, Factors influencing the loss of pesticides in drainage from a cracking clay soil: *J. Hydrology*, v. 159, nos. 1/4, pp. 235-253.

Harris, J.C., 1990, Rate of aqueous photolysis, *in* Lyman, W.J., Reehl, W.F., and Rosenblatt, D.H., eds., *Handbook of chemical property estimation methods: Environmental behavior of organic compounds:* American Chemical Society, Washington, D.C., pp. 8-1 to 8-43.

Harrison, S.A., Watschke, T.L., Mumma, R.O., Jarrett, A.R., and Hamilton, G.W., Jr., 1993, Nutrient and pesticide concentrations in water from chemically treated turfgrass, *in* Racke, K.D., and Leslie, A.R., eds., *Pesticides in urban environments: Fate and significance:* American Chemical Society Symposium Series 522, pp. 191-207.

Hassink, J., Klein, A., Kördel, W., and Klein, W., 1994, Behaviour of herbicides in non-cultivated soils: *Chemosphere*, v. 28, no. 2, pp. 285-295.

Hay, L.E., and Battaglin, W.A., 1990, Effects of land-use buffer size on Spearman's partial correlations of land use and shallow ground-water quality: U.S. Geological Survey Water-Resources Investigations Report 89-4163, 28 p.

Hayduk, W., and Laudie, M., 1974, Prediction of diffusion coefficients for nonelectrolytes in dilute aqueous solutions: *AIChE J.*, v. 20, pp. 611-615.

Heath, R.C., 1984, Ground-water regions of the United States: U.S. Geological Survey Water-Supply Paper No. 2242, 78 p.

Hebb, E.A., and Wheeler, W.B., 1978, Bromacil in lakeland soil ground water: *J. Environ. Qual.*, v. 7, no. 4, pp. 598-601.

Helgesen, J.O., and Rutledge, A.T., 1989, Relations between land use and water quality in the High Plains Aquifer of south-central Kansas, *in* Mallard, G.E., ed., U.S. Geological Survey Toxic Substances Hydrology Program—Proceedings of the technical meeting, Phoenix, Arizona, September 26-30, 1988: U.S. Geological Survey Water-Resources Investigations Report 88-4220, pp. 437-443.

Helling, C.S., 1987, Movement of agricultural chemicals into and within groundwater aquifers, in Proceedings of a meeting of the Great Plains Agricultural Council, Fort Collins, Colorado: Great Plains Agricultural Council, Lubbock, Tex., pp. 57-71.

Helling, C.S., and Gish, T.J., 1986, Soil characteristics affecting pesticide movement into ground water, in Garner, W.Y., Honeycutt, R.C., and Nigg, H.N., eds., Evaluation of pesticides in ground water: Developed from a symposium sponsored by the Division of Pesticide Chemistry at the 189th Meeting of the American Chemical Society, Miami Beach, Florida: American Chemical Society Symposium Series 315, pp. 14-38.

Helling, C.S., Zhuang, W., Gish, T.J., Coffman, C. B., Isensee, A.R., Kearney, P.C., Hoagland, D.R., and Woodward, D, 1988, Persistence and leaching of atrazine, alachlor, and cyanazine under no-tillage practices: Chemosphere, v. 17, no. 1, pp. 175-187.

Hemmamda, S., Calmon, M., and Calmon, J.P., 1994, Kinetics and hydrolysis mechanism of chlorsulfuron and metsulfuron-methyl: Pest. Sci., v. 40, pp. 71-76.

Herrchen, M., Kördel, W., Klein, W., and Hamm, R.T., 1990, Lysimeter studies of the experimental insecticide BAS 263 I: J. Environ. Sci. Health, v. B25, no. 1, pp. 31-53.

Hileman, B., 1994, Environmental estrogens linked to reproductive abnormalities, cancer: Chemical and Engineering News, v. 72, no. 5, pp. 19-23.

Hill, B.D., and Schaalje, G.B., 1985, A two-compartment model for the dissipation of deltamethrin in soil: J. Agric. Food Chem., v. 33, pp. 1001-1006.

Hill, B.D., Inaba, D.J., and Schaalje, G.B., 1991, Soil-pan method for studying pesticide dissipation in soil, in Nash, R.G., and Leslie, A.R., eds., Groundwater residue sampling design: American Chemical Society Symposium Series 465, pp. 358-366.

Hindall, S.M., 1978, Effects of irrigation on water quality in the sand plain of central Wisconsin: University of Wisconsin-Extension, Geological and Natural History Survey, Madison, Wis., 50 p.

Hodge, J.E., 1993, Pesticide trends in the professional and consumer markets, in Racke, K.D., and Leslie, A.R., eds., Pesticides in urban environments: Fate and significance: American Chemical Society Symposium Series 522, pp. 10-17.

Hogmire, H.W., Weaver, J.E., and Brooks, J.L., 1990, Survey for pesticides in wells associated with apple and peach orchards in West Virginia: Bull. Environ. Contam. Toxicol., v. 44, no. 1, pp. 81-86.

Holden, L.R., Graham, J.A., Whitmore, R.W., Alexander, W.J., Pratt, R.W., Liddle, S.K., and Piper, L.L., 1992, Results of the National Alachlor Well Water Survey: Environ. Sci. Technol., v. 26, no. 5, pp. 935-943.

Holden, P.W., 1986, Pesticides and groundwater quality: Issues and problems in four states: National Academy Press, Washington, D.C., 124 p.

Hollis, J.M., 1991, Mapping the vulnerability of aquifers and surface waters to pesticide contamination at the national/regional scale, in Walker, A., ed., Pesticides in soils and water: Current perspectives: Proceedings of a symposium organised by the British Crop Protection Council, held at the University of Warwick, Coventry, UK, British Crop Protection Council Monograph 47: British Crop Protection Council, Farnham, Surrey, UK, pp. 165-174.

Holm, J.V., Rügge, K., Bjerg. P.L., and Christensen, T.H., 1995, Occurrence and distribution of pharmaceutical organic compounds in the groundwater downgradient of a landfill (Grindsted, Denmark): Environ. Sci. Technol., v. 29 no. 5, pp. 1415-1420.

Hornsby, A.G, 1989, Relating pesticide management for water quality to health-based standards, in Proceedings of the Great Plains Agricultural Council, Lubbock, Texas: Great Plains Agricultural Council, Lubbock, Texas, pp. 95-103.

Hotzman, F.W., and Mitchell, W.H., 1977, Leaching of dicamba and Vel-4207 in modified soil, in *Proceedings of the 31st Annual Meeting of the Northeastern Weed Science Society*: Northeastern Weed Science Society, Salisbury, Md., pp. 364-370.

Howard, P.H., and Neal, M., 1992, *Dictionary of chemical names and synonyms*: Lewis Publishers, Chelsea, Mich., variously paged.

Hoyer, B.E., and Hallberg, G.R., 1991, Groundwater vulnerability regions of Iowa: Iowa Department of Natural Resources, Geological Survey Bureau Special Map Series 11, scale 1:500,000, 1 sheet.

Huang, L.Q., and Ahrens, J.F., 1991, Residues of alachlor in soil after application of controlled release and conventional formulations: *Bull. Environ. Contam. Toxicol.*, v. 47, no. 3, pp. 362-367.

Huang, L.Q., and Frink, C.R., 1989, Distribution of atrazine, simazine, alachlor, and metolachlor in soil profiles in Connecticut: *Bull. Environ. Contam. Toxicol.*, v. 43, no. 1, pp. 159-164.

Hurle, K., and Walker, A., 1980, Persistence and its prediction, in Hance, R.J., ed., *Interactions between herbicides and the soil*: Academic Press, N.Y., pp. 83-122.

Hutzler, N.J., Crittenden, J.C., Gierke, J.S., and Johnson, A.S., 1986, Transport of organic compounds with saturated groundwater flow: Experimental results: *Water Resour. Res.*, v. 22, no. 3, pp. 285-295.

Immerman, F.W., and Drummond, D.J, 1984. *National urban pesticide applicator survey: Final report—overview and results*: U.S. Environmental Protection Agency, Office of Pesticides Programs, Washington, D.C., variously paged.

Isabel, D., and Villeneuve, J., 1991, Significance of the dispersion coefficient in the stochastic modelling of pesticides transport [sic] in the unsaturated zone: *Ecolog. Model.*, v. 59, pp. 1-10.

Isensee, A.R., 1991, Dissipation of alachlor under *in situ* and simulated vadose zone conditions: *Bull. Environ. Contam. Toxicol.*, v. 46, no. 4, pp. 519-526.

Isensee, A.R., Helling, C.S., Gish, T.J., Kearney, P.C., Coffman, C.B., and Zhuang, W., 1988, Groundwater residues of atrazine, alachlor, and cyanazine under no-tillage practices: *Chemosphere*, v. 17, no. 1, pp. 165-174.

Isensee, A.R., Nash, R.G., and Helling, C.S., 1990, Effect of conventional vs. no-tillage on pesticide leaching to shallow groundwater: *J. Environ. Qual.*, v. 19, no. 3, pp. 434-440.

Isensee, A.R., and Sadeghi, A.M., 1994, Effects of tillage and rainfall on atrazine residue levels in soil: *Weed Sci.*, v. 42, no. 3, pp. 462-467.

Istok, J.D., Smyth, J.D., and Flint, A.L., 1993, Multivariate geostatistical analysis of ground-water contamination: A case history: *Ground Water*, v. 31, no. 1, pp. 63-74.

Jackson, G., Webendorfer, B., Harkin, J., and Shaw, B., 1983b, *Aldicarb and Wisconsin's groundwater*: University of Wisconsin-Extension, Cooperative Extension Service Programs, Madison, Wis., 8 p.

Jackson, G., Webendorfer, B., Shaw, B., and Harkin, J., 1983a, *Pesticides in groundwater—How they get there, what happens to them, how to keep them out*: University of Wisconsin-Extension, Cooperative Extension Service Programs, Madison, Wis., 6 p.

Jarczyk, H.J., 1987, Studies on the leaching characteristics of crop protection chemicals in a monolith lysimeter installation: *Pflanzenschutz-Nachrichten Bayer*, v. 40, no. 1, pp. 49-77.

Jayachandran, K., Steinheimer, T.R., Somasundaram, L., Moorman, T.B., Kanwar, R.S., and Coats, J.R., 1994, Occurrence of atrazine and degradates as contaminants of subsurface drainage and shallow groundwater: *J. Environ. Qual.*, v. 23, no. 2, pp. 311-319.

Jaynes, D.B., 1991, Field study of bromacil transport under continuous-flood irrigation: *Soil Sci. Soc. Am. J.*, v. 55, no. 3, pp. 658-664.

Jeffers, P.M., Coty, P., Luczak, S., and Wolfe, N.L., 1994, Halocarbon hydrolysis rates—A search for ionic strength and heterogeneous effects: *J. Environ. Sci. Health*, v. A29, no. 4, pp. 821-831.

Jeffers, P.M., Ward, L.M., Woytowitch, L.M., and Wolfe, N.L., 1989, Homogeneous hydrolysis rate constants for selected chlorinated methanes, ethanes, ethenes, and propanes: *Environ. Sci. Technol.*, v. 23, no. 8, pp. 965-969.

Jernlås, R., 1985, Leaching of 2,3,6-TBA [2,3,6-trichlorobenzoic acid] from a sandy soil, *in* Svenska Ograskonferensen, *Weeds and weed control: 26th Swedish Weed Conference, Uppsala*: Swedish University of Agricultural Sciences, Department of Plant Husbandry and Research Information Centre, Uppsala, Sweden, v. 1, pp. 197-208.

Jernlås, R., and Klingspor, P., 1981, Leaching of TCA [trichloroacetic acid] from arable fields, *in* Svenska Ograskonferensen, *Weeds and weed control: 22nd Swedish Weed Conference, Uppsala*: Swedish University of Agricultural Sciences, Department of Plant Husbandry and Research Information Centre, Uppsala, Sweden, v. 1, pp. 120-128.

Johnsen, T.N., Jr., and Morton, H.L., 1989, Tebuthiuron persistence and distribution in some semiarid soils: *J. Environ. Qual.*, v. 18, no. 4, pp. 433-438.

Johnson, S., 1988, 1988 forestry chemical update, *in* Tomascheski, J.H., ed., *Proceedings, tenth annual Forest Vegetation Management Conference, Eureka, California*: Forest Vegetation Management Conference, pp. 116-119.

Johnson, W.G., and Lavy, T.L., 1994, In-situ dissipation of benomyl, carbofuran, thiobencarb, and triclopyr at three soil depths: *J. Environ. Qual.*, v. 23, no. 3, pp. 556-562.

Johnson-Logan, L.R., Broshears, R.E., and Klaine, S.J., 1992, Partitioning behavior and the mobility of chlordane in groundwater: *Environ. Sci. Technol.*, v. 26, no. 11, pp. 2234-2239.

Johnston, W.R., Ittihadieh, F.T., Craig, K.R., and Pillsbury, A.F., 1967, Insecticides in tile drainage effluent: *Water Resour. Res.*, v. 3, no. 2, pp. 525-537.

Jones, F.A., Chesters, G., and Harkin, J.M., 1987, A site specific predictive method for regulating leachable pesticides, <u>in</u> *Proceedings of the Fourth Annual Eastern Regional Ground Water Conference, Burlington, Vermont*: National Water Well Association, Dublin, Ohio, pp. 529-543.

Jones, F.A., Fathulla, R.N., Chesters, G., and Harkin, J.M., 1988, Distribution and persistence of aldicarb residues in the sand-and-gravel aquifer of central Wisconsin. 2. Simulation of aldicarb residue distributions in groundwater using a combined leaching and groundwater transport model, *in* Marani, A, ed., *Advances in environmental modelling*: Elsevier, Amsterdam, The Netherlands, pp. 85-104.

Jones, R.L., 1986, Field, laboratory, and modeling studies on the degradation and transport of aldicarb residues in soil and ground water, *in* Garner, W.Y., Honeycutt, R.C., and Nigg, H.N., eds., *Evaluation of pesticides in ground water: Developed from a symposium sponsored by the Division of Pesticide Chemistry at the 189th Meeting of the American Chemical Society, Miami Beach, Florida*: American Chemical Society Symposium Series 315, pp. 197-218.

——1990, Pesticides in ground water: Conduct of field studies, *in* Hutson, D.H., and Roberts, T.R., eds., *Environmental fate of pesticides*: Wiley, Chichester, England, pp. 27-46.

Jones, R.L., Black, G.W., and Estes, T.L., 1986a, Comparison of computer model predictions with unsaturated zone field data for aldicarb and aldoxycarb: *Environ. Toxicol. Chem.*, v. 5, pp. 1027-1037.

Jones, R.L., Hansen, J.L., Romine, R.R., and Marquardt, T.E., 1986b, Unsaturated zone studies of the degradation and movement of aldicarb and aldoxycarb residues: *Environ. Toxicol. Chem.*, v. 5, pp. 361-372.

Jones, R.L., Hunt, T.W., Norris, F.A., and Harden, C.F., 1989, Field research studies on the movement and degradation of thiodicarb and its metabolite methomyl: *J. Contam. Hydrol.*, v. 4, pp. 359-371.

Jones, R.L., Rourke, R.V., and Hansen, J.L., 1986c, Effect of application methods on movement and degradation of aldicarb residues in Maine potato fields: *Environ. Toxicol. Chem.*, v. 5, pp. 167-173.

Junk, G.A., Richard, J.J., and Dahm, P.A., 1984, Degradation of pesticides in controlled water-soil systems, *in* Krueger, R.F., and Seiber, J.N., eds., *Treatment and disposal of pesticide wastes: Based on a symposium sponsored by the Division of Pesticide Chemistry at the 186th meeting of the American Chemical Society, Washington, D.C.:* American Chemical Society Symposium Series 259, pp. 37-67.

Junk, G.A., Spalding, R.F., and Richard, J.J., 1980, Areal, vertical, and temporal differences in ground water chemistry: II. Organic constituents: *J. Environ. Qual.*, v. 9, no. 3, pp. 479-483.

Jürgens, H.J., and Roth, R., 1989, Case study and proposed decontamination steps of the soil and groundwater beneath a closed herbicide plant in Germany: *Chemosphere*, v. 18, nos. 1-6, pp. 1163-1169.

Jury, W.A., Elabd, H., Clendening, L.D., and Resketo, M., 1986a, Evaluation of pesticide transport screening models under field conditions, *in* Garner, W.Y., Honeycutt, R.C., and Nigg, H.N., eds., *Evaluation of pesticides in ground water: Developed from a symposium sponsored by the Division of Pesticide Chemistry at the 189th Meeting of the American Chemical Society, Miami Beach, Florida:* American Chemical Society Symposium Series 315, pp. 384-395.

Jury, W.A., Elabd, H., and Resketo, M., 1986b, Field study of napropamide movement through unsaturated soil: *Water Resour. Res.*, v. 22, no. 5, pp. 749-755.

Jury, W.A., Farmer, W.J., and Spencer, W.F, 1984a, Behavior assessment model for trace organics in soil: II. Chemical classification and parameter sensitivity: *J. Environ. Qual.*, v. 13, no. 4, pp. 567-572.

Jury, W.A., Focht, D.D., and Farmer, W.J., 1987b, Evaluation of pesticide groundwater pollution potential from standard indices of soil-chemical adsorption and biodegradation: *J. Environ. Qual.*, v. 16, no. 4, pp. 422-428.

Jury, W.A., and Ghodrati, M., 1989. Overview of organic chemical environmental fate and transport modeling approaches, *in* Sawhney, B.L., and Brown, K., eds., *Reactions and movement of organic chemicals in soils*, Soil Science Society of America Special Publication 22: Soil Science Society of America and American Society of Agronomy, Madison, Wis., pp. 271-305.

Jury, W.A., Spencer, W.F., and Farmer, W.J., 1983, Behavior assessment model for trace organics in soil: I. Model description: *J. Environ. Qual.*, v. 12, no. 4, pp. 558-564.

——1984b, Behavior assessment model for trace organics in soil: III. Application of screening model: *J. Environ. Qual.*, v. 13, no. 4, pp. 573-579.

——1984c, Behavior assessment model for trace organics in soil: IV. Review of experimental evidence: *J. Environ. Qual.*, v. 13, no. 4, pp. 580-586.

Jury, W.A., Winer, A.M., Spencer, W.F., and Focht, D.D., 1987a, Transport and transformations of organic chemicals in the soil-air-water ecosystem: *Rev. Environ. Contam. Toxicol.*, v. 99, pp. 119-170.

Kalinski, R.J., Kelly, W.E., Bogardi, I., Ehrman, R.L., and Yamamoto, P.D., 1994, Correlation between DRASTIC vulnerabilities and incidents of VOC contamination of municipal wells in Nebraska: *Ground Water*, v. 32, no. 1, pp. 31-34.

Kalkhoff, S.J., and Schaap, B.D., 1995, Agricultural chemicals in ground and surface water in a small watershed in Clayton County, Iowa, 1988-91: U.S. Geological Survey Water-Resources Investigations Report 95-4158, 35 p.

Kalkhoff, S.J., Detroy, M.G., Cherryholmes, K.L., and Kuzniar, R.L., 1992, Herbicide and nitrate variation in alluvium underlying a corn field at a site in Iowa County, Iowa: *Water Resour. Bull.*, v. 28, no. 6, pp. 1001-1011.

Kanwar, R.S., and Baker, J.L., 1994, Tillage and chemical management effects on groundwater quality, in *Agricultural Research To Protect Water Quality: Proceedings of the conference, Minneapolis, Minnesota, USA:* Soil Water and Conservation Society, Arkeny, Iowa, v. 1, pp. 455-459.

Kanwar, R.S., Stoltenberg, D.E., Pfeiffer, R., Karlen, D., Colvin, T.S., and Simpkins, W.W., 1994, Transport of nitrate and pesticides to shallow groundwater system [sic] as affected by tillage and crop rotation practices, in *Agricultural Research To Protect Water Quality: Proceedings of the conference, Minneapolis, Minnesota, USA:* Soil Water and Conservation Society, Arkeny, Iowa, v. 1, pp. 270-273.

Karickhoff, S.W., Brown, D.S., and Scott, T.A., 1979, Sorption of hydrophobic pollutants on natural sediments: *Water Res.*, v. 13, pp. 241-248.

Karickhoff, S.W., and Morris, K.R., 1985, Sorption dynamics of hydrophobic pollutants in sediment suspensions: *Environ. Toxicol. Chem.*, v. 4, pp. 469-479.

Katz, B.G., 1993, Biogeochemical and hydrological processes controlling the transport and fate of 1,2-dibromoethane (EDB) in soil and ground water, central Florida: U.S. Geological Survey Water-Supply Paper 2402, 35 p.

Katz, B.G., Choquette, A.F., Orona, M.A., and Pendexter, W.S., 1990, Ethylene dibromide contamination in ground water and its implications for network design, in *Abstracts of papers from the 156th Annual National Meeting of the American Association for the Advancement of Science Annual Meeting, New Orleans, Louisiana*: American Association for the Advancement of Science, Washington, D.C., p. 30.

Katz, B.G., and Mallard, G.E., 1981, Chemical and microbiological monitoring of a sole-source aquifer intended for artificial recharge, Nassau County, New York, in Cooper, W.J., ed., *Chemistry in water reuse:* Ann Arbor Science Publishers, Ann Arbor, Mich., v. 1, pp. 165-183.

Kaufman, D.D., and Kearney, P.C., 1970, Microbial degradation of s-triazine herbicides: *Residue Rev.*, v. 32, pp. 235-265.

Keim, A.M., Ruedisili, L.C., Baker, D.B., and Gallagher, R.E., 1989, Herbicide monitoring of tile drainage and shallow groundwater in northwestern Ohio farm fields—A case study, in Weigmann, D.L., ed., *Pesticides in terrestrial and aquatic environments: Proceedings of a national research conference*: Virginia Polytechnic Institute and State University, Virginia Water Resources Research Center, Blacksburg, Va., pp. 62-78.

Kelley, R.D., 1985, *Synthetic organic compound sampling survey of public water supplies*: Iowa Department of Water, Air and Waste Management, Des Moines, Iowa, 38 p.

Kelley, R.D., and Wnuk, M., 1986, *Little Sioux River Synthetic Organic Compound Municipal Well Sampling Survey*: Iowa Department of Water, Air, and Waste Management, Des Moines, Iowa, 24 p.

Kellogg, R.L., Maizel, M., and Goss, D., 1992, Agricultural chemical use and ground water quality: Where are the potential problem areas? U.S. Department of Agriculture, Soil Conservation Service, Washington, D.C., variously paged.

Kenaga, E. E., 1980, Predicted bioconcentration factors and soil sorption coefficients of pesticides and other chemicals: *Ecotoxicol. Environ. Safety*, v. 4, pp. 26-38.

Kerdijk, H.N., 1981, Groundwater pollution by heavy metals and pesticides from a dredge spoil dump, *in* Glasbergen, P., and van Lelyveld, H., eds., *Quality of groundwater: Proceedings of an international symposium, Noordwijkerhout, the Netherlands*: Elsevier Scientific Publishing, Amsterdam, Holland, pp. 279-286.

Khan, M.A., and Liang, T., 1989, Mapping pesticide contamination potential: *Environ. Manag.*, v. 13, no. 2, pp. 233-242.

Khan, S.U., 1978, Kinetics of hydrolysis of atrazine in aqueous fulvic acid solution: *Pest. Sci.*, v. 9, pp. 39-43.

Kimball, C., and Goodman, J., 1989, Non-point source pesticide contamination of shallow ground water, *in* *1989 International Winter Meeting of the American Society of Agricultural Engineers*: American Society of Agricultural Engineers, St Joseph, Mich., pp. 7-9.

Kimball, C.G., 1988, Ground-water monitoring techniques for non-point-source pollution studies, *in* Collins, A.G., and Johnson, A.I., eds., *Ground-water monitoring techniques for non-point-source pollution studies*: American Society for Testing and Materials, Philadelphia, Pa., pp. 430-441.

King, P.H., and McCarty, P.L., 1966, The movement of pesticides in soils, *in* *Proc. 21st Industrial Wastes Conference*: Purdue University, West Lafayette, Ind, pp. 156-171.

Kladivko, E.J., Van Scoyoc, G.E., Monke, E.J., Oates, K.M., and Pask, W., 1991, Pesticide and nutrient movement into subsurface tile drains on a silt loam soil in Indiana: *J. Environ. Qual.*, v. 20, no. 1, pp. 264-270.

Klaine, S.J., Hinman, M.L., Winkelmann, D.A., Sauser, K.R., Martin, J.R., and Moore, L.W., 1988. Characterization of agricultural nonpoint pollution: Pesticide migration in a west Tennessee watershed: *Environ. Toxicol. Chem.*, v. 7, pp. 609-614.

Klaseus, T.G., Buzicky, G.C., and Schneider, E.C., 1988, *Pesticides and groundwater: Surveys of selected Minnesota wells*: Minnesota Department of Health and Minnesota Department of Agriculture, 95 p.

Klecka, G.M., and Gonsior, S.J., 1984, Reductive dechlorination of chorinated methanes and ethanes by reduced iron(II) porphyrins: *Chemosphere*, v. 13, no. 3, pp. 391-402.

Klein, A.J., 1993, personal communication, electronic data transfer (mailed diskette), November 12, Monsanto Corporation, St. Louis, Mo.

Klein, M., 1991, Application and validation of pesticide leaching models: *Pest. Sci.*, v. 31, pp. 389-398.

Kline and Company, 1990, *1990 Professional markets for pesticides and fertilizers*: Kline and Company, Fairfield, N.J.

Knisel, W.G., and Leonard, R.A., 1989, Irrigation impact on groundwater: Model study in humid region: *J. Irrig. Drain. Eng.*, v. 115, no. 5, pp. 823-838.

Knutson, H., Kadoum, A.M., Hopkins, T.L., Swoyer, G.F., and Harvey, T.L., 1971, Insecticide usage and residues in a newly developed Great Plains irrigation district: *J. Pest. Monitor.*, v. 5, no. 1, pp. 17-27.

Koelliker, J.K., Steichen, J.M., Yearout, R.D., Heiman, A.T., and Grosh, D.L., 1987, *Identification of factors affecting farmstead well water quality in Kansas*: Kansas State University, Kansas Water Resources Research Institute, 50 p.

Kögel-Knabner, I., Knabner, P., and Deschauer, H., 1990, Enhanced leaching of organic chemicals in soils due to binding to dissolved organic carbon, *in* Arendt, F., Hinsenveld, M., and van den Brink, W.J., eds., *Contaminated soil '90*: Kluwer Academic Publishers, The Netherlands, pp. 323-329.

Kolberg, R.L., Weiss, M.J., Prunty, L.D., and Fleeker, J.R., 1989, Influence of irrigation and rainfall on the movement of insecticides through a sandy loam soil, *in* Weigmann, D.L., ed., *Pesticides in terrestrial and aquatic environments: Proceedings of a national research conference*: Virginia Polytechnic Institute and State University, Virginia Water Resources Research Center, Blacksburg, Va., pp. 447-456.

Kolpin, D.W., 1995a, personal communication, peer review, U.S. Geological Survey, Iowa City, Iowa.

——1995b, personal communication, electronic mail, December 15, U.S. Geological Survey, Iowa City, Iowa.

Kolpin, D.W., and Burkart, M.R., 1991, Work plan for regional reconnaissance for selected herbicides and nitrate in ground water of the mid-continental United States, 1991: U.S. Geological Survey Open-File Report 91-59, 18 p.

Kolpin, D.W., Burkart, M.R., and Thurman, E.M., 1994, Herbicides and nitrate in near-surface aquifers in the midcontinental United States, 1991: U.S. Geological Survey Water-Supply Paper 2413, 34 p.

Kolpin, D.W., and Goolsby, D.A., 1995, A regional monitoring network to investigate the occurrence of agricultural chemicals in near-surface aquifers of the midcontinental USA, *in* Kovar, K., and Krasny, J., eds., *Groundwater quality: Remediation and protection*: International Association of Hydrological Sciences, Wallingford, Oxfordshire, pp. 13-20.

Kolpin, D.W., Goolsby, D.A., Aga, D.S., Iverson, J.L., and Thurman, E.M., 1993, Pesticides in near-surface aquifers: Results of the midcontinental United States ground-water reconnaissance, 1991-1992, *in* Goolsby, D.A., Boyer, L.L., and Mallard, G.E., compilers, Selected papers on agricultural chemicals in water resources of the midcontinental United States: U.S. Geological Survey Open File Report 93-0418, pp. 64-73.

Kolpin, D.W., Goolsby, D.A., and Thurman, E.M., 1995, Pesticides in near-surface aquifers. An assessment using highly sensitive analytical methods and tritium: *J. Environ. Qual.,* v. 24, no. 6, pp. 1125-1132.

Kolpin, D.W., and Thurman, E.M., 1995, Postflood occurrence of selected agricultural chemicals and volatile organic compounds in near-surface unconsolidated aquifers in the Upper Mississippi River Basin, 1993: U.S. Geological Survey Circular 1120-G, 20 p.

Kolpin, D.W., Thurman, E.M., and Goolsby, D.A., 1996, Occurrence of selected pesticides and their metabolites in near-surface aquifers of the midwestern United States: *Environ. Sci. Technol.*, v. 30, no. 1, pp. 335-340.

Komor, S.C., and Emerson, D.G., 1994, Movements of water, solutes, and stable isotopes in the unsaturated zones of two sand plains in the upper Midwest: *Water Resour. Res.*, v. 30, no. 2, pp. 253-267.

Koncal, J.J., Gorske, S.F., and Fretz, T.A., 1981, Leaching of EPTC, alachlor, and metolachlor through a nursery medium as influenced by herbicide formulations: *HortScience*, v. 16, no. 6, pp. 757-758.

Konrad, J.G., Armstrong, D.E., and Chesters, G., 1967, Soil degradation of diazinon, a phosphorothioate insecticide: *Agronomy J.*, v. 59, pp. 591-594.

Konrad, J.G., and Chesters, G., 1969, Degradation in soils of ciodrin, an organophosphate insecticide: *J. Agric. Food Chem.*, v. 17, no. 2, pp. 226-230.

Konrad, J.G., Chesters, G., and Armstrong, D.E., 1969, Soil degradation of malathion, a phosphorothioate insecticide: *Proc. Soil Sci. Soc. Am.*, v. 33, pp. 259-262.

Kördel, W., Herrchen, M., and Klein, W., 1991, Experimental assessment of pesticide leaching using undisturbed lysimeters: *Pest. Sci.*, v. 31, no. 3, pp. 337-348.

Koskinen, W.C., 1989, Analysis of pesticides in soil and water, *in* Nelson, D.W., and Dowdy, R.H., eds, *Methods for ground water quality studies: Proceedings of a national workshop held at Arlington, Virginia*: University of Nebraska-Lincoln, Agricultural Research Division, Lincoln, Nebr., pp 70-88.

Koterba, M.T., Banks, W.S.L., and Shedlock, R.J., 1993, Pesticides in shallow groundwater in the Delmarva Peninsula: *J. Environ. Qual.*, v. 22, no. 3, pp. 500-518.

Krapac, I.G., Roy, W.R., Smyth, C.A., and Barnhardt, M.L., 1993, Occurrence and distribution of pesticides in soil at agrichemical facilities in Illinois, *in* *Agrichemical facility site contamination study*: Illinois Department of Agriculture, Springfield, Ill., variously paged.

Krawchuk, B.P., and Webster, B.G.R., 1987, Movement of pesticides to ground water in an irrigated soil: *Water Poll. Res. J. Canada*, v. 22, no. 1, pp. 129-146.

Kray, W.C., Jr., and Castro, C.E., 1964, The cleavage of bonds by low-valent transition metal ions. The homogeneous dehalogenation of vicinal dihalides by chromous sulfate: *J. Am. Chem. Soc.*, v. 86, pp. 4603-4608.

Krider, J.N., 1986, Agricultural irrigation and groundwater quality in humid areas of the United States, *in* Summers, J.B., and Anderson, S.S., eds., *Toxic substances in agricultural water supply and drainage—Defining the problems: Proceedings from the 1986 regional meetings, Fresno, California*: U.S. Committee on Irrigation and Drainage, Denver, Colo., pp. 181-189.

Kriegman-King, M., and Reinhard, M., 1992, Transformation of carbon tetrachloride in the presence of sulfide, biotite and vermiculite: *Environ. Sci. Technol.*, v, 26, no. 11, pp. 2198-2206.

Krill, R.M., and Sonzogni, W.C., 1986, Chemical monitoring of Wisconsin's groundwater: *American Water Works Association Journal*, v. 78, no. 9, pp. 70-75.

Kross, B.C., Hallberg, G.R., Bruner, D.R., Libra, R.D., Rex, K.D., Weih, L.M.B., Vermace, M.E., Burmeister, L.F., Hall, N.H., Cherryholmes, K.L., Johnson, J.K., Selim, M.I., Nations, B.K., Seigley, L.S., Quade, D.J., Dudler, A.G., Sesker, K.D., Culp, M.A., Lynch, C.F., Nicholson, H.F., and Hughes, J.P., 1990, The Iowa state-wide rural well-water survey water-quality data: Initial analysis: Iowa Department of Natural Resources Technical Information Series 19, 142 p.

Kubiak, R., Fuhr, F., Mittelstaedt, W., Hansper, M., and Steffens, W., 1988, Transferability of lysimeter results to actual field situations: *Weed Sci.*, v. 36, no. 4, pp. 514-518.

Kuhn, E.P., and Suflita, J.M., 1989, Dehalogenation of pesticides by anaerobic microorganisms in soils and groundwater—A review, *in* Sawhney, B.L., and Brown, K., eds., *Reactions and movement of organic chemicals in soils*, Soil Science Society of America Special Publication 22: Soil Science Society of America and American Society of Agronomy, Madison, Wis., pp. 111-180.

Kuivila, K.M., 1993, Diazinon concentrations in the Sacramento and San Joaquin Rivers and San Francisco Bay, California, February 1993: U.S. Geological Survey Open-File Report 93-440, 2 p.

Kung, K.-J.S., 1990a, Preferential flow in a sandy vadose zone. 1. Field observation: *Geoderma*, v. 46, pp. 51-58.

——1990b, Preferential flow in a sandy vadose zone. 2. Mechanism and implications: *Geoderma*, v. 46, pp. 59-71.

——1993, Laboratory observation of funnel flow mechanism and its influence on solute transport: *J. Environ. Qual.*, v. 22, no. 1, pp. 91-102.

Kurtz, D.A., and Parizek, R.R., 1986, Complexity of contaminant dispersal in a karst geological system, *in* Garner, W.Y., Honeycutt, R.C., and Nigg, H.N., eds., *Evaluation of Pesticides in Water: Developed from a symposium sponsored by the Division of Pesticide Chemistry at the 189th Meeting of the American Chemical Society, Miami Beach, Florida*: American Chemical Society Symposium Series 315, pp. 256-281.

LaFleur, K.S., 1979, Sorption of pesticides by model soils and agronomic soils: *Soil Sci.*, v. 127, no. 2, pp. 94-101.

——1980, Loss of pesticides from Congaree sandy loam with time: Characterization: *Soil Sci.*, v. 130, no. 2, pp. 83-87.

LaFleur, K.S., Wojeck, G.A., and McCaskill, W.R., 1973, Movement of toxaphene and fluometuron through Dunbar soil to underlying ground water: *J. Environ. Qual.*, v. 2, no. 4, pp. 515-518.

Lagas, P., Verdam, B., and Loch, J.P.G., 1989, Threat to groundwater quality by pesticides in The Netherlands, *in* Sahuquillo, J., and O'Donnell, A.T., eds., *Groundwater management: Quantity and quality*: International Association of Hydrological Sciences, Wallingford, Oxfordshire, UK, p. 171-180.

Lambert, S.M., 1967, Functional relationship between sorption in soil and chemical structure: *J. Agric. Food Chem.*, v. 15, no. 4, pp. 572-576.

Lambert, S.M., Porter, P.E., and Schieferstein, H., 1965, Movement and sorption of chemicals applied to the soil: *Weeds*, v. 13, pp. 185-190.

Larson, S.J., 1995, personal communication, manuscript in preparation, University of Minnesota, Gray Freshwater Biological Institute, Navarre, Minn.

Lasaga, A.C., 1981, Transition state theory, *in* Lasaga, A.C., and Kirkpatrick, R.J., eds., *Kinetics of geochemical processes*: Mineralogical Society of America, Washington, D.C., pp. 135-169.

Lavy, T.L., 1989, Pesticide residue monitoring: *Arkansas Farm Research*, p. 9.

Lavy, T.L., and Fenster, C.R., 1974, Herbicide breakdown in soil: *Farm, Ranch and Home Quarterly*, Winter, pp. 1-3.

Lavy, T.L., Helling, C.S., and Cheng, H.H., 1985, Pesticides, in *Agriculture and groundwater quality:* Council for Agricultural Science and Technology, Ames, Iowa, pp. 38-62.

Lavy, T.L., Mattice, J.D., and Kochenderfer, J.N., 1989, Hexazinone persistence and mobility of a steep forested watershed [sic]: *J. Environ. Qual.*, v. 18, no. 4, pp. 507-514.

Lavy, T.L., Roeth, F.W., and Fenster, C.R., 1973, Degradation of 2,4-D and atrazine at three soil depths in the field: *J. Environ. Qual.*, v. 2, no. 1, pp. 132-137.

Lawrence, J.R., Eldan, M., and Sonzogni, W.C., 1993, Metribuzin and metabolites in Wisconsin (U.S.A.) well water: *Water Res.*, v. 27, no. 8, pp. 1263-1268.

Lawruk, T.S., Hottenstein, C.S., Herzog, D.P., and Rubio, F.M., 1992, Quantification of alachlor in water by a novel magnetic particle-based ELISA: *Bull. Environ. Contam. Toxicol.*, v. 48, no. 5, pp. 643-650.

Leake, C.R., 1991, Lysimeter studies: *Pest. Sci.*, v. 32, no. 3, pp. 363-373.

Leake, C.R., Arnold, D.J., Newby, S.E., and Somerville, L., 1987, Benazolin ethyl—A case study of herbicide degradation and leaching, in *1987 British Crop Protection Conference—Weeds: Proceedings of a conference held at Brighton Metropole, England*: British Crop Protection Council, v. 2, pp. 577-583.

Lear, B., Towson, A.J., and Miyagawa, S.T., 1983, The value of leaching for removing inorganic bromide residues from soil after application of methyl bromide: *Acta Horticulturae*, v. 152, pp. 305-313.

Lee, C.M., and Macalady, D.L., 1990, unpublished data, November 26, Colorado School of Mines, Golden, Colo.

Lee, J.K., Fuhr, F., and Kyung, K.S., 1994, Behaviour of carbofuran in a rice plant-grown lysimeter throughout four growing seasons: *Chemosphere*, v. 29, no. 4, pp. 747-758.

Lee, L.K., and Nielsen, E.G., 1988, Groundwater: Is it safe to drink? *Choices*, v. 3, no. 3, pp. 22-23.

Lee, P.W., Fukoto, J.M., Hernandez, H., and Stearns, S.M., 1990, Fate of monocrotophos in the environment: *J. Agric. Food Chem.*, v. 38, no. 2, pp. 567-573.

Le Grand, H.E., 1970, Movement of agricultural pollutants with groundwater, *in* Willrich, T.L., and Smith, G.E., eds., *Agricultural practices and water quality:* Iowa State University Press, Ames, Iowa, pp. 303-313.

Leidy, V.A., and Taylor, R.E., 1992, Overview of susceptibility of aquifers to contamination, Union County, Arkansas: U.S. Geological Survey Water-Resources Investigations Report 92-4094, 35 p.

Leistra, M., 1986, Modelling the behaviour of organic chemicals in soil and ground water: *Pest. Sci.*, v. 17, pp. 256-264.

——1988, Behaviour and significance of pesticide residues in ground water: *Aspects Appl. Biol.*, v. 17, pp. 223-229.

Leistra, M., and Boesten, J.J.T.I., 1989, Pesticide contamination of groundwater in western Europe: *Agric. Ecosys. Environ.*, v. 26, pp. 369-389.

Leistra, M., Dekker, A., and van der Burg, A.M.M., 1984a, Computed and measured leaching of the insecticide methomyl from greenhouse soils into water courses: *J. Water, Air, Soil Poll.*, v. 23, pp. 155-167.

Leistra, M., Groen, A.E., Crum, S.J., and van der Pas, L.J.T., 1991, Transformation rate of 1,3-dichloropropene and 3-chloroallyl alcohol in topsoil and subsoil material of flower-bulb fields: *Pest. Sci.*, v 31, pp. 197-207.

Leistra, M., Tuinstra, L.G.M.T., van der Burg, A.M.M., and Crum, S.J.H., 1984b, Contribution of leaching of diazinon, parathion, tetrachlorvinphos and triazophos from glasshouse soils to their concentrations in water courses: *Chemosphere*, v. 13, no. 3, pp. 403-413.

LeMasters, G., 1994, personal communication, telephone conference, December 9, Wisconsin Department of Agriculture, Trade and Consumer Protection, Madison, Wis.

LeMasters, G., and Doyle, D.J., 1989, *Grade A Dairy Farm Well Water Quality Survey*: Wisconsin Department of Agriculture and Wisconsin Agricultural Statistics Service, Madison, Wis., 36 p.

LeMasters, G., and Baldock, J., 1995, *A survey of atrazine in Wisconsin groundwater: Phase one report*: Wisconsin Department of Agriculture, Madison, Wis., 16 p.

Leonard, R.A., and Knisel, W.G., 1988, Evaluating groundwater contamination potential from herbicide use: *Weed Technology*, v.2, no. 2, pp. 207-216.

——1989, Groundwater loadings by controlled-release pesticides: A GLEAMS simulation: *Trans. Am. Soc. Agric. Eng.*, v. 32, no. 6, pp. 1915-1922.

Leonard, R.A., Knisel, W.G., Davis, F.M., and Johnson, A.W., 1988, Modeling pesticide metabolite transport with GLEAMS, *in* Hay, D.R., ed., *Planning now for irrigation and drainage in the 21st century: Proceedings of a conference: Lincoln, Nebraska*: American Society of Civil Engineers, N. Y., pp. 255-263.

——1990, Validating GLEAMS with field data for fenamiphos and its metabolites: *J. Irrig. Drain. Eng.*, v. 116, no. 1, pp. 24-35.

Leonard, R.A., Knisel, W.G., and Still, D.A., 1987, GLEAMS: Groundwater loading effects of agricultural management systems: *Trans. Am. Soc. Agric. Eng.*, v. 30, pp. 1403-1418.

Leonard, R.A., Shirmohammadi, A., Johnson, A.W., and Marti, L.R., 1988, Pesticide transport in shallow groundwater, *in* Collins, A.G., and Johnson, A.I., eds., Ground-water contamination: Field methods: A symposium: American Society for Testing and Materials Special Technical Publication 963, pp. 776-778.

Letey, J., and Pratt, P.F., 1983, Agricultural pollutants and groundwater quality, *in Proceedings from the International Workshop on Behaviour of Pollutants in Unsaturated Zones, Bet Dagan, Israel:* Springer-Verlag, Berlin, pp. 211-222.

Levy, J., Chesters, G., Read, H.W., and Gustafson, D.P., 1993, Distribution, sources and fate of atrazine in a sandy-till aquifer, *in* Chesters, G., Levy, J., Reed, H.W., and Gustafson, D.P., eds., *Distribution, transport and fate of major herbicides and their metabolites:* University of Wisconsin-Madison, Water Resources Center, Madison, Wis., pp. 1-47

Lewallen, M.J., 1971, Pesticide contamination of a shallow bored well in the southeastern Coastal Plains: *Ground Water,* v. 9, no. 6, pp. 45-48.

Li, G.-C., and Felbeck, G.T., Jr., 1972, Atrazine hydrolysis as catalyzed by humic acids: *Soil Sci.,* v. 114, no. 3, pp. 201-209.

Libra, R.D., and Hallberg, G.R., 1993, Agricultural drainage wells in Iowa: Hydrogeologic settings and water-quality implications: Iowa Department of Natural Resources Technical Information Series 24, 39 p.

Libra, R.D., Hallberg, G.R., and Hoyer, B.E., 1987, Impacts of agricultural chemicals on ground water quality in Iowa, *in* Fairchild, D.M., ed., *Ground water quality and agricultural practices:* Lewis Publishers, Chelsea, Mich., p. 185-215.

Libra, R.D., Hallberg, G.R., Hoyer, B.E., and Johnson, L.G., 1986, Agricultural impacts on ground water quality: The Big Spring basin study, *in Proceedings, the Agricultural Impacts on Ground Water—A conference, Omaha, Nebraska:* National Water Well Association, Omaha, Nebr., p. 253-273.

Libra, R.D., Hallberg, G.R., Littke, J.P., Nations, B.K., Quade, D.J., and Rowden, R.D., 1991, Groundwater monitoring in the Big Spring basin 1988-1989: A summary review: Iowa Department of Natural Resources Technical Information Series 21, 29 p.

Libra, R.D., Hallberg, G.R., Ressmeyer, G.G., and Hoyer, B.E., 1984, Groundwater quality and hydrogeology of Devonian-carbonate aquifers in Floyd and Mitchell Counties: Iowa Geological Survey Open File Report 84-2, 149 p.

Libra, R.D., Hallberg, G.R., Rex, K.D., Kross, B.C., Seigley, L.S., Culp, M.A., Johnson, J.K., Nicholson, H.F., Berberich, S.L., and Cherryholmes, K.L., 1993, The Iowa state-wide rural well-water survey: June 1991, repeat sampling of the 10% subset: Iowa Department of Natural Resources Technical Information Series 26, 30 p.

Lindberg, R.D., and Runnells, D.D., 1984, Ground water redox reactions: An analysis of equilibrium state applied to E_H measurements and geochemical modeling: *Science,* v. 225, no. 4665, pp. 925-927.

Liqiang, J., Shukui, H., Liansheng, W., Chao, L., and Deben, D., 1994, The influential factors of hydrolysis of organic pollutants in environment [sic]: *Chemosphere,* v. 28, no. 10, pp. 1749-1756.

Liszewski, M.J., and Squillace, P.J., 1991, The effect of surface-water and ground-water exchange on the transport and storage of atrazine in the Cedar River, Iowa, *in* Mallard, G.E., and Aronson, D.A., eds., U.S. Geological Survey Toxic Substances Hydrology Program—Proceedings of the technical meeting, Monterey, California, March 11-15, 1991: U.S. Geological Survey Water-Resources Investigations Report 91-4034, pp. 1-9.

Little, D.L., 1994, Don't buy the myths—Noted scientist challenges extremist environmental agenda: *Farm Chemicals,* v. 157, no. 5, p. 49.

Litwin, Y.J., Hantzsche, N.N., and George, N.A., 1983, Groundwater contamination by pesticides: A California assessment, in *Proceedings of the Fourteenth Biennial Conference on Ground Water, Sacramento, California*: University of California, California Water Resources Center, Davis, Calif.

Liu, L.C., 1974, Leaching of fluometuron and diuron in a vega alta soil: *J.Agr.Univ.P.R.*, v. 58, no. 4, pp. 473-482.

Loague, K.M., and Green, R.E., 1991, Statistical and graphical methods for evaluating solute transport models: Overview and application: *J. Contam. Hydrol.*, v. 7, nos. 1-2, pp. 51-73.

Loague, K.M., Green, R.E., Giambelluca, T.W., and Liang, T.C., 1990, Impact of uncertainty in soil, climatic, and chemical information in a pesticide leaching assessment: *J. Contam. Hydrol.*, v. 5, no. 2, pp. 171-194.

Loague, K.M., Miyahira, R.N., Green, R.E., Oki, D.S., Giambelluca, T.W., and Schneider, R.C., 1995, Chemical leaching near the Waiawa Shaft, Oahu, Hawaii: 2. Modeling results: *Ground Water*, v. 33, no. 1, pp. 124-138.

Loague, K.M., Miyahira, R.N., Oki, D.S., Green, R.E., Schneider, R.C., and Giambelluca, T.W., 1994, Chemical leaching near the Waiawa Shaft, Oahu, Hawaii: 1. Field experiments and laboratory analysis: *Ground Water*, v. 32, no. 6, pp. 986-996.

Loch, J.P.G., 1991, Effect of soil type on pesticide threat to the soil/groundwater environment, in Richardson, M.L., ed., *Chemistry, agriculture and the environment:* The Royal Society of Chemistry, Cambridge, England, pp. 291-307.

Logan, T.J., Eckert, D.J., and Beak, D.G., 1994, Tillage, crop and climatic effects on runoff and tile drainage losses of nitrate and four herbicides: *Soil Tillage Res.*, v. 30, pp. 75-103.

Long, T., 1989, Groundwater contamination in the vicinity of agrichemical mixing and loading facilities, in *Illinois Agricultural Pesticides Conference: Proceedings*: University of Illinois, College of Agricultural Consumer and Environmental Services, Urbana-Champaign, Ill., pp. 139-149.

Lorber, M.N., Cohen, S.Z., and DeBuchananne, G.D., 1990, A national evaluation of the leaching potential of aldicarb, Part 2. An evaluation of ground water monitoring data: *Ground Water Monitor. Rev.*, v. 10, no. 1, pp. 127-141.

Lorber, M.N., Cohen, S.Z., Noren, S.E., and DeBuchananne, G.D., 1989, A national evaluation of the leaching potential of aldicarb Part 1. An integrated assessment methodology: *Ground Water Monitor. Rev.*, v. 9, no. 4, pp. 109-125.

Lorber, M.N., and Offutt, C.K., 1986, A method for the assessment of ground water contamination potential using a Pesticide Root Zone Model (PRZM) for the unsaturated zone, in Garner, W.Y., Honeycutt, R.C., and Nigg, H.N., eds., *Evaluation of pesticides in ground water: Developed from a symposium sponsored by the Division of Pesticide Chemistry at the 189th Meeting of the American Chemical Society, Miami Beach, Florida:* American Chemical Society Symposium Series 315, pp. 342-365.

Louis, J.B., and Vowinkel, E., 1989, Effect of agricultural chemicals on groundwater quality in the New Jersey coastal plain, in D.L. Weigmann, D.L., ed., *Terrestrial and aquatic environments: Proceedings of a national conference*: Virginia Polytechnic Institute and State University, Virginia Water Resources Research Center, Blacksburg, Va., pp. 80-88.

Lovley, D.R., Chapelle, F.H., and Woodward, J.C., 1994, Use of dissolved H_2 concentrations to determine distribution of microbially catalyzed redox reactions in anoxic groundwater: *Environ. Sci. Technol.*, v. 28, no. 7, pp. 1205-1210.

Lucas, R., 1995, *1993 certified commercial pesticide applicator survey for non-agricultural pesticides*: Research Triangle Institute, Research Triangle Park, N.C.

Lutz, J.F., Byers, G.E., and Sheets, T.J., 1973, The persistence and movement of picloram and 2,4,5-T in soils: *J. Environ. Qual.*, v. 2, no. 4, pp. 485-488.

Lym, R.G., and Messersmith, C.G., 1988, Survey for picloram in North Dakota groundwater: *Weed Technology*, v. 2, no. 2, pp. 217-222.

Ma, L., and Spalding, R.F., 1995, Pesticide occurrence and persistence in recharge basin drainage area [sic] in York, Nebraska, *in* American Chemical Society, *Preprints of papers presented at the 209th National Meeting, Anaheim, California*: American Chemical Society, Division of Environmental Chemistry, v. 35, no. 1, pp. 270-273.

Maathuis, H., Wasiuta, V., Nicholaichuk, W., and Grover, R., 1988, Study of herbicides in shallow groundwater beneath three irrigated sites in Outlook Irrigation District, Saskatchewan: Saskatchewan Research Council Publication R-844-13-E-88, 94 p.

Mabey, W.R., and Mill, T., 1978, Critical review of hydrolysis of organic compounds in water under environmental conditions: *J. Phys. Chem. Ref. Data*, v. 7, no. 2, pp. 383-425.

Macalady, D.L., and Wolfe, N.L., 1985, Effect of sediment sorption and abiotic hydrolyses. 1. Organophosphorothioate esters: *J. Agric. Food Chem.*, v. 33, no. 2, pp. 167-173.

——1987, Influences of aquatic humic substances on the abiotic hydrolysis of organic contaminants: A critical review, *in* American Chemical Society, *Preprints of papers presented at the 194th National Meeting, New Orleans, Lousiana*: American Chemical Society, Division of Environmental Chemistry, v. 27, no. 2, pp. 12-15.

Mackay, D.M., 1988, Groundwater contamination by organic chemicals: Overview and consideration of the potential importance of pesticide inert ingredients, in *Proceedings of the Western Society Weed Science*: Utah State University Plant Science Department, Logan, v. 41, pp. 7-15.

Mackay, D., and Shiu, W.Y., 1981, A critical review of Henry's Law constants for chemicals of environmental interest: *J. Phys. Chem. Ref. Data*, v. 10, no. 4, pp. 1175-1199.

Mackay, D.M., and Smith, L.A., 1990, Agricultural chemicals in groundwater: Monitoring and management in California: *J. Soil Water Conserv.*, v. 45, no. 2, p. 253-255.

Mackay, D., and Paterson, S., 1982, Fugacity revisited: The fugacity approach to environmental transport: *Environ. Sci. Technol.*, v. 16, no. 12, pp. 654A-660A.

Maddy, K.T., Fong, H.R., Lowe, J.A., Conrad, D.W., and Fredrickson, A.S., 1982, A study of well water in selected California communities for residues of 1,3-dichloropropene, chloroallyl alcohol and 49 organophosphate or chlorinated hydrocarbon pesticides: *Bull. Environ. Contam. Toxicol.*, v. 29, no. 3, pp. 354-359.

Madhun, Y.A., Young, J.L., and Freed, V.H., 1986, Binding of herbicides by water-soluble organic materials from soil: *J. Environ. Qual.*, v. 15, no. 1, p. 64.

Madison, R.J., and Brunett, J.O., 1985, Overview of the occurrence of nitrate in ground water of the United States, *in* Moody, D.W., Fischer, J.N., and Chase, E.B., eds., National water summary 1984—Hydrologic events, selected water-quality trends, and ground-water resources: U.S. Geological Survey Water-Supply Paper 2275, pp. 93-105.

Maes, C.M., Pepple, M., Troiano, D., Weaver, W., and Kimaru, W., 1991, Sampling for pesticide residues in California well water: 1991 update well inventory data base: California Environmental Protection Agency Report EH 92-02, 129 p.

Majewski, M.S., 1993, personal communication, telephone conference, September 29, U.S. Geological Survey, Sacramento, Calif.

Majewski, M.S., and Capel, P.D., 1995, *Pesticides in the atmosphere: Distribution, trends, and governing factors*: Ann Arbor Press, Chelsea, Mich., 214 p.

Majka, J.T., and Lavy, T.L., 1977, Adsorption, mobility, and degradation of cyanazine and diuron in soils: *Weed Sci.*, v. 25, no. 5, pp. 401-406.

Mandelbaum, R.T., Wackett, L.P., and Allan, D.L., 1993, Rapid hydrolysis of atrazine to hydroxyatrazine by soil bacteria: *Environ. Sci. Technol.*, v. 27 no. 9, pp. 1943-1946.

Mansell, R.S., Calvert, D.V., Stewart, E.H., Wheeler, W.B., Rogers, J.S., Graetz, D.A., Allen, L.H., Overman, A.R., and Knipling, E.P., 1977, Fertilizer and pesticide movement from citrus groves in Florida flatwood soils: U.S. Environment Protection Agency Technical Report EPA-600/2-77-177, 156 p.

Mansell, R.S., Wheeler, W.B., and Calvert, D.V., 1980, Leaching losses of two nutrients and an herbicide [terbacil] from two sandy soils during transient drainage: *Soil Sci.*, v. 130, no. 3, pp. 140-150.

Marade, S.J., and Segawa, R.T., 1988, Sampling for residues of molinate and thiobencarb in well water and soil in the Central Valley: California Department of Food and Agriculture, Division of Pest Management, Environmental Protection and Worker Safety, variously paged.

Marani, A., and Chesters, G., 1990, Pesticide transformations and movement in soils, *in* Jorgensen, S.E., ed., *Modelling in ecotoxicology*: Elsevier, Amsterdam, The Netherlands, pp. 215-231.

Marin, P.A., and Droste, E.X., 1986, Contamination of ground water as a result of agricultural use of ethylene dibromide (EDB), in *Proceedings of the Third Annual Eastern Regional Ground Water Conference, Springfield, Massachusetts:* National Water Well Association, Dublin, Ohio, pp. 277-306.

Maritz Marketing Research, 1992, *1992 golf course pesticide use study*: Maritz Marketing Research, Fenton, Mo.

Marley, J.M.T., 1980, Persistence and leaching of picloram applied to a clay soil on the Darling Downs: *Queensland J. Agric. Animal Sci.*, v. 37, no. 1, pp. 15-25.

Martin, D.F., Norris, C.D., and Martin, B.B., 1991, Intrusion indices—A measure of groundwater quality: *J. Environ. Sci. Health*, v. A26, no. 6, pp. 899-911.

Matthies, M., 1987, Fate modelling of pesticides in groundwater, *in* Greenhalgh, R., and Roberts, T.R., eds., *Pesticide science and biotechnology: Proceedings of the Sixth International Congress of Pesticide Chemistry, held in Ottawa, Canada*: Blackwell Scientific Publications, Oxford, England, pp. 373-380.

Mayer, J.R., Lacher, T.E., Jr., Elkins, N.R., and Thorn, C.J., 1991, Temporal variation of ethylene dibromide (EDB) in an unconfined aquifer, Whatcom County, Washington, USA: A twenty-seven month study: *Bull. Environ. Contam. Toxicol.*, v. 47, no. 3, pp. 368-373.

McCarty, P.L., Reinhard, M., and Rittmann, B.E., 1981, Trace organics in groundwater: *Environ. Sci. Technol.*, v. 15, no. 1, pp. 40-51.

McConnell, J.B., 1988, Ethylene dibromide (EDB) trends in the Upper Floridan Aquifer, Seminole County, Georgia, October 1981 to November 1987: U.S. Geological Survey Water-Resources Investigations Report 89-4034, 11 p.

McKay, L.D., Gillham, R.W., and Cherry, J.A., 1993, Field experiments in a fractured clay till 2. Solute and colloid transport: *Water Resour. Res.*, v. 29, no. 12, pp. 3879-3890.

McKenna, D.P, 1990, Geologic Mapping for protection of groundwater resources, *in* Information exchange on models and data needs relating to the impact of agricultural practices on water quality: Workshop proceedings: U.S. Geological Survey, Office of Water Data Coordination, Reston, Va., pp. 99-100.

McKenna, D.P., Chou, J., Griffin, R.A., Valkenburg, J., Spencer, L.L., and Gilkeson, J.L., 1988, Assessment of the occurrence of agricultural chemicals in groundwater in a part of Mason County, Illinois, in *Proceedings, Agricultural Impacts on Groundwater—A conference, Des Moines, Iowa*: National Water Well Association, Dublin. Ohio, pp. 389-406.

McKenna, D.P., and Keefer, D.A., 1991, Potential for agricultural chemical contamination of aquifers in Illinois: Illinois State Geological Survey Open File Report 1991-7R, 16 p.

McKinley, R.S., and Arron, G.P., 1987, Distribution of 2,4-D and picloram residues in environmental components adjacent to a treated right-of-way: Canada Department of Energy and Mines Research Report OH/R-87/49/K, 33 p.

McLaughlin, D., Reid, L.B., Li, S., and Hyman, J., 1993, A stochastic method for characterizing ground-water contamination: *Ground Water*, v. 31, no. 2, pp. 237-259.

McLean, J.E., Sims, R.C., Doucette, W.J., Caupp, C.R., and Grenney, W.J., 1988, Evaluation of mobility of pesticides in soil using U.S. EPA methodology: *J. Environ. Eng.*, v. 114, no. 3, pp. 689-703.

McMahon, P.B., Chapelle, F.H., and Jagucki, M.L., 1992, Atrazine mineralization potential of alluvial-aquifer sediments under aerobic conditions: *Environ. Sci. Technol.*, v. 26, no. 8, pp. 1556-1559.

McMahon, P.B., Litke, D.W., Paschal, J.E., and Dennehy, K.F., 1994, Ground water as a source of nutrients and atrazine to streams in the South Platte River Basin: *Water Resour. Bull.*, v. 30. no. 3, pp. 521-530.

McNeill, A., 1990, Leaching of pesticides from treated timbers—A source of pollution: *J. Instit. Water Environ. Manag.*, v. 4, no. 4, pp. 330-334.

McVoy, C.W., 1988, Factors affecting groundwater vulnerability, in *Proceedings of the 42st Annual Meeting of the Northeastern Weed Science Society*: Northeastern Weed Science Society, Salisbury, Md., pp. 16-20.

Meeks, Y.J., and Dean, D., 1990, Evaluating ground-water vulnerability to pesticides: *J. Water Resour. Plan. Manag.*, v. 116, no. 5, pp. 693-707.

Mehnert, E., Schock, S.C., Barnhardt, M.L., Caughey, M.E., Chou, S.F.J., Dey, W.S., Dreher, G.B., and Ray, C., 1995, The occurrence of agricultural chemicals in Illinois' rural private wells: Results from the pilot study; *Ground Water Monitor. Remed.*, v. 15, no. 1, pp. 142-149.

Meikle, R.W., and Youngson, C.R., 1978, The hydrolysis rate of chlorpyrifos, *O,O*-diethyl *O*-(3,5,6-trichloro-2-pyridyl)phosphorothioate, and its dimethyl analog, chlorpyrifos-methyl, in dilute aqueous solution: *Arch. Environ. Contam. Toxicol.*, v. 7, pp. 13-22.

Meister Publishing Company, 1995, *Farm chemicals handbook '95*: Meister Publishing, Willoughby, Ohio, variously paged.

Messina, V., 1994, Chlorine and cancer: *Good Medicine*, Winter, pp. 6-9.

Meulenberg, E.P., Mulder, W.H., and Stoks, P.G., 1995, Immunoassays for pesticides: *Environ. Sci. Technol.*, v. 29, no. 3, pp. 553-561.

Meyer, M.T., and Thurman, E.M., 1994, The transport of cyanazine and atrazine metabolites in the unsaturated zone, *in* American Chemical Society, *Preprints of papers presented at the 208th National Meeting, San Diego, California*: American Chemical Society, Division of Environmental Chemistry, v. 34, no. 2, pp. 571-573.

———1995, The degradation and transport of cyanazine metabolites in surface water of the midwestern United States, *in* Proceedings of technical meeting, U.S. Geological Survey Toxic Substances Hydrology Program, Colorado Springs, Colorado, 1995: U.S. Geological Survey Water-Resources Investigations Report 94-4015.

Michael, J.L., Neary, D.G., and Wells, M.J.M., 1989, Picloram movement in soil solution and streamflow from a coastal plain forest: *J. Environ. Qual.*, v. 18, no. 1, pp. 89-95.

Milde, K., Milde, H., Ahlsdorf, B., Litz, N., Müller-Wegerner, U., and Stock, R., 1988, Protection of highly permeable aquifers against contamination by xenobiotics, in *Karst hydrogeology and karst environment protection*: International Association of Hydrogeology 21st Congress held in Guilin, China, v. 1, pp. 194-201.

Miles, C.J., Yanagihara, K., Ogata, S., Van De Verg, G., and Boesch, R., 1990, Soil and water contamination at pesticide mixing and loading sites on Oahu, Hawaii: *Bull. Environ. Contam. Toxicol.*, v. 44, no. 6, pp. 955-962.

Mill, T., 1980, Chemical and photo-oxidation, *in* Hutzinger, O, ed., *The handbook of environmental chemistry*: Springer-Verlag, Berlin, pp. 77-105.

———1989, Structure activity relationships for photooxidation processes in the environment: *Environ. Toxicol. Chem.*, v. 8, pp. 31-43.

Mill, T., and Mabey, W.R., 1985, Photochemical transformations, *in* Neely, W.B., and Blau, G., eds., *Environmental exposure from chemicals:* CRC Press, Boca Raton, Fla. v. 1, pp. 175-216.

Miller, C., Pepple, M., Troiano, J., Weaver, D., Kimaru, W., and State Water Resources Control Board, 1990, Sampling for pesticide residues in California well water: 1990 update well inventory data base: California Department of Food and Agriculture Report EH 90-11.

Miller, C.T., 1987, Groundwater quality: *J. Water Pollution Control Federation*, v. 59, no. 6, pp. 513-531.

Miller, C.T., and Mayer, A.S., 1990, Groundwater: A review of the 1989 literature: *Res. J. WPCF*, v. 62, no. 5, pp. 700-737.

Miller, C.T., and Pedit, J.A., 1992, Use of a reactive surface-diffusion model to describe apparent sorption-desorption hysteresis and abiotic degradation of lindane in a subsurface material: *Environ. Sci. Technol.*, v. 26, no. 7, pp. 1417-1427.

Miller, M.E., and Alexander, M., 1991, Kinetics of bacterial degradation of benzylamine in a montmorillonite suspension: *Environ. Sci. Technol.*, v. 25, no. 2, pp. 240-245.

Miller, W.L., Foran, J.A., Huber, W., Davidson, J.M., Moye, H.A., and Spangler, D.P., 1989, Ground water monitoring for temik (Aldicarb) in Florida: *Water Resour. Bull.*, v. 25, no. 1, pp. 79-86.

Millipore Corporation, 1990, Millipore Corporation product literature: Millipore Corporation, Bedford, Mass.

Mills, M.S., and Thurman, E.M., 1994a, Preferential dealkylation reactions of *s*-triazine herbicides in the unsaturated zone: *Environ. Sci. Technol.*, v. 28, no. 4, pp. 600-605.

———1994b, Reduction of nonpoint source contamination of surface water and groundwater by starch encapsulation of herbicides: *Environ. Sci. Technol.*, v. 28, no. 1, pp. 73-79.

Milne, G.W.A., ed., 1995, *CRC handbook of pesticides*: CRC Press, Boca Raton, Fla., 402 p.

Mink, F.L., Risher, J.F., and Stara, J.F., 1989, The environmental dynamics of the carbamate insecticide aldicarb in soil and water: *Environmental Pollution*, v. 61, pp. 127-155.

Minnesota Environmental Quality Board, 1992, *Silvicultural systems: A background paper for a generic environmental impact statement on timber harvesting and forest management in Minnesota*: Jaakko Poyry Consulting, St. Paul, Minn., 61 p.

Minton, N.A., Leonard, R.A., and Parker, M.B., 1990, Concentration of total fenamiphos residue over time in the profile of two sandy soils: *Applied Agricultural Research*, v. 5, no. 2, pp. 127-133.

Mischke, T., Brunetti, K., Acosta, V., Weaver, D., and Brown, M., 1985, Agricultural sources of DDT residues in California's environment: California Department of Food and Agriculture, Environmental Hazards Assessment Program, 42 p.

Mitchem, P.S., Hallberg, G.R., Hoyer, B.E., and Libra, R.D., 1988, Ground-water contamination and land management in the karst area of northeastern Iowa, *in* Collins, A.G., and Johnson, A.I., eds., Ground-water contamination: Field methods: A symposium: American Society for Testing and Materials Special Technical Publication 963, pp. 442-458.

Monohan, K., and Field, J. A., 1995, Analysis of dacthal and its metabolites in soil and water, *in* American Chemical Society, *Preprints of papers presented at the 209th National Meeting, Anaheim, California*: American Chemical Society, Division of Environmental Chemistry, v. 35, no. 1, pp. 214-217.

Monsanto Agricultural Company, 1990, The National Alachlor Well Water Survey (NAWWS): Data summary: *Monsanto Technical Bulletin*, July, 4 p.

Moody, D.W., Carr, J., Chase, E.B., and Paulson, R.W., compilers, 1988, National Water Summary 1986—Hydrologic Events and Ground-Water Quality: U.S. Geological Survey Water-Supply Paper 2325, 560 p.

Moorman, T.B., Cambardella, C.A., and Novak, J.M., 1994, Spatial variability of biological and chemical properties in Iowa soils and implications for water quality assessments, in *Agricultural Research To Protect Water Quality: Proceedings of the conference, Minneapolis, Minnesota, USA*: Soil Water and Conservation Society, Arkeny, Iowa, v. 1, pp. 424-431.

Mortland, M.M., and Raman, K.V., 1967, Catalytic hydrolysis of some organic phosphate pesticides by copper(II): *J. Agric. Food Chem.*, v. 15, no. 1, p. 163.

Moye, H.A., and Miles, C.J., 1988, Aldicarb contamination of groundwater: *Rev. Environ. Contam. Toxicol.*, v. 105, pp. 99-145.

Moyer, J.R., and Blackshaw, R.E., 1993, Effect of soil moisture on atrazine and cyanazine persistence and injury to subsequent cereal crops in southern Alberta: *Weed Technol.*, v. 7, no. 4, pp. 988-994.

Mueller, T.C., 1994, Comparison of PRZM computer model predictions with field lysimeter data for dichlorprop and bentazon leaching: *J. Environ. Sci. Health*, v. A29, no. 6, pp. 1183-1195.

Mueller, T.C., Jones, R.E., and Bush, P.B., 1992, Comparison of PRZM and GLEAMS computer model predictions with field data for alachlor, metribuzin and norflurazon leaching: *Environ. Toxicol. Chem.*, v. 11, no. 3, pp. 427-436.

Muir, D.C., and Baker, B.E., 1976, Detection of triazine herbicides and their degradation products in tile-drain water from fields under intensive corn (maize) production: *J. Agric. Food Chem.*, v. 42, no. 1, pp. 122-125.

Munch, D.J., and Frebis, C.P., 1992, Analyte stability studies conducted during the National Pesticide Survey: *Environ. Sci. Technol.*, v. 26, no. 5, pp. 921-925.

Myott, D.H., 1980, *Groundwater quality assessment for Nassau County 1978 report year*: Nassau County Department of Health, Nassau County, N.Y., 39 p.

Nair, D.R., and Schnoor, J.L., 1992, Effect of two electron acceptors on atrazine mineralization rates in soil: *Environ. Sci. Technol.*, v. 26, no. 11, pp. 2298-2300.

Nakamura, K., Shiba, H., and Hasegawa, H., 1983, Leaching of several herbicides with water in lysimeter soils: *J. Pest. Sci.*, v. 8, pp. 9-15.

Nash, R.G., 1988, Dissipation from soil, *in* Grover, R., ed., *Environmental chemistry of herbicides*: CRC Press, Boca Raton, Fla., pp. 132-169.

Nash, R.G., Helling, C.S., Ragone, S.E., and Leslie, A.R., 1991, Groundwater residue sampling: Overview of the approach taken by government agencies, *in* Nash, R.G., and Leslie, A.R., eds., *Groundwater residue sampling design*: American Chemical Society Symposium Series 465, pp. 1-13.

National Research Council, 1993, *Report of the committee on techniques for assessing ground water vulnerability:* National Research Council, Washington, D.C.

Nations, B.R., and Hallberg, G.R., 1992, Pesticides in Iowa precipitation: *J. Environ. Qual.*, v. 21, no. 3. pp. 486-492.

Neal, R.H., Teso, R.R., Younglove, Y., and Sheeks, D.L., III, 1991, Seasonal rainfall effects on pesticide leaching in Riverside, California: California Department of Pesticide Regulation Report EH 91-07, 40 p.

Nearpass, D.C., 1972, Hydrolysis of propazine by the surface acidity of organic matter: *Soil Sci. Soc. Am. J.*, v. 36, p. 606.

Neary, D.G., 1983, Monitoring herbicide residues in springflow after an operational application of hexazinone: *South. J. Appl. Forestry*, v. 7, no. 4, pp. 217-223.

Neary, D.G., Bush, P.B., and Douglass, J.E., 1983, Off-site movement of hexazinone in stormflow and baseflow from forest watersheds: *Weed Sci.*, v. 31, pp. 543-551.

Neary, D.G., Bush, P.B., Douglass, J.E., and Todd, R.L., 1985, Picloram movement in an Appalachian hardwood forest watershed: *J. Environ. Qual.*, v. 14, no. 4, pp. 585-592.

Neary, D.G., and Michael, J.L., 1989, Effect of sulfometuron methyl on ground water and stream quality in coastal plain forest watersheds: *Water Resour. Bull.*, v. *25*, no. 3, pp. 617-623.

Neil, C.D., Williams, J.S., and Weddle, T.K., 1989, Report to the 114th Maine State Legislature Energy and Natural Resources Committee: Pilot pesticides in ground water study, final report: Maine Geological Survey Open-File Report No. 89-2, 43 p.

Newland, L.W., Chesters, G., and Lee, G.B., 1969, Degradation of γ-BHC in simulated lake impoundments as affected by aeration: *J. Water Pollution Control Federation*, v. 41, no. 5, pp. R174-R188.

Newton, M., Horner, L.M., Cowell, J.E., White, D.E., and Cole, E.C., 1994, Dissipation of glyphosate and aminomethylphosphonic acid in North American forests: *J. Agric. Food Chem.*, v. 42, no. 8. pp. 1795-1802.

Ngabe, B., Bidleman, T.F., and Falconer, R.L., 1993, Base hydrolysis of α- and γ-hexachloro-cyclohexanes: *Environ. Sci. Technol.*, v. 27, no. 9, pp. 1930-1933.

Nicholls, P.H., 1988, Factors influencing entry of pesticides into soil water: *Pest. Sci.*, v. 22, pp. 123-137.

Nofziger, D.L., and Hornsby, A.G., 1986, A microcomputer-based management tool for chemical movement in soil: *Appl.Agr.Res.*, v. 1, pp. 50-56.

Nome, F., Rubira, A.F., Franco, C., and Ionescu, L.G., 1982, Limitations of the pseudophase model of micellar catalysis. The dehydrochlorination of 1,1,1-trichloro-2,2-bis(p-chlorophenyl)ethane and some of its derivatives: *J. Phys. Chem.*, v. 86, no. 10, pp. 1881-1885.

Norris, F.A., Noling, J.W., Jones, R.L., Kirkland, S.D., Overman, A.J., and Stanley, C.D., 1991, Field studies of ethoprop movement and degradation in two Florida soils: *J. Contam. Hydrol.*, v. 8, pp. 299-315.

Nose, K., 1984, Some problems on dispersion coefficient [sic] and simulation for pesticide leaching through soil: *J. Pest. Sci.*, v. 9, pp. 1-6.

Novak, J.M., Moorman, T.B., and Karlen, D.L., 1994, Influence of soil aggregate size on atrazine sorption kinetics: *J. Agricultural Food Chemicals*, v. 42, no. 8, pp. 1809-1812.

Nowell, L.H., and Resek, E.A., 1994, Summary of national standards and guidelines for pesticides in water, bed sediment, and aquatic organisms and their application to water-quality assessments: U.S. Geological Survey Open-File Report No. 94-44, 115 p.

O'Neill, H.J., Pollock, T.L., Bailey, H.S., Milburn, P., Gartley, C., and Richards, J.E., 1989, Dinoseb presence in agricultural subsurface drainage from potato fields in northwestern New Brunswick, Canada: *Bull. Environ. Contam. Toxicol.*, v. 43, no. 6, pp. 935-940.

Obreza, T.A., and Ontermaa, E.O., 1991, Small-scale ground water monitoring for 1,3-dichloropropene in southwest Florida: *Soil and Crop Sci. Soc. Florida Proc.*, v. 50, pp. 94-98.

Odanaka, Y., Taniguchi, T., Shimamura, Y., Iijima, K., Koma, Y., Takechi, T., and Matano, O., 1994, Runoff and leaching of pesticides in golf course [sic]: *J. Pest. Sci.*, v. 19, no. 1, pp. 1-10.

Ogram, A.V., Jessup, R.E., Ou, L.T., and Rao, P.S.C., 1985, Effects of sorption on biological degradation rates of (2,4-dichlorophenoxy)acetic acid [2,4-D] in soils: *Appl. Environ. Microbiol.*, v. 49, no. 3, pp. 582-587.

Ojima, M., 1995, Experiments on pollutants runoff [sic] in a golf course under land site development: *Acta Universitatis Carolinae Geologica*, v. 39, pp. 359-370.

Ojima, M., Matsura, F., and Hori, K., 1993, Experimental analysis on the moving characteristics of pesticides scattered on putting-greens of golf courses [sic], in *Proceedings of XXV Congress of International Association for Hydraulic Research, Tokyo, Japan:* Local Organizing Committee of the 25th IAHR Congress and Japan Society of Civil Engineers, Tokyo, Japan, D-10, pp. 328-335.

Oki, D.S., and Giambelluca, T.W., 1987, DBCP, EDB, and TCP contamination of ground water in Hawaii: *Ground Water*, v. 25, no. 6, pp. 693-702.

Ontario Ministry of the Environment, 1987a, *Pesticides in Ontario drinking water, 1985*: Ontario Ministry of the Environment, Water Resources Branch, Drinking Water Section, Toronto, Canada, 31 p.

———1987b, Pesticides in Ontario drinking water—1986: Ontario Ministry of the Environment, Water Resources Branch, Drinking Water Section, Toronto, Canada, 51 p.

———1990, Pesticides in Ontario drinking water—1987 groundwater sampling program: Ontario Ministry of the Environment, Water Resources Branch, Toronto, Canada, 49 p.

Oreskes, N., Shrader-Frechette, K., and Belitz, K., 1994, Verification, validation, and confirmation of numerical models in the earth sciences: *Science*, v. 263, no. 5147, pp. 641-646.

Osgerby, J.M., 1973, An approach to the prediction of the leaching of herbicides in soils, in *Proceedings of the Eleventh British Weed Control Conference, Hotel Metropole, Brighton:* A.W. Billit, Droitwich, England, pp. 792-799.

Oshima, R.J., Torres, G., Nelson, S.J., and Mischke, T.M., 1980, Monitoring selected ground water basins for the presence of aldicarb: A cooperative California interagency study: California Department of Food and Agriculture, State Water Resources Control Board, Department of Health Services Report EH 80-03, 9 p.

Otero, C., and Rodenas, E., 1986, Influence of n-butyl and n-hexyl alcohols in the dehydrohalogenation of DDT in cationic micelles of *N*-cetyl-*N,N,N*-trimethylammonium bromide, chloride, and hydroxide: *J. Phys. Chem.*, v. 90, no. 22, pp. 5771-5775.

Pacenka, S., Porter, K.S., Jones, R.L., Zecharias, Y.B., and Hughes, H.B.F., 1987, Changing aldicarb residue levels in soil and groundwater, eastern Long Island, New York: *J. Contam. Hydrol.*, v. 2, pp. 73-91.

Pacenka, S., and Steenhuis, T., 1984, *User's guide for the MOUSE computer program*: Cornell University, Agricultural Engineering Department, Ithaca, N.Y., 47 p.

Parrish, R., and Smith, C.N., 1990, A method for testing whether model predictions fall within a prescribed factor of true values, with an application to pesticide leaching: *Ecolog. Model.*, v. 51, pp. 59-72.

Parsons, D.W., and Witt, J.M., 1989, Pesticides in groundwater of the United States of America: A report of a 1988 survey of state lead agencies: Oregon State University Extension Service Report EM 8406, variously paged.

Pavlostathis, S.P., and Mathavan, G.N., 1992, Desorption kinetics of selected volatile organic compounds from field contaminated soils: *Environ. Sci. Technol.*, v. 26, no. 3, pp. 532-538.

Pennell, K.D., Hornsby, A.G., Jessup, R.E., and Rao, P.S.C., 1990, Evaluation of five simulation models for predicting aldicarb and bromide behavior under field conditions: *Water Resour. Res.*, v. 26, no. 11, pp. 2679-2693.

Pennington, K.L., Harper, S., and Koskinen, W.C., 1991, Interactions of herbicides with water-soluble soil organic matter. *Weed Sci.*, v. 39, pp. 667-672.

Peoples, S.A., Maddy, K.T., Cusick, W., Jackson, T., Cooper, C., and Frederickson, A.S., 1980, A study of samples of well water collected from selected areas in California to determine the presence of DBCP and certain other pesticide residues: *Bull. Environ. Contam. Toxicol.*, v. 24, no. 4, pp. 611-618.

Perdue, E.J., and Wolfe, N.L., 1982, Modification of pollutant hydrolysis kinetics in the presence of humic substances: *Environ. Sci. Technol.*, v. 16, no. 12, pp. 847-852.

Pereira, W.E., and Hostettler, F.D., 1993, Nonpoint source contamination of the Mississippi River and its tributaries by herbicides: *Environ. Sci. Technol.*, v. 27, no. 8, pp. 1542-1552.

Perry, C.A., 1990, Source, extent, and degradation of herbicides in a shallow aquifer near Hesston, Kansas: U.S. Geological Survey Water-Resources Investigations Report 90-4019, 24 p.

Perry, C.A., and Anderson, M.R., 1991, Statistical comparison of selected chemical constituents in water from chemigation and conventional irrigation wells in Kansas, 1987: U.S. Geological Survey Water-Resources Investigations Report 91-4049, 12 p.

Perry, C.A., Robbins, F.V., and Barnes, P.L., 1988, Factors affecting leaching in agricultural areas and an assessment of agricultural chemicals in the ground water of Kansas: U.S. Geological Survey Water-Resources Investigations Report 88-4104, 55 p.

Peryea, F.J, 1989, Leaching of lead and arsenic in soils contaminated with lead arsenate pesticide residues: Washington State University, Tree Fruit Research and Extension Center Project A-158-WASH, 59 p.

Peterson, F.L., and Hargis, D.R., 1971, Effect of storm runoff disposal and other artificial recharge to Hawaiian Ghyben-Herzberg aquifers: University of Hawaii Water Resources Research Center Technical Report 54, 51 p.

Petrovic, A.M., and Hummel, N.W., Jr., 1987, The grass may be greener, but...: *New York's Food and Life Sci.*, v. 17, no. 1, pp. 25-26.

Petrovic, A.M., Young, R.A., Sanchirico, C.A., and Lisk, D.J., 1994a, Triadimenol in turfgrass lysimeter leachates after fall application of triadimefon and overwintering: *Chemosphere*, v. 29, no. 2, pp. 415-419.

——1994b, Migration of isazofos nematocide in irrigated turfgrass soils: *Chemosphere*, v. 28, no. 4, pp. 721-724.

Pettit, G., 1988, Assessment of Oregon's groundwater for agricultural chemicals, in *Agricultural Impacts on Ground Water Conference,* Des Moines, Iowa, Association of Ground Water Scientists and Engineers: National Well Water Association, Dublin, Ohio, pp. 279-295.

Pettyjohn, W.A., 1981, Introduction to artificial ground water recharge: U.S. Environmental Protection Agency. Office of Research and Development, Robert S. Kerr Environmental Research Laboratory, 44 p.

Phillips, L.J., and Birchard, G.F., 1991, Use of STORET data to evaluate variations in environmental contamination by Census Division: *Chemosphere*, v. 22, nos. 9-10, pp. 835-848.

Phillips, P.J., and Shedlock, R.J., 1993, Hydrology and chemistry of groundwater and seasonal ponds in the Atlantic Coastal Plain in Delaware, USA: *J. Hydrology*, v. 141, nos. 1-4, pp. 157-178, 40 p.

Pickett, C.H., Hawkins, L.S., Pehrson, J.E., and O'Connell, N.V., 1990, Herbicide use in citrus production and ground water contamination in Tulare County: California Department of Food and Agriculture Report PM 90-1, 40 p.

——1992, Irrigation practices, herbicide use and ground-water contamination in citrus production: A case study in California: *Agric. Ecosys. Environ.*, v. 41, pp. 1-17.

Pignatello, J.J., 1991, Desorption of tetrachloroethene and 1,2-dibromo-3-chloropropane from aquifer sediments: *Environ. Toxicol. Chem.*, v. 10, no. 11, pp. 1399-1404.

Pignatello, J.J., Ferrandino, F.J., and Huang, L.Q., 1993, Elution of aged and freshly added herbicides from a soil. *Environ. Sci. Technol.*, v. 27, no. 8, pp. 1563-1571.

Pignatello, J.J., Frink, C.R., Marin, P.A., and Droste, E.X., 1990, Field-observed ethylene dibromide in an aquifer after two decades: *J. Contam. Hydrol.*, 5, no. 2, pp. 195-214.

Pignatello, J.J., and Huang, L.Q., 1991, Sorptive reversibility of atrazine and metolachlor residues in field soil samples: *J. Environ. Qual.*, v. 20, no. 1, pp. 222-228.

Pinholster, G., 1995, Drinking recycled wastewater: *Environ. Sci. Technol.*, v. 29, no. 4, pp. 174A-179A.

Pinto, E., 1980, *Report of groundwater contamination study in Wicomico County, Maryland*: Wicomico County Health Department. Salisbury, Md., 9 p.

Pionke, H.B., and Glotfelty, D.E., 1989, Nature and extent of groundwater contamination by pesticides in an agricultural watershed: *Water Res*, v. 23, no. 8, pp. 1031-1037.

Pionke, H.B., Glotfelty, D.E., Lucas, A.D., and Urban, J.B., 1988, Pesticide contamination of groundwaters in the Mahantango Creek watershed: *J. Environ. Qual.*, v. 17, no. 1, pp. 76-84.

Plumb, R.H., Jr., 1985, Disposal site monitoring data: Observations and strategy implications, *in* Hitchon, B., and Trudell, M.R., eds., *Proceedings, Second Canadian/American Conference on Hydrogeology: Hazardous wastes in ground water, a soluble dilemma: Banff, Alberta, Canada*: National Water Well Association, Dublin, Ohio, pp. 69-77.

——1991, The occurrence of appendix IX organic constituents in disposal site ground water: *Ground Water Monitor. Rev.*, v. 11, no. 2, pp. 157-164.

Plummer, L.N., Michel, R.L., Thurman, E.M., and Glynn, P.D., 1993, Environmental tracers for age dating young ground water, *in* Alley, W.M., ed., *Regional ground-water quality:* Van Nostrand Reinhold, N.Y., p. 255-294.

Poletika, N.N., Jury, W.A., and Yates, M.V., 1995, Transport of bromide, simazine, and MS-2 coliphage in a lysimeter containing undisturbed, unsaturated soil: *Water Resour. Res.*, v. 31, no. 4, pp. 801-810.

Porter, K.S., Wagenet, R.J., Jones, R.L., and Marquardt, T.E., 1990, Field research on aldicarb management practices for upstate New York: *Environ. Toxicol. Chem.*, v. 9, no. 3, pp. 279-287.

Pothuluri, J.V., Moorman, T.B., Obenhuber, D.C., and Wauchope, R.D., 1990, Aerobic and anaerobic degradation of alachlor in samples from a surface-to-groundwater profile: *J. Environ. Qual.*, v. 19, pp. 525-530.

Potter, T.L., and Carpenter, T.L., 1995, Occurrence of alachlor environmental degradation products in groundwater: *Environ. Sci. Technol.*, v. 29, no. 6, pp. 1557-1563.

Priddle, M.W., Jackson, R.E., and Mutch, J.P., 1989, Contamination of the sandstone aquifer of Prince Edward Island, Canada, by aldicarb and nitrogen residues: *Ground Water Monitor. Rev.*, v. 9, no. 4, pp. 134-140.

Priddle, M.W., Jackson, R.E., Novakowski, K.S., Denhoed, S., Graham, B.W., Patterson, R.J., Chaput, D., and Jardine, D., 1987, Migration and fate of aldicarb in the sandstone aquifer of Prince Edward Island: *Water Pollution Research J. Canada*, v. 22, no. 1, pp. 173-185.

Priddle, M.W., Mutch, J.P., and Jackson, R.E., 1992, Long-term monitoring of aldicarb residues in groundwater beneath a Canadian potato field: *Arch. Environ. Contam. Toxicol.*, v. 22, no. 2, pp. 183-189.

Priebe, D.L., and Blackmer, A.M., 1989, Preferential movement of oxygen-18-labeled water and nitrogen-15-labeled urea through macropores in a Nicollet soil: *J. Environ. Qual.*, v. 18, no. 1, pp. 66-72.

Puri, R.K., Orazio, C.E., Kapila, S., Clevenger, T.E., Yanders, A.F., McGrath, K.E., Buchanan, A., Czarnezki, J., and Bush, J., 1990, Studies on the transport and fate of chlordane in the environment, *in* Kurtz, D.A., ed., *Long range transport of pesticides:* Lewis Publishers, Chelsea, Mich., pp. 271-289.

Racke, K.D., Lubinski, R.N., Fontaine, D.D., Miller, J.R., McCall, P.J., and Oliver, G.R., 1993, Comparative fate of chlorpyrifos insecticide in urban and agricultural environments, *in* Racke, K.D., and Leslie, A.R., eds., *Pesticides in urban environments: Fate and significance*: American Chemical Society Symposium Series 522, pp. 70-85.

Radcliffe, D.E., Tollner, E.Q., Hargrove, W.L., Clark, R.L., and Golabi, M.H., 1988, Effect of tillage practices on infiltration and soil strength of a typic hapludult soil after ten years: *Soil Sci. Soc. Am. J.*, v. 52, pp. 798-804.

Raloff, J., 1994, The gender benders: *Science News*, v. 145, pp. 24-27.

Ramanand, K., Sharmila, M., Panda, D., and Sethunathan, N., 1988, Leaching of carbofuran in flooded field under puddled and nonpuddled conditions: *J. Environ. Sci. Health*, v. B23, no. 3, pp. 225-234.

Randhawa, S.K., and Gill, H.S., 1984, Leaching behaviour of herbicides in soils: *J. Res. Punjab Agric. Univ.*, v. 21, no. 2, pp. 159-162.

Rao, P.S.C., and Alley, W.M., 1993, Pesticides, *in* Alley, W.M., ed., *Regional ground-water quality*: Van Nostrand Reinhold, N.Y., pp. 345-382.

Rao, P.S.C., Edvardsson, K.S.V., Ou, L.T., Jessup, R.E., Nkedi-Kizza, and Hornsby, A.G., 1986, Spatial variability of pesticide sorption and degradation parameters, *in* Garner, W.Y., Honeycutt, R.C., and Nigg, H.N, eds., *Evaluation of Pesticides in Water: Developed from a symposium sponsored by the Division of Pesticide Chemistry at the 189th Meeting of the American Chemical Society, Miami Beach, Florida:* American Chemical Society Symposium Series 315, pp. 100-115.

Rao, P.S.C., Green, R.E., Balasubramanian, V., and Kanehiro, Y., 1974, Field study of solute movement in a highly aggregated oxisol with intermittent flooding: II. Picloram: *J. Environ. Qual.*, v. 3, no. 3, pp. 197-202.

Rao, P.S.C., Hornsby, A.G., and Jessup, R.E., 1985, Indices for ranking the potential for pesticide contamination of groundwater: *Soil and Crop Sci. Soc. Florida Proc.*, v. 44, pp. 1-8.

Rao, P.S.C., and Jessup, R.E., 1983, Sorption and movement of pesticides and other toxic organic substances in soils, *in* Nelson, D.W., Tanji, K.K., and Elrick, D.E., eds., *Chemical mobility and reactivity in soil systems: Proceedings of a symposium:* Soil Science Society of America and American Society of Agronomy, Madison, Wis. pp. 183-201.

Reiml, D., Scheunert, I., and Korte, F., 1989, Leaching of conversion products of [^{14}C]buturon from soil during 12 years after application: *J. Agric. Food Chem.*, v. 37, no. 1, pp. 244-248.

Rex, K.D., Libra, R.D., Hallberg, G.R., Kross, B.C., Field, R.W., Johnson, J.K., Nicholson, H.F., Cherryholmes, K.L., and Hall, N.H., 1993, The Iowa state-wide rural well-water survey: October 1990, repeat sampling of the 10% subset: Iowa Department of Natural Resources Technical Information Series 25, 26 p.

Rezende, M.C., Rubira, A.F., Franco, C., and Nome, F., 1983, Effect of normal and functional micelles in elimination reactions of polyhalogenated pesticides: *J. Chem. Soc., Perkin Trans.*, v. 2, pp. 1075-1078.

Rice, R.C., Jaynes, D.B., and Bowman, R.S., 1991, Preferential flow of solutes and herbicide under irrigated fields: *Trans. Am. Soc. Agric. Eng.*, v. 34, no. 3, pp. 914-918.

Richard, J.J., Junk, G.A., Avery, M.J., Nehring, N.L., Fritz, J.S., and Svec, H.J., 1975, Analysis of various Iowa waters for selected pesticides: Atrazine, DDE, and dieldrin—1974: *J. Pest. Monit.*, v. 9, no. 3, pp. 117-123.

Richard, T.L., and Steenhuis, T.S., 1988, Tile drain sampling of preferential flow on a field scale: *J. Contam. Hydrol.*, v. 3, pp. 307-325.

Richards, R.P., 1992a, *Water quality in private rural wells in the Limestone Ridge area near Carey, Ohio*: Heidelberg College, Water Quality Laboratory, Tiffin, Ohio, 79 p.

——1992b, *Temporal variability of chemistry in selected private rural wells in Ohio*: Heidelberg College, Water Quality Laboratory, Tiffin, Ohio, 82 p.

——1994, personal communication, electronic data transfer (magnetic tape), June 15, Heidelberg College, Water Quality Laboratory, Tiffin, Ohio.

Richards, R.P., and Baker, D.B., 1992, Pesticide concentration patterns in agricultural drainage networks in the Lake Erie Basin: *Environ. Toxicol. Chem.*, v. 12, no. 1, pp. 13-26.

Riley, D., 1976, Physical loss and redistribution of pesticides in the liquid phase, *in* Beynon, K.I., ed., *Persistence of insecticides and herbicides: Papers presented at a symposium held at the University of Reading:* British Crop Protection Council, pp. 109-115.

Risch, M.R., 1994, A summary of pesticides in ground-water data collected by government agencies in Indiana, December 1985 to April 1991: U.S. Geological Survey Open-File Report 93-133, 30 p.

Ritter, W.F., 1990, Pesticide contamination of ground water in the United States—A review: *J. Environ. Sci. Health*, v. B25, no. 1, pp. 1-29.

Ritter, W.F., Chirnside, A.E.M., and Scarborough, R.W., 1989, Pesticide movement in a coastal plain soil under irrigation, *in* Summers, J.B., and Anderson, S.S., eds., *Toxic substances in agricultural water supply and drainage: An international environmental perspective: Papers fron the Second Pan-American Regional Conference of the International Commission on Irrigation and Drainage, Ottawa, Canada*: U.S. Committee on Irrigation and Drainage, Denver, Colo. pp. 389-400.

Ritter, W.F., Scarborough, R.W., and Chirnside, A.E.M., 1994, Contamination of groundwater by triazines, metolachlor and alachlor: *J. Contam. Hydrol.*, v. 15, nos. 1-2, pp. 73-92.

Rittmann, B.E., McCarty, P.L., and Roberts, P.V., 1980, Trace-organics biodegradation in aquifer recharge: *Ground Water*, v 18, no. 3, pp. 236-243.

Roaza, H.P., Pratt, T.R., and Moore, B.W., 1989, *Hydrogeology and nonpoint source contamination of ground water by ethylene dibromide in northeast Jackson County, Florida*: Northwest Florida Water Management District Water Resources, Havana, Fla., 96 p.

Robbins, J.W.D., and Kriz, G.J., 1969, Relation of agriculture to groundwater pollution: A review: *Trans. Am. Soc. Agric. Eng.*, v. 12, no. 3, pp. 397-403.

Roberts, A. L., Sanborn, P. N., and Gschwend, P. M., 1992, Nucleophilic substitution reactions of dihalomethanes with hydrogen sulfide species: *Environ. Sci. Technol.*, v. 26, no. 11, pp. 2263-2274.

Roberts, P.V., Goltz, M.N., Summers, R.S., Crittenden, J.C., and Nkedi-Kizza, P., 1987, The influence of mass transfer on solute transport in column experiments with an aggregated soil: *J. Contam. Hydrol.*, v. 1, pp. 375-393.

Roberts, P.V., Hopkins, G.D., Mackay, D.M., and Semprini, L., 1990, A field evaluation of in-situ biodegradation of chlorinated ethenes: Part I, Methodology and field site characterization: *Ground Water*, v. 28, no. 4, pp. 591-604.

Roberts, P.V., McCarty, P.L., Reinhard, M., and Schreiner, J., 1980, Organic contaminant behavior during groundwater recharge: *J. Water Pollution Control Federation*, v. 52, no. 1, pp. 161-172.

Roberts, P.V., Reinhard, M., and Valocchi, A.J., 1982, Movement of organic contaminants in groundwater: Implications for water supply: *J. Am. Water Works Assoc.*, August, pp. 408-413.

Roberts, T.R., 1990, Environmental fate of pesticides: A perspective, *in* Hutson, D.H., and Roberts, T.R., eds., *Environmental fate of pesticides:* Wiley, Chichester, England, pp. 1-12.

Roberts, T.R., and Stoydin, G., 1976, The degradation of (Z)- and (E)-1,3-dichloropropenes and 1,2-dichloropropane in soil: *Pest. Sci.*, v. 7, pp. 325-335.

Rodgers, E.G., 1968, Leaching of seven *s*-triazines: *Weed Sci.*, v. 16, no. 2, pp. 117-120.

Roeth, F.W., Lavy, T.L., and Burnside, O.C., 1969, Atrazine degradation in two soil profiles: *Weed Sci.*, v. 17, pp. 202-205.

Rogers, N.K., and Talbert, R.E., 1981, Dissipation and leaching of metriflufen under field and controlled conditions: *Weed Sci.*, v. 29, no. 5, pp. 561-565.

Rose, C., 1990, A hidden pesticide peril: Inert, but toxic, additives: *Des Moines Sunday Register,* January 28, pp. 3A+.

Rose, S.C., Harris, G.L., Armstrong, A.C., Williams, J.R., Howse, K.R., and Tranter, N., 1991, The leaching of agrochemicals under different agricultural land uses and its effect on water quality, *in* Peters, N.E., and Walling, D.E., eds., *Sediment and stream water quality in a changing environment: Trends and explanation:* International Association of Hydrological Sciences, Wallingford, Oxfordshire, UK, pp. 249-257.

Rosen, L., 1994, A study of DRASTIC methodology with emphasis on Swedish conditions: *Ground Water*, v. 32, no. 2, pp. 278-285.

Rossi, P., de Carvalho-Dill, A., Muller, I., and Aragno, M., 1994, Comparative tracing experiments in a porous aquifer using bacteriophages and fluorescent dye on a test field located at Wilerwald (Switzerland) and simultaneously surveyed in detail on a local scale by radio-magneto-tellury (12-240 kHz): *Environ. Geol.*, v. 23, no. 3, pp. 192-200.

Rothschild, E.R., Manser, R.J., and Anderson, M.P., 1982, Investigation of aldicarb in ground water in selected areas of the central sand plain of Wisconsin: *Ground Water*, v. 20, no. 4, pp. 437-445.

Rouchaud, J., Gustin, F., and Wauters, A., 1994, Soil biodegradation and leaf transfer of insecticide imidacloprid [sic] applied in seed dressing in sugar beet crops: *Bull. Environ. Contam. Toxicol.*, v. 53, no. 3, pp. 344-350.

Roux, P.H., Balu, K., and Bennett, R., 1991a, A large-scale retrospective ground water monitoring study for metolachlor: *Ground Water Monitor. Rev.*, v. 11, no. 3, pp. 104-114.

Roux, P.H., Hall, R.L., and Ross, R.H., Jr., 1991b, Small-scale retrospective ground water monitoring study for simazine in different hydrogeological settings: *Ground Water Monitor. Rev.*, v. 11, no. 3, pp. 173-181.

Rowden, R.D., 1995, unpublished data, Iowa Department of Natural Resources, Geological Survey Bureau, Iowa City, Iowa.

——1995, personal communication, written, May 8, Iowa Department of Natural Resources, Iowa City, Iowa.

Rowden, R.D., Libra, R.D., Hallberg, G.R., and Nations, B., 1993, Groundwater monitoring in the Big Spring Basin 1990-1991: A summary review: A report of the Big Spring Basin Demonstration Project: Iowa Department of Natural Resources Technical Information Series 27, 36 p.

Roy, W.R., Krapac, I.G., Dey, W.S., and Mehnert, E., 1993, Pesticides in geologic materials at agrichemical facilities in Illinois: Definitions of "contamination," in *Agrichemical facility site contamination study*: Illinois Department of Agriculture, Springfield, Ill., variously paged.

Rudolph, D., Goss, J., Graham, A., Kachanoski, G., Scafe, M., Aspinall, D., van den Broek, R., Clegg, S., Barry, D., and Stimson, J., 1992, *Ontario farm groundwater quality survey winter 1991/92*: University of Waterloo, Waterloo Centre for Ground Water Research., 152 p.

——1993, *Ontario farm groundwater quality survey summer 1992:* Waterloo Center for Groundwater Research, University of Waterloo, 162 p.

Rügge, K., Bjerg, P.L., and Christensen, T.H., 1995, Distribution of organic compounds from municipal solid waste in the groundwater downgradient of a landfill (Grindsted, Denmark): *Environ. Sci. Technol.*, v. 29, no. 5, pp. 1395-1400.

Russell, J.D., Cruz, M., White, J.L., Bailey, G.W., Payne, W.R., Jr., Pope, J.D., Jr., and Teasley, J.I., 1968, Mode of chemical degradation of *s*-triazines by montmorillonite: *Science*, v. 160, pp. 1340-1342.

Rutledge, A.T., 1987, Effects of land use on ground-water quality in central Florida— Preliminary results, U.S. Geological Survey Toxic Waste-Ground-Water Contamination Program: U.S. Geological Survey Water Resources Investigations Report 86-4163, 49 p.

Sabatini, D.A., and Austin, T.A., 1990, Sorption and transport of pesticides in ground water: Critical review: *J. Irrig. Drain. Eng.*, v. 116, no. 1, pp. 3-15.

Sabbagh, G.J., Geleta, S., Elliott, R.L., Williams, J.R., and Griggs, R.H., 1991. Modification of EPIC to simulate pesticide activities: EPIC-PST: *Trans. Am. Soc. Civ. Eng.*, v. 34, no. 4, pp. 1683-1692.

Sadeghi, A.M., and Isensee, A.R., 1994, Spatial distribution of atrazine residues in soil and shallow groundwater: Effect of tillage and rainfall timing: *Agric. Ecosys. Environ.*, v. 48, pp. 67-76.

Salo, J.E., Harrison, D., and Archibald, E.M., 1986, Removing contaminants by groundwater recharge basins: *J. American Water Works Association*, v. 78, no. 9, pp. 76-81.

Sauer, T.J., Fermanich, K.J., and Daniel, T.C., 1990, Comparison of the pesticide root zone model simulated and measured pesticide mobility under two tillage systems: *J. Environ. Qual.*, v. 19, no. 4, pp. 727-734.

Sawhney, B.L., Pignatello, J.J., and Steinberg, S.M., 1988, Determination of 1,2-dibromoethane (EDB) in field soils: Implications for volatile organic compounds: *J. Environ. Qual.*, v. 17, no. 1, pp. 149-152.

Scarano, 1986, The Massachusetts aldicarb well water survey, in *Proceedings of the third annual Eastern Regional Ground Water Conference, Springfield, Massachusetts:* National Water Well Association, Dublin, Ohio, pp. 261-276.

Scheibe, T.D., and Lettenmaier, D.P., 1989, Risk-based selection of monitoring wells for assessing agricultural chemical contamination of ground water: *Ground Water Monitor. Rev.*, v. 9, no. 4, pp. 98-108.

Scheuerman, P.R., Bitton, G., Overman, A.R., and Gifford, G.E., 1979, Transport of viruses through organic soils and sediments: *J. Envir. Eng. Div. Am. Soc. Civ. Eng.*, v. 105, no. EE4, p. 629.

Schiavon, M., 1988, Studies of the leaching of atrazine, of its chlorinated derivatives, and of hydroxyatrazine from soil using ^{14}C ring-labeled compounds under outdoor conditions: *Ecotoxicol. Environ. Safety*, v. 15, pp. 46-54.

Schiner, G.R., and German, E.R., 1983, Effects of recharge from drainage wells on quality of water in the Floridan Aquifer in the Orlando area, central Florida: U.S. Geological Survey Water Resources Investigations Report No. 82-4094, 130 p.

Schmaland, G., 1983, Auswaschung einiger Insektizide aus landwirtschaftlich genutzten Boden im Lysimeterversuch: *Acta hydrochim. et hydrobiol.*, v. 11, no. 3, p. 269-278.

Schmidt, K.D., 1986a, DBCP in ground water of the Fresno-Dinuba area, California, in *Proceedings, the Agricultural Impacts on Ground Water—A conference, Omaha, Nebraska*: National Water Well Association, Dublin, Ohio, pp. 511-529.

——1986b, Effect of irrigation on groundwater quality in the southwest, *in* Summers, J.B., and Anderson, S.S., eds., *Toxic substances in agricultural water supply and drainage—Defining the problems*: *Proceedings from the 1986 regional meetings, Fresno, California*: U.S. Committee on Irrigation and Drainage, Denver, Colo., pp. 273-289.

Schmidt, K.D., and Sherman, I., 1987, Effect of irrigation on groundwater quality in California: *J. Irrig. Drain. Eng.*, v 113, no. 1, pp. 16-29.

Schmitz, G.L., and Witt, W.W., 1988, Environmental factors affecting chlorimuron degradation: *Weed Sci. Soc. Am. Abstr.*, v. 28, p. 80.

Schneider, A.D., Wiese, A.F., and Jones, O.R., 1970, Movement and recovery of herbicides in the Ogallala Aquifer, *in* Mattox, R.B., and Miller, W.D., eds. The Ogallala Aquifer: A symposium: Texas Tech University International Center for Arid and Semi-Arid Land Studies Special Report 39, pp. 219-226.

——1977, Movement of three herbicides in a fine sand aquifer: *Agronomy J.*, v. 69, pp. 432-436.

Schneider, R.C., Green, R.E., Apt, W.J., Bartholomew, D.P., and Caswell, E.P., 1990, Field movement and persistence of fenamiphos in drip-irrigated pineapple soils: *Pest. Sci.*, v. 30, pp. 243-257.

Schottler, S.P., Eisenreich, S.J., and Capel, P.D., 1994, Atrazine, alachlor, and cyanazine in a large agricultural river system: *Environ. Sci. Technol.*, v. 28, no. 6, pp. 1079-1089.

Schowanek, D., and Verstraete, W., 1991, Hydrolysis and free radial [sic] mediated degradation of phosphonates: *J. Environ. Qual.*, v. 20, no. 4, pp. 769-776.

Schwarzenbach, R.P., Giger, W., Hoehn, E., and Schneider, J.K., 1983, Behavior of organic compounds during infiltration of river water to groundwater. Field studies: *Environ. Sci. Technol.*, v. 17, no. 8, pp. 472-479.

Schwarzenbach, R.P., and Westall, J., 1981, Transport of nonpolar organic compounds from surface water to groundwater. Laboratory sorption studies: *Environ. Sci. Technol.*, v. 15, no. 11, pp. 1360-1367.

Seaburn, G.E., and Aronson, D.A., 1974, Influence of recharge basins on the hydrology of Nassau and Suffolk Counties, Long Island, N.Y.: U.S. Geological Survey Water-Supply Paper 2031, 66 p.

Segal, D.S., Neary, D.G., Best, G.R., and Michael, J.L., 1987, Effect of ditching, fertilization, and herbicide application on groundwater levels and groundwater quality in a flatwood spodosol: *Soil and Crop Sci. Soc. Florida Proc.*, v. 46, pp. 107-112.

Segawa, R.T., Maykoski, R., and Sava, R.J., 1986, Survey for triazine herbicides in well water, Glenn County, 1986: California Department of Food and Agriculture Report EH 86-03.

Seigley, L., 1993, personal communication, memorandum, July 7, Department of Natural Resources, Geological Survey Bureau, Iowa City, Iowa.

Seigley, L.S., and Hallberg, G.R., 1991, *Groundwater quality observations from the Bluegrass watershed Audobon County, Iowa: A report by the Integrated Farm Management Demostration Project*: Iowa Department of Natural Resources Technical Information Series 20, 50 p.

Seitz, H.R., La Sala, A.M., Jr., and Moreland, J.A., 1977, Effects of drain wells on the ground-water quality of the Western Snake Plain Aquifer, Idaho: U.S. Geological Survey Open-File Report No. 76-673, 83 p.

Senseman, S.A., 1993, personal communication, telephone conference, August 9, University of Arkansas, Department of Agronomy, Fayetteville, Ark.

Senseman, S.A., Lavy, T.L., and Daniel, T.C., 1990, Survey of Arkansas groundwater at pesticide mixing and loading sites, in *Arkansas Agricultural Pesticide Association, Proceedings of the annual meeting*: Arkansas Agricultural Pesticide Association, Little Rock, Ark., v. 29, p. 15.

Sevcik, P., and Khir, M., 1980, Kinetics and mechanism of oxidation of Cr(II) ions by tetrachloromethane: *Collection of Czechoslovakia Chemical Communications*, v. 45, pp. 21-25.

Shaffer, M.J., Halvorson, A.D., and Pierce, F.J., 1991, Nitrate leaching and economic analysis package (NLEAP): Model description and application, in Follett, R.F., Kenney, D.R., and Cruse, R.M., eds., *Managing nitrogen for groundwater quality and farm profitability: Proceedings of a symposium*: Soil Science Society of America, Madison, Wis., pp. 285-322.

Shaffer, R.D., and Penner, D., 1991, Evaluation of leaching prediction models for herbicide movement in the soil vadose zone, in *Pesticides in the next decade: Proceedings of the Third National Research Conference on Pesticides:* Virginia Polytechnic Institute and State University, Virginia Water Resources Research Center, Blacksburg, Va., pp. 751-771.

Shestopalov, V.M., and Molozhanova, H.G., 1992, Prediction of pesticide behavior in subsurface water, in Schnoor, J.L., ed., *Fate of pesticides and chemicals in the environment:* Wiley, N.Y., pp. 371-384.

Shirmohammadi, A., and Knisel, W.G., 1989, Irrigated agriculture and water quality in south [sic]: *J. Irrig. Drain. Eng.*, v. 115, no. 5, pp. 791-806.

Shirmohammadi, A., Magette, W.L., Brinsfield, R.B., and Staver, K., 1989, Ground water loading of pesticides in the Atlantic Coastal Plain: *Ground Water Monitor. Rev.*, v. 9, no. 4, pp. 141-148.

Shoemaker, L.L., Magette, W.L., and Shirmohammadi, A., 1990, Modeling management practice effects on pesticide movement to ground water: *Ground Water Monitor. Rev.*, v. 10, no. 1, pp. 109-115.

Shrestha, S.P., and Loganathan, G.V., 1994, Monte Carlo simulation and effective medium approximation in subsurface flow modeling: *Ground Water*, v. 32, no. 6, pp. 929-936.

Sichani, S.A., Engel, B.A., Monke, E.J., Eigel, J.D., and Kladivko, E.J., 1991, Validating GLEAMS with pesticide field data on a Clermont silt loam soil: *Trans. Am. Soc. Civ. Eng.*, v. 34, no. 4, pp. 1732-1737.

Simmleit, N., and Herrmann, R., 1987a, The behavior of hydrophobic, organic micropollutants in different karst water systems I. Transport of micropollutants and contaminant balances during the melting of snow: *J. Water, Air, Soil Poll.*, v. 34, pp. 79-95.

——1987b, The behavior of hydrophobic, organic micropollutants in different karst water systems II. Filtration capacity of karst systems and pollutant sinks: *J. Water, Air, Soil Poll.*, v. 34, pp. 97-109.

Sirons, G.J., Frank, R., and Sawyer, T., 1973, Residues of atrazine, cyanazine, and their phytotoxic metabolites in a clay loam soil: *J. Agric. Food Chem.*, v. 21, no. 6, pp. 1016-1020.

Sitts, J.A., 1989, Survey for bentazon in well water of 15 California counties, December 1988-May 1989: California Department of Food and Agriculture, Division of Pest Management, Environmental Protection and Worker Safety, Environmental Monitoring and Pest Management Branch Report EH 89-10, 106 p.

Skipper, H.D., Gilmour, C.M., and Furtick, W.R., 1967, Microbial versus chemical degradation of atrazine in soils: *Proc. Soil Sci. Soc. Am.*, v. 31, no. 5, pp. 653-656.

Skipper, H.D., and Volk, V.V., 1972, Biological and chemical degradation of atrazine in three Oregon soils: *Weed Sci.*, v. 20, no. 4, pp. 344-347.

Smelt, J.H., Dekker, A., Leisra, M., and Houx, N.W.H., 1983, Conversion of four carbamoyloximes in soil samples from above and below the soil water table: *Pest. Sci.*, v. 14, pp. 173-181.

Smith, A.E., and Aubin, A.J., 1993, Degradation of [^{14}C]amidosulfuron in aqueous buffers and in an acidic soil: *J. Agric. Food Chem.*, v. 41, no. 12, pp. 2400-2403.

——1994, Loss of enhanced biodegradation of 2,4-D and MCPA in a field soil following cessation of repeated herbicide applications: *Bull. Environ. Contam. Toxicol.*, v. 53, no. 1, pp. 7-11.

Smith, A.E., and Bridges, D.C., 1995, Potential movement of certain herbicides following application to golf courses, *in* American Chemical Society, *Preprints of papers presented at the 209th National Meeting, Anaheim, California*: American Chemical Society, Division of Environmental Chemistry, v. 35, no. 1, pp. 233-236.

Smith, A.E., Grover, R., Emmond, G.S., and Korven, H.C., 1975, Persistence and movement of atrazine, bromacil, monuron, and simazine in intermittently-filled irrigation ditches: *Can.J.Plant Sci.*, v. 55, no. 3, pp. 809-816.

Smith, A.E., and Tillotson, W.R., 1993, Potential leaching of herbicides applied to golf course greens, *in* Racke, K.D., and Leslie, A.R., eds., *Pesticides in urban environments: Fate and significance*: American Chemical Society Symposium Series 522, pp. 168-181.

Smith, A.E., Weldon, O., Slaughter, W., Peeler, H., and Mantripragada, N., 1993, A greenhouse system for determining pesticide movement from golf course greens: *J. Environ. Qual.*, v. 22, no. 4, pp. 864-867.

Smith, C.J., 1990, Hydrogeology with respect to underground contamination, *in* Hutson, D.H., and Roberts, T.R., eds., *Environmental fate of pesticides:* Wiley, Chichester, England, pp. 47-99.

Smith, C.N., Brown, D.S., Parrish, R.S., Asmussen, L.E., Leonard, R.A., Hicks, D.W., Payne, W.R., and Fletcher, R.S., 1991, Field testing pesticide transport models at a cooperative test site near Plains, Georgia, *in* Hotcher, K.J., ed., *Proceedings of the 1991 Georgia Resources Conference*: University of Georgia, Institute of Natural Resources, Athens, Ga., pp 298-301.

Smith, M.C., Bottcher, A.B., Campbell, K.L., and Thomas, D.L., 1991b, Field testing and comparison of the PRZM and GLEAMS models: *Trans. Am. Soc. Agric. Eng.*, v. 34, no. 3, pp. 838-847.

Smith, M.C., Thomas, D.L., Bottcher, A.B., and Campbell, K.L., 1990, Measurement of pesticide transport to shallow ground water: *Trans. Am. Soc. Agric. Eng.*, v. 33, no. 5, pp. 1573-1581.

Smolen, J.M., and Stone, A.T., 1994, Phosphorothioate ester hydrolysis catalyzed by dissolved metals and metal containing mineral surfaces, *in* American Chemical Society, *Preprints of papers presented at the 208th National Meeting, San Diego, California*: American Chemical Society, Division of Environmental Chemistry, v. 34, no. 2, pp. 534-547.

Snethen, D., and Robbins, V., 1988, Farmstead well contamination factor study: Kansas Department of Health and Environment, unpublished project summary VR85, 5 p.

Soller, D.R., 1992, Applying the DRASTIC model—A review of county-scale maps: U.S. Geological Survey Open-File Report 92-297, 36 p.

Somasundaram, L., and Coats, J.R., 1991, Pesticide transformation products in the environment, *in* Somasundaram, L., and Coats, J.R., eds., *Pesticide transformation products: Fate and significance in the environment:* American Chemical Society Symposium Series 459, pp. 2-9.

Somasundaram, M.V., Ravindran, G., and Tellam, J.H., 1993, Ground-water pollution of the Madras urban aquifer, India: *Ground Water*, v. 31, no. 1, pp. 4-11.

Sophocleous, M., Townsend, M.A., and Whittemore, D.O., 1990, Movement and fate of atrazine and bromide in central Kansas croplands: *J. Hydrology*, v. 115, pp. 115-137.

Soren, J., and Stelz, W.G., 1985. Ground water contamination by aldicarb pesticide in eastern Suffolk County, Long Island, in Perspectives on Nonpoint Source Pollution: Proceedings of a national conference, Kansas City, Missouri, May 19-22, 1985: U.S. Environmental Protection Agency Report EPA 440/5-85-001, pp. 101-108.

Spalding, R.F., 1989, Complexities associated with the interpretation of trace pesticide levels in ground water: *Ground Water Monitor. Rev.*, v. 9, no. 4, pp. 79-88.

———1992, Assessment of statewide groundwater quality data from domestic wells in rural Nebraska: Nebraska Health Department Open File Report, 49 p.

Spalding, R.F., Burbach, M.E., and Exner, M.E., 1989, Pesticides in Nebraska's ground water: *Ground Water Monitor. Rev.*, v. 9, no. 4, pp. 126-133.

Spalding, R.F., Junk, G.A., and Richard, J.J., 1980, Pesticides in ground water beneath irrigated farmland in Nebraska, August 1978: *J. Pest. Monit.*, v. 14, no. 2, pp. 70-73.

Spillmann, P., 1989, Decomposition and elimination of typical pollutants from sanitary landfills in porous aquifers, *in* Sahuquillo, J., and O'Donnell, A.T., eds., *Groundwater management: Quantity and quality*: International Association of Hydrological Sciences, Wallingford, Oxfordshire, UK, pp. 227-233.

Sposito, G., 1984, *The surface chemistry of soils*: Oxford University Press, N.Y, 234 p.

Spruill, T.B., 1983, Statistical summaries of selected chemical constituents in Kansas ground-water supplies, 1976-81: U.S. Geological Survey Open-File Report 83-263, 33 p.

Squillace, P.J., and Thurman, E.M., 1992, Herbicide transport in rivers: Importance of hydrology and geochemistry in nonpoint-source contamination: *Environ. Sci. Technol.*, v. 26, no. 3, pp. 538-545.

Squillace, P.J., Thurman, E.M., and Furlong, E.T., 1993, Groundwater as a nonpoint source of atrazine and deethylatrazine in a river during base flow conditions: *Water Resour. Res.*, v. 29, no. 6, pp. 1719-1729.

Stahnke, G.K., Shea, P.J., Tupy, D.R., Stougaard, R.N., and Shearman, R.C., 1991, Pendimethalin dissipation in Kentucky bluegrass turf: *Weed Sci.*, v. 39, no. 1, pp. 97-103.

Stark, J.R., Strudell, J.D., Bloomgren, P.A., and Eger, P., 1987, Ground-water and soil contamination near two pesticide-burial sites in Minnesota: U.S. Geological Survey Water-Resources Investigations Report 87-4115, 48 p.

Starr, R.C., Gillham, R.W., and Sudicky, E.A., 1985, Experimental investigation of solute transport in stratified porous media 2. The reactive case: *Water Resour. Res.*, v. 21, no. 7, pp. 1043-1050.

Stecher, L.S., and Rainwater, K., 1988, Quantity and quality of recharge to the Ogallala Aquifer from urban runoff, in *Proceedings of FOCUS Conference on Southwestern Ground Water Issues, Albuquerque, New Mexico*: National Water Well Association, Dublin, Ohio, pp. 145-164.

Steenhuis, T.S., Boll, J., Shalit, G., Selker, J.S., and Merwin, I.A., 1994, A simple equation for predicting preferential flow solute concentrations: *J. Environ. Qual.*, v. 23, no. 5, pp. 1058-1064.

Steenhuis, T.S., Paulsen, R., Richard, T., Staubitz, W., Andreini, M., and Surface, J., 1988, Pesticide and nitrate movement under conservation and conventional [sic] tilled plots, *in* Hay, D.R., ed., *Planning now for irrigation and drainage in the 21st century: Proceedings of a conference: Lincoln, Nebraska*: American Society of Civil Engineers, N.Y., pp. 585-595.

Steenhuis, T.S., Staubitz, W., Andreini, M.S., Surface, J., Richard, T., Paulsen, R., Pickering, N.B., Hagerman, J.R., and Geohring, L.D., 1990, Preferential movement of pesticides and tracers in agricultural soils: *J. Irrig. Drain. Eng.*, v. 116, no. 1, pp. 50-66.

Steenhuis, T.S., van der Marel, M., and Pacenka, S., 1984, A pragmatic model for diagnosing and forecasting ground water contamination, *in* National Water Well Association Conference on Practical Applications of Ground Water Models, *Proceedings of the National Water Well Association Conference on the practical applications of ground water models: Fawcett Center, Columbus, Ohio*: National Water Well Association, Worthington, Ohio.

Steichen, J., Koelliker, J., Grosh, D., Heiman, A., Yearout, R., and Robbins, V., 1988, Contamination of farmstead wells by pesticides, volatile organics, and inorganic chemicals in Kansas: *Ground Water Monitor. Rev.*, v. 8, no. 3, pp. 153-159.

Stover, J.A., and Guitjens, J.C., 1990, Aldicarb in vadose zone: Review: *J. Irrig. Drain. Eng.*, v. 116, no. 1, pp. 36-49.

Stuart, C.G., and Demas, C.R., 1990, Organic chemical analyses of ground water in Louisiana, Water years 1984-1988: Louisiana Department of Transportation and Development Water Resources Basic Records Report 18, 80 p.

Subramaniam, V., and Hoggard, P.E., 1988, Metal complexes of glyphosate: *J. Agric. Food Chem.*, v. 36, no. 6, pp. 1326-1329.

Sudicky, E.A., Gillham, R.W., and Frind, E.O., 1985, Experimental investigation of solute transport in stratified porous media 1. The nonreactive case: *Water Resour. Res.*, v. 21, no. 7, pp. 1035-1041.

Suffolk County Department of Health Services, 1989, *Status of pesticide sampling programs 1980-1988:* County of Suffolk, Department of Health Services, Hauppauge, N. Y., 23 p.

Summit, G.D., Dupont, R.R., Parker, T.D.T., and Deer, H.M., 1989, Modeling and measurement of tebuthiuron (Spike) mobility in intermountain soils, *in* Weigmann, D.L., ed., *Pesticides in terrestrial and aquatic environments: Proceedings of a national research conference:* Virginia Polytechnic Institute and State University, Virginia Water Resources Research Center, Blacksburg, Va., pp. 491-500.

Sundaram, K.M.S., and Nott, R., 1989, Mobility of diflubenzuron in two types of forest soils: *J. Environ. Sci. Health*, v. B24, no. 1, pp. 65-86.

Suntio, L.R., Shiu, W.Y., Mackay, D., Seiber, J.N., and Glotfelty, D., 1988, Critical review of Henry's Law Constants for pesticides: *Rev. Environ. Contam. Toxicol.*, v. 103, pp. 1-59.

Szabo, Z., Rice, D.E., Ivahnenko, T., and Vowinkel, E.F., 1994, Delineation of the distribution of pesticides and nitrate in an unconfined aquifer in the New Jersey coastal plain by flow-path analysis, in *New directions in pesticide research, development, management, and policy: Proceedings of the Fourth National Conference on Pesticides:* Virginia Polytechnic Institute and State University, Virginia Water Resources Center, Blacksburg, Va., pp. 100-119.

Taboada, E.R., Dekker, A., Van Kammen-Polman, A., Smelt, J.H., Boesten, J.J.T.I., and Leistra, M., 1994, Adsorption, degradation and leaching of pirimicarb in orchard soils: *Sci. Total Environ.*, v. 153, pp. 253-260.

Tamblyn, T.A., and Beck, L.A., 1968, Distribution of pesticides in California: *J. Phys. Chem.*, v. 2, no. 7, pp. 728-732.

Teso, R.R., Younglove, T., Peterson, M.R., Sheeks III, D.L., and Gallavan, R.E., 1988, Soil taxonomy and surveys: Classification of areal sensitivity to pesticide contamination of groundwater: *J. Soil Water Conserv.*, July-August, p. 348-352.

Thamke, J.N., and Clark, M.L., 1988, Populations affected by pesticides in public ground-water supplies in Iowa, *in* Waterstone, M., and Burt, R.J., eds., *Proceedings of the Symposium on Water-Use Data for Water Resources Management*: American Water Resources Association, Bethesda, Md., pp. 381-388.

Thomas, C.R., and Robinson, W.H., 1994, Dispersion of chlorpyrifos in soil beneath concrete slabs: *Bull. Environ. Contam. Toxicol.*, v. 53, no. 1, pp. 1-6.

Thomas, R.G., 1990, Volatilization from water, *in* Lyman, W.J., Reehl, W.F., and Rosenblatt, D.H., eds., *Handbook of chemical property estimation methods: Environmental behavior of organic compounds:* American Chemical Society, Washington, D.C., pp. 15-1 to 15-34.

Thompson, C.A., 1990, Nitrate and pesticide distribution in the west fork Des Moines River alluvial aquifer: Iowa Department of Natural Resources Technical Information Series 18, 43 p.

Thompson, C.A., Libra, R.D., and Hallberg, G.R., 1986, Water quality related to ag-chemicals in alluvial aquifers in Iowa, in *Proceedings of Agricultural Impacts on Groundwater—A conference, Omaha, Nebraska*: National Water Well Association, Dublin, Ohio, pp. 224-242.

Thorbjarnarson, K.W., and Mackay, D.M., 1994, A forced-gradient experiment on solute transport in the Borden aquifer 3. Nonequilibrium transport of the sorbing organic compounds: *Water Resour. Res.*, v. 30, no. 2, pp. 401-419.

Thorstenson, D.C., 1984, The concept of electron activity and its relation to redox potentials in aqueous geochemical systems: U.S. Geological Survey Open-File Report 84-072, 66 p.

Thurman, E.M., 1985, *Organic geochemistry of natural waters*: M. Nijhoff, Boston, Mass., 497 p.

Thurman, E.M., Goolsby, D.A., Meyer, M.T., and Kolpin, D.W., 1991, Herbicides in surface waters of the midwestern United States: The effect of spring flush: *Environ. Sci. Technol.*, v. 25, no. 10, pp. 1794-1796.

Thurman, E.M., Goolsby, D.A., Meyer, M.T., Mills, M.S., Pomes, M.L., and Kolpin, D.W., 1992, A reconnaissance study of herbicides and their metabolites in surface water of the midwestern United States using immunoassay and gas chromatography/mass-spectrometry: *Environ. Sci. Technol.*, v. 26, no. 12, pp. 2440-2447.

Thurman, E.M., Meyer, M.T., Mills, M.S., Zimmerman, L.R., Perry, C.A., and Goolsby, D.A., 1994, Formation and transport of deethylatrazine and deisopropylatrazine in surface water: *Environ. Sci. Technol.*, v. 28, no. 13, pp. 2267-2277.

Timlin, D.J., Ahuja, L.R., and Ankeny, M.D., 1994, Comparison of three field methods to characterize apparent macropore conductivity: *Soil Sci. Soc. Am. J.*, v. 58, pp. 278-284.

Todd, D.K., 1974, Salt-water intrusion and its control: *J. Am. Water Works Assoc.*, March, pp. 180-187.

Tooby, T.E., and Marsden, P.K., 1991, Interpretation of environmental fate and behaviour data for regulatory purposes, *in* Walker, A., ed., *Pesticides in soils and water: Current perspectives: Proceedings of a symposium organised by the British Crop Protection Council, held at the University of Warwick, Coventry, UK,* British Crop Protection Council Monograph 47: British Crop Protection Council, Farnham, Surrey, UK, pp. 3-10.

Torstensson, L., 1981, A method to indicate leaching of TCA [trichlorobenzoic acid] during field conditions, *in* Svenska Ograskonferensen, *Weeds and weed control: 22nd Swedish Weed Conference, Uppsala*: Swedish University of Agricultural Sciences, Department of Plant Husbandry and Research Information Centre, Uppsala, Sweden, v. 1, pp. 129-132.

Traub-Eberhard, U., Kördel, W., and Klein, W., 1994, Pesticide movement into subsurface drains on a loamy silt soil: *Chemosphere*, v. 28, no. 2, pp. 273-284.

Trevisan, M., Capri, E., and Del Re, A.A.M., 1993, Pesticide soil transport models: Model comparisons and field evaluation: *Toxicol. Environ. Chem.*, v. 40, nos. 1-4, pp. 71-82.

Trevisan, M., Montepiani, C., Ghebbioni, C., and Del Re, A. A. M., 1991, Evaluation of potential hazard of propanil to groundwater: *Chemosphere*, v. 22, no. 7, pp. 637-643.

Troiano, J., Garretson, C., Krauter, C., and Brownell, J., 1990, Atrazine leaching and its relation to percolation of water as influenced by three rates and four methods of irrigation water application: California Department of Food and Agriculture, Division of Pest Management, Environmental Protection and Worker Safety, Environmental Monitoring and Pest Management Branch Report EH 90-7, variously paged.

Troiano, J., and Sitts, J., 1990, Survey for alachlor, atrazine, metolachlor and nitrate residues in well water in Merced County and their relation to soil and well characteristics: California Department of Food and Agriculture, Division of Pest Management, Environmental Protection and Worker Safety, Environmental Monitoring and Pest Management Branch Report EH 90-3, 83 p.

Troiano, J., and Segawa, R.T., 1987, Survey for herbicides in well water in Tulare County: California Department of Food and Agriculture, Division of Pest Management, Environmental Protection and Worker Safety, Environmental Monitoring and Pest Management Branch Report EH 87-01, 13 p.

Troiano, J., Turner, B., and Miller, N., 1987, Sampling for residues of fenamiphos, fenamiphos sulfoxide and fenamiphos sulfone in well water: California Department of Food and Agriculture, Division of Pest Management, Environmental Protection and Worker Safety, Environmental Monitoring and Pest Management Branch Report EH 87-8, 31 p.

Tucker, W.A., and Nelken, L.H., 1990, Diffusion coefficients in air and water, in Lyman, W.J., Reehl, W.F., and Rosenblatt, D.H., eds., Handbook of chemical property estimation methods: Environmental behavior of organic compounds: American Chemical Society, Washington, D.C., pp. 17-1 to 17-25.

Turco, R.F., and Konopka, A.E., 1988, Agricultural impact on groundwater quality: Purdue University, Water Resources Research Center, Technical Report 185, 56 p.

Uchrin, C.G., and Katz, J., 1985, Reversible and resistant components of hexachlorocyclohexane (lindane) sorption to New Jersey coastal plain aquifer solids: J. Environ. Sci. Health., v. A20, no. 2, pp. 205-218.

U.S. Department of Agriculture, 1991, State Soil Geographic Data Base (STATSGO): Data Users Guide: U.S. Department of Agriculture, Soil Conservation Service Miscellaneous Publication 1492, 88 p.

U.S. Department of Commerce, 1990, 1987 Census of Agriculture: U.S. Department of Commerce, Bureau of the Census, Agricultural Atlas of the United States, Part 1, 199 p. (plus Appendix).

U.S. Environmental Protection Agency, 1975, Manual of water well construction practices: U.S. Environmental Protection Agency, Office of Water Supply Report EPA-570/9-75-100.

——1986, Pesticides in ground water: Background document: U.S. Environmental Protection Agency, Office of Ground-Water Protection Report EPA 440/6-86-002, 72 p.

——1987, FIFRA scientific advisory approval of proposed changes to inert ingredient lists 1 and 2: U.S. Environmental Protection Agency internal memorandum from F.S. Bishop, 5 p.

——1990a, National survey of pesticides in drinking water wells: Phase I report: Environmental Protection Agency, Office of Pesticides and Toxic Substances Report EPA 570/9-90-015, variously paged.

——1990b, Suspended, cancelled, and restricted pesticides: U.S. Environmental Protection Agency, Office of Pesticides and Toxic Substances, 93 p.

——1990c, Pesticide Fact Handbook: Noyes Data Corporation, Park Ridge, N.J., v. 1, 666 p.; v. 2, 827 p.

——1992a, Another look: National survey of pesticides in drinking water wells, Phase II report: U.S. Environmental Protection Agency, Office of Pesticides and Toxic Substances Report EPA 579/09-91-020, 166 p.

——1992b, Pesticides in ground water database: A compilation of monitoring studies, 1971-1991: U.S. Environmental Protection Agency Report EPA 734-12-92-001, variously paged.

——1994a, Memorandum from Linda J. Fisher (Assistant Administrator, U.S. Environmental Protection Agency) to Douglas Campt (Director, Office of Pesticide Programs), 16 p.

——1994b, Database documentation for the National Pesticide Survey, Section Three, 97 p.

——1995, Water resources impact analysis for the triazine herbicides, unpublished report, U.S. Environmental Protection Agency, Office of Pesticides and Toxic Substances, 155 p.

U.S. Forest Service, 1978, *Report of the Forest Service, fiscal year 1977*: U.S. Department of Agriculture—Forest Service, Washington, D.C., variously paged.

——1985, *Report of the Forest Service, fiscal year 1984*: U.S. Department of Agriculture—Forest Service, Washington, D.C., variously paged.

——1989, *Report of the Forest Service, fiscal year 1988*: U.S. Department of Agriculture—Forest Service, Washington, D.C., variously paged.

——1990, *Report of the Forest Service, fiscal year 1989*: U.S. Department of Agriculture—Forest Service, Washington, D.C., variously paged.

——1991, *Report of the Forest Service, fiscal year 1990*: U.S. Department of Agriculture—Forest Service, Washington, D.C., variously paged.

——1992, *Report of the Forest Service, fiscal year 1991*: U.S. Department of Agriculture—Forest Service, Washington, D.C., variously paged.

——1993, *Report of the Forest Service, fiscal year 1992*: U.S. Department of Agriculture—Forest Service, Washington, D.C., variously paged.

——1994, *Report of the Forest Service, fiscal year 1993*: U.S. Department of Agriculture—Forest Service, Washington, D.C., variously paged.

Utermann, J., Kladivko, E.J., and Jury, W.A., 1990, Evaluating pesticide migration in tile-drained soils with a transfer function model: *J. Environ. Qual.*, v. 19, no. 4, pp. 707-714.

Valo, R., Kitunen, V., Salkinoja-Salonen, M., and Raisanen, S., 1984, Chlorinated phenols as contaminants of soil and water in the vicinity of two Finnish sawmills: *Chemosphere*, v. 13, no. 8, pp. 835-844.

Valocchi, A.J., and Roberts, P.V., 1983, Attenuation of groundwater contaminant pulses: *J. Hydrol. Eng.*, v. 109, no. 12, pp. 1665-1682.

van Biljon, J.J., Groeneveld, H.T., and Nel, P.C., 1988, Leaching depth of metolachlor in a drift sand soil: *Appl. Plant Sci.*, v. 2, no. 2, pp. 77-80.

——1990, Leaching depth of metolachlor in different soils: *Appl. Plant Sci.*, v. 4, no. 2, p. 46-49.

van de Weerd, H., and van der Linden, A.M.A., 1991, *Behaviour of pesticides in aquifer materials: Interpretation of an in situ experiment*: National Institute of Public Health and Environmental Protection, Bilthoven, The Netherlands, 84 p.

van der Zee, S.E.A.T.M., and Boesten, J.J.T.I., 1991, Effects of soil heterogeneity on pesticide leaching to groundwater: *Water Resour. Res.*, v. 27, no. 12, pp. 3051-3063.

van Genuchten, M.T., Davidson, J.M., and Wierenga, P.J., 1974, An evaluation of kinetic and equilibrium equations for the prediction of pesticide movement through porous media: *Soil Sci. Soc. Am.*, v. 38, pp. 29-35.

van Genuchten, M.T., and Wierenga, P.J., 1976, Mass transfer studies in sorbing porous media I. Analytical Solutions: *Soil Sci. Soc. Am. J.*, v. 40, no. 4, pp. 473-480.

van Genuchten, M.T., Wierenga, P.J., and O'Connor, G.A., 1977, Mass transfer studies in sorbing porous media: III. Experimental evaluation with 2,4,5-T: *Soil Sci. Soc. Am. J.*, v. 41, pp. 278-285.

Vanachter, A., van Pee, G., van Wambeke, E., and van Assche, C., 1981, Bromide concentration in water after methyl bromide soil disinfestation 1. Relation between soil type, efficiency of leaching and bromide concentration in the leaching water: *Med. Fac. Landboww. Rijksuniv. Gent.*, v. 46, no. 1, pp. 343-349.

Vicari, A., Catizone, P., and Zimdahl, R.L., 1994, Persistence and mobility of chlorsulfuron and metsulfuron under different soil and climatic conditions: *Weed Research*, v. 34, pp. 147-155.

Villeneuve, J., Banton, O., and LaFrance, P., 1990, A probabilistic approach for the groundwater vulnerability to contamination by pesticides: The VULPEST model: *Ecolog. Model.*, v. 51, pp. 47-58.

Voelker, D.C., 1989, Quality of water from public-supply wells in principal aquifers of Illinois, 1984-87: U.S. Geological Survey Water-Resources Investigations Report 88-4111, 29 p.

Voss, C.I., 1984, SUTRA—Saturated-Unsaturated Transport: A finite-element simulation model for saturated-unsaturated, fluid-density-dependent ground-water flow with energy transport or chemically-reactive single-species solute transport: U.S. Geological Survey Water-Resources Investigations Report 84-4369, 429 p.

Vowinkel, E.F., 1991, Comparison of relations between shallow ground-water quality and land use in two New Jersey Coastal Plain aquifer systems, *in* G. E. Mallard, G.E., and Aronson, D.A., eds., U.S. Geological Survey Toxic Substances Hydrology Program—Proceedings of the technical meeting, Monterey, California, March 11-15, 1991: U.S. Geological Survey Water-Resources Investigations Report 91-4034, pp. 307-313.

Vowinkel, E.F., and Battaglin, W.A., 1989, Methods of evaluating the relation of ground-water quality to land use in a New Jersey Coastal Plain aquifer system, *in* Mallard, G.E., ed., U.S. Geological Survey Toxic Substances Hydrology Program—Proceedings of the technical meeting, Phoenix, Arizona, September 26-30, 1988: U.S. Geological Survey Water-Resources Investigations Report 88-4220, pp. 405-410.

Vowinkel, E.F., Clawges, R.M., and Uchrin, C.G., 1994, Evaluation of the vulnerability of water from public supply wells in New Jersey to contamination by pesticides, in *New directions in pesticide research, development, management, and policy: Proceedings of the Fourth National Conference on Pesticides*: Virginia Polytechnic Institute and State University, Virginia Water Resources Center, Blacksburg, Va., p. 495-510.

Wadd, D.J., and Drennan, D.S.H., 1989, Field study of the persistence and leaching of chlorsulfuron and metsulfuron-methyl, in *Brighton Crop Protection Conference—Weeds—1989: Proceedings of an international conference organised by the British Crop Protection Council, held at Brighton Centre and Brighton Metropole, Brighton, England*: British Crop Protection Council, v. *3*, pp. 1133-1138.

Wade, R.S., and Castro, C.E., 1973, Oxidation of iron(II) porphyrins by alkyl halides: *J. Am. Chem. Soc.*, v. 95, no. 1, pp. 226-230.

Wagenet, R.J., 1986a, Chemical processes that influence pesticide fate in soil, *in* Dale, D., ed., *Groundwater protection: Proceedings of the Northeastern States Cooperative Extension Agent Training Conference, Chicopee, Massachusetts*: Massachusetts Cooperative Extension, Amherst, Mass., pp. 18-26.

——1986b, Principles of modeling pesticide movement in the unsaturated zone, *in* Garner, W.Y., Honeycutt, R.C., and Nigg, H.N., eds., *Evaluation of pesticides in water: Developed from a symposium sponsored by the Division of Pesticide Chemistry at the 189th Meeting of the American Chemical Society, Miami Beach, Florida*: American Chemical Society Symposium Series 315, pp. 330-341.

Wagenet, R.J., and Hutson, J.L, 1986, Predicting the fate of nonvolatile pesticides in the unsaturated zone: *J. Environ. Qual.*, v. 15, no. 4, pp. 315-322.

——1990, Quantifying pesticide behavior in soil: *Ann. Rev. Phytopathology*, v. 28, pp. 295-319.

Waite, DT., Grover, R., Westcott, N.D., Sommerstad, H., and Kerr, L., 1992, Pesticides in ground water, surface water and spring runoff in a small Saskatchewan watershed: *Environ. Toxicol. Chem.*, v. 11, no. 6, pp. 741-748.

Walker, M.J., and Porter, K.S., 1990, Assessment of pesticides in upstate New York ground water: Results of a 1985-1987 sampling survey: *Ground Water Monitor. Rev.*, v. 10, no. 1, pp. 116-126.

Wallach, D., and Goffinet, B., 1991, Comment on "A method for testing whether model predictions fall within a prescribed factor of true values, with an application to pesticide leaching" by Parrish and Smith (1990): *Ecolog. Model.*, v. 54, pp. 137-141.

Waller, R.M., 1988, Ground water and the rural homeowner: U.S. Geological Survey general interest pamphlet, 37 p.

Wallrabenstein, L.K., and Baker, D.B., 1992, Agrichemical contamination in private water supplies, in *Proceedings of the FOCUS Conference on Eastern Regional Ground Water Issues: Boston Marriott Newton, Newton, Massachusetts*: Water Well Journal Publishing, Dublin, Ohio, pp. 697-711.

Wang, W., and Squillace, P., 1994, Herbicide interchange between a stream and an adjacent alluvial aquifer: *Environ. Sci. Technol.*, v. 28, no. 13, pp. 2336-2344.

Washington, J.W., 1995, Hydrolysis rates of dissolved volatile organic compounds: Principles, temperature effects, and literature review: *Ground Water*, v. 33, no. 3, pp. 415-424.

Watschke, T.L., and Mumma, R.O., 1989, The effect of nutrients and pesticides applied to turf on the quality of runoff and percolating water: Final report for the U.S. Department of the Interior, Geological Survey: Pennsylvania State Environmental Resources Research Institute Report ER 8904, 64 p.

Watson, V.J., Rice, P.M., and Monnig, E.C., 1989, Environmental fate of picloram used for roadside weed control: *J. Environ. Qual.*, v. 18, no. 2, pp. 198-205.

Watts, G.B., and Brown, N.A., 1985, Riviera Beach wellfield contamination: Florida Department of Environmental Regulation Report 85-10, 184 p.

Wauchope, R.D., 1978, The pesticide content of surface water draining from agricultural fields—A review: *J. Environ. Qual.*, v. 7, no. 4, pp. 459-472.

——1987, Effects of conservation tillage on pesticide loss with water, in Logan, T.J., Davidson, J.M., Baker, J.L., and Overcash, M.R., eds., *Effects of conservation tillage on groundwater quality: Nitrates and pesticides*: Lewis Publishers, Chelsea, Mich., pp. 205-215.

Wauchope, R.D., Buttler, T.M., Hornsby, A.G., Augustijn-Beckers, P.W.M., and Burt, J.P., 1992, The SCS/ARS/CES Pesticide Properties Database for Environmental Decision-Making: *Rev. Environ. Contam. Toxicol.*, v. 123, pp. 1-156.

Wauchope, R.D., Williams, R.G., and Marti, L.R., 1990, Runoff of sulfometuron-methyl and cyanazine from small plots: Effects of formulation and grass cover: *J. Environ. Qual.*, v. 19, no. 1, pp. 119-125.

Weaver, D.J., Marade, S.J., Miller, N., and Monier, M., 1988b, Studies on the persistence and leaching in soil of nematicides having use in flower bulb production in Humboldt and Del Norte Counties, California: California Department of Food and Agriculture, Division of Pest Management, Environmental Protection and Worker Safety, Environmental Monitoring and Pest Management Branch Report EH 88-14, 61 p.

Weaver, D.J., Quan, V., Collison, C.N., Saini, N., and Marade, S.J., 1988a, Monitoring the persistence and movement of fenamiphos in soils of lily bulb fields in Del Norte County, 1986: California Environmental Protection Agency, Environmental Hazards Assessment Program Report EH 88-1, 42 p.

Weaver, D.J., Sava, R.J., Zalkin, F., and Oshima, R.J., 1983, *Survey of ground water basins for DBCP, EDB, simazine and carbofuran*, v. 1 of *Pesticide movement to ground water:* California Department of Food and Agriculture, Division of Pest Management, Environmental Protection and Worker Safety, Sacramento, Calif., 67 p.

Weaver, J.E., Hogmire, H.W., Brooks, J.L., Sencindiver, J.C., and Bissonnette, G.K., 1987, Assessment of pesticide residues in surface and groundwater from a commercial apple orchard: West Virginia University, Agricultural and Forestry Experiment Station final report for the U.S. Department of Agriculture, contract TPSU-WVU-2057-330, 12 p.

Weed, D.A.J., Kanwar, R.S., Stoltenberg, D.E., and Pfeiffer, R.L., 1995, Dissipation and distribution of herbicides in the soil profile: *J. Environ. Qual.*, v. 24, no. 1, pp. 68-79.

Wegman, R.C.C., Greve, P.A., de Heer, H., and Hamaker, P., 1981, Methyl bromide and bromide-ion in drainage water after leaching of glasshouse soils: *J. Water, Air, Soil Poll.*, v. 16, pp. 3-11.

Wehr, M.A., Mattson, J.A., Bofinger, R.W., and Sajdak, R.L., 1992, Ground-application trial of hexazinone on the Ottawa National Forest: U.S. Department of Agriculture, Forest Service, North Central Forest Experiment Station Research Paper NC-308, 34 p.

Wehtje, G.R, Leavitt, J.R.C., Spalding, R.F., Mielke, L.N., and Schepers, J.S., 1981, Atrazine contamination of groundwater in the Platte Valley of Nebraska from non-point sources: *Sci. Total Environ.*, v. 21, pp. 47-51.

Wehtje, G.R., Mielke, L.N., Leavitt, J.R.C., and Schepers, J.S., 1984, Leaching of atrazine in the root zone of an alluvial soil in Nebraska: *J. Environ. Qual.*, v. 13, no. 4, pp. 507-513.

Wehtje, G.R., Spalding, R.F., Burnside, O.C., Lowry, S.R., and Leavitt, J.R.C., 1983, Biological significance and fate of atrazine under aquifer conditions: *Weed Sci.*, v. 31, no. 5, pp. 610-618.

Weintraub, R.A., Jex, G.W., and Moye, H.A., 1986, Chemical and microbial degradation of 1,2-dibromoethane (EDB) in Florida ground water, soil, and sludge, *in* Garner, W.Y., Honeycutt, R.C., and Nigg, H.N., eds., *Evaluation of pesticides in ground water: Developed from a symposium sponsored by the Division of Pesticide Chemistry at the 189th Meeting of the American Chemical Society, Miami Beach, Florida:* American Chemical Society Symposium Series 315, p. 294-310.

Welling, R., and Nicosia, S., 1986, Report on monitoring for alachlor in well water: I. Sampling in the Sacramento Valley: California Department of Food and Agriculture, Environmental Monitoring and Pest Management, Sacramento, Calif., 8 p.

Welling, R., Troiano, J., Maykoski, R., and Loughner, G., 1986, Effects of agronomic and geologic factors on pesticide movement in soil: Comparison of two ground water basins in California, *in Proceedings, the Agricultural Impacts on Ground Water—A conference, Omaha, Nebraska*: National Water Well Association, Omaha, Nebr., pp. 666-685.

Wells, D., and Waldman, E., 1991, *County level assessment of aldicarb leaching potential*: U.S. Environmental Protection Agency, Office of Pesticide Programs, Washington, D.C., 88 p.

White, R.E., Dyson, J.S., Gerstl, Z., and Yaron, B., 1986, Leaching of herbicides through undisturbed cores of a structured clay soil: *Soil Sci. Soc. Am. J.*, v. 50, pp. 277-283.

White, W.B., 1989, Introduction to the karst hydrology of the Mammoth Cave area, *in* White, W.B., and White, E.L., eds., *Karst hydrology: Concepts from the Mammoth Cave area*: Van Nostrand Reinhold, N.Y., pp. 1-13.

Whitehead, R.L., 1974, Chemical and physical data for disposal wells, Eastern Snake River Plain, Idaho: Idaho Department of Water Resources Water Information Bulletin 39, 31 p.

Whitmore, R.W., Kelly, J.E., and Reading, P.L., 1992, Executive summary, results, and recommendations, v. 1 of National Home and Garden Pesticide Use Survey: U.S. Environmental Protection Agency Report EPA RTI/5100/17-01F68-WO-0032, 140 p.

Wietersen, R.C., Daniel, T.C., Fermanich, K.J., Lowery, B., and McSweeney, K., 1993, Irrigation and polymer effects on herbicide transport through the unsaturated zone of a Sparta sand: *J. Environ. Qual.*, v. 22, no. 4, pp. 819-824.

Wildenschild, D., Jensen, K.H., Villholth, K., and Illangasekare, T.H., 1994, A laboratory analysis of the effect of macropores on solute transport: *Ground Water*, v. 32, no. 3, pp. 381-389.

Wilkerson, M., Oshima, R.J., Younglove, T., Margolis, H., and Marade, S.J., 1985, *Use of agronomic variables to predict ground water contamination in the San Joaquin valley, California*, v. 3 of *Pesticide movement to ground water*: California Department of Food and Agriculture, Division of Pest Management, Environmental Protection and Worker Safety, Sacramento, Calif., 63 p.

Wilkins, R.M., 1989, Controlled pesticide availability in soil environments, *in* Jamet, P., ed., *Methodological aspects of the study of pesticide behavior in soil:* INRA, Paris, France, pp. 57-62.

Williams, J.S., Neil, C.D., and Weddle, T.K., 1987, The influence of agricultural practices on ground water quality in Maine, in *Proceedings of the Fourth Annual Eastern Regional Ground Water Conference, Burlington, Vermont*: National Water Well Assocation, Dublin, Ohio, pp. 329-341.

Williams, J.S., and Tolman, A.L., 1985, Water-quality in sand and gravel aquifers in Maine: The influence of acid deposition, agriculture, and other non-point contamination sources, in *Proceedings of the Association of Ground Water Scientists and Engineers, Eastern Regional Ground Water Conference, Portland, Maine*: National Water Well Association, Worthington, Ohio, pp. 123-135.

Williams, W.M., Holden, P.W., Parsons, D.W., and Lorber, M.N., 1988, *Pesticides in ground water data base: 1988 interim report*: U.S. Environmental Protection Agency, Office of Pesticide Programs, Environmental Fate and Effects Division, Washington, D.C., variously paged.

Williamson, A.K., 1994, personal communication, electronic mail, April 28, U.S, Geological Survey, Tacoma, Wash.

Willis, G.H., and Hamilton, R.A., 1973, Agricultural chemicals in surface runoff, ground water, and soil: I. Endrin: *J. Environ. Qual.*,v. 2, no. 4, pp. 463-466.

Wilson, J.T., McNabb, J.F., Balkwill, D.L., and Ghiorse, W.C., 1983, Enumeration and characterization of bacteria indigenous to a shallow water-table aquifer: *Ground Water*, v. 21, no. 2, pp. 134-142.

Wilson, L.G., Osborn, M.D., Olson, K.L., Maida, S.M., and Katz, L.T., 1990, The ground water recharge and pollution potential of dry wells in Pima County, Arizona: *Ground Water Monitor. Rev.*, v. 10, no. 3, pp. 114-121.

Wilson, M.P., Savage, E.P., Adrian, D.D., Aaronson, M.J., Keefe, T.J., Hamar, D.H., and Tessan, J. T., 1987, Groundwater transport of the herbicide, atrazine, Weld County, Colorado: *Bull. Environ. Contam. Toxicol.*, v. 39, pp. 807-814.

Windholz, M., ed., 1976, *The Merck Index* (9th ed.): Merck & Co., Rahway, N.J., variously paged.

Winkelmann, D.A., and Klaine, S.J., 1991, Atrazine metabolite behavior in soil-core microcosms: Formation, disappearance, and bound residues, *in* Somasundaram, L., and Coats, J.R., eds., *Pesticide transformation products: Fate and significance in the environment*: American Chemical Society Symposium Series 459, pp. 75-92.

Winnett, G., Marucci, P., Reduker, S., and Uchrin, C.G., 1990a, Fate of parathion in ground water in commercial cranberry culture in the New Jersey pinelands: *Bull. Environ. Contam. Toxicol.*, v. 45, no. 3, pp. 382-388.

——1990b, The fate of chlorothalonil in ground water in commercial cranberry culture in the New Jersey pine barrens: *J. Environ. Sci. Health*, v. A25, no. 6, pp. 587-595.

Wittmann, C., and Hock, B., 1991, Development of an ELISA for the analysis of atrazine metabolites deethylatrazine and deisopropylatrazine: *J. Agric. Food Chem.*, v. 39, no. 6, pp. 1194-1200.

Wolfe, N.L., Kitchens, B.E., Macalady, D.L., and Grundl, T.J., 1986, Physical and chemical factors that influence the anaerobic degradation of methyl parathion in sediment systems: *Environ. Toxicol. Chem.*, v. 5, pp. 1019-1026.

Wolfe, N.L., and Macalady, D.L., 1992, New perspectives in aquatic redox chemistry: Abiotic transformations of pollutants in groundwater and sediments: *J. Contam. Hydrol.*, v. 9, nos. 1-2, pp. 17-34.

Worthing, C.R., and Walker, S.B., eds., 1987, *The pesticide manual* (8th ed.): The British Crop Protection Council, Thornton Heath, United Kingdom, 1081 p.

Wu, L., Baker, J.M., Allmaras, R.R., Dowdy, R.H., Lamb, J.A., and Anderson, J.L., 1994, Modeling approaches for infiltration and preferential-flow: A review, in *Agricultural Research To Protect Water Quality: Proceedings of the conference, Minneapolis, Minnesota, USA:* Soil Water and Conservation Society, Arkeny, Iowa, v. 1, pp. 370-374.

Wylie, B.K., Shaffer, M.J., Brodahl, M.K., Dubois, D., and Wagner, D.G., 1993, Predicting spatial distributions of nitrate leaching in northeastern Colorado: *J. Soil Water Conserv.*, v. 49, no. 3, pp. 288-293.

Wyman, J.A., Jensen, J.O., Curwen, D., Jones, R.L., and Marquardt, T.E., 1985, Effects of application procedures and irrigation on degradation and movement of aldicarb residues in soil: *Environ. Toxicol. Chem.*, v. 4, pp. 641-651.

Yen, P.Y., Koskinen, W.C., and Schweizer, E.E., 1994, Dissipation of alachlor in four soils as influenced by degradation and sorption processes: *Weed Sci.*, v. 42, no. 2, pp. 233-240.

Yiacoumi, S., and Tien, C., 1994, A model of organic solute uptake from aqueous solutions by soils: *Water Resour. Res.*, v. 30, no. 2, pp. 571-580.

Young, C.P., 1981, The impact of point source pollution on groundwater quality: *Sci. Total Environ.*, v. 21, pp. 61-70.

Zaki, M.H., 1986, Groundwater contamination with synthetic organic compounds and pesticides in Suffolk County: *Northeastern Environ. Sci.*, v. 5, nos. 1/2, pp. 15-22.

Zaki, M.H., Moran, D., and Harris, D., 1982, Pesticides in groundwater: The aldicarb story in Suffolk County, NY: *Am. J. Public Health*, v. 72, no. 12, pp. 1391-1395.

Zalkin, F., Wilkerson, M., and Oshima, R.J., 1984, *Pesticide contamination in the soil profile at DBCP, EDB, simazine and carbofuran application sites*, v. 2 of *Pesticide movement to ground water*: California Department of Food and Agriculture, Division of Pest Management, Environmental Protection and Worker Safety, Sacramento, Calif., 168 p.

Zandvoort, R., van den Born, G.W., Braber, J.M., and Smelt, J.H., 1980, Leaching of the herbicide bromacil after application on railroads in The Netherlands: *J. Water, Air, Soil Poll.*, v. 13, pp. 363-372.

Zheng, S.Q., Cooper, J.F., Fontanel, P.V., Coste, C.M., and Deat, M., 1993, Distribution and dissipation of metolachlor in soil columns: *J. Environ. Sci. Health*, v. 28, no. 6, pp. 641-653.

Index

Note: Page numbers followed by f refer to figures; page numbers followed by t refer to tables.